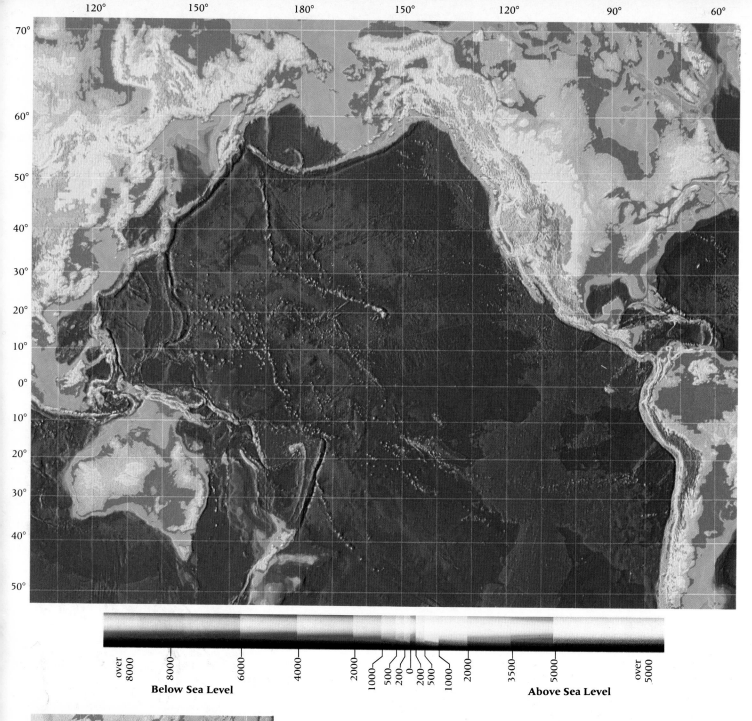

Below Sea Level | over 8000 | 8000 | 6000 | 4000 | 2000 | 1000 | 500 | 200 | 0 | 200 | 500 | 1000 | 2000 | 3500 | 5000 | over 5000 | Above Sea Level

RELIEF OF THE SURFACE OF THE EARTH

Maps of the continental surfaces and ocean floors (above and left) are based on the Mercator projection. Relief is shown by plastic shading with illumination from a simulated sun 20° above the western horizon. Depth or elevation zones, labeled in meters on the color scale, are shown by varying the hue and saturation of colors. The hemispheric images (right) use the same color system.

These maps were produced from digital data on a 5-minute by 5-minute latitude-longitude grid, representing the best available information from the following sources:

United States, Central America, Southern Canada, Western Europe, Japan: *U.S. Defense Mapping Agency.* ▪ Australia, New Zealand: *Branch of Mineral Resources CSIRO (Australia), Department of Scientific and Industrial Research (New Zealand).* ▪ Rest of the world land masses: *Fleet Numerical Oceanography Center (U.S. Navy) from 10-minute grid.* ▪ Oceanic areas: *U.S. Naval Oceanographic Office.*

Digital imagery by Dr. Peter W. Sloss, NOAA-National Geophysical Data Center, Boulder, Colorado, with image processing equipment at the National Snow and Ice Data Center, Boulder, Colorado, 1986. Data used in producing these images are available from the National Geophysical Data Center, Code E/GC, 325 Broadway, Boulder, Colorado, 80303-3328.

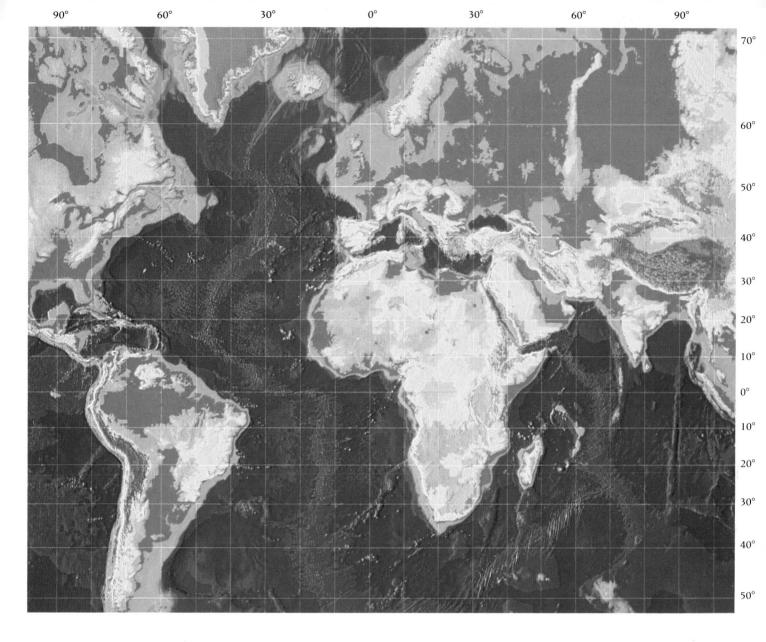

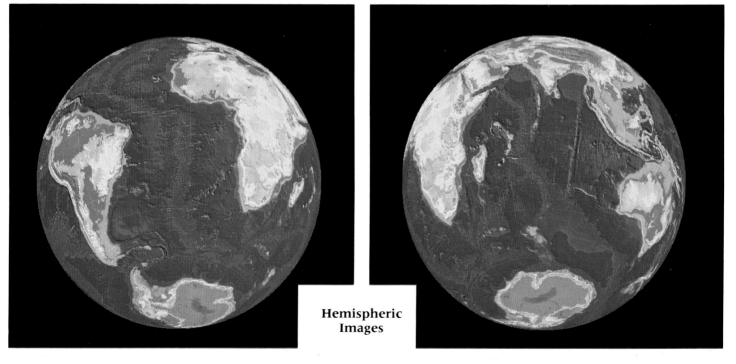

Hemispheric Images

Introducing
Physical Geography

INTRODUCING PHYSICAL GEOGRAPHY

ALAN STRAHLER
Boston University

ARTHUR STRAHLER
Columbia University (emeritus)

JOHN WILEY & SONS, INC.
New York • Chichester • Brisbane • Toronto • Singapore

Acquisitions Editor	Barry Harmon
Marketing Manager	Catherine Faduska
Production Editor	Sandra Russell
Designer	Pedro A. Noa
Manufacturing Manager	Andrea Price
Photo Research Manager	Stella Kupferberg
Photo Researcher	Alexandra Truitt
Illustration/Map Coordinator	Edward Starr
Illustrations/Maps	J/B Woolsey Associates/Maryland Cartographics
Cover Photo	Chris Harris/West Light

This book was set in New Baskerville by Ruttle Graphics, Inc., and printed and bound by Von Hoffmann Press, Inc. The cover was printed by The Lehigh Press, Inc. The color separations were prepared by Lehigh Press Colortronics.

Recognizing the importance of preserving what has been written, it is a policy of John Wiley & Sons, Inc. to have books of enduring value published in the United States printed on acid-free paper, and we exert our best efforts to that end. The paper in this book was manufactured by a mill whose forest management programs include sustained yield harvesting of its timberlands. Sustained yield harvesting principles ensure that the number of trees cut each year does not exceed the amount of new growth.

Strahler, Alan H.
Introducing Physical Geography/Alan Strahler, Arthur Strahler.
Includes index.
ISBN 0-471-57667-0 (pbk.)
1. Physical geography. I. Strahler, Arthur Newell, 1918– . II. Title.
GB54.5.S77 1994
910.02—dc20

93-26814
CIP

Printed in the United States of America

10 9 8 7 6 5 4 3 2 1

BRIEF TABLE OF CONTENTS

PREFACE

Introducing Physical Geography is the sixth major collaboration between the authors under the Wiley imprint, but it is a truly new book that departs from its predecessors in many ways. In the 40-plus years since the publication of *Physical Geography* by Arthur Strahler in 1951, the college population has changed markedly. Today, science must be taught in an open, accessible manner to reach students, and written material needs to communicate directly and simply. Moreover, today's students are accustomed to visual learning and to an interactive methodology.

These changes in the student body have challenged us to provide a new physical geography text that draws on the Strahler heritage of accurate science while being genuinely accessible to the student. With *Introducing Physical Geography*, we believe we have achieved this goal. The text has been specifically crafted not only to provide the essential science of physical geography and its component disciplines, but also to optimize and reinforce the student's learning.

Introducing Physical Geography is also our first new title in which Alan Strahler has served as senior author. Having taught undergraduates in physical geography continuously for more than 20 years in four major educational institutions, Alan is not only intimately familiar with the science topics covered in this book, but he is also very aware of the needs of today's students. In preparing this book, he has taken the lead in devising entirely new treatments of topics in physical geography that are accessible and inviting, and yet still provide the essential principles of science without sacrificing accuracy or stooping to superficiality. Each sentence has been worded to ensure that it communicates its meaning clearly and simply. Every map and line drawing has been carefully designed and styled to make sure that its message is clear, obvious, and direct. Every photo has been selected not only to provide a fine illustration of the relevant concept, but also to demonstrate it effectively and strikingly.

COURSE GOALS

Introducing Physical Geography has been written for a one-semester or one-quarter course in which a majority of students are nonmajors fulfilling a college science requirement. These students need to learn how to rea-son logically, to understand the nature of cause and effect, and to become acquainted with fundamental scientific principles such as the conservation of energy and matter. Physical geography can easily provide the context for this learning, and it has the added advantage of being relevant to many aspects of students' everyday experiences. For this audience, the appropriate focus is on the broad concepts of physical geography and their application, and this is the focus we have adopted for this book.

In designing the content of *Introducing Physical Geography*, we have emphasized the standard curriculum for physical geography that has evolved over the years. The sequence of topics is deliberately conventional and eminently teachable. However, as experienced instructors of physical geography, we know that the conventional curriculum is difficult to cover in a one-semester or one-quarter course. Our approach in this book has been to provide a balanced level of coverage of the field, so that when an instructor chooses to emphasize a particular area of science—climatology or geomorphology, for example—a firm foundation is there. If a topic is deemphasized, the text will provide a basic level of understanding that can stand alone without lecture support.

Physical geography is a very broad field that draws freely from many component sciences. In developing the topical coverage for *Introducing Physical Geography*, we made difficult decisions about what to include and what to omit. In our selections, we have tried to emphasize topics that are truly important. We have constantly asked, "Is this topic really necessary? Do other topics depend on it? Is it essential for understanding the big picture? Will it be useful and important to a nonscience major a year from now?" In producing a book of reasonable size and length, we have not had the luxury of including topics and concepts that are secondary to the book's main mission.

THE LEARNING ENVIRONMENT

Content

Students need orientation as they read a textbook. Where are we in the scheme of this book? What topics will be covered next? Why are they important? How do they relate to topics that have already been presented? Years of classroom lecturing have convinced us of the importance of helping students understand the structure of a course as they learn, and we have carried this idea forward specifically in *Introducing Physical Geography.*

We begin the book with a Prologue, in which the student becomes an astronaut stationed on the moon and embarks on a journey home to earth. On this journey, the student sees the planet at ever-increasing scales, first as a whole, then as a realm of the four great spheres, then as regions differentiated by climate, vegetation, and soils, and finally as a collection of individual landforms as the return spacecraft proceeds to a touchdown. The journey is both a metaphor and a model for the learning to come. The attention to orientation continues throughout the book in transitional paragraphs provided at the beginning and end of each chapter, and in introductions to major topics within chapters. The book concludes with an Epilogue that raises the environmental implications of the physical geography that the student has learned.

"What Is This Chapter All About?"

Students learn more effectively if what they are learning can be related to their life experience. This is one of the great strengths of physical geography, in that the subject matter concerns processes and phenomena that students encounter on a daily basis. To enhance this linkage, we provide a simple one-page introduction to each chapter, entitled, "What Is This Chapter All About?" Rather than simply preview the chapter's contents, this essay personalizes the chapter with an experience or topic that directly involves the student. Most of these essays are written in the second person to involve the student more directly. For example, in Chapter 12, the student gains a respect for earthquakes by experiencing a tsunami from the deck of a hotel room on Waikiki beach. In Chapter 15, the student is introduced to the erosive power of rivers by riding a raft down the Colorado River through the Grand Canyon.

Eye on the Environment

Students also enjoy learning that their new knowledge is important and useful. To help reinforce text material and provide relevance, we have included a short feature within each chapter labeled "Eye on the Environment." These boxed features focus the concepts of the chapter specifically on an environmental topic. Many are concerned with global climate and climatic change, which is a frequent concern of news and documentary coverage on television and appears often in the popular press. Accordingly, our goal for these features is to reinforce the relevance of the chapter material.

Key Terms and Key Statements

Another problem created by the breadth of physical geography is the large number of terms. Vocabulary is an essential part of any science, but we must guard against overloading the student with arcane words and soon-to-be-forgotten phrases. For each term, from "ablational till" to "yazoo stream," we have asked, "Is it used more than once? Is it related to a central concept? Is it just a label? Do we really need it?" To help organize the terms that we have included, we have set off in boldface the key terms—those dozen or more terms in each chapter that are most important—and listed them at the close of each chapter. Less important terms are set in italics. All terms, whether bold or italic, are defined in the glossary at the close of the book.

As another aid to learning, we have embedded in the text a number of key statements, one or two sentences long, that focus attention on key concepts. They are set apart from the text in a slightly larger type face to catch student attention.

End of Chapter Student Aids

Students also need help in mastering the content of the chapters. The text provides chapter summaries, key term lists, review questions, and in-depth questions at the end of each chapter. The chapter summary is worded like a scientific abstract, succinctly covering all the major concepts of the chapter. The list of key terms indicates the most important concepts that must be studied to understand the chapter material. The review questions are designed as oral or written exercises that require description or explanation of important ideas and concepts. Some questions utilize sketching or graphing as a way of motivating students to visualize key cencepts. Also provided are in-depth questions that require more synthesis or the reorganization of knowledge in a new context.

Illustrations

Because students learn visually as well as from the written text, illustrations are of major importance. An entirely new art program for *Introducing Physical Geography* provides scientific accuracy with visual appeal. The content of each graph or drawing has been devised with direct communication in mind. Photos are equally important. Whenever possible we selected our photos for visual impact—a method of holding student interest and drawing attention to the relevant material.

SUPPLEMENTS

Study Guide. By Don L. Gary of Nicholls State University. This extensive guide contains chapter summaries, learning objectives, vocabulary development, short-answer questions, visualization exercises, and critical thinking questions, all of which are designed to reinforce and challenge students' understanding of text principles.

Exercise Manual. By Arthur Strahler and Alan Strahler of Boston University. Designed to build on students' knowledge of physical geography developed through class lecture and textbook study, the manual puts this knowledge to use by solving problems based on course topics. The Exercise Manual to accompany *Introducing Physical Geography* provides all data, figures, and maps needed for the exercises, which are keyed into the parent textbook.

Instructor's Manual and Test Bank. By Jim Westernik of Concord College. This combination Instructor's Manual/Test Bank provides the instructor with chapter outlines, summaries, supplementary topics, tips for the first-time lecturer, current discussion topics, alternative syllabi, additional resources, and supplementary topics. The test questions supplied consist of multiple-choice, matching, and short-answer as well as essay questions.

Full-color Overhead Transparencies. Over 70 full-color transparencies from the text are provided in a form suitable for projection in the classroom. These illustrations will also be available as slides.

Supplementary Slide Set. Approximately 150 slides from the text and the authors' personal collection are available.

CD-Rom. Approximately 300 additional photographs from the authors' personal collections are available on CD-Rom with an easy to use photo manager (Photo Prowler by Xing Corporation) which will facilitate your ability to customize the electronic presentation of these images.

ACKNOWLEDGMENTS

The preparation of *Introducing Physical Geography* was greatly aided by many reviewers who read and evaluated the various parts of the manuscript. They include Kevin Anderson, Augustana College; Don Gary, Nicholls State University; Roland L. Grant, Eastern Montana College; Richard Hackett, Oklahoma State University; Vatche Tchakerian, Texas A&M University; Ted Alsop, Utah State University; Joseph Ashley, Montana State University; Randall Cerveny, Arizona State University; Andrew Marcus, Montana State University; John Watkins, University of Kentucky; Jim Westerik, Concord College; Richard Crooker, Kutztown University; Jeffrey Lee, Texas Tech; Bernard McGonigle, Community College of Philadelphia; Thomas Wikle, Oklahoma State; Kenneth White, Nicholls State University; George Aspbury, Illinois State; David Castillon, South West Missouri State; Vance Halliday, University of Wisconsin; Julie Laity, California State University, Northridge; Maurice Hackett, Southwest Missouri State University; Richard Mackinnon, Allan Hancock College; Marilyn Raphael, University of California at Los Angeles; and Peter Herron, University of Calgary.

We are also most grateful to the many instructors who returned our extensive questionnaire on the revision of *Elements of Physical Geography*, which was very helpful in preparing our new book. They include Claire C. Correale, Burlington County College; Richard Pollak, Notre Dame College–St. Johns University; Richard R. Knabel, Westchester Community College; Lawrence E. O'Brien, Orange County Community College; Richard A. Frank, Monroe Community College; Michael Bikerman, University of Pittsburgh; Brian Hanson, University of Delaware; Michael M. Folkoff, Salisbury State University; Brent R. Skeeter, Salisbury State University; Kenneth Bick, College of William & Mary; Curtis Breeding, Southwest Virginia Community College; Margaret M. Gripshover, Marshall University; Sidney Fletcher, Wingate College; Clarence M. Head, University of Central Florida; Dewey M. Stowers, University of South Florida; Allan L. Lippert, Manatee Community College; Michael Marchioni, East Tennessee State University; David A. Howarth, University of Louisville; Clara A. Leuthart, University of Louisville; Robert Swanson, Union College; Kimberly Medley, Miami University; Sharron Beigel, Edison State Community College; Janice Michael, Edison State Community College; Anthony J. Dzik, Shawnee State University; Roger L. Jenkinson, Taylor University; David R. Williams, Delta College; Julie A. Winkler, Michigan State University; John D. Kapter, University of Wisconsin; Terry Larson, Brainerd Community College; Larry League, Dickinson State University; Joseph M. Ashley, Jr., Montana State University; John J. Donahue, University of Montana; Darshan S. Kang, University of Montana; Noel L. Stirrat, College of Lake County; Kenneth L. Bowden, Northern Illinois University; Gary E. Greenwood, Lewis & Clark Community College; Paul D. Nelson, St. Louis Community College at Forest Park; William A. Noble, University of Missouri; Walter A. Schroeder, University of Missouri; Rose M. Sauder, University of New

Orleans; Don Gary, Nicholls State University; David R. Legates, University of Oklahoma; Peter W. Knightes, Central Texas College; Richard Giardino, Texas A&M University; Steven Jennings, Texas A&M University; Vatche Tchakerian, Texas A&M University; Frederick Lohrengel II, Southern Utah University; John C. Kimura, California State University; Rodney Steiner, California State University; Angus MacDonald, Los Angeles Valley College; David Bixler, Chaffey College; Peter Konovnitzine, Chaffey College; Harold Throckmorton, San Diego Mesa College; Richard MacKinnon, Allan Hancock College; James Feng, De Anza College; Ted Hamilton, Modesto Junior College; Michael L. Talbott, Bellevue Community College; Kenneth J. Langran; Thomas A. Terich, Western Washington University; and David Gosling, Glen Oaks Community College.

It is with particular pleasure that we thank the staff at Wiley for their careful work, encouragement, and sense of humor in the preparation and production of *Introducing Physical Geography*. They include our editor, Barry Harmon, our developmental editor Christine Peckaitis, our director of photo research, Stella Kupferberg, our photo researcher, Alexandra Truitt, our photo research associate, Alissa Mend, our designer, Pete Noa, our illustrator, Ed Starr, our marketing manager Catherine Faduska, our supplements editor Joan Kalkut, and our production editor Sandra Russell. We give special thanks to John Woolsey and his studio, J/B Woolsey Associates, and to Maryland Cartographics, who produced the final illustrations with skill, care, and artistry.

We are deeply indebted to Kristi Strahler for her help in so many phases of the preparation of the book, and for her support and endurance through the many long months of work this book required from start to finish. It would have been impossible without her.

Alan Strahler
Cambridge, Massachusetts

Arthur Strahler
Santa Barbara, California

October 1, 1993

ABOUT THE AUTHORS

Alan Strahler (b. 1943) received his B.A. degree in 1964 and his Ph.D. degree in 1969 from The Johns Hopkins University, Department of Geography and Environmental Engineering. He has held academic positions at the University of Virginia, the University of California at Santa Barbara, and Hunter College of the City University of New York, and is now Professor of Geography at Boston University. With Arthur Strahler, he is a coauthor of six textbook titles with six revised editions on physical geography and environmental science. He has also published over fifty articles in the refereed scientific literature, largely on the theory of remote sensing of vegetation. He has also contributed to the fields of plant geography, forest ecology, and quantitative methods. His work has been supported by over $2 million in grant and contract funds, primarily from NASA. In 1993, he was awarded the Association of American Geographers/Remote Sensing Specialty Group Medal for Outstanding Contributions to Remote Sensing.

Arthur Strahler (b. 1918) received his B.A. degree in 1938 from the college of Wooster, Ohio, and his Ph.D. degree in geology from Columbia University in 1944. He was appointed to the Columbia University faculty in 1941, serving as Professor of Geomorphology from 1958 to 1967 and as Chair of the Department of Geology from 1959 to 1962. He was elected as a Fellow

of both the Geological Society of America and the Association of American Geographers for his pioneering contributions to quantitative and dynamic geomorphology, contained in over thirty major papers in leading scientific journals. He is the author or coauthor with Alan Strahler of fifteen textbook titles with twelve revised editions in physical geography, environmental science, the earth sciences, and geology. His most recent new title, *Understanding Science: An Introduction to Concepts and Issues*, published by Prometheus Books in 1992, has been widely reviewed as an introductory college text on philosophy of science.

CONTENTS

Introducing Physical Geography

Introducing Earth

What Is Physical Geography All About?

Physical geography is the study of the surface of our home planet, Earth. It is not a catalog of mountains and rivers, but a branch of science that investigates how and why the surface changes from day to day, from year to year, and over millions of years. For the physical geographer, the surface includes four parts. First is the land surface itself, which is covered by soil and vegetation. Second is the upper layer of the earth's crust, which provides rock materials that can be shaped into landforms by running water, wind, waves, and ice. Third is the lower layer of the atmosphere, in which clouds and weather occur. Last is the upper layer of the oceans, where ocean waters influence land near the coast.

Perhaps the easiest way to introduce physical geography is to examine the features of our planet—first as a whole and then by zooming in on continents, regions, and, finally, local areas. But to make it more interesting, imagine that the year is 2050 and that you are an astronaut on a lunar base. Relax and follow along as you move from the lunar base to a space station orbiting the earth, and finally to the descent and landing of a space shuttle carrying you safely home.

It's eight hundred hours on December 17, Houston time, and your duty watch is just beginning. Taking your place at the lab's environmental control desk, you check the oxygen and carbon dioxide balances and review the pressure and water vapor concentration log from the previous watch. All is well. You rise from your desk and turn. A small push on your right foot produces a giant but graceful step. You are standing in front of the viewport, looking out across the moonscape. Not much scenery here at Lunar Base Alpha, at least not when you're five earth days into the long lunar night. But in the black sky above, there is a truly majestic sight—a huge blue, green, and brown disk veiled by swirls and films of white. It is Earth, your home.

From your lunar viewpoint, the earth's environmental realms are easily identified by their colors. The blue of the world oceans—the *hydrosphere*—dominates the area of the disk. Rich browns characterize the *lithosphere*—the solid rocky portion of the planet—where it rises above the seas. Vast white cloud expanses form and fade in the gaseous portion of the planet—the *atmosphere*. The lush greens of plants are the visible portion of the *biosphere*, the realm of life on earth, which dominates the *life layer*—the interface between air, land, and ocean.

The right-hand edge of the disk is in shadow, and you see the dramatic contrast between the day and night portions of the globe. The land and water masses of the illuminated side bask in the sunlight, and here the temperature of the life layer is increasing. On the shadowed side of the planet, however, the heat of the oceans, atmosphere, and continents is continuously being lost—radiated into outer space. Without sunlight, the earth's surface cools.

A Meteosat photo of earth, showing cloud and water vapor patterns. Taken from geostationary orbit.

This classic photo of the earth from space was taken by astronauts returning from a lunar voyage in December, 1972. At this time of year, the south polar region is tilted toward the sun and much of Antarctica is visible.

Over the next few hours, you notice the planet's slow rotation. The continent of South America glides across the center of the disk to its edge, then disappears. Australia emerges from behind the curtain of night and makes its way out onto the illuminated portion of the disk. As you watch, each land and water mass slowly takes its turn in the sun, absorbing and reflecting its quota of solar energy.

Africa and South America receive the lion's share of the sunlight striking the earth. These continents lie near the equator, where they meet the strong, vertical rays of the sun. North America and Europe, on the other hand, are short-changed because they intercept the sun's rays at a low angle. At the lower rim of the earth's disk, the stark, white continent of Antarctica rotates slowly but remains sunlit, experiencing 24-hour daylight.

As the earth days go by, you often gaze at your planet, marking the changes. Away from the equator, huge swirls of clouds form, dissolve, and reform in its atmosphere, making their way eastward across the disk. First one portion of a continent, and then another, is obscured. These swirls mark the passage of weather fronts and storms that are moved eastward by persistent westerly winds above the surface.

Near the earth's equator in Africa you notice a band of patchy, persistent cloudiness bracketed by reddish-brown areas of earth that are normally clear. The clouds result when warm, moist air is heated by excess solar energy and rises. As the air rises, it cools, and the moisture it contains condenses, forming clouds and producing rain. The lush, green landscape that is occasionally visible underneath the cloud belt seems to thrive on the warm temperature and abundant rainfall it receives. The reddish-brown areas are vast deserts. These receive the air that rises over the equator and becomes depleted of moisture. As the air descends toward the desert, it warms. Showing the colors of rock and soil, the hot, dry deserts are barren of plants.

During the remainder of your six-month tour of duty at the lunar base, you watch the slow changes of the planet with the season. By June, the earth has changed its position in the sun. Antarctica has moved below the southern rim. The sunlit northern rim of the disk now includes northern Canada, Siberia, and the Arctic Ocean. In previous months, you watched the green wave of vegetation sweep northward, up and across North America, Europe, and eastern Asia, following the warming temperatures of spring. The band of tropical cloudiness has also moved northward, bringing a green wave in Africa along with it as well. Clearly, each region of the earth has its unique climate, responding to the rhythms of the season in different ways. You remember fondly the days of June on earth—the smell of warm soil mixed with the scent of flowers, the hot sidewalks and streets at noon, the long evenings and warm nights alive with quite rustlings. As the end of your tour draws near, you count the days eagerly, thinking of your return to your true home, planet Earth.

It's sixteen hundred hours, June 18, Houston time, aboard the Space Station *Enterprise,* in low earth orbit. Your duty watch is just over. Being an environmental specialist in orbit around the earth is

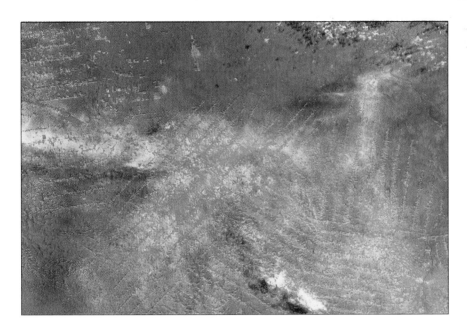

A Landsat satellite image of deforestation in the Rondonia region, Brazil. In this false-color image, vegetation appears red and forest clearings appear blue.

not the hardest job on the space station, but it isn't the easiest either. Today's shift required changing the filter on the nutrient pump for the algae, and in the weightlessness of orbit, it was a ticklish, delicate job.

It's been quite a trip so far. You recall the events of the last few weeks—the arrival of the new crew on the lunar shuttle at Moon Base Alpha's spaceport, the preparations for the departure of your crew, the relaunch of the lunar shuttle, the slow days in transit to earth orbit, and, at long last, the docking of the lunar craft at the space station. But the most memorable part of the return trip was approaching the earth ever closer. With each day and hour, it grew larger and larger. Continents became regions of forests and deserts, mountain ranges, and inland seas. At last the planet became a sphere, its edges falling away and out of view as the lunar shuttle neared the space station. Now, in low earth orbit, finer detail is visible—systems of rivers, agricultural patterns, large cities.

Still drawn to the majesty and serenity of the earth turning below

you, you float gently to the viewport for yet another few minutes of observation. Beneath you stretches Amazonia, its lush green rainforests visible through gaps in the clouds. The verdant pattern is interrupted by brownish patches of recently cleared land, some with hazy smoke plumes at their edges. The rainforest is but one of the many types of vegetation covers you have noted from space, ranging from spruce forests to tropical grasslands. The vegetation patterns have fit the cloud patterns well and have also marked the seasonal changes of solar illumination you noted from your lunar base.

Some hours later, your view is of China. Finally, the clouds have cleared, and a disaster is in view—the Yangtze River is in flood. The river is now a wide, shining ribbon, filling its broad valley from edge to edge. A few tiny green islands are visible, but you know that from this distance they are actually ranges of hills or diked cities. Where the Yangtze meets the sea, a plume of sediment-laden river water curls lazily across the ocean waters. In this view, the work of rivers is especially evident—moving sediment from highlands to lower lands and the ocean, under the power of rains generated ultimately by unequal solar heating of the globe.

The following day, your view is of the San Francisco Bay area. The San Andreas fault is clearly visible on the east side of the peninsula as a line marked by valleys and linear lakes. Crossing the Golden Gate, it heads northwest, near the ocean's edge. As you have surveyed the world's landforms from your orbital vantage point, you have noted how

The San Francisco Bay area, viewed by the Landsat satellite from earth orbit.

many seem to be related to the great structural features of the earth's surface—its great upwarps and downwarps creating mountain zones and inland seas, its folded and faulted crust producing mountains, valleys, and plains.

It's nine hundred hours, June 26, Houston time, aboard the Space Shuttle *Discovery*. The tension is

building, strongly now. Gravity is starting to return, and the earth shuttle is beginning to feel the slowing of friction with the atmosphere. The communications blackout period is about to begin. Strapped firmly into your seat, you contemplate the bumpy ride ahead with a racing heart. Fortunately, you are just a passenger on this seg-

ment and can leave the tricky duties of navigating the shuttle to a safe landing to the flight crew and their computers. You'll be landing at Edwards Spaceport, in the California desert, since the two other spaceports are experiencing winds and rain. For now, there is nothing much to see out of the nearby viewport—it is night over the Pacific. The minutes pass slowly—too slowly. The attitude-control jets fire from time to time, as the shuttle moves through its slowing maneuvers.

Soon the line of the horizon begins to redden. The sun rises quickly, spectacularly, moving rapidly skyward. Below, the California coast approaches, the white stripe of its beaches gleaming in the sun. The light brown scars of recent debris avalanches on the steep marine cliffs dissect the gray-green of the chaparral. Off to the right, is Los Angeles. The pink bubble of polluted air over the city is plainly visible at this early hour.

Still descending, you cross the Tehachapi Mountains, which are now very close, and drop toward the Mojave Desert. A patch of stripes flashes by—lines of sand dunes silhouetted in the early light. The landing gear descends, shaking your craft with a thud as it locks in place. Your grip on the handrest tightens as your body tenses in anticipation. The shuttle flares out over the flat, salty soil of the dry lake. Bang! The wheels hit the ground. Touchdown! The brakes and drogue chute press you hard against the seat. You're home at last!

The Space Shuttle runway at Edwards Air Force Base, California, as photographed from the Space Shuttle Challenger.

THE PHYSICAL GEOGRAPHY OF OUR PLANET

Our opening essay touches on a number of subjects that are part of physical geography, ranging from earth—sun relationships to evolution of landforms. These are a few of the important topics in physical geography that we will develop further in this book. However, there are some essential ideas and themes that need to be discussed first. We begin with the "spheres"—the four great realms of earth—mentioned in the essay. Next is an exploration of scale—in space and in time. Finally, we examine the earth as the human habitat and introduce physical geography as the science that studies it.

At its most basic level, physical geography is concerned with the atmosphere, the lithosphere, the hydrosphere, and the biosphere and their interactions. The four realms have a zone of interaction called the life layer. Let's discuss the four realms in more depth.

The Four Great Realms

The **atmosphere** is a gaseous layer that surrounds the earth. It receives and returns flows of heat and moisture from and to the surface. The atmosphere also supplies vital elements—carbon, hydrogen, oxygen, and nitrogen—needed to sustain the life-forms on the lands.

The solid earth, or **lithosphere,** forms the stable platform for the life layer. The solid rock of the lithosphere bears a shallow layer of soil in which nutrient elements become available to organisms. The lithosphere is sculpted into landforms. These relief features, like mountains, hills, and plains, provide varied habitats for plants, animals, and humans.

The earth's four great realms are the atmosphere, lithosphere, hydrosphere, and biosphere.

Water, in all its forms, constitutes the **hydrosphere.** The main mass of the hydrosphere lies in the world's oceans, but water also occurs in the atmosphere as gaseous vapor, liquid droplets, and solid ice crystals. Water is found in or atop the uppermost layers of the lithosphere, forming ground water reservoirs, lakes, streams, and rivers. Water is essential to life and required for the survival of organisms.

The **biosphere** encompasses all living organisms of the earth. Life-forms on earth utilize the gases of the atmosphere, the water of the hydrosphere, and the nutrients of the lithosphere, and so the biosphere is dependent on all three of the other great realms. Figure P.1 diagrams this relationship.

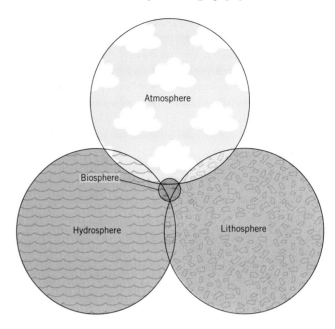

Figure P.1   The earth realms, shown as intersecting circles.

The Life Layer

The **life layer** is the shallow surface zone that contains most of the biosphere (Figure P.2). It includes the surface of the lands and the upper 100 meters (300 ft) or so of the ocean. On land, the life layer is the zone of interaction among the biosphere, lithosphere, and atmosphere. The hydrosphere is represented on land by rain, snow, still water in ponds and lakes, or running water in rivers. In the ocean, the life layer is the zone of interactions among the hydrosphere, biosphere, and atmosphere. The lithosphere is represented by nutrients dissolved in the upper layer of sea water.

Figure P.2   As this sketch shows, the life layer is the layer of the earth's surface that supports nearly all of the earth's life. This includes the land and ocean surfaces and the atmosphere in contact with them.

Throughout our exploration of physical geography, we will often refer to the life layer and the four realms that interact within it. We will find that the themes and subjects of physical geography concern processes that involve the atmosphere, hydrosphere, lithosphere, or biosphere—and usually two or three of these realms simultaneously.

SCALE

Think again about your recent trip home to earth from the lunar base. While making your trip, you noted aspects of several themes in physical geography. These include earth–sun relationships, global atmospheric circulation, precipitation and weather systems, world climates, natural vegetation, soils, the earth's surface structure, and landforms and the agents that produce them.

But what is interesting is that you observed them in about that order as you made your return from moon to earth. This relates to the **scale** with which each subject is concerned. Earth–sun relationships come first, because the sun is the energy source that powers most of the important processes that occur within the life layer. To study these processes, we need to consider the planet and its global energy balance as a whole.

The sun's energy is not distributed evenly across the earth's surface. This unequal solar heating produces currents of air and water, just as a current of hot air rises from a candle flame, and these currents constitute the global atmospheric circulation. When we move still closer to the earth, the cloud patterns of weather systems become obvious. By watching the earth over time, we see the regular movements of weather systems. Taking these movements along with solar control of surface temperature, we have the basis for climates of the world.

The nature of the earth's cover of plants, and the soil layers beneath them, are strongly related to climate. Local factors are also important in determining the exact patterns of vegetation and soils. Is the region a mountain or a valley? Is it being farmed or do sheep graze there? This specific information is not useful when we consider the planet as a whole, but it is important to collect when we consider the nature of the land's vegetation cover and soils at a middle scale. We look at a still closer scale to see features such as a sand dune on a beach or a rocky cliff cut away by a river. These individual landforms are produced by local action of wind or water.

The point here is that the processes acting in nature have a wide range of scales in time and space. Some processes—for example, the global circulation of the atmosphere—are best understood by considering the earth as a whole. Still other examples are concerned with very local phenomena, such as sand grains being blown along a beach. Time scales vary widely, too. It may take millions of years to produce a chain of mountains like the Himalayas, but a fracturing of the earth's crust and the resulting earthquake may last only a few minutes.

If you look at the table of contents for this book, you will note that our study of physical geography moves from the coarse scale to the fine, from global phenomena to local phenomena. There will be exceptions to this overall plan, but understanding the focus of this organization will be helpful in learning about the systems and processes of physical geography.

PHYSICAL GEOGRAPHY AND THE HUMAN HABITAT

The lands of the earth are the habitat of the human species, as well as other forms of terrestrial life. Since physical geography is the study of the earth's surface and the processes that act on it, physical geography is also the study of the **human habitat.** This personal involvement helps make physical geography interesting and rewarding to learn. Everyday concerns like weather systems are a good example. Since the weather is always changing, physical geography will help you understand how and why it changes.

> **Because physical geography focuses on the life layer, it is also the study of the human habitat.**

Still other aspects of physical geography help explain your local landscape. The shapes of hills and valleys normally change too slowly to be noticed on a daily basis. But if you watch soil being washed down a slope during a sudden rain shower, you can extend this phenomenon to the land around you and come to understand how years of rainfall slowly carve the landscape into unique landforms.

If you have the opportunity to travel, studying physical geography will help you understand more about the landscapes you visit. While considering a Peace Corps assignment to the country of Mali, on the African continent, you might expect to find a tropical jungle. But your study of physical geography will tell you that much of Mali looks like our American desert!

Physical geography also encompasses a concern for our environment. For example, global climate change is a widespread concern today. Is the earth getting warmer? Is it getting cooler? Will there be another ice age? Scientists offer us conflicting opinions. To understand their conclusions, you need some background about the processes and systems that influ ence global climate. While our survey of physical geography focuses on natural processes, it will often explain environmental problems. We will encourage you to sift the data and the opinions drawn from current studies.

Humans are now the dominant species on the planet. Nearly every part of the earth has felt our impact in some way. As our population continues to grow, human impacts on natural systems will continue to increase. Each of us is charged with the responsibility to treat the earth well and respect it. Understanding the processes that shape the human habitat as they are described by physical geography will help us carry out this mission.

THE PROLOGUE IN REVIEW

Physical geography is a branch of science that investigates the earth's surface, and how and why it changes. It focuses on the life layer in which the four great realms—atmosphere, hydrosphere, lithosphere, and biosphere—interact. The processes of interaction between the realms can be examined at different scales. For example, earth-sun relationships control global heating and in this way influence global wind systems. Vegetation and soils are strongly related to climate, which varies at a regional scale. And individual landforms, such as a sand dune, are produced by processes acting at local scales. Since physical geography focuses on the land surface, it includes the study of the human habitat—the local environment in which we live and interact with natural processes.

KEY TERMS

physical geography	atmosphere	lithosphere
hydrosphere	biosphere	life layer
scale	human habitat	

REVIEW QUESTIONS

1. What are the four great realms of earth? Name and describe each realm.

2. What is the life layer? Where is it located?

3. Provide two examples of processes that operate at global, middle, and local scales.

4. What relation does physical geography have to the human habitat?

IN-DEPTH QUESTION

Imagine that you are planning a mountain backpacking trip—for example, along the Appalachian or Pacific Coast Trails. Which aspects of physical geography might be useful to know about? Why?

Chapter 1

Our Place in the Sun

Brrring! The alarm at the side of your bed goes off. It's 7:30 A.M. Already the sun is streaming in the window, reflecting brightly off your white pillowcase. You place your head under the pillow to bring back the darkness, then kick off the sheet, letting the ceiling fan cool your sweaty skin. It's going to be a hot one today, all right.

Brrring! The alarm at the side of your bed goes off. It's 7:30 A.M. You grope for the alarm, but it is difficult to find it in the pale predawn light. You pull the goosedown quilt up and over your head to block out the chilly air, and retrieve a wayward foot that has inadvertently tested the temperature. It's going to be a cold one today, all right.

Some people like hot summers on the beach, while others like cold winters on the ski slopes. How about you? If you live on a part of our planet that experiences strong seasonal changes, you may have a preference. However, many parts of our planet experience only small temperature variations throughout the year. What causes the seasons? Why do some areas experience the seasons more intensely than others?

Two eternal cycles are involved. In the first, the earth slowly turns on its axis, making a complete rotation every 24 hours. This cycle determines our daily time system. The plan of rotation on an axis also led naturally to the system of latitude and longitude—our method for locating places on the surface of the earth.

The second eternal cycle is the complete revolution of the earth about the sun every 365 days. This cycle determines our yearly time system. As the earth revolves about the sun, the axis of the earth's rotation remains pointed toward the same place in the heavens, no matter where the earth is located. This means that sometimes the northern hemisphere is tilted toward the sun, while the southern hemisphere is tilted away.

If the northern hemisphere is tilted toward the sun, days are longer in the north and the sun's rays strike the northern hemisphere at a more direct angle. These effects produce more intense solar heating, so temperatures are generally warmer. The effect: summer in the northern hemisphere. At another time of year, the southern hemisphere is tilted toward the sun, and the northern hemisphere is tilted away. Northern hemisphere days are then shorter and the sun's rays strike the hemisphere at a lower angle. This produces winter in the northern hemisphere.

These two great cycles continue, day after day, year after year, regulating the processes of earthly life. Without these cycles, life as we know it would not exist. They are truly an appropriate starting point for our study of the physical geography of our planet.

Stonehenge, England

We all know that the "earth is round"—that is, our planet is shaped like a sphere, or ball. In fact, one of our first lessons in school is usually to locate our country, state, and city on the globe. Later on, we see pictures taken from space by astronauts or by orbiting satellites, confirming the roundness of our planet. Thus, we build up a mental picture of the earth and our location on it.

Many of our ancestors, however, were not aware of our planet's spherical shape. To sailors of the Mediterranean Sea in ancient times, the shape and breadth of the earth's oceans and lands were unknown. On their ships and out of sight of land, the sea surface looked perfectly flat and bounded by a circular horizon. Given this view, many sailors concluded that the earth had the form of a flat disk, and that their ships would fall off, if they traveled to its edge.

High in a jet airplane at cruising altitude, you could come to the same conclusion. Peering out the window, you see below a vast expanse of earth or ocean fading into a distant and level horizon. At sunset, you notice an interesting phenomenon (Figure 1.1). While the sun is still visible to you, illuminating the clouds at your altitude with an orange glow, the ground below you is in shadow. This means that the sun is below the horizon to a ground observer but above the horizon to an air traveler. This only happens when the curvature of the earth blocks the sun for the ground observer, but not for you, the air traveler.

Actually, the earth is not perfectly spherical. Its true shape is described as an *oblate ellipsoid*. The outward force of the earth's rotation causes the earth to bulge slightly at the equator, producing a shape in which the

Figure 1.1   This magnificent sunset, seen from the passenger cabin of a high flying jet aircraft over the southern Pacific Ocean, demonstrates the earth's curvature. Since the sun is still visible above the horizon at the level of the aircraft, it lights the clouds and sky. At the same moment, the ocean below is in shadow. The sun has already left the sky for an observer on the surface.

earth is about three-tenths of a percent wider at the equator than at the poles.

Scientists still study the earth's shape and attempt to measure it as precisely as possible. This is important since the information is needed by satellite navigation systems for aircraft, ocean vessels, and ground vehicles seeking to determine their exact location. The more precisely the earth's shape is known, the more accurately location may be determined.

EARTH ROTATION

Another fact about our planet, which we learn early in life, is that the earth spins slowly, making a full turn with respect to the sun every day. We use the term **rotation** to describe this motion. One complete rotation with respect to the sun defines the *solar day*. By convention, the solar day is divided into exactly 24 hours.

The earth rotates on its *axis*, an imaginary straight line through the earth's center. The intersections of the axis of rotation and the earth's surface are defined as the poles. To distinguish between the two poles, one is called the *north pole* and the other, the *south pole*.

Direction of Rotation

The direction of earth rotation can be determined by using one of the following guidelines (Figure 1.2):

- Imagine yourself looking down on the north pole of the earth. From this position, the earth is turning in a counterclockwise direction (Figure 1.2*a*).
- Imagine yourself off in space, viewing the planet much as you would view a globe in the library, with the north pole on top. The earth is rotating from left to right, or in an eastward direction (Figure 1.2*b*).

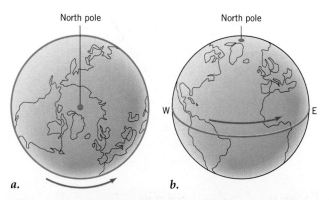

a.   *b.*

Figure 1.2 The direction of rotation of the earth can be thought of as (*a*) counterclockwise at the north pole, or (*b*) from left to right (eastward) at the equator.

> **The direction of the earth's rotation is counterclockwise, when we look down on the north pole from above.**

The earth's rotation is important for two reasons. First, it provides the framework for locating position on the earth's surface and for measuring time. Second, the rotation influences physical and life processes on earth, as we will now describe.

Physical Effects of Earth Rotation

The physical effects of the earth's rotation are of great importance to us and our environment. The first, and perhaps most obvious, effect of rotation is that it imposes a daily, or *diurnal*, rhythm to which plants and animals respond. The phenomena affected include light, heat, air humidity, and air motion. Green plants respond to the daily rhythm by storing energy during the day and consuming some of it at night. Animals adjust their activities to the daily rhythm, some preferring the day, others the night, for food-gathering. The daily cycle of incoming solar energy and the corresponding cycle of fluctuating air temperature will be topics for analysis in chapters 2 and 3.

Second, both air and water flow paths are turned consistently in one direction because of the earth's rotation. In the northern hemisphere, flows are turned toward the right and, in the southern hemisphere, toward the left, when viewed from the starting point. This phenomenon is called the *Coriolis effect*. It is of great importance when studying the earth's systems of winds and ocean currents. We will investigate both the Coriolis effect and its influence on winds and currents in Chapter 5.

A third physical effect of the earth's rotation is the movement of the tides. The moon exerts its gravitational attraction on the earth, while at the same time the earth is turning with respect to the moon. These forces induce a rhythmic rise and fall of the ocean surface, known as the *tide*. These motions in turn cause water currents of alternating direction to flow in the shallow salty waters of the coastal zone. The ocean tide may have no importance for a grain farmer in Kansas, but for the clam digger and charter boat captain on Cape Cod, the tidal cycle is a clock regulating daily activities. The ebb and flow of tidal currents is a lifegiving pulse for many plants and animals that live in coastal saltwater environments. The tide and its currents are discussed further in Chapter 17.

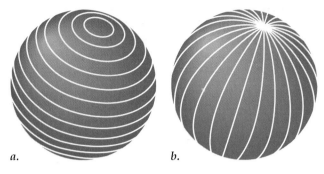

Figure 1.3 (*a*) Parallels of latitude divide the globe cross-wise into rings. (*b*) Meridians of longitude divide the globe from pole to pole.

THE GEOGRAPHIC GRID

The geographic grid provides a method for describing locations on the earth's surface. Without such a system, human society would literally be lost. Because the earth's surface is curved, and not flat, we cannot divide it into a rectangular grid, like a sheet of graph paper. Instead, we divide it using imaginary circles set on the surface that are perpendicular to the axis of rotation in one direction and parallel to the axis of rotation in the other direction.

Parallels and Meridians

Imagine a point on the earth's surface. As the earth rotates, the point traces out a path in space, following an *arc* —that is, a curved line that forms a portion of a circle. With the completion of one rotation, the line forms a full circle. This is known as a parallel of latitude, or a **parallel** (Figure 1.3*a*). Parallels cut the globe much as you might slice an onion to produce onion rings—that is, perpendicular to the main axis. The largest parallel of latitude lies midway between the two poles and is designated the **equator.** The equator is a fundamental reference line for measuring the position of points on the globe.

Imagine now slicing the earth with a plane that passes through the axis of rotation, instead of across it. This is the way you might cut up a lemon to produce wedges. The cut traces a circle on the globe passing through both poles. A half-part of this circle, connecting one pole to the other, is known as a meridian of longitude, or, more simply, a **meridian** (Figure 1.3*b*).

Meridians and parallels define compass directions. Meridians are north-south lines, so you are following a

---

The geographic grid consists of an orderly system of circles—meridians and parallels—that are used to locate position on the globe.

---

meridian if you walk north or south. Parallels are east-west lines, and so you are following a parallel if you walk east or west. There can be any number of parallels and meridians. Every point on the globe is associated with a unique combination of one parallel and one meridian. The position of the point is defined by their intersection. The total system of parallels and meridians forms a network of intersecting circles called the **geographic grid.**

Latitude and Longitude

We use a special system to label parallels and meridians—latitude and longitude. Parallels are identified by latitude, and meridians by longitude. Since parallels are circles and meridians are half-circles, we use degrees of arc as latitude and longitude measures to mark our place along particular parallels and meridians. What do we mean by *degrees of arc?* Figure 1.4 illustrates this concept. A full circle consists of 360°. An arc, or portion of a circle, has a lesser angle associated with it. Thus, we can measure the arc by measuring the degrees of that angle.

Latitude measures the position of a given point in terms of its angular distance from the equator (Figure 1.5). That is, latitude is an indicator of how far north or south of the equator a given point is situated. Latitude is measured in degrees of arc from the equator (0°) toward either pole, where the value reaches 90°. The equator divides the globe into two equal portions, or *hemispheres.* All points north of the equator—that is, in the *northern hemisphere*—are designated as north latitude, and all points south of the equator—in the *southern hemisphere*—are designated as south latitude.

Figure 1.5 shows how latitude is measured. The point *P* lies at the latitude of 50 degrees north, which we can abbreviate as lat. 50°N. Notice that latitude

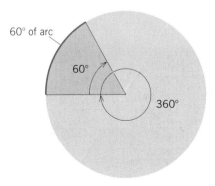

Figure 1.4 A circle consists of 360°, measured as an angle from its center. An arc is part of a circle. An angle is associated with an arc measuring the arc's length in degrees. We refer to this measure as degrees of arc.

angle arc is actually measured along the meridian that passes through the point.

Longitude is a measure of the position of a point eastward or westward from a reference meridian, called the *prime meridian.* As Figure 1.5 shows, longitude is the angle of arc, measured in degrees, between the meridian of a given point and the prime meridian. This arc is measured east to west along the parallel that passes through the point. In this example, the point *P* lies at longitude 60 degrees west (long. 60°W).

The prime meridian passes through the old location of the Royal Observatory at Greenwich, near London, England (Figure 1.6). For this reason it is also referred to as the Greenwich meridian. This meridian has the value long. 0°. The longitude of any given point on the globe is measured eastward or westward from this meridian, depending on which is the shorter arc. Longitude thus ranges from 0° to 180°, east or west. When both the latitude and longitude of a place are known, it can be accurately and precisely located on the geographic grid.

When arcs of latitude or longitude are measured other than in full-degree increments, *minutes* and *seconds* can be used. A minute is one-sixtieth of a degree, and a second is one-sixtieth of a minute, or one thirty-six-hundredth of a degree. Thus, a latitude like 41

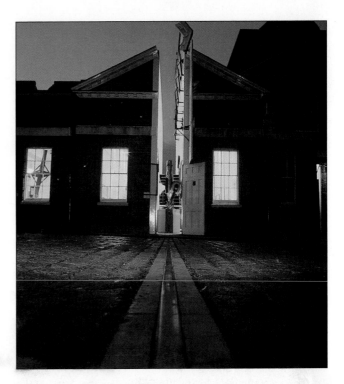

Figure 1.6 This photograph, taken at dusk at the old Royal Observatory at Greenwich, England, shows the prime meridian, which has been marked as a stripe on the fore-court paving. The dome covering the observatory is partly opened and the telescope, located exactly on the meridian, is pointed toward the sky as if making an observation.

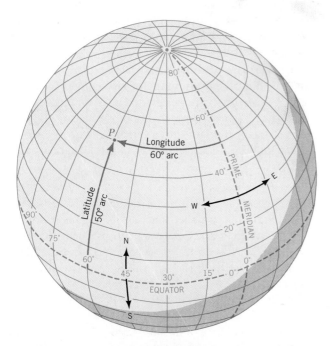

Figure 1.5 The point *P* has a latitude of 50°. This means that the arc between *P*'s parallel and the equator has an angle of 50°. We say that P is located on the 50th north parallel. *P* also has a longitude of 60°. Therefore, the arc between *P*'s meridian and the prime meridian, in Greenwich, England, is 60°. We speak of *P* as being on the 60th west meridian.

degrees, 27 minutes, and 41 seconds north (lat. 41°27' 41" N) means 41 degrees north plus 27/60 of a degree plus 41/3600 of a degree. This cumbersome system has now largely been replaced by decimal notation. In this example, the latitude 41°27'41" N translates to 41.4614°N.

Statements of latitude and longitude do not describe distances in kilometers or miles directly. However, for latitude, you can estimate conversions from degrees into kilometers. One degree of latitude is approximately equivalent to 111 km (69 mi) of surface distance in the north-south direction. This value can be rounded off to 110 km (70 mi) for multiplying in your head. For example, if you live at lat. 40°N (on the 40th parallel north), you are located about 40 × 110 = 4400 km (40 × 70 = 2800 mi) north of the equator.

East-west distances cannot be converted so easily from degrees of longitude into kilometers or miles because the meridians converge toward the poles (Figure 1.3*b*). Only at the equator is a degree of longitude equivalent to 111 km (69 mi). At lat. 60°N or S, meridians are twice as close as at the equator. So, one degree of longitude is reduced to half its equatorial length, or about 56 km (35 mi).

Map Projections

With a working understanding of the geographic grid, we can now consider how to display the locations of continents, rivers, cities, islands, and other geographic features on maps. This will take us briefly into the realm of *cartography*, the art and science of making maps. Our discussion will focus on a few simple types of maps used in this text.

As we observed earlier, the earth's surface is nearly spherical. However, maps are flat. It is impossible to copy a curved surface onto a flat surface without cutting, stretching, or otherwise distorting the curved surface in some way. So, making a map means devising an orderly way of changing the globe's geographic grid of curved parallels and meridians into a grid that lies flat. We refer to a system for changing the geographic grid to a flat grid as a **map projection.**

A map projection is a system for displaying the curved surface of the geographic grid on a flat surface.

Associated with every map is a *scale fraction*—a ratio that relates distance on the map to distance on the earth's surface. For example, a map with a scale fraction of 1:50,000 means that one unit of map distance equals 50,000 units of distance on the earth. Because a curved surface cannot be projected onto a flat surface without some distortion, the scale fraction of a map holds only for one point or a single line on the map. Away from that point or line, the scale fraction will be different.

We will concentrate on the three most useful map projections. The first is the polar projection, which is essential today for scientific uses like weather maps of the polar regions. Second is the Mercator projection, a navigator's map invented in 1569 by Gerhardus Mercator. It is a classic that has never gone out of style. Third is the Goode projection, named for its designer, Dr. J. Paul Goode. It has special qualities not found in the other two projections.

Polar Projection

The *polar projection* (Figure 1.7) can be centered on either the north or south pole. Meridians are straight lines radiating outward from the pole, and parallels are nested circles centered on the pole. Spacing of the parallels increases outward from the center, so that the map is usually cut off to show only one hemisphere. In that case, the equator forms the outer edge of the map. Because the intersections of the parallels with the meridians always form true right angles, this projection shows the true shapes of all small areas. If you drew a

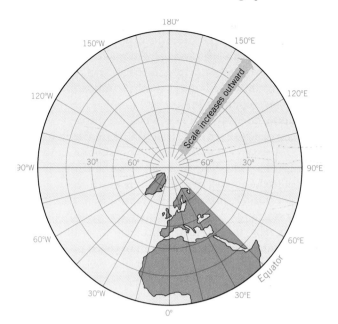

Figure 1.7 A polar projection. The map is centered on the north pole. All meridians are straight lines radiating from the center point, and all parallels are concentric circles. The scale fraction increases in an outward direction, making shapes appear larger toward the edges of the map.

small circle anywhere on the globe, it would reproduce as a perfectly shaped circle on this map. However, because the scale fraction increases in an outward direction, the circle would appear larger toward the edge than near the center.

Mercator Projection

The *Mercator projection* (Figure 1.8) is a rectangular grid with meridians as straight vertical lines and parallels as straight horizontal lines. Meridians are evenly spaced, but the spacing between parallels increases at higher latitude so that at 60° the spacing is double that at the equator. With closer movement to the poles the spacing increases even more, and the map must be cut off at some arbitrary parallel, such as 80°N. This produces a change of scale that enlarges features when they near the pole, as can easily be seen in Figure 1.8. There Greenland appears larger than Australia and is nearly the size of Africa!

The Mercator projection has special properties. The first is that a straight line drawn anywhere on the map is a line of constant compass direction. A navigator can therefore simply draw a line between any two points on the map and measure the direction angle or bearing of the line with a protractor. Once aimed in that compass direction, a ship or an airplane can be held to the same compass bearing to reach the final point or destination. However, this line will not necessarily follow the shortest actual distance between two points. The shortest path on the globe follows the arc of a *great circle*. (A great circle connects two points on the globe and has the earth's center as the circle's center.) On a Mercator projection, a great circle line will usually curve and can (falsely) seem to be a much longer distance than a compass line.

Because the Mercator projection shows the true compass direction of any straight line on the map, it is used to show many types of straight-line features. Among these are flow lines of winds and ocean currents, directions of crustal features (such as chains of volcanoes), and lines of equal values, such as lines of equal air temperature or equal air pressure. This explains why the Mercator projection is chosen for maps of temperatures, winds, and pressures.

Goode Projection

The *Goode projection* (Figure 1.9) uses two sets of mathematical curves (sine curves and ellipses) to form its meridians. Between the 40th parallels, sine curves are used, and beyond the 40th parallel toward the poles, ellipses are used. Since the ellipses converge to meet at the pole, the entire globe can be shown. The straight, horizontal parallels make it easy to scan across the map

at any given level to compare regions most likely to be similar in climate.

The Goode projection has one very important property—it presents areas of the earth's surface indicating their true sizes. That is, if we drew a small circle on a sheet of clear plastic and moved it over all parts of the Goode world map, the circle would always enclose an area with a constant value in square kilometers or square miles. Because of this property, we use a similar map to show geographical features that occupy surface areas. Examples of useful Goode projections include maps of the world's climate, soils, and vegetation.

The Goode projection displays the relative areas of land masses correctly but distorts their shape—especially near the poles.

The Goode map suffers from a serious defect. The Goode projection distorts the shapes of areas, particularly in high latitudes and at the far right and left edges. To minimize this defect, Dr. Goode split his map apart into separate, smaller sectors, each centered on a different vertical meridian. These were then assembled at the equator. This type of split map is

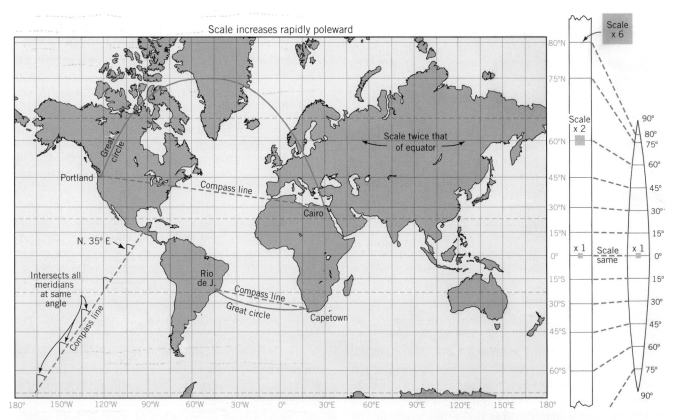

Figure 1.8　The Mercator projection. The diagram at the right shows how the map scale increases rapidly at higher latitudes. At lat. 60°, the scale is double the equatorial scale. At lat. 80°, the scale is six times greater than at the equator.

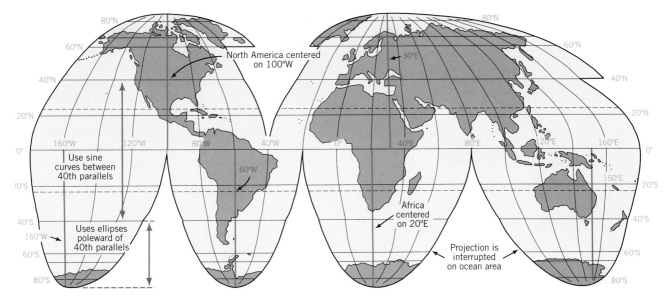

Figure 1.9   The Goode projection. The meridians in this projection follow sine curves between lat. 40° N and 40° S, and ellipses between lat. 40° and the poles. Although the shapes of continents are distorted, their areas are properly shown. (Copyright © by the University of Chicago. Used by permission of the Committee on Geographical Studies, University of Chicago.)

called an *interrupted projection*. Although the interrupted projection greatly reduces shape distortion, it separates parts of the earth's surface that actually lie close together, particularly in the high latitudes.

Maps and map projections are a practical application of the earth's geographic grid. Another practical application that involves both the grid and the rotation is global time, a subject we turn to next.

GLOBAL TIME

Our planet requires 24 hours for a full rotation with respect to the sun. Put another way, humans long ago decided to divide the solar day into 24 units, called hours, and devised clocks to keep track of hours in groups of 12. Yet, different regions set their clocks differently. For example, when it is 10:03 A.M. in New York, it is 9:03 A.M. in Chicago, 8:03 A.M. in Denver, and 7:03 A.M. in Los Angeles. Note that these times differ by exactly one hour. How did this system come about? How does it work?

Our global time system is oriented to the sun. Think for a moment about how the sun appears to move across the sky. In the morning, the sun is low on the eastern horizon, and as the day progresses, it rises higher until at *solar noon*, the sun reaches its highest point in the sky. If you check your watch at that moment, it will usually read a time somewhere near twelve o'clock (12:00 noon). After solar noon, the sun's elevation in the sky decreases. By late afternoon,

the sun appears low in the sky, and at sunset it rests on the western horizon.

Imagine for a moment that you are in Chicago, the time is noon, and the sun is at or near its highest point in the sky. Further imagine that you call a friend in New York, and ask about the time there and the position of the sun. You will receive a report that the time is 1:00 P.M. and that the sun is already past solar noon, its highest point. Calling a friend in Los Angeles, you hear that it is 10:00 A.M. there and that the sun is still working its way up to its highest point. Consider calling Mobile, Alabama. Your friend there will tell you that the time in Mobile is the same as in Chicago and that the sun is at about solar noon. What is happening here?

The difference in time between Chicago, New York, and Los Angeles makes sense because solar noon can occur simultaneously only at locations with the same longitude. In other words, only one meridian can be directly under the sun and experience solar noon at a given moment. At this given moment, locations on other meridians lie either to the east or to the west of Chicago. Locations to the east of Chicago like New York already will have passed solar noon, and locations to the west of Chicago like Los Angeles will not yet have reached solar noon. Since Mobile and Chicago have nearly the same longitude, they experience solar noon at approximately the same time.

Figure 1.10 indicates how time varies with longitude. In this figure, the inner disk shows a polar projection of the world, centered on the north pole. Meridians

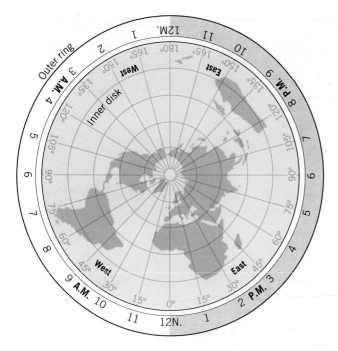

Figure 1.10 This diagram shows how longitude is related to time. The outer ring shows the time at locations identified by longitude meridians on the central map. The diagram is set to show noon conditions in Greenwich, England, that is, on the prime meridian. Clock time is earlier to the west of Greenwich and later to the east.

are straight lines (radii) ranging out from the pole. The outer ring indicates the time in hours. The figure shows the moment in time when the prime meridian is directly under the sun—that is, the 0° meridian is directly on the 12:00 noon mark. This means that, at this instant, the sun is at the highest point of its path in the sky in Greenwich, England. The alignment of meridians with hour numbers tells us the time in other locations around the globe. For example, the time in New York, which lies on roughly the 75°W meridian, is about 7 A.M. In Los Angeles, which lies roughly on the 120°W meridian, the time is about 4 A.M.

Standard Time

Locations with different longitudes experience solar noon at different times. Consider the result if every town and city set its clocks to read 12:00 at its own local solar noon. This would mean that every town or city that was not on exactly the same meridian would have a different time! In these days of instantaneous global communication, chaos would soon result.

The use of standard time simplifies the global time-keeping problem. In the **standard time system,** the globe is divided into 24 **time zones.** All inhabitants within the zone agree to keep time according to a *standard meridian* that passes through their zone. Since the

standard meridians are usually 15 degrees apart, the difference in time between adjacent zones is normally one hour.

> **In the standard time system, we keep global time according to nearby standard meridians that normally differ by one hour from each other.**

Six time zones occur in the United States. Their names and standard meridians of west longitude are

Eastern	75°
Central	90°
Mountain	105°
Pacific	120°
Alaska-Hawaii	150°
Bering	165°

If carried out strictly, the standard time system would consist of belts exactly 15 degrees in width, extending to meridians 7 1/2 degrees east and west of each standard meridian. However, this system could still be inconvenient, since the boundary meridians could divide a state, county, or city into two different time zones. As a result, time zone boundaries are often routed to follow agreed-upon natural or political boundaries.

Figure 1.11 presents a map of time zones for the contiguous United States and southern Canada. From this map, you can see that most time zone boundaries are conveniently located along an already existing and widely recognized line. For example, the Eastern time—Central time boundary line follows Lake Michigan down its center, and the Mountain time—Pacific time boundary follows a ridge-crest line also used by the Idaho–Montana state boundary.

International Date Line

When we take a world map or globe with 15° meridians and count them in an eastward direction, starting with the Greenwich meridian as 0, we find that 180th meridian is number 12 and that the time of this meridian is, therefore, 12 hours fast. Counting in a similar manner westward from the Greenwich meridian, we find that the 180th meridian is again number 12 but that the time is 12 hours slow. How can the same meridian be both 12 hours ahead of Greenwich time and 12 hours behind? This paradox is explained by the fact that the 180th meridian is the *International Date Line.* Except for the exact moment of midnight at this meridian, the calendar day observed to the west of the date line will be one ahead of that to the east.

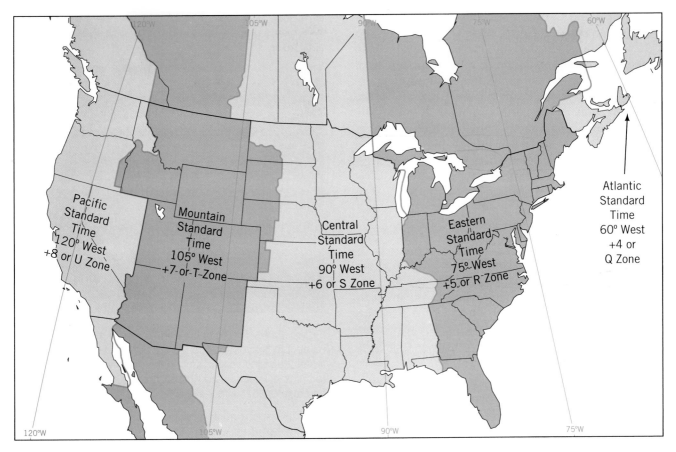

Figure 1.11 Time zones of the contiguous United States and southern Canada. The name and standard meridian are shown for each time zone. Time zones are sometimes distinguished by number or letter codes, which are also given. Note that time zone boundaries often follow preexisting natural or political boundaries.

Crossing the date line requires adjusting the calendar. For example, flying from Los Angeles to Sydney, Australia, you may depart on a Tuesday evening and arrive on a Thursday morning after a flight of only 14 hours duration. On a flight from Tokyo to San Francisco, you may actually arrive before you take off, taking the date change into account!

Daylight Saving Time

Especially in urban areas, many human activities begin well after sunrise and continue long after sunset. Therefore, we adjust our clocks during the part of the year that has a longer daylight period to correspond more closely with the modern pace of society. This adjusted time system, called *daylight saving time*, is obtained by setting ahead all clocks by one hour. The effect of the time change is to transfer the early morning daylight period, theoretically wasted while schools, offices, and factories are closed, to the early evening, when most people are awake and busy. Daylight saving time also yields a considerable savings in power used for electric lights.

THE EARTH'S REVOLUTION AROUND THE SUN

So far, we have discussed the importance of the earth's rotation on its axis. Another important motion of the earth is its **revolution,** or its movement in orbit around the sun.

The earth completes a revolution around the sun in about $365\frac{1}{4}$ days. Every four years the extra one-fourth day difference between the period of revolution and the calendar year of 365 days adds up to about one whole day. By inserting a 29th day in February in leap years, we correct the calendar for this effect. Further minor corrections are necessary to perfect this calendar system.

The earth's orbit around the sun is shaped like an ellipse, or oval. This means that the distance between the earth and sun varies somewhat through the year. At *perihelion*, which occurs on or about January 3, the earth is nearest to the sun. At *aphelion*, on or about July 4, the earth is farthest away from the sun. However, the elliptical orbit is shaped very much like a circle, and the distance between sun and earth varies only by about 3 percent during one revolution. For our purpose, we can regard the orbit as circular.

In which direction does the earth revolve? Imagine yourself in space, looking down on the north pole of the earth. From this viewpoint, the earth travels counterclockwise around the sun (Figure 1.12). This is the same direction as the earth's rotation. It is also the same direction as the moon's rotation and revolution around the earth. In fact, nearly all the planets and major satellites in our solar system have the same direction of rotation and revolution.

Tilt of the Earth's Axis

The seasons we experience on earth are related to the orientation of the earth's axis of rotation and the position of the sun. We usually describe this situation by stating that "the earth's axis is tilted." What does this really mean? Let's refer to Figure 1.13. The plane containing the earth's orbit around the sun is called the *plane of the ecliptic.* Notice that the earth's orbit is shown lying within the plane.

The axis of the earth's rotation is tilted by $23\frac{1}{2}°$ away from a right angle with the plane of the ecliptic.

Next, consider the axis of the earth's rotation. Rather than being at a right angle to the plane of the ecliptic, the axis is tilted at an angle of $23\frac{1}{2}°$ away from a right angle. That is, the angle between the axis and the plane of the ecliptic is $66\frac{1}{2}°$, not 90°. In addition, the direction the axis points toward is fixed in space. The axis is aimed toward Polaris, the north star. The direction of the axis does not change as the earth revolves, so that, for one part of the year, the north

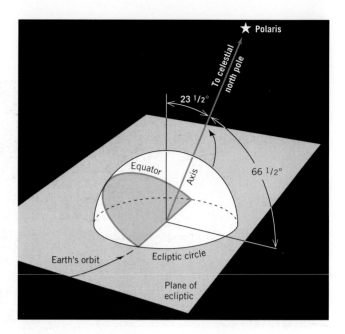

Figure 1.13 The tilt of the earth's axis of rotation with respect to its orbital plane. As the earth moves in its orbit on the plane of the ecliptic around the sun, its rotational axis remains pointed toward Polaris, the north star, and makes an angle of $66\frac{1}{2}°$ with the ecliptic plane.

pole is tilted away from the sun. This is the situation shown in the diagram. Later in the year, the north pole is tilted toward the sun by the same amount. Let's investigate this phenomenon in more detail.

Solstice and Equinox

Figure 1.14 diagrams the earth as it revolves in its orbit through the four seasons. Consider first the event on December 22, which is pictured on the far right. On this day, the earth is positioned so that the north polar end of its axis leans at the maximum angle away from the sun, $23\frac{1}{2}°$. This event is called the **winter solstice.** (While it is winter in the northern hemisphere, it is summer in the southern hemisphere, so you can use the term *December solstice* to avoid any confusion.) At this time, the southern hemisphere is tilted toward the sun and enjoys strong solar heating.

Six months later, on June 21, the earth is on the opposite side of its orbit in an equivalent position. At this event, known as the **summer solstice** (*June solstice*), the north polar end of the axis is tilted at its maximum angle of $23\frac{1}{2}°$ toward the sun. Thus, the north pole and northern hemisphere are tilted toward the sun, while the south pole and southern hemisphere are tilted away.

Figure 1.12 Viewed from a point over the earth's north pole, the earth both rotates and revolves in a counterclockwise direction. From this point, the moon also rotates counterclockwise.

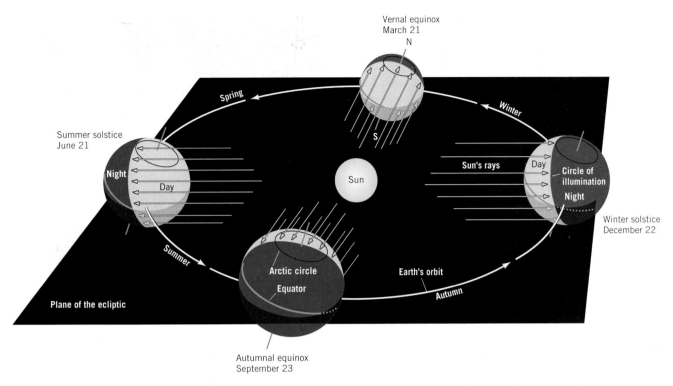

Figure 1.14 The four seasons result because the earth's tilted axis keeps a constant orientation in space as the earth revolves about the sun. This tips the northern hemisphere toward the sun for the summer solstice and away from the sun for the winter solstice. Both hemispheres are illuminated equally at the spring equinox and the fall equinox

Midway between the solstice dates, the equinoxes occur. At an **equinox,** the earth's axis makes a right angle with a line drawn to the sun, and neither the north nor south pole is tilted toward the sun. The *vernal equinox* occurs on March 21, and the *autumnal equinox* occurs on September 23. Conditions are identical as far as earth—sun relationships are concerned on each of the two equinoxes. We should also note that the date of any solstice or equinox in a particular year may vary by a day or so, since the revolution period is not exactly 365 days.

Equinox Conditions

Let's now look at equinoxes and solstices in more detail. The conditions at an equinox form the simplest case. Figure 1.15 illustrates two important concepts of global illumination that we use for describing equinoxes and solstices. The first concept is the *circle of illumination.* Note that the earth is always divided into two hemispheres with respect to the sun's rays. One hemisphere (day) is lit by the sun, and the other (night) lies in the darkness of the earth's shadow. The circle of illumination is the circle that separates the day hemisphere from the night hemisphere. The sec-

ond concept is the *subsolar point.* The sun is directly overhead at this single point on the earth's surface.

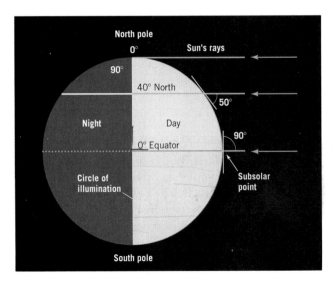

Figure 1.15 Equinox conditions. At this time, the earth's axis of rotation is exactly at right angles to the direction of solar illumination. The subsolar point lies on the equator. At both poles, the sun is seen at the horizon.

Eye on the Environment • Earth–Sun Cycles and Ice Ages

Stand back! The glaciers are coming! Well, perhaps not today or tomorrow, but during the past two to three million years, great sheets of ice—continental glaciers—have come and gone many times, covering large expanses of the northern hemisphere. A period when continental glaciers grow and cover the landscape is referred to as a *glaciation*. Right now, we are in an interglacial period that started 12,000 years ago, when the last continental ice sheets melted away. When will the ice sheets return?

A glaciation sets in when global climates become colder and high-latitude continents receive more snow. Each year, more snow accumulates than can be melted by warm summer weather, and eventually, an ice cap is formed. As the process continues over thousands of years, the ice cap can grow into a great continental ice sheet spreading over millions of square kilometers and covering an extremely large area of land.

What causes our climate to change so that glaciers form? Scientists have puzzled over this question since at least the middle of the nineteenth century, and many theories have been proposed. One theory that many scientists find reasonable states that glaciation is caused by slight changes in the amount of solar energy the earth receives. These slight changes are produced by minor changes in the earth's orbit and in the tilt of its axis.

The distance between the earth and sun at the time of summer solstice in the northern hemisphere varies slowly over time. This cycle of change lasts about 21,000 years. During the cycle, the average distance between earth and sun may vary from 1 to 5 percent greater to 1 to 5 percent less than the normal long-term average. When the distance is greater, less solar energy reaches earth and summers are colder.

Another cycle slowly changes the tilt of the earth's axis. Just as a spinning top sometimes develops a wobble, so the earth's axis also has a slight wobble that changes the earth's tilt from 24° to 22° and back again. This cycle takes about 40,000 years. When the tilt is smaller (22°), it tends to make the polar regions slightly colder and the equatorial regions slightly warmer. When the tilt is greater (24°), we have the opposite effect—warmer polar regions and a cooler equatorial belt.

Since both cycles are going on simultaneously, they can either cancel or reinforce one another. When both cycles combine to make the arctic regions colder and reduce summer snowmelt, ice cap formation is enhanced. When one cycle warms the poles while the other cools the arctic, snowmelt keeps the ice caps from growing quickly, and continental glaciers form more slowly or begin to waste away. If both cycles combine to warm the north polar regions, the ice cap melts and shrinks away.

The figure at right shows the effect of the combined cycles. The vertical axis shows the change in solar energy reaching the earth at the top of the atmosphere. The amount of solar energy is calculated for lat. 65°N on the summer solstice, June 21. The horizontal axis shows the time, thousands of years before the present date. The graph is known as the Milankovitch curve, named for Milutin Milankovich, the astronomer who first calculated it in 1938.

At equinox, the circle of illumination passes through the north and south poles, as we see in Figure 1.15. The sun's rays graze the surface at either pole, and the surface receives little or no solar energy. The subsolar point falls on the equator. Here the angle between the sun's rays and the earth surface is 90°, and solar illumination is received in full force. At an interme-diate latitude, such as 40°N, the rays of the sun at noon strike the surface at a lesser angle. Some simple geom-etry shows that the noon sun angle with the surface is equal to 90° minus the latitude, or 50°, for equinox conditions.

Imagine yourself at a single point on the earth, say, at a latitude of 40° north. Visualize the earth rotating from left to right, so that you turn with the globe, completing a full circuit in 24 hours. At the equinox, you spend 12 hours in darkness and 12 hours in sunlight. This is because the circle of illumination passes through the poles, dividing every parallel exactly in two. Thus, one important feature of the equinox is that day and night are of equal length everywhere on the globe.

At the equinox, all locations on earth experience a day and night of equal length.

As you can see, in the last 500,000 years about a dozen peaks of increased solar energy have alternated with dips to lower values. Some of these dips and peaks seem to correspond with the timings of glaciations and interglacial periods, as scientists attempt to date them using other evidence. The graph shows that our planet's solar energy is now decreasing from a peak occurring about 20,000 years ago.

One strongly supported theory is that these cycles in earth–sun relationships control the timing of our planet's ice ages. Other scientists however, have proposed that quite different mechanisms influence the onset of ice ages. These mechanisms include volcanic eruptions and cycles of change in the sun's brightness. Other factors, such as ocean current patterns, are also deemed important. We will touch on some of these in later chapters and *Eye on the Environment* features.

Questions

1. What two cycles are known to affect the amount of solar radiation received by polar regions of the earth?

2. What is the Milankovitch curve? What does it show about warm and cold periods during the last 500,000 years?

The Milankovitch curve. The vertical axis shows fluctuations in incoming solar energy at lat. 65° N on June 21, the summer solstice. These are calculated from mathematical models of the change in earth–sun distance and change in axial tilt with time. The zero value represents the present value. (Based on calculations by A. D. Vernekar, 1968. Copyright © A. N. Strahler.)

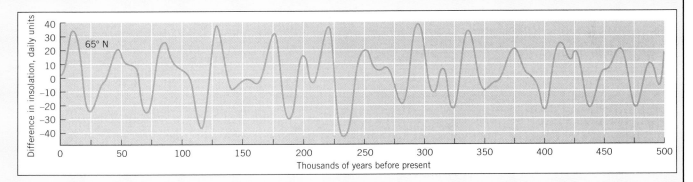

Solstice Conditions

Examine the solstice conditions, as shown in Figure 1.16. Summer solstice is shown on the left. Consider yourself back at a point on the lat. 40°N parallel. The circle of illumination does not divide your parallel in equal halves because of the tilt of the northern hemisphere toward the sun. Instead, the larger part is in daylight. For you, the day is now considerably longer (about 15 hours) than the night (about 9 hours).

The farther north you go, the more the effect increases. In fact, the entire area of the globe north of lat. 66 1/2° is on the daylight side of the circle of illumination. This parallel is known as the **arctic circle.** Even though the earth rotates through a full cycle during a 24-hour period, the area north of the arctic circle remains in continuous daylight. We also see that the subsolar point is at a latitude of 23 1/2°N. This parallel is known as the **tropic of cancer.** Because the sun is directly over the tropic of cancer at this solstice, solar energy is most intense here.

At the winter solstice, conditions are exactly reversed from those of the summer solstice. If you imagine yourself back at lat. 40°N, you find that the night is now about 15 hours long while the day is 6 hours. All the area south of lat. 66 1/2°S lies under the sun's rays, inundated with 24 hours of daylight. This parallel is known as the **antarctic circle.** The subsolar point has shifted to a point on the parallel at lat. $23\frac{1}{2}°$S, known as the **tropic of capricorn.**

The solstices and equinoxes represent conditions on only 4 of 365 days in the year. In between these

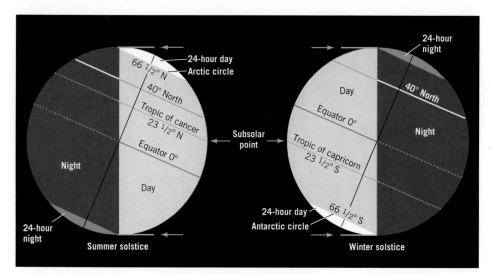

Figure 1.16 Solstice conditions. At the solstice, the north end of the earth's axis of rotation is fully tilted either toward or away from the sun. Because of the tilt, polar regions experience either 24-hour day or 24-hour night. The subsolar point lies on one of the tropics, at lat. 23½° N or S.

times, the subsolar point travels northward and southward, moving the latitude of maximum solar intensity across the belt between the two tropics. Areas of 24-hour daylight or 24-hour night shrink and then grow as the seasonal cycle progresses. The length of each daylight period is a little different from that of the next, as the earth continues its revolution around the sun.

An understanding of the causes for the seasons is essential to the topic we treat next—the global energy balance. In Chapter 2 we will examine in detail solar energy—the source that powers life processes—and its interaction with the earth's atmosphere and surface. This discussion will include an explanation of why the solar energy available varies through the seasons and with latitude.

OUR PLACE IN THE SUN IN REVIEW

This chapter introduced some very fundamental topics in physical geography—the rotation of the earth on its axis and the revolution of the earth in its orbit about the sun. The rotation provides the first great rhythm of our planet—the daily alternation of sunlight and darkness. The tides, and a sideward turning of ocean and air currents, are further effects of the earth's rotation, as we will see in more detail later.

The earth's rotation also leads to a natural system of location on the earth's surface by providing meridians and parallels—that is, the geographic grid. This system is indexed by our system of latitude and longitude, which uses the equator and the prime meridian as references. We require a map projection to display the earth's curved surface on a flat map. The polar projection is centered on either pole and pictures the globe as we might view it from the top or bottom. The Mercator projection converts the geographic grid into a flat, rectangular one and best displays directional features. The Goode projection distorts the shapes of continents and coastlines but preserves the areas of land masses in their correct proportion.

We monitor the earth's rotation by daily timekeeping. Each hour the earth rotates by 15 degrees. In the standard time system, we keep time according to a nearby standard meridian. Since standard meridians are normally 15 degrees apart, clocks around the globe usually differ by even hours.

The seasons are the second great earthly rhythm. They arise from the revolution of the earth in its orbit, combined with the fact that the earth's rotational axis is tilted with respect to its orbital plane. The solstices and equinoxes mark the cycle of this revolution. At the summer (June) solstice, the northern hemisphere is fully tilted toward the sun. At the winter (December) solstice, it is the southern hemisphere that is tilted toward the sun. At the equinoxes, day and night are of equal length.

KEY TERMS

rotation	parallel	equator
meridian	geographic grid	latitude
longitude	map projection	standard time system
time zone	revolution	winter solstice
summer solstice	equinox	arctic circle
tropic of cancer	antarctic circle	tropic of capricorn

REVIEW QUESTIONS

1. What is the shape of the earth? How do you know?

2. What is meant by earth rotation? Describe three physical effects of the earth's rotation.

3. Describe the geographic grid, including parallels and meridians.

4. How do latitude and longitude determine position on the globe? In what units are they measured?

5. Name three types of map projections and describe each briefly. Give reasons why you might choose different map projections to display different types of geographical information.

6. Explain the global timekeeping system. Define and use the terms standard time, standard meridian, and time zone in your answer.

7. What is meant by the "tilt of the earth's axis"? How is the tilt responsible for the seasons?

8. Sketch a diagram of the earth at an equinox. Show the north and south poles, the equator, and the circle of illumination. Indicate the direction of the sun's incoming rays and shade the night portion of the globe.

9. Sketch a diagram of the earth at the summer (June) solstice, showing the same features. Also include the tropics of cancer and capricorn, and the arctic and antarctic circles.

IN-DEPTH QUESTION

Suppose that the earth's axis was tilted at $40°$ to the plane of the ecliptic instead of $23\frac{1}{2}°$. What would be the global effects of this change? How would the seasons change at your location?

Chapter 2

The Earth's Global Energy Balance

The sun is deliciously warm on your back. The sound of the surf is subdued as you lie stretched full-length on your towel. The beach here at Cancun is everything you thought it would be—soft fine sand and clear blue water, with hazy tropical skies overhead. What a place to spend spring break!

It's not long before the sun on your back seems a little too warm, and so you roll over, pulling the brim of your hat down over your eyes to screen out the bright sunlight. Soon you are hot all over, and the sound of the gently breaking waves seems very inviting. Shaking off a few clinging grains of sand, you head for the surf. A wave breaks over your feet, splashing cool drops of foam on your legs. A few quick steps, a shallow dive, and your skin is tingling with the shock of the Caribbean waters. What a marvelous feeling!

Thoroughly refreshed, you emerge from the waves and walk back up to your place on the beach. As you towel the water from your skin, the tropical breeze takes on a cool edge. You lie down again, keeping low, and exposing your body to the direct rays of the sun overhead. Soon your skin dries, and you are warm once more.

Later, in the evening, you return to the beach. The stars are out now, and the white crescent is framed by palm trees behind you and the dark ocean in front. The gentle breakers splash phosphorescent foam up the beach, which quickly sinks into the fine sand. After a few minutes, you notice that your shoulders and arms are a bit cold. The sand, however, is still warm to your feet. You lie down, burrowing slightly into the soft surface to create a smooth support for your back and legs. Comfortable now, you relax and watch the sky overhead. Slowly, the stars grow brighter and the night behind them deepens to a velvet black. A shooting star flashes across the sky. This will be a spring break you'll always remember.

This description of a beach holiday probably seems far from the subject of this chapter—the earth's global energy balance. But, in fact, many of the principles involved in the global radiation balance are the same as those that warm or cool your skin on a Caribbean beach. Like your skin, the earth's surface absorbs sunlight and is warmed. At night, lacking sunlight, it loses heat and is cooled. The surface, as well as your skin, also loses heat when cooled by evaporation. And, the surface can be cooled or warmed by direct contact with a cool or warm atmosphere—just as your skin can be cooled or warmed by direct contact with the sand of the beach.

No matter whether the surface is your skin or a layer of grass-covered soil, it is constantly emitting radiation. In fact, all objects that receive and absorb radiation must also give off radiation. Our bodies emit radiation in the form of heat. Very hot objects, such as the sun or the filament of a light bulb, give off radiation that is nearly all in the form of light.

The temperature of a surface depends on the balance between the amount of energy it absorbs and the amount it loses. Lying on the beach at Cancun, you may enjoy absorbing the light radiation from the tropical sun and the heat from the coral sand. But if you absorb energy faster than you can emit it, your skin temperature will rise and you'll feel hot! However, a dip in the cool ocean quickly conducts away the excess energy, and your skin's temperature is comfortable again.

The same is true of the earth. Our planet is bathed in radiation emitted by the sun. But because the only source of energy warming the earth is the sun, over time the earth must emit energy at the same rate as it is received from the sun. If the sun's energy output increased, what would happen? You might guess that the earth would begin to warm up. But the earth would soon begin to emit more radiation because of its hotter temperature, and the outgoing radiation would eventually balance the incoming radiation. The only difference would be a warmer earth. We might expect the earth to cool a bit, if the sun produced less energy or if the earth moved farther away, for the same reason. So the earth's overall temperature really depends on the balance of solar energy that it absorbs and heat energy that it emits.

In this chapter, we examine the energy balance of the earth, including land and ocean surfaces and the atmosphere. This balance controls the seasonal and daily changes in the earth's surface temperature. It also produces differences in energy flow rates from place to place that drive currents of air and ocean water. These, in turn, produce the changing weather and rich diversity of climates we experience on the earth's surface.

Sunset, eastern Washington

This chapter explains how solar radiation is intercepted by our planet, flows through the earth's atmosphere, and interacts with the earth's land and ocean surfaces. Solar radiation, nearly all in the form of light energy, is the driving power source for wind, waves, weather, rivers, and ocean currents when absorbed at or near the surface of the earth. Most natural phenomena you see at the earth's surface are directly or indirectly solar-powered—from the downhill flow of rivers, to the movement of a sand dune, to the growth of a forest.

The sun's energy powers the processes of the life layer like the constant flow of gasoline powers a car cruising down the highway. But just as friction and air resistance constantly hold the car's speed in check, so the radiation of heat energy from our earth and atmosphere out to space constantly checks the buildup of energy absorbed from the sun. Thus, a balance is maintained between incoming solar radiation and outgoing terrestrial radiation.

Human activities now dominate many regions of the earth, and we have irreversibly modified our planet by farming much of its surface and by adding carbon dioxide to its atmosphere. Have we shifted the balance of energy flows? Is our earth absorbing more solar energy and becoming warmer? Is it absorbing less and becoming cooler? We must examine the global energy balance in detail before we can understand human impact on the earth–atmosphere system.

ELECTROMAGNETIC RADIATION

The primary subject of this chapter is radiation—that is, **electromagnetic radiation,** a form of energy emitted by all objects. Think of this form of energy as a collection, or spectrum, of waves of a wide range of wavelengths traveling quickly away from the surface of the object. *Wavelength* describes the distance separating one wave crest from the next wave crest (Figure 2.1). The unit used to measure wavelength is the *micrometer* (formerly called the micron). A micrometer is one-millionth of a meter, or ten-thousandth of a centimeter. This is such a small unit that the tip of your little finger is about 15,000 micrometers wide. In this text, we use the abbreviation μm for the micrometer. The first letter is the Greek letter μ, or mu. It is used in metric units to denote micro, meaning one-millionth.

The wavelength of the radiation determines its characteristics. For example, light and heat are familiar forms of electromagnetic radiation. Light has shorter wavelengths of about 0.4 to 3 μm, while heat has longer ones about 3 to 30 μm. In this chapter, we are primarily concerned with electromagnetic energy in the form of light and heat.

Radiation and Temperature

An inverse relationship exists between the wavelength of the radiation that an object emits and the temperature of the object. That is, the higher the temperature, the shorter the wavelengths emitted. For example, the radiation emitted by the sun's hot surface is mostly in the form of light at wavelengths of 0.4 to 0.7 μm. The earth's much cooler surface emits heat radiation mostly at wavelengths of 7 to 15 μm. Our eyes are sensitive to light radiation, while they do not respond directly to heat. Scientists refer to the energy emitted by the sun as **shortwave radiation,** because it consists primarily of the shorter ultraviolet, visible light, and shortwave infrared wavelengths. Shortwave radiation has wavelengths of less than 3 μm. In contrast, the earth emits **longwave radiation,** defined as radiation at wavelengths of greater than 3 μm.

Another important principle is that hot objects radiate more energy than cooler objects—much more. In fact, the amount of energy radiated by a surface is

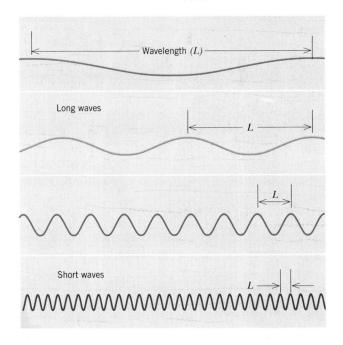

Figure 2.1 Electromagnetic radiation can be described as a collection of energy waves with different wavelengths. Wavelength L is the crest-to-crest distance between successive wave crests.

31

directly related to its absolute temperature raised to the fourth power. (Absolute temperature is temperature measured on the Kelvin scale, with zero degrees the absence of all heat.) So, a small increase in temperature can mean a large increase in the amount of radiation given off by an object. For example, water at the freezing point emits about one-third more radiation when warmed to room temperature.

Understanding these two principles—that hotter objects radiate more energy and that the energy is of shorter wavelengths—will help you understand the global energy balance.

Hotter objects radiate more energy, and at shorter wavelengths, than cooler objects.

Solar Radiation

Earlier, we noted that the sun's radiation is the main power source for many of the earth's processes. What are the characteristics of solar radiation?

Our sun is a ball of constantly churning gases that are heated by continuous nuclear reactions. In comparison with other stars, it is about average in size, and our sun has a surface temperature of about 6000°C (11,000°F). Like all objects it emits energy in the form of electromagnetic radiation. The energy travels in straight lines outward from the sun at a speed of about 300,000 km (186,000 mi) per second. At that rate, it takes the energy about $8\frac{1}{3}$ minutes to travel the 150 million km (93 million mi) from the sun to the earth.

Solar radiation travels through space without losing energy. However, the rays spread apart as they move away from the sun. This means that a planet farther from the sun, like Mars, will receive less radiation than one located nearer to the sun, like Venus. The earth intercepts only about half of one-billionth of the sun's total energy output.

The electromagnetic spectrum can be divided into four major portions, based on wavelengths. These regions are shown in Figure 2.2. At the left lies the *ultraviolet* radiation portion of the spectrum. Wavelengths range from about 0.2 to 0.4 μm. These short waves are high in energy, and they are nearly all absorbed by the gas molecules of the earth's atmosphere. However, when these rays penetrate to the surface of the earth, they can damage living tissues, such as human skin.

Toward the center of the solar spectrum is the **visible light** portion of the spectrum. Wavelengths range from 0.4 to 0.7 μm. Our eyes are sensitive to these waves as light energy. The color of the light is determined by its wavelength. The shorter wavelengths are perceived as violet, while the longer ones are perceived as red. In between is the rest of our familiar color spectrum, including blue, green, yellow, and orange. These wavelengths easily penetrate the earth's atmosphere.

Next is the *shortwave infrared* radiation region, ranging from 0.7 to 3 μm. This radiation comes from the sun and acts much like visible light, although our eyes are not sensitive to it. Shortwave infrared radiation also penetrates the atmosphere easily. At the right are the *thermal infrared* wavelengths, longer than 3 μm, which are emitted by cooler objects. We perceive this type of radiation as heat, such as you might feel sitting beside the glowing coals of a campfire.

Characteristics of Solar Energy

Although solar radiation includes ultraviolet, visible, and shortwave infrared radiation, the sun does not provide all types of energy equally. The tall curve at the left on the graph of Figure 2.2 shows the relative energy intensity, as it enters our outer atmosphere, from one end of the sun's spectrum to the other. Energy intensity is shown on the graph on the vertical scale. The ultraviolet portion of the spectrum accounts for about 9 percent of the total incoming solar energy. Also included in the ultraviolet division are X rays and gamma rays, forms of radiation that have shorter wavelengths than the ultraviolet. They are not shown on the graph. The solar energy curve peaks in the visible light range, which accounts for about 41 percent of the total energy received by the earth. The shortwave infrared portion of the spectrum accounts for 50 percent of the total energy. Very little energy arrives in wavelengths longer than 2 μm.

The intensity of solar energy is strongest in visible wavelengths.

The sun's interior is the source generating solar energy. Here, hydrogen is converted to helium under enormous confining pressure and at very high temperatures. In this nuclear fusion process, a vast quantity of energy is generated and finds its way to the sun's surface. Because the rate of production of nuclear energy is constant, the output of solar radiation is also nearly constant. So, given the average distance of earth from the sun, the amount of solar energy received on a small, fixed area of surface held at right angles to the sun's rays is almost constant. This rate of incoming shortwave energy is known as the *solar constant* and has a value of 1400 watts per square meter (W/m²). It is measured beyond the outer limits of the earth's atmosphere, before energy has been lost in passing through the earth's atmosphere.

The *watt* (W), which describes a rate of energy flow, is a familiar measure of electrical power consumption, being applied to stereo amplifiers, microwave ovens, and light bulbs. When describing the intensity of solar electromagnetic energy flow being received, we must specify the unit cross section of the beam. This is assigned the area of 1 square meter (1 m²). Thus, the measure of intensity of received (or emitted) radiation is given as watts per square meter (W/m²). Because there are no common equivalents for this energy flow rate in the English system, we will use only metric units.

Longwave Radiation from the Earth

Land, ocean, and atmosphere possess heat that is derived by absorbing the sun's rays. Since the earth's surface and atmosphere are much colder than the sun's surface, we can predict that our planet radiates less energy and that the energy released has longer wavelengths. This is also shown on the right side of Figure 2.2, where the curve indicates earth-to-space outgoing radiation. Notice that the overall height is much lower than for the solar curve on the left, which confirms our expectation. What about wavelength? The curve shows that the energy is emitted at wavelengths of about 3 to 30 μm, verifying our observation that a cooler surface emits radiation with longer wavelengths.

The curve of outgoing longwave radiation shows three distinct peaks at wavelengths of about 5, 10, and 20 μm. In between these wavelengths, atmospheric gases, in the form primarily of water vapor and carbon dioxide, absorb much of the radiation leaving the surface.

As Figure 2.2 shows, the flow of longwave energy leaving our planet is only a fraction of the flow of incoming shortwave solar energy. Why so low? One reason why is because solar rays fall on only one hemisphere, whereas longwave radiation is constantly emitted from the entire spherical surface of the earth.

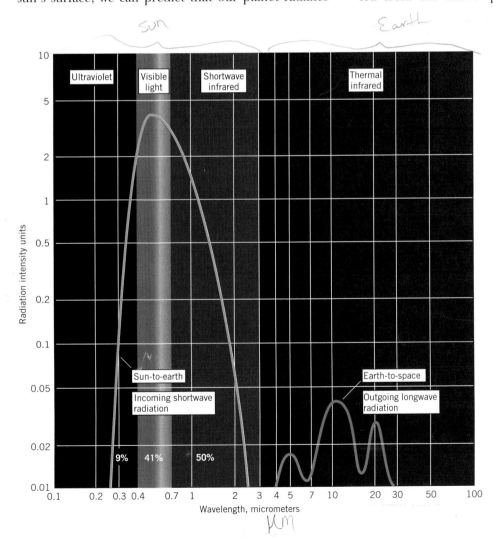

Figure 2.2 There are four important divisions within the electromagnetic spectrum—ultraviolet, visible light, short-wave infrared, and thermal infrared. The curve on the left shows the intensity of incoming solar shortwave radiation as it reaches the surface. The curve on the right shows the outgoing longwave radiation from earth and atmosphere to outer space.

Figure 2.3 A thermal infrared image of a suburban scene at night. Black and blue tones show lower temperatures, while yellow and red tones show higher temperatures. Ground and sky are coldest, while the heated windows of the homes are warmest.

Thus, outgoing radiation needs only to be one-fourth as strong. In addition, about one-third of the incoming solar radiation is reflected directly back to space. This means that only two-thirds remains to be absorbed by the atmosphere and surface and eventually reradiated as heat.

Because longwave radiation continues throughout the night, by using special sensors mounted on aircraft it is possible to obtain thermal infrared images of ground features. Figure 2.3 shows an image obtained in total darkness. Houses, windows and pavement appear bright because they are warm and radiate more intensely. In contrast, the moist surfaces of lawns are cooler and appear darker.

THE GLOBAL RADIATION BALANCE

As we have seen, the earth constantly absorbs solar shortwave radiation and emits longwave radiation. These two energy flows must balance. Figure 2.4 presents a diagram of this energy flow process, which we refer to as the earth's **radiation balance**.

Input of solar energy is in the form of shortwave radiation. However, part of this radiation is reflected back into space before it can be absorbed and stored as heat. Clouds and dust particles in the atmosphere contribute to this scattering, as do land and ocean surfaces. The remaining portion of shortwave energy from the sun is absorbed by either atmosphere, land, or ocean. Once absorbed, solar energy raises the temperature of the atmosphere, as well as the surfaces of the oceans and lands. However, the atmosphere, land, and ocean emit energy in the form of longwave radiation. This radiation ultimately leaves the planet, headed for outer space. The longwave radiation out-

flow tends to lower the temperature of the atmosphere, ocean, and land, and thus cool the planet.

In the long run, these flows balance—incoming energy absorbed and outgoing radiation emitted are equal. Since the temperature of a surface is determined by the amount of energy it absorbs and emits, the earth's overall temperature tends to remain constant. Shortwave absorption and longwave emissions balance for the earth as a whole, but they do not always necessarily balance within a single system, such as the atmosphere. For example, the atmosphere radiates much more longwave energy than it receives by absorption of shortwave energy.

> **In the long run, the rate of solar energy absorption by the earth is balanced by the rate of longwave energy emission from the earth.**

Insolation Variation with Sun Angle and Day Length

Although solar radiation received at the outer edge of the earth's atmosphere is nearly constant over time, the solar energy received at any point on the earth's surface can vary greatly. First, there is a daily variation—there is plenty of solar radiation during the day but none at night. Further changes in solar radiation are evident during different seasons. In summer, the sun is high in the sky and thus heats the surface of the earth more intensely. Also, the days are longer. Let's investigate these variations in more detail.

Refer back to Chapter 1 and Figure 1.15, showing equinox conditions. Only at the subsolar point does the earth's spherical surface present itself at a right

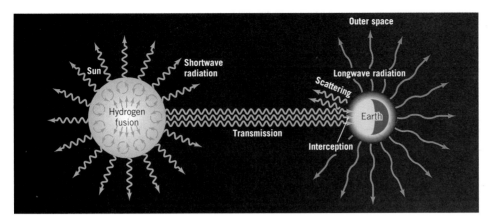

angle to the sun's rays. To a viewer on the earth at the subsolar point, the sun is directly overhead. But as we move away from the subsolar point toward either pole, the earth's curved surface becomes turned at an angle with respect to the sun's rays. To the earthbound viewer at a new latitude, the sun appears to be nearer the horizon. When the circle of illumination is reached, the sun's rays are parallel with the surface. That is, the sun is viewed just at the horizon. The point is that the angle of the sun in the sky at a particular location depends on both the latitude of the location and the time of day.

The angle of the sun in the sky is important, for it determines the intensity of solar radiation on the surface. Figure 2.5 shows that insolation is greatest where the sun's rays strike vertically. When the angle diminishes, the same amount of solar energy spreads over a greater area of ground surface. So when the sun is high in the sky overhead, solar radiation is more intense.

How does the angle of the sun vary during the day? Figure 2.6 shows how the sun's path in the sky changes from season to season. The diagram is drawn for latitude 40° N and is typical of conditions in middle latitudes in the northern hemisphere, for example, at New York or Denver. The diagram shows a small area of the earth's surface bounded by a circular horizon. This is the way things appear to an observer standing on a wide plain. The earth's surface appears flat, and the sun seems to travel inside a vast dome in the sky.

At equinox, the sun rises directly to the east and sets directly to the west. At noon, the sun rests at an angle of 50° above the horizon in the southern sky. At equinox, the sun is above the horizon for exactly 12 hours, as shown by the hour numbers on its path. At summer solstice, the sun is above the horizon for about 15 hours. In addition, the sun's path rises much higher in the sky, for at noon it will be 73° above the horizon. At winter solstice, the sun's path is low, reaching only 26° above the horizon, and the sun is visible

for only about 9 hours. Clearly, the solar radiation reaching the surface is much greater during the day at summer solstice than at winter solstice.

Insolation, Latitude, and the Seasons

Assume that the earth is a perfectly uniform sphere with no atmosphere. That is, there are no clouds or dust particles to intercept solar radiation aiming toward the surface. We use the term insolation (incoming solar radiation) to mean the amount of solar short-wave energy intercepted by an exposed surface. The insolation received in one day depends on two factors: (1) the angle at which the sun's rays strike the earth, and (2) the length of time of exposure to the rays.

Figure 2.7 shows annual insolation by latitude. As we might expect from the general effect of latitude on sun

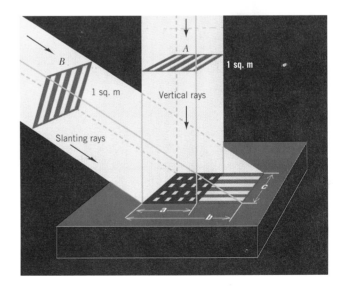

Figure 2.5 The angle of the sun's rays determines the intensity of insolation. The energy of vertical rays *A* is concentrated in square *a* by *c*, but the same energy in the slanting rays *B* is spread over a larger rectangle, *b* by *c*.

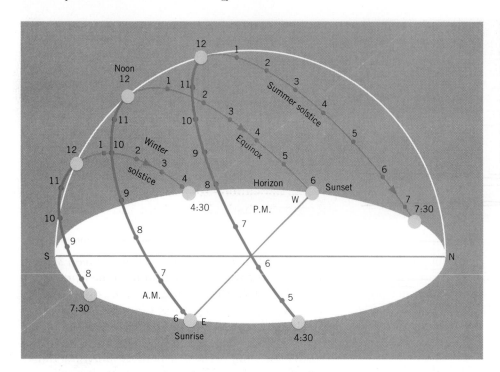

Figure 2.6 The sun's path in the sky changes greatly in position and height above the horizon from summer to winter. This diagram is for a location at lat. 40° N (New York or Denver). At the winter solstice, the sun is low in the sky and the period of daylight is short. At the summer solstice, the sun is high and the daylight period is long.

Incoming solar radiation, or insolation, depends on the sun's angle and the duration of daylight.

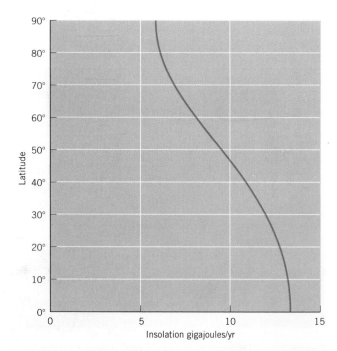

Figure 2.7 Total annual insolation from equator to pole. At the pole, annual insolation is about 40 percent of that at the equator.

angle, annual insolation is greatest at the equator and decreases toward the poles. However, annual insolation at the poles is still nearly 40 percent of that of the equator! The polar day is six months long, and clearly a large amount of insolation can be received during this lengthy day.

When the daily insolation values are plotted for a full year, they form a wavelike curve on a graph. Figure 2.8 is just such a graph. The curve for latitude 40° N shows the features just discussed—much greater daily insolation at the summer solstice (June) and much less daily insolation at the winter solstice (December).

Not every latitude shows a simple progression from low to high, back to low insolation through the course of the year. Notice that the equator has two periods of maximum insolation. These periods correspond to the equinoxes, when the sun is overhead at the equator. There are also two minimum periods corresponding to the solstices, when the subsolar point moves farthest north and south from the equator. All latitudes between the tropic of cancer ($23\frac{1}{2}°$ N) and the tropic of capricorn ($23\frac{1}{2}°$ S) have two maximum and minimum values. However, as either tropic is approached, the two maximum periods get closer and closer in time, and then merge into a single maximum.

Poleward of the two tropics is a single insolation cycle with a maximum at one solstice and a minimum at the other. At the arctic circle ($66\frac{1}{2}°$ N), insolation is reduced to zero on the day of the winter solstice. At latitudes further poleward, this seasonal period of no insolation increases from 1 day to 182 days.

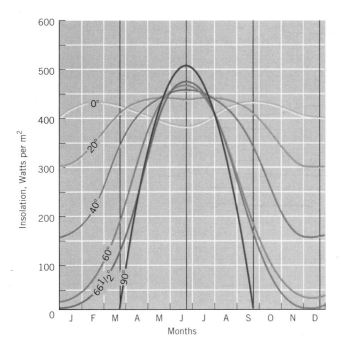

Figure 2.8 Insolation curves at various latitudes in the northern hemisphere (assuming no atmosphere). Dashed lines mark the equinoxes and solstices. Latitudes between the equator (0°) and tropic of cancer ($23\frac{1}{2}$° N) show two maximum values; others show only one. Poleward of the arctic circle ($66\frac{1}{2}$° N), insolation is zero for at least some period of the year. (Copyright © A. N. Strahler.)

The seasonal pattern of insolation is directly related to latitude. This pattern is important because it is the driving force for the annual cycle of climate. Nowhere on earth is insolation constant from day to day. Even at the equator, Figure 2.8 shows that there are significant seasonal differences in the solar energy available to warm the earth and atmosphere. Poleward of the tropics, the large differences produce the large variations in temperature from winter to summer.

World Latitude Zones

The seasonal pattern of insolation can be used as a basis for dividing the globe into broad latitude zones (Figure 2.9). The zone limits either shown in the figure or specified below should not be taken as absolute and binding, however. Rather, this system of names is a convenient way to identify general world geographic belts throughout this book.

The *equatorial zone* encompasses the equator and covers the latitude belt roughly 10° north to 10° south. The sun provides intense insolation throughout the year, and days and nights are of roughly equal length within this zone. Spanning the tropics of cancer and capricorn are the *tropical zones*, ranging from latitudes

10° to 25° north and south. A marked seasonal cycle exists in these zones, but it is still combined with a large total annual insolation.

Moving toward the poles from each of the tropical zones are transitional regions called the *subtropical zones*. For convenience, we assign these zones the latitude belts 25° to 35° north and south. At times we may extend "subtropical" a few degrees farther poleward or equatorward of these parallels.

The midlatitude zones lie between 35° and 55° latitudes in the northern and southern hemispheres.

The *midlatitude zones* are next, lying between 35° and 55° north and south latitude. In these belts, the sun's height in the sky shifts through a wide range annually. Differences in day length from winter to summer are also large. Thus, seasonal contrasts in insolation are also quite strong. In turn, these regions can experience a large range in annual surface temperature.

Bordering the midlatitude zones on the poleward side are the *subarctic zone* and *subantarctic zone*, 55° to 60° north and south latitudes. Astride the arctic and antarctic circles, $66\frac{1}{2}$° north and south latitudes, lie the

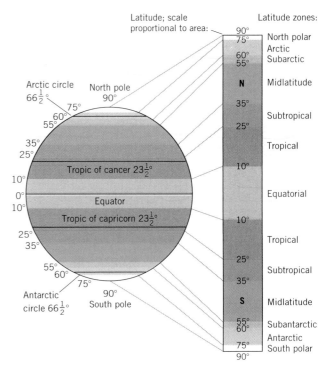

Figure 2.9 A geographer's system of latitude zones. These are based on the seasonal patterns of insolation observed over the globe.

arctic and *antarctic zones*. These zones have an extremely large yearly variation in day lengths, yielding enormous contrasts in insolation from solstice to solstice.

The *polar zones,* north and south, are circular areas between about 75° latitude and the poles. Here the polar regime of a six-month day and six-month night is predominant. These zones experience the greatest seasonal contrasts of insolation of any area on earth.

COMPOSITION OF THE ATMOSPHERE

A main function of this chapter is to explain the flows of energy within the atmosphere and between the atmosphere and the earth's surface. An understanding of this topic requires first some knowledge of basic facts about the atmosphere.

The earth's atmosphere consists of a mixture of various gases surrounding the earth to a height of many kilometers. This envelope of air is held to the earth by gravitational attraction. Since air is compressible, it is densest at sea level and thins rapidly upward. Almost all the atmosphere (97 percent) lies within 30 km (18 mi) of the earth's surface. The upper limit of the atmosphere is at a height of approximately 10,000 km (6000 mi), a distance approaching the diameter of the earth itself. From the earth's surface upward to an altitude of about 80 km (50 mi), the chemical composition of air is highly uniform in terms of the proportions of its gases.

Pure, dry air consists largely of nitrogen, about 78 percent by volume, and oxygen, about 21 percent (Figure 2.10). Other gases account for the remaining 1 percent. *Nitrogen* in the atmosphere exists as a molecule consisting of two nitrogen atoms (N_2). Nitrogen gas does not enter easily into chemical union with other substances and can be thought of mainly as a neutral substance. Very small amounts of nitrogen are extracted by soil bacteria and made available for use by plants. Otherwise, it is largely a "filler," adding inert bulk to the atmosphere.

In contrast to nitrogen, *oxygen gas* (O_2) is highly active chemically. It combines readily with other elements in the process of oxidation. Combustion of fuels represents a rapid form of oxidation, whereas certain forms of rock decay (weathering) represent very slow forms of oxidation. Living tissues require oxygen to convert foods into energy.

These two main component gases of the lower atmosphere are perfectly mixed so as to give the pure, dry air a definite set of physical properties, as though it were a single gas.

The remaining 1 percent of dry air is mostly argon, an inactive gas of little importance in natural processes. In addition, there is a very small amount of *carbon dioxide* (CO_2), amounting to about 0.033 percent. This gas is of great importance in atmospheric processes because of its ability to absorb radiant heat passing through the atmosphere from the earth. Carbon dioxide thus adds to the warming of the lower atmosphere. Carbon dioxide is also used by green plants in the photosynthesis process. During photosynthesis, CO_2 is converted into chemical compounds that build up the plant's tissues, organs, and supporting structures.

Ozone (O_3) is found mostly in the upper part of the atmosphere. Although its concentration is very low, ozone serves the essential function of absorbing ultraviolet radiation from the sun as the radiation passes through the atmosphere. Since this type of radiation can be damaging to plants and animals, earthly life depends on this ozone shield.

Water vapor is an important component of the atmosphere that varies in concentration from place to place and time to time.

Another important component of the atmosphere is *water vapor.* This is the gaseous form of water, in which individual water molecules are mixed freely throughout in the atmosphere, just like the other atmospheric component gases. Unlike the other component gases, however, the concentration of water vapor can vary highly. Usually, water vapor makes up less than 1 percent of the atmosphere. Under very warm, moist conditions, as much as 2 percent of the air can be water vapor. We are all familiar with water vapor as the component of the atmosphere that makes humid days feel so oppressive. Since water vapor, like carbon dioxide, is a good absorber of heat radiation, it also plays a role in warming the lower atmosphere

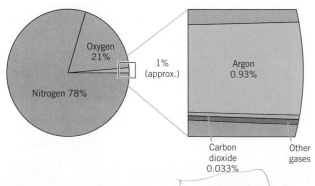

Figure 2.10 Components of the lower atmosphere. Values show percentage by volume. Nitrogen and oxygen form 99 percent of our air, with other gases, principally argon and carbon dioxide, accounting for the final 1 percent.

SENSIBLE HEAT AND LATENT HEAT TRANSFER

Thus far, we have discussed energy in the form of shortwave and longwave radiation rather extensively. However, energy can be transported in other ways, two of which are extremely important when discussing the global energy balance—sensible heat and latent heat transfer.

The most familiar form of heat storage and transport is known as **sensible heat**—the quantity of heat held by an object that can be sensed by touching or feeling. This kind of heat is measured by a thermometer. When the temperature of an object or a gas increases, its store of sensible heat increases. When two objects of unlike temperature contact each other, heat energy moves by *conduction* from the warmer to the cooler. Molecules of a solid or liquid are in constant motion (vibration)—like dancers moving on a crowded dance floor. During conduction, the faster moving molecules of the warmer substance pass along some of their energy of motion to adjacent molecules, in a sort of chain reaction. The warmer object cools, and the cooler object is warmed. This type of heat flow is referred to as **sensible heat transfer.**

One very important property of water vapor is that it contains **latent heat.** For liquid water to make the transition to water vapor, energy is required. For example, consider the cooling you feel in a breeze when your skin is wet. As the water evaporates from liquid to gas, it draws up the heat required to make the change of state and carries it away from your skin, keeping you cool. This is the latent heat. Latent means "hidden," and here the heat is carried in the form of the fast random motion of the free water vapor molecules as they enter the gaseous state during the process of evaporation.

The latent heat of water vapor in air cannot be measured by a thermometer, but energy is stored there just the same. When water vapor turns back to a liquid, the latent heat is released and becomes sensible heat. In the earth–atmosphere system, **latent heat transfer** occurs when water evaporates from a moist land surface or open water surface, transferring latent heat to the atmosphere. On a global scale, latent heat transfer provides a mechanism for moving large amounts of heat from one region of the earth to another.

THE GLOBAL ENERGY SYSTEM

The flow of energy from the sun to the earth and then back out into space is a complex system involving not only radiation, but also energy storage and transport. We will investigate this complex system in more detail to fully appreciate global weather and climate. Keep in mind that we are really studying the way the sun powers the earth's surface processes, ranging from wind systems to the cycle of precipitation.

Insolation Losses in the Atmosphere

Consider the flow of solar radiation through the atmosphere on its way toward the surface. As the shortwave radiation penetrates the atmosphere, its energy is absorbed or diverted in various ways. This is shown in Figure 2.11. The typical values for losses of incoming solar energy are taken to the nearest 5 percent.

At an altitude of 150 km (95 mi), the solar radiation spectrum possesses almost 100 percent of its original energy. However, by the time radiation penetrates to an altitude of 88 km (55 mi), absorption of X rays is nearly complete, and some of the ultraviolet radiation has been absorbed as well.

Molecules and particles in the air scatter radiation in all directions. When the rays are scattered back toward space, this process is referred to as reflection.

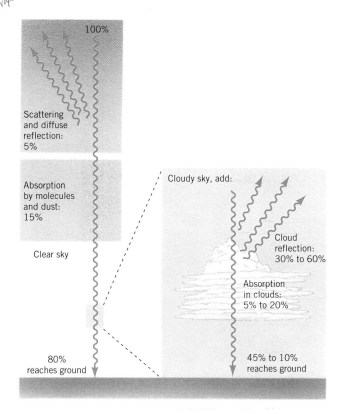

Figure 2.11 Losses of incoming solar energy are much lower with clear skies (*left*) than with cloud cover (*right*). (Copyright © A. N. Strahler.)

Eye on the Environment • The Ozone Layer–Shield to Life

When you hear about the ozone on the nightly news, the news story is probably referring to damage to the atmosphere's *ozone layer*. The ozone layer serves as a shield against harmful solar ultraviolet radiation. The ozone layer begins at an altitude of about 15 km (9 mi) and extends to about 55 km (35 mi) above the earth.

The action of ultraviolet rays on ordinary oxygen molecules produces ozone. When the rays are absorbed, the oxygen molecule (O_2) splits into two oxygen atoms (O + O). Each of these atoms can in turn combine with O_2 molecules to yield O_3, or ozone. Like O_2, ozone also absorbs ultraviolet rays, and it splits back to O_2 + O. Thus, ozone is constantly formed, destroyed, and then reformed in the ozone layer, absorbing ultraviolet radiation with each transformation.

This absorption protects the earth's surface from most of the sun's ultraviolet radiation. If these ultraviolet rays were to reach the earth's surface in their full intensity, all bacteria exposed on the earth's surface would be destroyed and animal tissues would be severely damaged. The presence of the ozone layer is an essential protection in maintaining a viable environment for life on this planet.

The release into the atmosphere of *chlorofluorocarbons* or *CFCs* poses a serious threat to the ozone layer. CFCs are synthetic chemical compounds containing chlorine, fluorine, and carbon atoms. Before 1976, when the Environmental Protection Agency in the United States issued a ban on aerosol spray cans charged with CFCs, many households used such aerosols. CFCs continue to be widely used, however, as the cooling fluid in refrigeration systems so that when these appliances are disposed of, their CFCs are released to the air.

Molecules of CFCs in the atmosphere are very stable close to the earth, and so they diffuse upward without chemical change until they eventually reach the ozone layer. As these compounds absorb ultraviolet radiation, they are decomposed and chlorine atoms are released. The chlorine atoms in turn attack molecules of ozone, converting them in large numbers into ordinary oxygen molecules by a chain reaction. In this way the ozone concentration within the stratospheric ozone layer is reduced. The intensity of ultraviolet radiation reaching the earth's surface can increase. A marked increase in the incidence of skin cancer in humans is one of the predicted effects. Other possible effects include reduction of crop yields and the death of some forms of aquatic life.

As studies of the ozone layer based on satellite data accumulated during the 1980s, a substantial decline in the total global ozone was noted. By 1987, following an 8 percent ozone decline in the preceding eight years, scientists agreed that the decrease in the ozone layer was escalating far faster than had been predicted.

Compounding the problem of global ozone decrease was the discovery in the mid-1970s of a "hole" in the ozone layer over the continent of Antarctica. (See figure at right.) Here, seasonal thinning of the layer occurs during the early spring of that hemisphere, and ozone reaches a minimum during the month of October. Most scientists now confidently attribute the seasonal thinning to the presence of CFCs.

In 1992 researchers reanalyzed 12 years of weather satellite data and compared the results with ground measurements. Researchers concluded that at various points in the northern hemisphere the ozone shield had thinned by as much as 6 percent. Damage to the ozone layer was noted as far south as Florida, and some loss persisted from the early spring into the early summer.

Volcanic dust, inserted into the stratosphere, also can act to reduce ozone concentrations. The June, 1991 eruption of Mt. Pinatubo, in the Philippines, reduced global ozone in the stratosphere by 4 percent during the following year, with reductions in the layer over midlatitudes of up to 9 percent.

If the global ozone layer thins substantially, the rate of incoming ultraviolet solar radiation will be expected to increase. Scientists estimate that for each 1 percent decrease in global ozone, ultraviolet radiation should increase by 2 percent. Measurements of ultraviolet radiation reaching the earth's surface, however, have not always shown increases in recent years. Several other factors effect year-to-year changes in recorded ultraviolet radiation. These factors include variations in average cloud cover and air pollution. Volcanic dust suspended in the upper atmosphere can also reduce ozone concentration and increase surface ultraviolet radiation. Because of these other contributing factors, we need to use caution before linking any increase in ultraviolet radiation to ozone depletion.

Responding to the global threat of ozone depletion, and its anticipated impact on the biosphere, 23 nations endorsed a United Nations plan in 1987 for cutting global CFC consumption by 50 percent by the year 1999. In 1988, under an agreement known as the Montreal Protocol, the United States joined 31 other nations in approving a similar goal. In 1990 the international agreement was expanded and strengthened to a goal of 50 percent reduction of CFCs by 1995 and 85 percent by 1997. That goal was further strengthened in 1992, when the United States announced a phaseout of all American CFC manufacturing by the end of 1995.

Although the manufacture and consumption of CFCs has now been halted or greatly reduced, the ozone layer will continue to thin for some time as CFCs already in the lower atmosphere continue to work their way upward to the stratosphere. However, the annual rate of increase in concentration of CFCs is already slowing in response to the production cutbacks. In 1993, scientists reported increases of only 1–3 percent, compared to values of 10–20 percent in previous years.

Questions

1. How does the ozone layer protect the life layer?
2. What are CFCs and how do they impact the ozone layer?
3. When and where have ozone reductions been reported? Have corresponding reductions in ultraviolet radiation been noted?
4. What actions have governments around the world taken on the ozone layer problem?

Map of ozone concentration over the South Pole on October 6, 1991. Colors indicate ozone levels. High ozone concentrations are in red and orange. Purple, pink and white values are lowest. Outlines of the continents are shown in black. The ozone hole over Antarctica shows very clearly in the center of the image. Produced from data of the Nimbus-7 satellite by NASA and NOAA researchers.

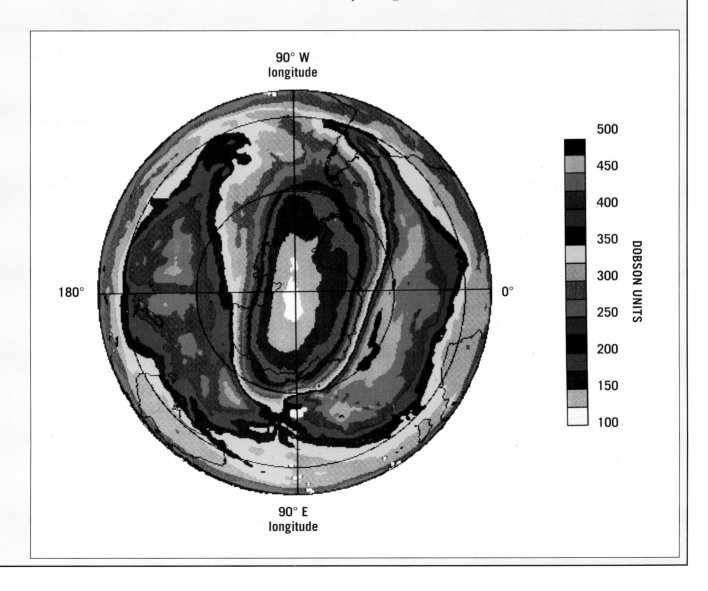

90° W longitude

180°

0°

90° E longitude

500
450
400
350
300
250
200
150
100

DOBSON UNITS

As radiation moves through deeper and denser atmospheric layers, gas molecules cause some visible light rays to be turned aside in all possible directions, a process known as **scattering**. Dust and other particles in the air cause further scattering. Scattered radiation is unchanged, except for the direction of its path. Scattered radiation moving in all directions through the atmosphere is known as *diffuse radiation*. Because scattering occurs in all directions, some energy is sent back into outer space and some flows down to the earth's surface. When the scattering turns radiation back to space, we call this process *reflection*. Under clear sky conditions, this diffuse reflection sends about 5 percent of the incoming solar radiation back to space.

Another form of energy loss in the atmosphere is *absorption*. In this process, solar rays strike gas molecules and dust, and their energy is absorbed as sensible heat. This accounts for about 15 percent of the incoming solar radiation. Thus, when skies are clear, reflection and absorption combined total about 20 percent, leaving as much as 80 percent of the solar radiation to reach the ground.

Both carbon dioxide and water vapor are capable of directly absorbing some wavelengths of solar radiation. Carbon dioxide is present fairly uniformly in the air, but, as we noted earlier, the water vapor content of the air can vary greatly. Thus, absorption can vary from one global environment to another. Absorption results in a rise in air temperature. In this way, some direct heating of the lower atmosphere takes place during incoming solar radiation.

When clouds are present, they can greatly increase the amount of incoming solar radiation reflected back to space. The bright white surfaces of clouds are extremely good reflectors of shortwave radiation. Cloud reflection can account for a direct turning back to space of 30 to 60 percent of incoming radiation (Figure 2.11). Clouds also absorb radiation, perhaps as much as 5 to 20 percent. Under conditions of a heavy cloud layer, as little as 10 percent of incoming solar radiation can actually reach the ground.

Albedo

The percentage of shortwave radiant energy scattered upward by a surface is termed its **albedo.** For example, a surface that reflects 40 percent of the shortwave radiation it receives has an albedo of 40 percent. Albedo is an important property of a surface, because it determines how fast the surface heats up when exposed to insolation. That is, a surface with a high albedo, such as snow or ice (45 to 85 percent), reflects much or most of the solar radiation and so heats up slowly. A surface with a low albedo, such as black pavement (3 percent), absorbs nearly all the solar radiation and so heats up quickly. The albedo of a water surface is very low (2 percent) for nearly vertical rays. Thus, water is a very good absorber of shortwave radiation. However, when the sun shines on a calm water surface

a.

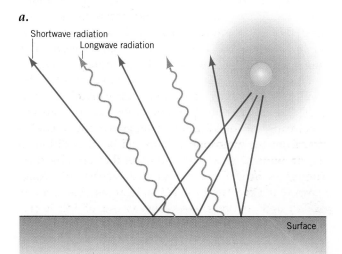

Shortwave radiation
Longwave radiation

Surface

b.

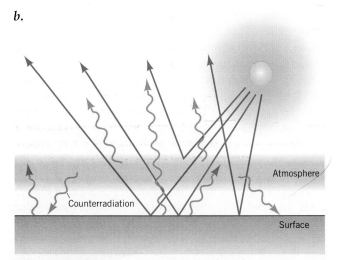

Atmosphere

Counterradiation

Surface

Figure 2.12 Effect of the atmosphere on the surface energy budget. Without an atmosphere, (*a*) solar radiation is reflected by surface scattering back to space, and the surface emits longwave radiation to space. In (*b*), the atmosphere absorbs much upwelling longwave energy and emits some back to the surface as counterradiation. The atmosphere also scatters some shortwave radiation back to space.

at a low angle, much of the radiation is directly reflected as sun glint, producing a higher albedo. For fields, forests, and bare ground, albedos are of intermediate value, ranging from as low as 3 percent to as high as 25 percent.

The earth's overall albedo lies in the range of 29 to 34 percent.

Certain orbiting satellites are equipped with instruments to measure the energy levels of shortwave and infrared radiation. Radiation, both incoming from the sun and outgoing from the atmosphere and earth's surfaces below, is measured. Data from these satellites have been used to estimate the earth's average albedo. This albedo includes reflection by the earth's atmosphere as well as its surfaces. Therefore, scattering by clouds, dust, and atmospheric molecules is included. The albedo values obtained in this way vary between 29 and 34 percent. This means that the earth—atmosphere system directly returns back to space slightly less than one-third of the solar radiation it receives.

Counterradiation and the Greenhouse Effect

Although the ground can only radiate longwave energy upward, the atmosphere can radiate longwave energy both upward to space, where it is lost, and downward to the ground. Recall that water vapor and carbon dioxide in the atmosphere continuously absorb much of the longwave radiation emitted upward from the earth's surface. This means that part of the ground radiation absorbed by the atmosphere is radiated back down toward the earth surface. This process is called counterradiation. For this reason, the lower atmosphere, with its longwave-absorbing gases, acts like a blanket that returns heat to the earth. This mechanism helps to keep surface temperatures from dropping excessively during the night or in winter at middle and high latitudes. Cloud layers are even more important than carbon dioxide and water vapor in producing a blanketing effect to retain heat in the lower atmosphere, since clouds are excellent absorbers and emitters of longwave radiation.

The principle of counterradiation is shown in Figure 2.12. In part (*a*), no atmosphere is present. Solar energy is reflected upward by scattering, and longwave energy is emitted by the surface directly to space. In part (*b*), an atmospheric layer absorbs the upwelling longwave radiation. The portion that is absorbed and reemitted downward is the counterradiation. Somewhat the same principle is used in greenhouses and in homes using large windows to entrap solar heat.

Here, the glass windows permit entry of shortwave energy but block the exit of longwave energy. This phenomenon is even better demonstrated by the intense heating of air in a closed automobile left parked in the sun. We use the expression **greenhouse effect** to describe this atmospheric heating principle.

Global Energy Budgets of the Atmosphere and Surface

The global energy budget for the earth's surface and atmosphere is diagrammed in Figure 2.13. We begin with a discussion of shortwave radiation, shown in the upper part of the figure. Since shortwave radiation comes only from the sun, we are concerned only with the downward flow of solar radiation and its fate.

As shown in the left part of the figure, reflection by molecules and dust, clouds, and the surface (including the oceans) totals 31 percent. Note that the value of 31 percent given in Figure 2.13 lies between the satellite observation values of 29 to 34 percent. The right side of the upper figure shows values for absorption in the atmosphere. The combined losses through absorption by molecules, dust, and clouds average 21 percent. With 31 percent of incoming solar energy reflected and 21 percent absorbed in the atmosphere, 48 percent is left. This amount is absorbed by the earth's land and water surfaces. We have now accounted for all the insolation energy.

The right part of Figure 2.13 shows the components of outgoing longwave radiation for the earth's surface, the atmosphere, and the planet as a whole, using the same percentage units as for the incoming radiation. The large arrow on the left shows total longwave radiation leaving the earth's land and ocean surface. The long, thin arrow indicates that 6 units are lost to space, while 107 units are absorbed by the atmosphere. The loss is therefore equivalent to 113 percentage units.

This means that the surface emits 113 percent of the total incoming solar radiation! How can this be correct? To answer this question, note that the surface receives a flow of 97 units of longwave counterradiation from the atmosphere. This will be in addition to the 48 units of shortwave radiation that are absorbed (upper figure). Thus, the surface receives 97 (longwave) + 48 (shortwave) = 145 units in all. Thus, a loss of 113 units by longwave radiation out of 145 is certainly possible.

On the far right of the figure are two smaller arrows. These show the flow of energy away from the surface as latent heat and sensible heat. Also recall that latent heat is absorbed when water evaporates. Therefore, evaporation of water from moist soil or ocean transfers latent heat from the surface to the atmosphere. Also

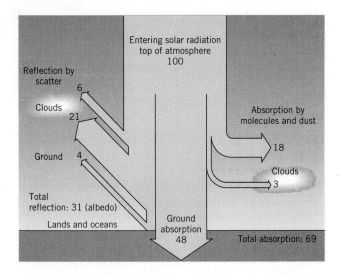

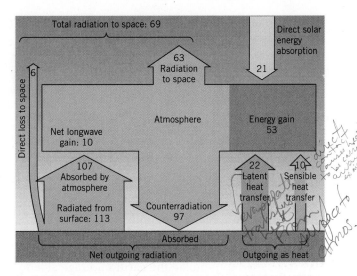

Figure 2.13 Diagram of the global energy balance. Values are percentage units based on total insolation as 100. The left figure shows the fate of incoming solar radiation. The right figure shows longwave energy flows occurring between the surface and atmosphere and space on the left. On the right are the transfers of latent heat, sensible heat, and direct solar absorption that balance the budget for earth and atmosphere.

recall that the transfer of sensible heat refers to heat transferred by conduction. When air is warmed by direct contact with a surface and rises, it takes its heat along with it. In this process, heat flows from the surface to the atmosphere.

Latent heat flows from a moist surface to the atmosphere when water evaporates.

These last two heat flows are not part of the radiation balance, since they are not in the form of radiation. However, they are a very important part of the total energy budget of the surface, which includes all forms of energy. Taken together, these two flows account for 32 units leaving the surface. With their contribution, the surface energy balance is complete. Total gains are 97 (longwave) + 48 (shortwave) = 145. Total losses are 113 (longwave) + 22 (latent heat) + 10 (sensible heat) = 145.

Like the earth's surface, the atmosphere also emits longwave radiation (Figure 2.12*b*). However, this radiation is divided into two parts. One part is directed into space (63 units), while the other part is counterradiation directed back to the earth's surface (97 units). For the box representing the atmosphere, the total of this emitted radiation is 63 + 97 = 160 percentage units. This figure seems absurdly large until we note that the atmosphere receives not only 107 units of longwave radiation, but also 22 units of latent heat, 10 units of sensible heat from the surface, and 21 units of shortwave radiation by direct absorption. These units total

to 160. So the energy budget balances for the atmosphere as well as for the surface.

This analysis helps us to understand the vital role of the atmosphere in warming the surface though the greenhouse effect. Because of counterradiation, the surface receives 145 units and radiates away 113 of them. In order to do so, the surface must be warm indeed. Without the counterradiation of the atmosphere, the surface would have only 48 units of absorbed shortwave radiation to lose. The temperature needed to radiate so few units is well below freezing. So without the greenhouse effect, our planet would be a very cold, forbidding place.

NET RADIATION, LATITUDE, AND THE ENERGY BALANCE

Solar energy is intercepted by our planet, and because some of it is absorbed, the heat level of our planet tends to rise. At the same time, our planet radiates energy into outer space, a process that tends to reduce its level of heat energy. Over time, these flows must balance for the earth as a whole. However, incoming and outgoing radiation flows for any given surface do not have to be in balance. At night, for example, there is no incoming radiation, yet the earth's surface and atmosphere still emit outgoing radiation. In any one place and time, more radiant energy is being gained than lost, or vice versa.

Net radiation expresses the difference between all incoming radiation and all outgoing radiation. In places where energy is coming in faster than it is going out, net radiation will be a positive quantity, providing a surplus. In other places where energy is going out faster than it is coming in, net radiation will be a negative quantity, yielding a radiant energy deficit. Our analysis of the radiation balance has already shown that for the entire globe as a unit, the net radiation is zero on an annual basis.

Net radiation at a surface measures the balance between incoming and outgoing radiation.

In Figure 2.8, we saw that solar energy input varies strongly with latitude. What is the effect of this difference? A study of the net radiation profile spanning the entire latitude range 90° N to 90° S will answer this question. In this analysis we will use yearly averages for each latitude, so that the effect of seasons is concealed.

The lower part of Figure 2.14 presents a global profile of net radiation from pole to pole. Between about lat. 40° N and S there is a net radiant energy gain labeled "surplus," and as a result, incoming solar radiation exceeds outgoing longwave radiation throughout the year. This phenomenon occurs only if there are

other means to carry away the excess heat energy. Similarly, poleward of 40° N and S, the net radiation is negative and is labeled "deficit"—meaning that outgoing longwave radiation exceeds incoming shortwave radiation. This can happen only if energy is brought into these middle latitudes from the equatorial and tropical zones.

What mechanism brings energy from the lower latitudes poleward? Actually, two processes are responsible. First, the circulation pattern of the oceans brings warm water poleward, moving heat energy across latitude boundaries. Because the heat is carried by the water, this is sensible heat transfer. Second, the movement of warm, moist air poleward carries latent heat energy to higher latitudes, as a latent heat transfer. These two energy transport processes taken together are called **poleward heat transport,** since the transport moves heat energy from lower to higher latitudes. The upper part of Figure 2.14 diagrams this flow.

If you carefully examine the lower part of Figure 2.14, you will find that the area on the graph labeled "surplus" is equal in size to the combined areas labeled "deficit." The matching of these areas confirms the fact that net radiation for the whole earth is zero and that global incoming shortwave radiation exactly balances global outgoing longwave radiation.

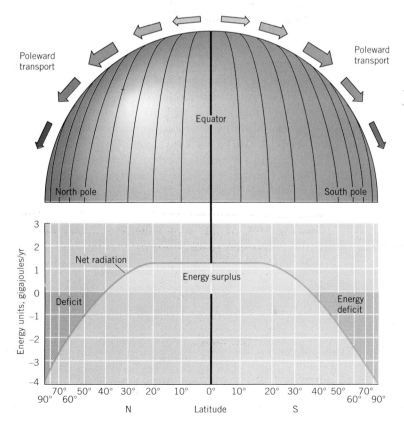

Figure 2.14 Annual surface net radiation from pole to pole. Where net radiation is positive, incoming solar radiation exceeds outgoing longwave radiation. There is an energy surplus, and so energy moves poleward as latent heat and sensible heat.

Although our analysis of the earth's energy balance is far from complete, it is clear that the balance is a sensitive one, involving a number of factors that determine how energy is transmitted and absorbed. Has industrial activity already altered the components of the planetary radiation balance? An increase in carbon dioxide will increase the absorption of longwave radiation by the atmosphere, enhancing the greenhouse effect. Will this change cause a steady and permanent rise in average atmospheric temperature?

An increase in atmospheric dust particles at upper levels of the atmosphere will increase the scattering of incoming shortwave radiation and thus reduce the shortwave energy available to warm the surface. On the other hand, increased dust content at low levels will act to absorb more longwave radiation and so raise the surface temperature. Has either change occurred? Human habitation, through cultivation and urbanization, has profoundly altered the earth's land surfaces. Have these changes in surface albedo and in the capacity of the ground to absorb and to emit longwave radiation changed the energy balance? Answers to such questions require further study of the processes of heating and cooling of the earth's atmosphere, lands, and oceans.

THE EARTH'S GLOBAL ENERGY BALANCE IN REVIEW

Electromagnetic radiation is a form of wave energy emitted by all objects. The wavelength of the radiation determines its characteristics. Shortwave radiation, emitted by the sun, includes ultraviolet, visible, and shortwave infrared radiation. Longwave radiation, emitted by earth surfaces, takes the form of heat. The hotter an object, the shorter the wavelength of the radiation and the greater the amount of radiation that it emits. Radiation flows are measured in watts per square meter. The earth continuously absorbs solar (shortwave) radiation and emits longwave radiation. In the long run, the gain and loss of energy balances, and the earth's average temperature remains constant.

Insolation, the amount of solar radiation available at a location, is greater when the sun is higher in the sky. Daily insolation is also larger when the period of daylight is longer. Annual insolation is greatest at the equator and least at the poles. However, the poles still receive 40 percent of the radiation received at the equator. The pattern of annual insolation with latitude leads to a natural naming convention for latitude zones: equatorial, tropical, subtropical, midlatitude, subarctic (subantarctic), arctic (antarctic), and polar.

The earth's atmosphere is dominated by nitrogen and oxygen gases. Carbon dioxide and water vapor, though minor constituents by volume, are very important because they absorb longwave radiation and enhance the greenhouse effect.

Part of the solar radiation passing through the atmosphere is absorbed or scattered by molecules, dust, and larger particles. Some of the scattered radiation returns to space as diffuse reflection. The land and ocean surfaces also reflect some solar radiation back to space, leaving the earth's surface to absorb from about 10 to 40 percent of the solar radiation, depending on the presence and nature of clouds. The proportion of radiation that a surface absorbs is termed its albedo. The albedo of the earth as a whole is about 30 percent. Heat energy can also be transmitted to or from a surface as latent heat or sensible heat.

The atmosphere absorbs longwave energy emitted by the earth's surface, causing the atmosphere to counterradiate some of that longwave radiation back to earth, thereby creating the greenhouse effect. Because of this heat trapping, the earth's surface temperature is considerably warmer than we might expect for an earth without an atmosphere.

Although absorbed shortwave radiation balances longwave radiation emitted by the earth's surface and atmosphere for the earth as a whole, this net radiation does not have to balance at every surface location on the globe. At latitudes lower than 40 degrees, annual net radiation is positive, while it is negative at higher latitudes. This imbalance can be sustained only if heat is exported from low to high latitudes by poleward heat transport. This heat is transported poleward as sensible heat by currents of warm water and as latent heat by flows of warm, moist air.

KEY TERMS	electromagnetic radiation	shortwave radiation	longwave radiation
	visible light	radiation balance	insolation
	ozone	sensible heat	sensible heat transfer
	latent heat	latent heat transfer	scattering
	albedo	counterradiation	greenhouse effect
	net radiation	poleward heat transport	

REVIEW QUESTIONS

1. What is electromagnetic radiation? How does the temperature of the hood of a car influence the nature of the electromagnetic radiation emitted?

2. Describe the types of radiation emitted by the sun and their characteristics.

3. Compare the terms shortwave radiation and longwave radiation. What are their sources?

4. How does the sun's path in the sky influence the amount of insolation a location receives in a day?

5. What is the influence of latitude on daily and annual insolation?

6. Sketch the world latitude zones on a circle representing the globe and give their approximate latitude ranges.

7. Why are carbon dioxide, ozone, and water vapor important atmospheric constituents?

8. Describe latent heat transfer and sensible heat transfer.

9. What is the fate of incoming insolation? Define scattering and absorption, including the role of clouds.

10. Define albedo and give two examples.

11. Describe the counterradiation process and how it relates to the greenhouse effect.

12. Discuss the energy balance of the earth's surface. Identify the types and sources of energy that the surface receives and do the same for energy that it loses.

13. Discuss the energy balance of the atmosphere. Identify the types and sources of energy that the atmosphere receives and do the same for energy that it loses.

14. What is net radiation? How does it vary with latitude?

IN-DEPTH QUESTIONS

1. Suppose the earth's axis of rotation was perpendicular to the orbital plane instead of tilted at $23\frac{1}{2}°$. How would global insolation be affected? How would insolation vary with latitude? How would the path of the sun in the sky change with the seasons?

2. Imagine that you are following a beam of either (a) shortwave solar radiation entering the earth's atmosphere heading toward the surface, or (b) a beam of longwave radiation emitted from the surface heading toward space. How will the atmosphere influence the beam?

Chapter 3

Air Temperature

The stars are bright and clear tonight. It's easy to trace the outline of the Big Dipper with the North Star at the edge of its rim. In fact, you can even see the white blur of the Milky Way, arching across the sky and disappearing behind the peaks of the tall fir trees that surround the meadow. Brrr! It's getting chilly now. With regret, you decide to abandon your stargazing and return to the campsite, just a few minutes away in the midst of the conifer forest. As you approach, the campfire is burning brightly. You step close, enjoying its warm glow. A friend offers you a mug of hot cocoa. You hold it with both hands, feeling its warmth seep into your cold fingers. "I'll bet we'll have frost tonight," you think as you sip the brown, sweet froth from the steaming cup. Sure enough, the next morning the grass in the meadow sparkles with white crystals that slowly turn to glistening beads when they are hit by the sun's first rays.

The trail is steep and the sun is hot. Though it was cool early this morning in the forest below, the low shrubs and scattered grasses don't relieve the force of the sun, now high overhead. The stark granite rocks are warm to the touch as you scramble over a low ledge. Straightening up, you see the summit ahead, shimmering in the rising currents of hot air. "I'm really glad I brought that extra water bottle," you think as you continue upward on the rocky path. "I'm going to need it by the time I make it to the top."

The view from the summit is absolutely spectacular. In the thin, clear air, the surrounding peaks stand majestically against the deep blue sky, some with patches of white snow glistening in the intense sunlight. Below the peaks, the gray-brown tones of the barren summit slopes grade into the deep green mantle of conifer forests that line the valleys and lower slopes. It's been a long, hard climb, especially the last part up the crusted surface of the snow field. The view, however, is worth every moment of exertion. You wouldn't have missed it for anything!

The subject of this chapter is air temperature, primarily the air temperature we experience every day within a few feet from the ground surface. What influences air temperature? Three factors are important. The first is the nature of the ground surface. In our example above, the grassy meadow loses heat more rapidly at night than the forest does. Air temperatures in the meadow drop more quickly, and frost forms first on the meadow grass. The dry, rocky surface of the mountain trail heats up more quickly than the forest, and so air temperatures rise more rapidly on the barren ground near the summit.

The second factor is the cyclic nature of solar radiation. The two motions of the earth—its daily rotation on its axis and its annual revolution in its solar orbit—create the two great cycles of solar radiation that produce day and night, and winter and summer. In turn, these cycles produce cycles of air temperature that follow daily and annual patterns. The daily cycle is obvious and natural in our example—the cool night spent in the forest followed by the warm day on the trail to the top. The annual cycle varies with latitude. In most parts of North America we experience a strong seasonal temperature variation, but near the equator the variation will be much smaller.

The third factor is elevation. At high elevation, there is less atmosphere above the surface. This means that the greenhouse effect provides a less effective insulating blanket. Average temperatures are cooler, since more surface heat is lost to space. This allows snow to accumulate and remain longer on the peaks in our example. Understanding air temperatures will be easy if you keep these three factors in mind.

Snow-covered lodgepole pines, Yellowstone National Park, Wyoming.

Temperature is a familiar concept. Essentially, it is a measure of the level of sensible heat of matter, whether it is gaseous (air), liquid (water), or solid (rock or dry soil). We know from experience that when a substance receives heat its temperature rises. As we saw in Chapter 2, the earth as a whole receives a constant flow of radiant shortwave energy from the sun. The earth also radiates longwave energy to space. During the day, the flow of shortwave radiation absorbed exceeds longwave energy emitted, and the surface temperature increases. At night, this net radiation balance reverses. No shortwave radiation strikes the darkened side of the earth, but longwave energy is still emitted from the surface. Therefore, surface temperatures decrease.

The temperature of a surface can be raised or lowered by processes other than absorption or emission of radiant energy. **Conduction** describes the flow of heat from a warmer substance to a colder one when the two are touching. In this process, the temperature of air in direct contact with the ground is raised when the ground receives strong sunlight. **Evaporation,** the process by which water changes from a liquid to a gas by absorbing heat, tends to lower the temperature of a wet surface. In the case of a patch of wet soil, the latent heat absorbed by the liquid water is taken away from the surface as the water vapor molecules leave the soil and move into the air layer above.

Air temperature is measured for recordkeeping at 1.2 m (4 ft) above the surface. However, the temperature of a surface can be somewhat different than the temperature of the air above it. For example, if you've ever walked barefoot across a grassy lawn on a summer day, you've probably noticed that the surface temperature can be a bit cooler than the air temperature at head height. Although such differences in temperature occur, they are generally not large because the lowermost layer of air tends to be well mixed. (The mixing process, or convection, will be covered in Chapter 5.) This coupling between the temperature of the surface and temperature at standard height means that we can explain measured air temperatures by understanding the principles that govern the temperatures of surfaces.

In most places, air temperatures follow natural cycles. First, there is the daily rhythm of rise and fall of air temperature, which is caused by the rhythm of incoming solar energy (insolation), varying from day to night. Second is the seasonal rhythm. As we saw in Chapter 2, this variation is caused by the tilt of the earth's axis, which exposes the northern and southern hemispheres to differing lengths of daylight and solar intensities as the year progresses.

Because the daily and annual cycles of insolation vary systematically with latitude, we will see that some temperature cycles are closely related to latitude. For example, temperatures generally fall as we move poleward. Since the sun at higher latitudes does not rise as high in the sky, on average it provides less energy to warm earth surfaces there. Latitude also enhances seasonal contrasts. As much solar energy may be received at high latitudes in the summer as at the equator. In winter, however, much less energy is received. Because of this annual variation, high latitudes experience a much greater range in temperatures through the year.

Another important temperature difference occurs between oceanic and continental locations in mid- and high latitudes. It arises because water bodies heat and cool more slowly than continents. In the summer, we head for the beach when the weather gets hot because air temperatures there are cooler. Why? Because the water is cooler than the land, the air near the coast also tends to be cooler. In contrast, during the winter, coastal regions tend to have warmer temperatures, because the ocean water is often warmer than the land.

Air temperature cycles—daily and seasonal—and the influence of latitude and location on air temperature are the primary subject of this chapter. Keep in mind as you study that the main cause of air temperature differences is the difference in the amount of the sun's energy that strikes the earth's surface. Because the amount of energy from the sun reaching the earth varies from day to day, from season to season, and from latitude to latitude, temperatures also vary.

MEASUREMENT OF AIR TEMPERATURE

Air temperature is an item of weather information that we encounter daily. In the United States, temperature is still widely measured and reported using the Fahrenheit scale. The freezing point of water on the Fahrenheit scale is 32°F, and the boiling point is 212°F. This represents a range of 180°F. In this book, we use the Celsius temperature scale, which is the international standard. On the Celsius scale, the freezing point of water is 0°C and the boiling point is 100°C. Thus, 100 Celsius degrees are the equivalent of 180 Fahrenheit degrees ($1°C = 1.8°F$; $1°F = 0.56°C$). Conversion formulas between these two scales are given in Figure 3.1.

You are probably familiar with the *thermometer* as an instrument for measuring temperature. In this device, a liquid inside a glass tube expands and contracts with temperature, and the position of the liquid surface indicates the temperature. Since air temperature can vary with height, it is measured at a standard level—1.2 m (4 ft) above the ground. A *thermometer shelter* (Figure

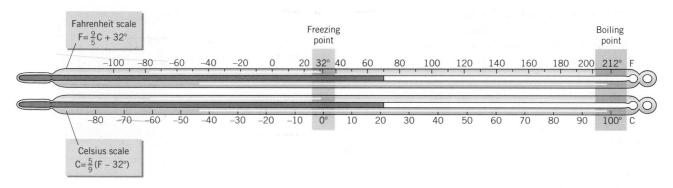

Figure 3.1 A comparison of the Celsius and Fahrenheit temperature scales. At sea level, the freezing point of water is at Celsius temperature (C) 0°, while it is 32° on the Fahrenheit (F) scale. Boiling occurs at 100°C, or 212°F.

3.2*a*) holds thermometers or other simple weather instruments at this height while sheltering them from the direct rays of the sun. Air circulates freely through the louvers, ensuring that temperatures inside the shelter are the same as the outside air.

Air temperatures are now automatically recorded by thermistors at a uniform height above the ground.

Thermometers are being replaced by newer instruments for the routine measurement of temperatures.

Perhaps you have used a digital fever thermometer, which reads out your body temperature directly as a number. It uses a device called a *thermistor*, which changes its electrical resistance with temperature. By measuring this resistance, the temperature may be obtained automatically. Many weather stations are now equipped with temperature measurement systems that use thermistors (Figure 3.2*a*).

Temperature Statistics

Although some weather stations report temperatures hourly, most stations only report the highest and low-

a.

b.

Figure 3.2 (*a*) Maximum-minimum temperature instrument system (left). To the right is a large recording rain gauge. (*b*) A thermometer shelter. This white wooden louvered box houses maximum-minimum thermometers and other instruments.

est temperatures recorded during a 24-hour period. These are most important values in observing long-term trends in temperature. They are recorded daily by more than 5000 stations in the United States.

Daily, monthly, and yearly temperature statistics for a station are produced using the daily maximum and minimum temperature. The mean daily temperature is defined as the average of the maximum and minimum daily values. Records of mean daily temperatures averaged for monthly and yearly intervals of many years' duration are compiled for each observing station. These averages are used to describe the climate of the station and its surrounding area.

THE DAILY CYCLE OF AIR TEMPERATURE

Because the earth rotates on its axis, incoming solar energy can vary widely throughout the 24-hour period. Insolation is greatest in the middle of the daylight period, when the sun is high in the sky, and falls to zero at night. Recall from Chapter 2 that net radiation is the balance between incoming energy absorbed by a surface and outgoing radiation that is emitted from that surface. Since received solar radiation varies greatly, so does net radiation. During the day, net radiation is positive, and the surface gains heat. At night, net radiation is negative, and the surface loses heat by radiating it to the sky and space. Since the air next to the surface is warmed or cooled as well, air temperatures follow the same cycle. This results in the daily cycle of rising and falling air temperatures.

> **The daily cycle of temperature is controlled by the daily cycle of net radiation.**

Daily Insolation and Net Radiation

Let's look in more detail at how insolation, net radiation, and air temperature are linked in this daily cycle. The three graphs in Figure 3.3 show average curves of daily insolation, net radiation, and air temperature that we might expect for a typical observing station at lat. 40° to 45°N in the interior United States. The time scale is set so that 12:00 noon occurs when the sun is at its highest elevation in the sky.

Graph (*a*) shows daily insolation. At the equinox (middle curve), insolation begins at about sunrise (6 A.M.), rises to a peak value at noon, and declines to zero at sunset (6 P.M.). At the June solstice, insolation begins about 2 hours earlier (4 A.M.) and ends about 2 hours later (8 P.M.). The June peak is much greater than at equinox, and the total insolation for the day is

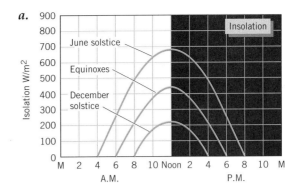

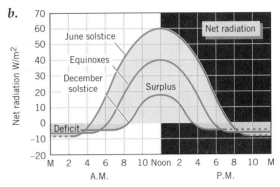

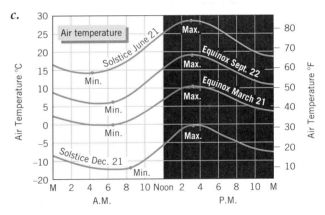

Figure 3.3 Idealized cycles of insolation, net radiation, and air temperature for a midlatitude station in the interior United States. Insolation (*a*) is a strong determiner of net radiation (*b*). Air temperatures (*c*) respond by generally increasing while net radiation is positive and decreasing when net radiation is negative.

also much greater. At the December solstice, insolation begins about 2 hours later than the equinox curve (8 A.M.) and ends about 2 hours earlier (4 P.M.). Both the peak intensity and daily total insolation are greatly reduced in the winter solstice season.

Graph (*b*) shows net radiation for the surface. Recall that when net radiation is positive, the surface gains heat, and when negative, it loses heat. The curves for the solstices and equinox generally resemble those of insolation. Unlike insolation, net radiation begins the 24-hour day as a negative value—a deficit—at midnight. The deficit continues into the early morning hours. Net radiation shows a positive value—a surplus—shortly after sunrise and rises sharply to a peak at noon. In the afternoon, net radiation decreases as insolation decreases. A value of zero is reached shortly before sunset. With no incoming insolation, net radiation then becomes negative, showing a deficit, where it remains until midnight.

Although the three net radiation curves show the same general daily pattern—negative to positive to negative—they differ greatly in height. For the June solstice, the positive values are quite large, and careful inspection of the figure will show that the area of surplus is much larger than the area of deficit. This means that for the day as a whole, net radiation is positive. At the December solstice, the surplus period is short and the surplus is small. The total deficit, which extends nearly 18 hours, outweighs the surplus. This means that the net radiation for the entire day is negative. As we will see, this pattern of positive daily net radiation in the summer and negative daily net radiation in the winter drives the annual cycle of temperatures.

Daily Temperature

Graph (*c*) shows the typical, or average, daily air temperature cycle. The minimum daily temperature usually occurs about a half hour after sunrise. Since net radiation has been negative during the night, heat has flowed from the ground surface, and the ground has cooled the air to its lowest temperature. As net radiation becomes positive, the surface warms quickly and transfers heat to the air above. Air temperature rises sharply in the morning hours and continues to rise long after the noon peak of net radiation.

We should expect the air temperature to rise as long as a radiation surplus is in effect, and in theory this should produce a temperature maximum just before sunset. However, another process begins in the early afternoon. Mixing of the lower air by vertical currents distributes heat upward, offsetting the temperature rise. Therefore, the temperature peak usually occurs in the midafternoon. It is shown in the figure at about 3 P.M., but it usually occurs between 2 and 4 P.M., depending on local climatic conditions. By sunset, air temperature is falling rapidly. It continues to fall, but at a decreasing rate, throughout the night.

Temperatures are lowest just after sunrise and highest in midafternoon.

Note also that the general level of the temperature curves varies with the seasons. In the summer, the daily curve is high, showing warm temperatures. In winter, the curve is low, reflecting cold temperatures. In between are the equinoxes, with their intermediate temperatures. Since the temperatures lag behind the seasonal changes in net radiation, the September equinox shows temperatures considerably warmer than the March equinox. Even though net radiation is the same for the two, each curve lags behind, reflecting earlier seasonal conditions.

Temperatures Close to the Ground

The temperatures we have discussed thus far are measured under standard conditions, at a short distance from the ground surface. As we get closer to the ground surface, the temperature trends get more

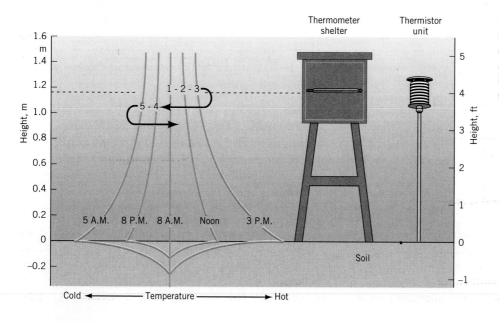

Figure 3.4 A simplified diagram of the temperature profile close to the ground surface. The soil surface becomes very hot by midafternoon but cools greatly at night.

Figure 3.5 Urban surfaces (left) are composed of asphalt, concrete, building stone, and similar materials. Sewers drain away rain water, keeping urban surfaces dry. Rural surfaces (right) are composed of moist soil, largely covered by vegetation.

extreme. Under the direct rays of the sun, a pavement surface heats to much higher temperatures than the air at standard height. At night, this surface layer cools to temperatures that are lower than the air temperature at standard height. These effects are weaker under a forest cover where the ground is shaded and moist. On a dry desert floor or on the pavements of city streets and parking lots, however, the surface temperature extremes can be large.

Figure 3.4 presents a series of temperature profiles as they might be observed throughout the day close to the ground. At 8 A.M. (curve 1), the temperature of air and soil is uniform, producing a vertical line on the graph. By noon (curve 2), the surface is considerably warmer than the air at standard height, and the soil below the surface has been warmed as well. By 3 P.M. (curve 3), the soil surface is very much warmer than the air at standard height. By 8 P.M. (curve 4), the surface is cooler than the air, and by 5 A.M. (curve 5), it is much colder. The daily cycle shows its greatest range near the soil surface and gradually weakens with depth.

Urban and Rural Surface Characteristics

Human activity has altered much of the earth's land surface. Cities are an obvious example. To build a city, vegetation is removed, and soils are covered with pavement or structures. Using the principles of the surface energy balance we have learned, we can understand the effects of these changes on the temperatures of urban areas as compared to rural regions.

Figure 3.5 diagrams some of the differences between urban and rural surfaces. In rural areas, the land surface is normally covered with a layer of vegetation. A natural activity of plants is **transpiration.** In this process, water is taken up by plant roots and moved to the leaves, where it evaporates. Since evaporation cools

a surface by removing heat, we would expect the rural surface to be cooler.

The cooling effect of vegetation is even greater in a forest. Not only are transpiring leaves abundant, but also solar radiation is intercepted by thick layers of leaves from the tops of the trees to the forest floor. Thus, solar warming is not concentrated intensely at the ground surface, but rather serves to warm the whole forest layer.

Another feature of the rural environment is that the soil surfaces can be moist. During a rainstorm, precipitation seeps into the soil. When sunlight reaches the soil surface, the water evaporates, thus removing heat and keeping the soil cool.

In contrast are the typical surfaces of the city. Rain runs easily off the impervious surfaces of roofs, sidewalks, and streets. The rainwater is channeled into sewer systems, where it flows directly into rivers, lakes, or oceans, instead of soaking into the soil beneath the city surfaces. Because the surfaces are dry, the full energy of insolation warms the surfaces, with no energy expended to evaporate water. Therefore, city surfaces heat the air to higher temperatures during the daytime.

Urban surfaces lack moisture and so are warmer than rural surfaces during the day. At night, urban materials conduct stored heat to the surface, also keeping temperatures warmer.

Another factor contributing to warmer urban temperature is the ability of concrete, stone, and asphalt to conduct and hold heat better than soil, even when the soil is dry. Therefore, during the day, more heat can be stored by the city's building materials. At night, the heat is conducted back to the surface, keeping nighttime temperatures warmer. Heat absorption is enhanced by the many vertical surfaces in cities, which reflect radiation from surface to surface. Since some radiation is absorbed with each reflection, the network of vertical surfaces tends to trap heat more effectively than a single flat surface.

The Urban Heat Island

As a result of these effects, air temperatures in the central region of a city are typically several degrees warmer than those of the surrounding suburbs and countryside. Figure 3.6 shows city air temperatures on a typical summer evening. The lines of equal air temperature delineate a **heat island.** The heat island persists through the night because of the availability of a large quantity of heat stored in the ground during the daytime hours.

Another important factor in warming the city is fuel consumption. In winter, furnaces warm buildings. Interior heat is conducted to the outside walls and roofs, which radiate heat into the urban environment. In summer, city temperatures are raised through the use of air conditioning. The machinery pumps heat out of buildings, releasing the heat to the air. The power used to run the air conditioning systems is also released as heat.

TEMPERATURE STRUCTURE OF THE ATMOSPHERE

Thus far, we have discussed temperatures in the range of the atmosphere that surrounds us—up to about 2 m (6 ft) above the ground surface. How different are air temperatures at increasing heights through the atmosphere? In general, temperatures are lower. The decrease in measured temperature is described by the **lapse rate,** which indicates the drop in temperature in degrees C per 1000 m (or degrees F per 1000 ft).

Why is air cooler at higher altitudes? Recall from Chapter 2 that most of the incoming solar radiation penetrates the atmosphere and is absorbed at the surface. In general, the further the air is away from that warm surface, the cooler the air will be.

> **In the lower atmosphere, air temperatures decrease with increasing altitude.**

Figure 3.7 shows how temperature varies with altitude for a typical summer day in the midlatitudes. Altitude is plotted on the vertical axis and temperature on the horizontal axis. The curve is an average one—if

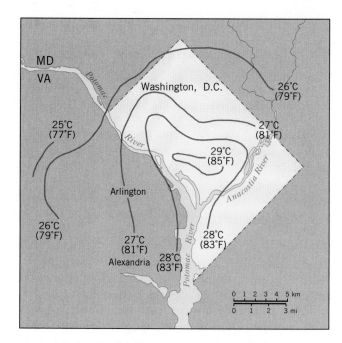

Figure 3.6 A heat island over Washington, D.C., and surrounding areas. Air temperatures were taken at 10:00 P.M. on an evening in early August. (Data of H. E. Landsberg.)

we sent up a balloon equipped to radio the temperature of the air every minute or two, and sent up many balloons over a long period of time, we would obtain an average profile of temperature very much like that shown in the figure. Temperature drops with altitude at an average rate of 6.4°C/1000 m (3.5°F/1000 ft). This average value is known as the **environmental temperature lapse rate.** For example, when the air temperature near the surface is a pleasant 21°C (70°F), the air at an altitude of 12 km (40,000 ft) will be a bone-chilling –55°C (–70°F). Keep in mind that the environmental temperature lapse rate is an average value and that on any given day the observed lapse rate might be quite different.

Figure 3.7 shows another important feature of the atmosphere. For the first 12 km (7 mi) or so, temperature falls with increasing elevation. However, at 12 to 15 km in height, the temperature stops decreasing. In fact, above that height, temperature slowly increases with elevation. This feature has led atmospheric scientists to distinguish two different layers in the lower atmosphere—the troposphere and the stratosphere.

Troposphere

The **troposphere** is the lowest atmospheric layer, in which temperature decreases with increasing elevation. It extends from the ground up to 12–15 km (7–9 mi). Since almost all human activity occurs in this layer, it is of primary importance to us. Everyday weather phenomena, such as clouds or storms, occur mainly in this layer.

One important feature of the troposphere is that it contains significant amounts of water vapor. When the water vapor content is high, vapor can condense into water droplets, forming low clouds and fog, or be deposited as ice crystals, forming high clouds. When condensation or deposition is rapid, rain, snow, hail, or sleet—collectively termed *precipitation*—may be produced and fall to earth. Regions where water vapor content is high throughout the year will therefore have moist climates. In desert regions, where water vapor is present only in small amounts, precipitation is infrequent. In Chapter 2, we also described the important role of water vapor in absorbing and counterradiating heat emitted by the earth's surface. In this way, water vapor helps to create the greenhouse effect, which warms the earth to habitable temperatures.

The troposphere contains countless tiny dust particles, so small and light that the slightest movements of the air keep them aloft. These are called *aerosols.* They are swept into the air from dry desert plains, lakebeds, and cultivated fields, or they are released by volcanoes. Oceans are also a source of aerosols. Strong winds blowing over the ocean lift droplets of spray into the air. These droplets of spray lose most of their moisture by evaporation, leaving tiny particles of salt as residues that are carried high into the air. Forest fires and

brushfires are another important source of aerosols, contributing particles of soot as smoke. Meteors contribute dust particles as they vaporize from the heat of friction upon entering the upper layers of air. Industrial processes that burn coal or fuel oil incompletely release aerosols to the air as well.

The most important function of aerosols is that certain types serve as nuclei, or centers, around which water vapor condenses to form tiny droplets. When these droplets occur in high concentration, they are visible to the eye as clouds or fog. Aerosols in the troposphere also scatter sunlight, thus brightening the whole sky while reducing slightly the intensity of the solar beam. This scattering behavior is strongest for red light, and thus, the red colors of sunrise and sunset are accentuated by dust.

The height at which the troposphere gives way to the stratosphere above is known as the *tropopause.* Here, temperatures stop decreasing with altitude and start to increase. The altitude of the tropopause varies somewhat with latitude and season. This means that the troposphere is not uniformly thick.

The troposphere and stratosphere are the two lowermost atmospheric layers.

Stratosphere

Above the troposphere lies the **stratosphere** in which the air becomes slightly warmer as altitude increases. The stratosphere extends to a height of roughly 50 km (30 mi) above the earth's surface. It is the home of strong, persistent winds that blow from west to east. There is little mixing of air between the troposphere and stratosphere, and the stratosphere holds very little water vapor or dust.

One important feature of the stratosphere is that it contains the ozone layer. As we saw in Chapter 2, the ozone layer absorbs solar ultraviolet radiation and thus shields earthly life from this intense, harmful form of energy. In fact, the warming of the stratosphere is caused by the absorption of solar energy by ozone molecules.

High-Mountain Environments

Did you ever hike or drive to the top of a high mountain? Mount Whitney, in California, perhaps (Figure 3.8)? Or New Hampshire's Mount Washington? Or North Carolina's Mount Mitchell? If so, you have probably made the simple observation that temperatures generally drop as you go up in elevation. We've already noted this fact in the discussion of the environmental

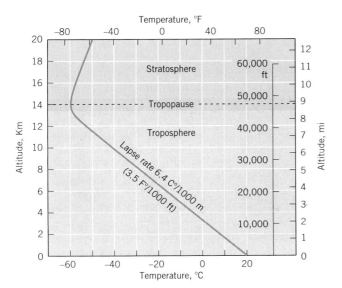

Figure 3.7 A typical environmental temperature lapse-rate curve for a summer day in the midlatitudes. The rate of temperature decrease with elevation, or lapse rate, is shown at the average value of 6.4°C/1000 m (3.5°F/1000 ft). The tropopause separates the troposphere, where temperature decreases with increasing elevation, from the stratosphere, where temperature is constant or slightly increases with elevation.

Figure 3.8 Mt. Whitney (elevation 4418 m, 14,495 ft) and surrounding peaks of the Southern Sierra Nevada. The excellent visibility and dark blue sky are products of the thin air at this elevation.

temperature lapse rate. But what else might you notice as you climb? You might notice that you sunburn a bit faster. Perhaps you also find yourself getting out of breath more quickly. And if you camp out, you may notice that the nighttime temperature gets lower than you might expect, even given that temperatures are generally cooler.

All these effects are related to the fact that air density decreases with elevation. At high elevations, the air is thin—that is, there are fewer gas molecules in a unit volume of air. Therefore, it makes sense that you can feel short of breath on a high mountain simply because the oxygen pressure in your lungs is lower than you are used to at lower elevations. With fewer air molecules and dust particles to scatter and absorb the sun's light, the sun's rays will feel stronger. And there is also less carbon dioxide and water vapor, and so the greenhouse effect is reduced. With the reduced warming effect, temperatures will tend to drop lower at night.

At high elevations, air temperatures are generally cooler and show a greater day-to-night range.

What is the effect of elevation on the daily temperature cycle? Figure 3.9 shows temperature graphs for five stations ascending the Andes Mountain range in Peru. The July monthly mean for each station is shown as a horizontal line. Mean temperatures clearly decrease with elevation, from 16°C (61°F) at sea level to −1°C (30°F) at 4380 m (14,370 ft). The daily range clearly increases with elevation, except for Cuzco. This large city does not experience nighttime temperatures as low as you might expect because of its urban heat island and accompanying air pollution effects.

Temperature Inversion and Frost

So far, air temperatures seem to decrease with height above the surface. Is this always true? Think about what happens on a clear, calm night. Under clear conditions, the ground surface radiates longwave energy to the sky, net radiation becomes negative, and the surface cools. This means that the air near the surface will also be cooled, so as you move away from the cool surface, the air becomes warmer.

This situation is illustrated in Figure 3.10. The straight, slanting line of the normal environmental

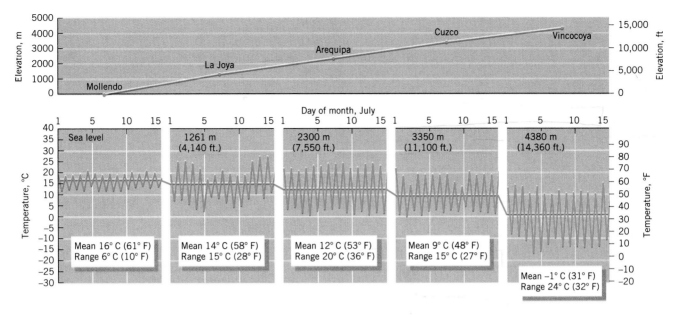

Figure 3.9 Daily maximum and minimum air temperatures for mountain stations in Peru, lat. 15° S. All data cover the same 15-day observation period in July. As elevation increases, the mean daily temperature decreases and the temperature range increases. (Data from Mark Jefferson.)

lapse rate bends to the left in a "J" hook. In this example, the air temperature at the surface, point A, has dropped to –1°C (30°F). This value is the same as at point B, some 750 m (2500 ft) aloft. As we move up from ground level, temperatures become warmer up to about 300 m (1000 ft). Here the curve reverses itself, and normal lapse rate takes over. The lower portion of the lapse rate curve is called a low-level **temperature inversion.** Here, the normal cooling trend is reversed and temperatures increase with height.

> **In a temperature inversion, air temperature does not decrease with altitude but increases instead.**

In the case shown, the temperature of the lowermost air has fallen below the freezing point, 0°C (32°F). For sensitive plants, this temperature condition is called a killing frost when it occurs during the growing season. Perhaps you have seen news reports about killing frosts damaging fruit trees or crops in Florida or California. Growers commonly use several methods to reduce the inversion. Oil-burning heaters are used to warm the surface air layer and create air circulation. Or large fans are used to mix the cool air at the surface with the warmer air above (Figure 3.11).

Low-level temperature inversions often occur over snow-covered surfaces in winter. Inversions of this type are very intense and can extend thousands of meters into the air. They build up over many long nights in arctic and polar regions, where the solar heat

of the short winter day cannot completely compensate for nighttime cooling. Inversions can also result when a warm air layer overlies a colder one. This type of inversion is often found along the west coasts of major continents, and we will discuss it in more detail in Chapter 4.

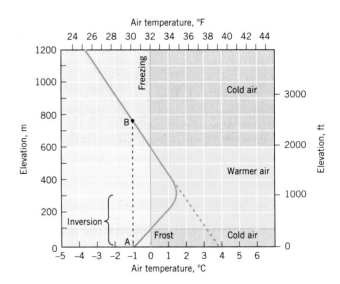

Figure 3.10 A low-level temperature inversion with frost. Instead of temperature decreasing with elevation (dashed line) from, say, 4°C (38°F), the surface temperature is at –1°C (30°F) and temperature increases with elevation (solid line) for several hundred meters (1,000 ft or so) above the ground. Since the surface is below the freezing point, frost forms.

Figure 3.11 Smudge pots burning kerosene are used to warm the cold air of an inversion, protecting delicate pear blossoms from frost. Hood River, Oregon.

THE ANNUAL CYCLE OF AIR TEMPERATURE

As the earth revolves around the sun, the tilt of the earth's axis causes an annual cycle of variation in insolation. This cycle produces an annual cycle of net radiation, which, in turn, causes an annual cycle to occur in mean monthly air temperatures. Although the annual cycle of net radiation is the most important factor in determining the annual temperature cycle, another important consideration is location—maritime or continental. That is, places located well inland and far from oceans generally experience a stronger temperature contrast from winter to summer. We will begin our study of the annual temperature cycle, however, with the relationship between net radiation and temperature for four examples, ranging from the equator almost to the arctic circle.

> **The annual cycle of net radiation, which results from the variation of insolation with the seasons, drives the annual cycle of air temperatures.**

Net Radiation and Temperature

Graph (*a*) of Figure 3.12 shows the yearly cycle of the net radiation rate for four stations. The average value of net radiation flow rate for the month is plotted, in units of w/m². Graph (*b*) shows mean monthly air temperatures for these same stations. We will compare the net radiation graph with the air temperature graph for each station, beginning with Manaus, a city on the Amazon River in Brazil.

At Manaus, located nearly on the equator, the average net radiation rate is strongly positive in every month. However, there are two minor peaks. These coincide approximately with the equinoxes, when the sun is nearly straight overhead. A look at the temperature graph of Manaus shows uniform air temperatures, averaging about 27°C (81°F) for the year. The annual temperature range, or difference between the highest and lowest mean monthly temperature, is only 1.7°C (3°F). In other words, near the equator the temperature is similar each month. There are no temperature seasons.

We go next to Aswan, Egypt, a very dry desert location on the Nile River at lat. 24°N. The positive net radiation rate curve shows that a large radiation surplus exists for every month. Furthermore, the net radiation rate curve has a much stronger annual cycle, with values for June and July that are almost double those of December and January. The temperature graph shows a corresponding annual cycle, with an annual range of about 17°C (30°F). June, July, and August are terribly hot, averaging over 32°C (90°F).

Moving farther north, we come to Hamburg, Germany, lat. 54°N. The net radiation rate cycle here is also strongly developed. The rate is positive for nine months, providing a radiation surplus. When the rate becomes negative for three winter months, a deficit is produced. The temperature cycle reflects the reduced total insolation at this latitude. Summer months reach a maximum of just over 16°C (60°F), while winter months reach a minimum of just about freezing (0°C or 32°F). The annual range is about 17°C (30°F), the same as at Aswan.

Finally, we travel to Yakutsk, Siberia, lat. 62°N. During the long, dark winters, the net radiation rate is

a.

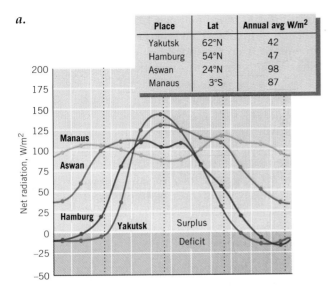

Place	Lat	Annual avg W/m²
Yakutsk	62°N	42
Hamburg	54°N	47
Aswan	24°N	98
Manaus	3°S	87

b.

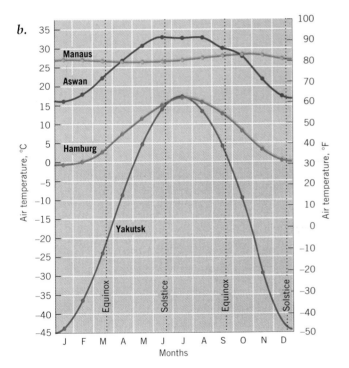

Figure 3.12 Net radiation (*a*) and temperature (*b*) cycles for four stations ranging from the equatorial zone to the arctic zone. Aswan (Egypt), Hamburg (Germany), and Yakutsk (Siberia) all show strong cycles with summer maxima and winter minima. In contrast is Manaus (Brazil), very near the equator, which shows two net radiation peaks and a nearly uniform temperature. (Data courtesy of David H. Miller.)

negative, and there is a radiation deficit that lasts about six months. During this time, air temperatures drop to extremely low levels. For three of the winter months, monthly mean temperatures are between −35 and −45°C (−30 and −50°F). This is actually one of the

coldest places on earth. In summer, when daylight lasts most of a 24-hour day, the net radiation rate rises to a strong peak. In fact, this peak value is higher than those of the other three stations. As a result, air temperatures show a phenomenal spring rise to summer-month values of over 13°C (55°F). In July, the temperature is about the same as for Hamburg. The annual range at Yakutsk is enormous—over 60°C (110°F). No other place on earth, even the south pole, has so great an annual range.

Land and Water Contrasts

Have you ever visited San Francisco? If so, you probably noticed that this magnificent city has quite a unique climate. Fog is frequent, and cool, damp weather prevails for most of the year (see Figure 4.13). In fact, Mark Twain is reputed to have said, "The coldest winter I ever spent in my life was summer in San Francisco!" The cool climate is due to its location—on the tip of a peninsula, with the Pacific Ocean on one side and San Francisco Bay on the other. Ocean and bay water temperatures are quite cool, since a southward-flowing current sweeps cold water from Alaska down along the northern California coast. Winds from the west move cool, moist ocean air, as well as clouds and fog, across the peninsula, keeping summer air temperatures low and winter temperatures above freezing. Figure 3.13 shows a typical record of temperatures for San Francisco for a week in the summer. Temperatures hover around 13°C (60°F) and change only a little from day to night.

In contrast is a location far from the water, like Yuma, Arizona, also shown in Figure 3.13. Located in the Sonoran Desert, air temperatures here are much warmer on the average—about 28°C (82°F). Clearly, no ocean cooling is felt in Yuma! The daily range is also much greater, nearly 20°C (36°F). Hot desert days become cool desert nights, with clear, dry air allowing the ground to lose heat rapidly.

Land—water contrasts keep air temperatures at coastal locations more constant than at interior continental locations.

It seems only natural that coastal locations should be cooler in summer and warmer in winter, and have a more limited day-to-night temperature range, than stations in the interior of continents. But did you ever wonder why? The important principle is this: the surface of any extensive, deep body of water heats more slowly and cools more slowly than the surface of a large body of land when both are subjected to the same intensity of insolation. Because of this principle, daily

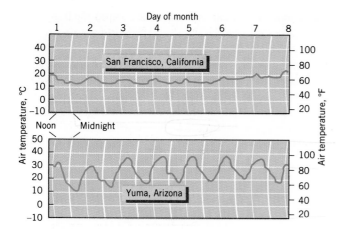

Figure 3.13 A recording thermometer made these continuous records of the rise and fall of air temperature over a period of a week in summer at San Francisco, California, and at Yuma, Arizona. The daily cycle is strongly developed at Yuma, a station in the desert. In contrast, the graph for San Francisco, on the Pacific Ocean, shows a very weak daily cycle.

and annual air temperature cycles will be quite different at coastal locations than at interior locations.

Four important thermal differences between land and water surfaces account for the land–water contrast (see Figure 3.14). The most important difference is that solar radiation penetrates water, distributing the absorbed heat throughout a substantial water layer. In contrast, solar radiation does not penetrate soil or rock, so its heating effect is concentrated at the surface. The radiation therefore warms a thick water surface layer only slightly, while a thin land surface layer is warmed more intensely.

Oceans heat and cool more slowly than continents.

A second thermal factor is that water is slower to heat than dry soil or rock. The *specific heat* of a substance describes how the temperature of a substance changes with a given input of heat. Consider an experiment using a small volume of water and the same volume of rock. Let's suppose we warm each volume by the same amount—perhaps 1°C (1.8°F). If we measure the amount of heat required, we learn that it takes about five times as much heat to raise the temperature of the water one degree as it does to raise the temperature of the rock one degree. That is, the specific heat of water is about five times greater than that of rock. The same will be true for cooling—after losing the same amount of heat, the temperature of water falls less than the temperature of rock.

A third difference between land and water surfaces is related to mixing. That is, a warm water surface layer can mix with cooler water below, producing a more uniform temperature throughout. For the open ocean, the mixing is produced by wind-generated waves. Clearly, no such mixing occurs on land surfaces, unless we mix the soil with a shovel!

A fourth thermal difference is that an open water surface can be cooled easily by evaporation. Land surfaces can also be cooled by evaporation, but only if water is present at or near the soil surface. When the surface dries, evaporation stops. In contrast, a free water surface can always provide evaporation.

Daily Temperature Cycle

What is the effect of the contrast in thermal characteristics of land and water on temperatures? Let's examine the daily cycle of air temperature first. Two sets of daily air temperature curves are shown in Figure 3.15—El Paso, Texas, and North Head, Washington—and four months are presented—January, April, July, and October.

The El Paso curves show the temperature environment of an interior desert in midlatitudes. Because soil moisture content is low and vegetation is sparse, evaporation and transpiration will not be important cooling effects. Cloud cover is generally light. Under these circumstances, the ground surface heats intensely during the day and cools rapidly at night. Air temperatures show an average daily range of 11 to 14°C (20 to 25°F). This type of variation represents an interior temperature environment—typical of a station located in the interior of a continent, far from the ocean's influence.

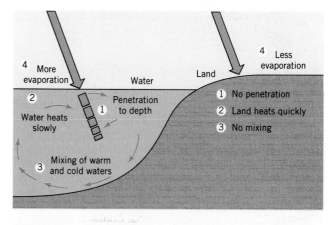

Figure 3.14 Four differences that illustrate why a land surface heats more rapidly and more intensely than the surface of a deep water body. Locations near the ocean have more uniform air temperatures—cooler in summer and warmer in winter—because of these differences.

Figure 3.15 The average daily cycle of air temperature for four different months shows the effect of continental and maritime location. Daily and seasonal ranges are great at El Paso, a station in the continental interior, but are only weakly developed at North Head, Washington, which is on the Pacific coast. The seasonal effect on overall temperatures is stronger at El Paso.

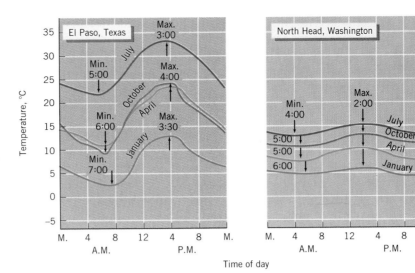

North Head is located on the Washington coast. Here, prevailing westerly winds sweep cool, moist air off the adjacent Pacific Ocean. The average daily range at North Head is a mere 3°C (5°F) or less. Persistent fogs and cloud cover also contribute to the minimal daily range. Note further that the annual range is much restricted, especially when compared to El Paso. North Head exemplifies a coastal temperature environment, typical of a station located in the path of oceanic air.

Annual Temperature Cycle

Let's now turn to the effects of land-water surface contrasts on the temperature cycle for the year. We have already noted for El Paso and North Head that the temperature cycle for the four months plotted shows a greater range for the interior station than for the coastal one. Let's look in more detail at the annual cycle for another pair of stations—Winnipeg, Manitoba, located in the heart of the North American continent, and the Scilly Islands, off the southwestern tip of England, which are surrounded by the waters of the Atlantic Ocean. This time, the two stations chosen are at the same latitude, 50°N. As a result, they have the same insolation cycle and receive the same potential amount of solar energy for surface warming. Figure 3.16 shows their annual cycles of temperature as well as the insolation curve common to both.

Monthly temperature maximums and minimums occur later at coastal stations than at interior stations.

The temperature graphs for the Scilly Islands and Winnipeg confirm the effects we have already noted for North Head and El Paso—that the annual range in

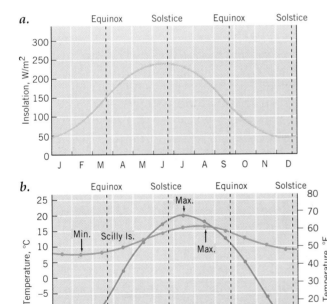

Figure 3.16 Annual cycles of insolation (*a*) and monthly mean air temperature (*b*) for two stations at lat. 50° N: Winnipeg, Canada, and Scilly Islands, England. Insolation is identical for the two stations. Winnipeg temperatures clearly show the large annual range and earlier maximum and minimum that are characteristic of its continental location. Scilly Islands temperatures show its maritime location in the small annual range and delayed maximum and minimum.

temperature is much larger for the interior station (39°C, 70°F) than for the coastal station (8°C, 15°F). Note that the nearby ocean waters keep the air temperature at the Scilly Islands well above freezing in the winter, while January temperatures at Winnipeg fall to near –20°C (–5°F)!

Another important effect concerns the timing of maximum and minimum temperatures. Insolation reaches a maximum at summer solstice, but it is still strong for a long period afterward. This means that heat energy continues to flow into the ground well into August. Therefore, the hottest month of the year for interior regions is July, the month following the solstice. Similarly, the coldest month of the year for large land areas is January, the month after the winter solstice. This is because the ground continues to lose heat even after insolation begins to increase.

Over the oceans and at coastal locations, maximum and minimum air temperatures are reached a month later than on land—in August and February, respectively. Because water bodies heat or cool more slowly than land areas, the air temperature changes more slowly. This effect is clearly shown in the Scilly Islands graph, where February is slightly colder than January.

WORLD PATTERNS OF AIR TEMPERATURE

We have learned some important principles about air temperatures in this chapter. Some of these principles are local—such as the effect of urban and rural surfaces on temperatures—and still others are global—such as the effect of latitude or elevation on temperature, or coastal-interior location. We now turn to world temperature patterns, tying our discussion closely to these principles. First, however, we will need a quick explanation of air temperature maps and their meaning.

Maps of isotherms show centers of high and low temperatures as well as temperature gradients.

The distribution of air temperatures is often shown on a map by **isotherms**—lines drawn to connect locations having the same temperature. Figure 3.17 shows a map on which the observed air temperatures have been recorded and placed at their proper location on the map. These may be single readings, such as a daily maximum or minimum, or they may be averages of many years of records for a particular day or month of a year, depending on the purposes of the map. The isotherms are constructed by drawing smooth lines

through and among the points in a way that best indicates a uniform temperature, given the observations at hand. Usually, isotherms representing 5- or 10-degree differences are chosen, but they can be drawn for any range of selected temperatures.

Isothermal maps are valuable because they make the important features of the temperature pattern clearly visible. Centers of high or low temperatures are clearly outlined. Also visible are *temperature gradients*—directions along which temperature changes. Centers and gradients give the broad pattern of temperature, as shown in our discussion of world patterns below.

Factors Controlling Air Temperature Patterns

World patterns of isotherms are largely explained by three factors that we have already discussed. The first of these is latitude. As latitude increases, average insolation decreases, and so temperatures decrease as well, making the poles colder than the equator. The effect of latitude on seasonal variation is also important. For example, more solar energy is received at the poles at the summer solstice than at the equator. We must therefore consider the time of year as well as the latitude in understanding world temperature patterns.

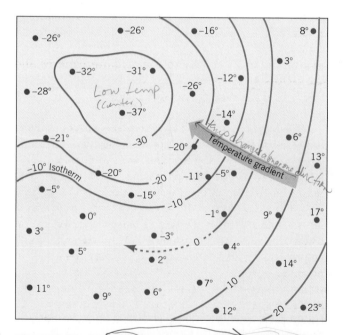

Figure 3.17 Isotherms are used to make temperature maps. Each line connects points that have the same temperature. Where temperature changes along one direction, a temperature gradient exists. Where isotherms close in a tight circle, a center exists. This example shows a center of low temperature.

The second factor is that of coastal-interior contrasts. Coastal stations that receive marine air from prevailing winds have more uniform temperatures—
cooler in summer and warmer in winter. Interior stations, on the other hand, show a much larger annual
variation in temperature. Ocean currents, discussed
further in Chapter 5, can also have an effect. By keeping coastal waters somewhat warmer or cooler than
expected, temperatures at maritime stations will be
influenced in a similar manner.

Elevation is the third important factor. At higher elevations, temperatures will be cooler. Therefore, we
expect world temperature maps to show the presence
of mountain ranges, which will be cooler than surrounding regions.

World Air Temperature Patterns for January and July

With these factors in mind, let's look at some world
temperature maps in more detail. Figure 3.18 consists
of maps of world temperatures for two months—
January and July. The Mercator projections show temperature trends from the equator to the midlatitude
zones, and the polar projections give the best picture
for high latitudes. From the maps, we can make six
important points about the temperature patterns and
the factors that produce them. Be sure to follow along
by examining the maps carefully.

**1. Temperatures decrease from the equator to the
poles.** Annual insolation decreases from the equator to
the poles, thus causing temperatures to decrease. This
temperature gradient is most clearly seen in the polar
maps for the southern hemisphere in January and
July. On these maps, the isotherms are nearly circular, decreasing to a center of low temperature on
Antarctica near the south pole. The center is much
colder in July, when most of the polar region is in
perpetual night. We can also see this same general
trend in the north polar maps, but the continents com-
plicate the pattern. The general temperature gradient
from the equator poleward is also evident on the
Mercator maps.

**2. Large landmasses located in the subarctic and
arctic zones develop centers of extremely low temperatures in winter.** The two large landmasses we have in
mind are North America and Eurasia. The January
north polar map shows these low temperature centers

very well. The cold center in Siberia, reaching –50°C
(–58°F), is strong and well defined. The cold center
over northern Canada is also quite cold (–35°C, –32°F)
but is not as well defined. Both features are visible on
the January Mercator map. Greenland shows a low
temperature center as well, but it has a high dome of
glacial ice, as discussed in point 6.

**3. Temperatures in equatorial regions change little
from January to July.** Note the broad space between
25°C (78°F) isotherms, which is evident on both
January and July Mercator maps. In this region, the
temperature is greater than 25°C (78°F) but less than
30°C (86°F). Although the two isotherms move a bit
from winter to summer, the equator always falls
between them. This demonstrates the uniformity of
equatorial temperatures. (An exception is the northern end of the Andes Mountains in South America,
where high elevations, and thus cooler temperatures,
exist at the equator.) Equatorial temperatures are uniform primarily because insolation at the equator does
not change greatly with the seasons.

**4. Isotherms make a large north-south shift from
January to July over continents in the midlatitude and
subarctic zones.** Figure 3.19 demonstrates this principle. In the winter, isotherms dip equatorward, while in
the summer, they arch poleward. This effect is shown
in North America and Eurasia in the January and July
Mercator maps. In January the isotherms drop down
over these continents, and in June they curve upward.
For example, the 15°C (58°F) isotherm lies over central Florida in January. But by July this same isotherm
has moved far north, cutting the southern shore of
Hudson Bay and then looping far up into northwestern Canada. In contrast are the isotherms over oceans,
which shift much less. This striking difference is due to
the contrast between oceanic and continental surface
properties, which cause continents to heat and cool
more rapidly than oceans.

**5. Highlands are always colder than surrounding
lowlands.** You can see this by looking at the pattern of
isotherms around the Rocky Mountain chain, in the
western United States, on the Mercator maps. In winter, the –5°C (23°F) and –10°C (13°F) isotherms dip
down around the mountains, indicating that the center of the range is colder. In summer, the 20°C (68°F)
and 25°C (78°F) isotherms also dip down, showing the
same effect even though temperatures are much
warmer. The Andes Mountains in South America show
the effect even more strongly. The principle at work
here is that temperatures decrease with an increase
in elevation.

**6. Areas of perpetual ice and snow are always
intensely cold.** Greenland and Antarctica contain our
planet's two great ice sheets. Notice how they stand out

(continued on page 22)

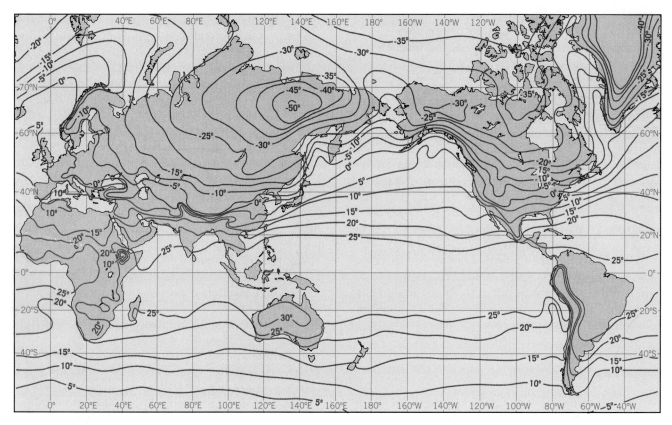

JANUARY

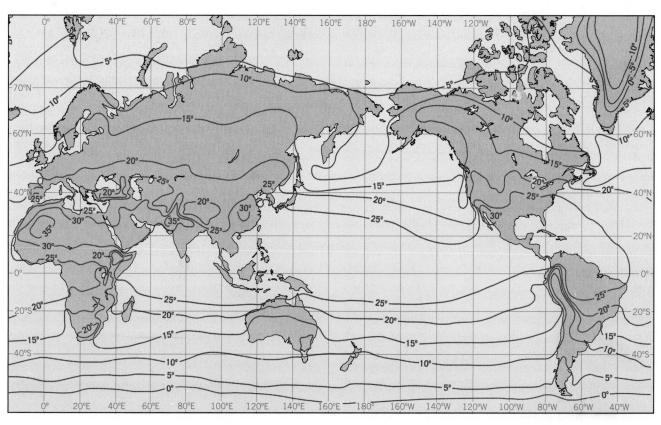

JULY

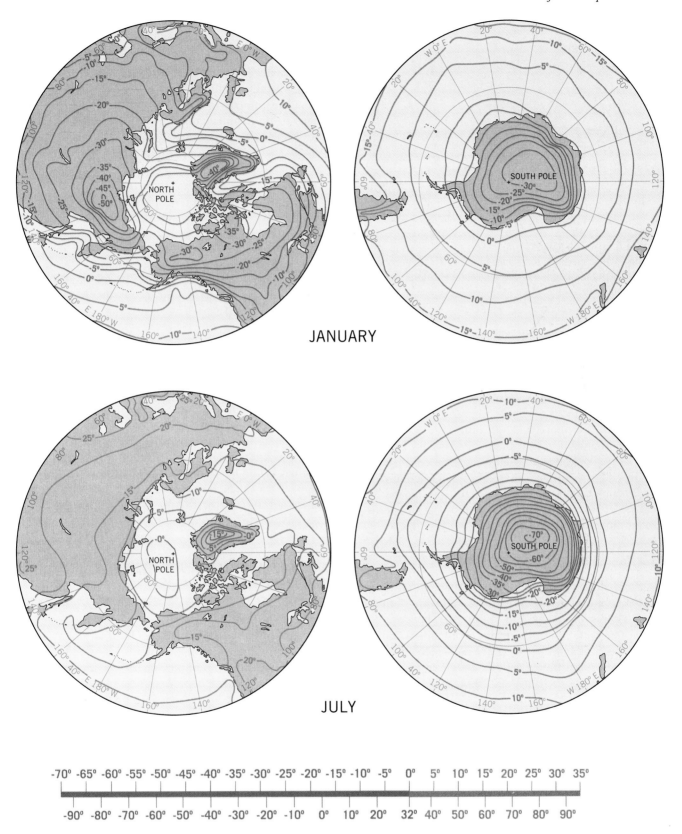

JANUARY

JULY

-70° -65° -60° -55° -50° -45° -40° -35° -30° -25° -20° -15° -10° -5° 0° 5° 10° 15° 20° 25° 30° 35°

-90° -80° -70° -60° -50° -40° -30° -20° -10° 0° 10° 20° 32° 40° 50° 60° 70° 80° 90°

Figure 3.18   Mean monthly air temperatures (°C) for January and July, Mercator and polar projections. (Compiled by John E. Oliver.)

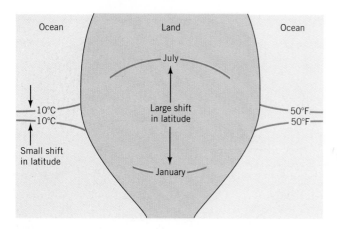

Figure 3.19 The seasonal migration of isotherms is much greater over continents than over oceans. This difference occurs because oceans heat and cool much more slowly than continents.

on the polar maps as cold centers in both January and July. They are cold for two reasons. First, their surfaces are high in elevation, rising to over 3000 m (10,000 ft) in their centers. Second, the white snow surfaces reflect much of the insolation. Since little solar energy is absorbed, little is available to warm the snow surface and the air above it. The Arctic Ocean, bearing a cover of floating ice, also maintains its cold temperatures throughout the year. However, the cold is much less intense in January than on the Greenland Ice Sheet, since ocean water underneath the ice acts as a heat reservoir to keep the ice above from getting extremely cold.

The Annual Range of Air Temperatures

Figure 3.20 is a world Mercator map showing the annual range of air temperatures. The lines, resembling isotherms, show the difference between the January and July monthly means. We can explain the features of this map by using the same effects of latitude, interior-maritime location, and elevation.

The global pattern of annual temperature range shows the smallest range for tropical oceans and the largest for northern hemisphere continental interiors.

---

1. The annual range increases with latitude, especially over northern hemisphere continents. This trend is most clearly shown for North America and Asia. This is due to the increasing contrast between summer and winter insolation, which increases with latitude.

2. The greatest ranges occur in the subarctic and arctic zones of Asia and North America. The map clearly shows two very strong centers of large annual range—one in northeast Siberia and the other in northwest Canada–eastern Alaska. In these regions, summer insolation is nearly the same as at the equator, while winter insolation is very low.

3. The annual range is moderately large on land areas in the tropical zone, near the tropics of cancer and capricorn. These are regions of large deserts—North Africa (Sahara), southern Africa (Kalahari), and Australia (interior desert) are examples. Dry air and the absence of clouds and moisture allow these continental locations to cool strongly in winter and warm strongly in summer, even though insolation contrasts with the season are not as great as at higher latitudes.

4. The annual range over oceans is less than that over land at the same latitude. This can be clearly seen by following a parallel of latitude—40°N, for example. Starting from the right, we see that the range is between 5 and 10°C (9 and 18°F) over the Atlantic but increases to about 30°C (54°F) in the interior of North America. In the Pacific, the range falls to 5°C (9°F) just off the California coast and increases to 15°C (27°F) near Japan. In Central Asia, the range is near 35°C (63°F). Again, these major differences are due to the contrast between land and water surfaces. Since water heats and cools much more slowly than land, a narrower range of temperatures is experienced.

5. The annual range is very small over oceans in the tropical zone. As shown on the map, the range is less than 3°C (5°F) since insolation varies little with the seasons near the equator and water heats and cools slowly.

GLOBAL WARMING AND THE GREENHOUSE EFFECT

Tune in to the evening news, and one of the recurring scientific subjects you will hear about is global warming. Is our climate changing? Many scientists have concluded that the temperature of the planet is warming significantly because of human activities. Other scientists argue that the warming of the last few years is part of a natural global cycle.
the last few years is part of a natural global cycle.

You may also hear that carbon dioxide, produced by human activities, is a major cause of warming. This gas is released to the atmosphere in large quantities by fossil fuel burning. Why is this a problem? Recall that the greenhouse effect is caused by atmospheric absorption of longwave radiation, largely by carbon dioxide and water vapor, that is emitted from the earth. Because some of this absorbed energy is counterradiated back from the atmosphere, the earth's surface is warmed by an extra amount.

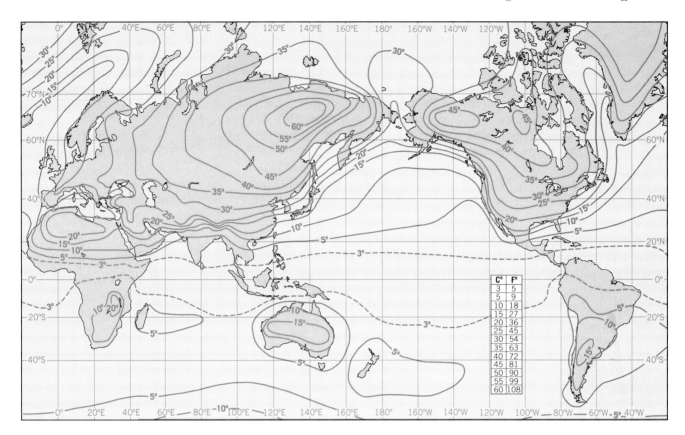

Figure 3.20 Annual range of air temperature in Celsius degrees. Data show differences between January and July means. The inset box shows the conversion of Celsius degrees to Fahrenheit for each isotherm value. (Compiled by John E. Oliver.)

Because human activities now continuously increase the amount of carbon dioxide in the air, many scientists are concerned that the greenhouse effect is escalating and that global temperatures are rising—or will rise soon—in response. (See Eye on the Environment: Carbon Dioxide—On the Increase, in this chapter.) Also of concern are other gases that are normally present in very small concentrations—methane, nitrous oxide, ozone, and the chlorofluorocarbons. (See Eye on the Environment: The Ozone Layer—Shield to Life, Chapter 2.) These also enhance the greenhouse effect, even though they are even less abundant than CO_2. Taken together with CO_2, they are referred to as **greenhouse gases.**

The Temperature Record

Has the buildup of greenhouse gases caused global temperatures to rise? If not, will this buildup warm the earth in the near future? There seems to be no doubt that an increased concentration of greenhouse gases can lead to global warming. The question seems to be whether or not warming will actually take place, given the cyclic nature of climate. If so, we also wish to know how soon it will occur and how great the warming will be.

Figure 3.21 shows the earth's mean annual surface temperature from 1856 to 1992. The temperature is expressed as a difference from the average annual temperature for the period 1951–1980. Two curves are shown—yearly data and smoothed data, taken from a 10-year moving average. Although temperature has increased since the last century, there have been wide swings in the mean annual surface temperature.

Several theories have been proposed to explain the variations. For example, some measures of solar activity seem to suggest that varying solar output induces similarly varying temperatures. If this theory is correct, then the present warming may not be the result of greenhouse gas accumulation. Instead, solar cycles may be responsible.

Another factor to consider when discussing global warming is volcanic activity. Volcanic dust and gases emitted in major eruptions are typically carried upward into the stratosphere, where strong winds spread them quickly throughout the entire layer. These particles have a cooling effect on the troposphere because they scatter more solar energy back to space. For example, the eruption of Mount Pinatubo in the Philippines in the spring of 1991 lofted 15 to 20 million tons of sulfuric acid particles into the

(continued on page 26)

Eye on the Environment • Carbon Dioxide — On the Increase

In the centuries before global industrialization, carbon dioxide concentration in the atmosphere was at a level slightly less than 300 parts per million (ppm), or about 3/100ths of a percent by volume. During the last hundred years or so, that amount has been substantially increased. Why? The answer is fossil fuel burning. When fuels like coal, oil, or natural gas are burned, they yield water vapor and carbon dioxide. The release of water vapor does not present a problem, because a large amount of water vapor is normally present in the global atmosphere. But because the natural amount of CO_2 is so small, fossil fuel burning has raised the level to about 350 ppm. This is a 22 percent increase.

The figure below shows how CO_2 has increased with time since 1860. Until 1940 or so, the level remained nearly stable. But after 1940 CO_2 began a rapid rise—so rapid that at the present rate of increase (about 4 percent per year), the amount of CO_2 in the air will double by 2030! Even if worldwide fossil fuel combustion is cut in half, the prediction is that doubling

will occur by 2050, only about a half-century away.

Not all the carbon dioxide emitted into the air by fossil fuel burning remains there. Instead, a complex cycle moves CO_2 throughout the life layer. Photosynthesis is one part of this carbon cycle. Plants use CO_2 in photosynthesis, the process by which they use light energy to build their tissues. So, plants take in carbon dioxide, removing it from the atmosphere. When plants die, their remains are digested by decomposing organisms, which release CO_2. This process returns CO_2 to the atmosphere. Normally, these processes are in balance.

Humans have tilted this balance by clearing land and burning the vegetation cover as new areas of forest are opened for development. This practice increases the amount of CO_2 in the air. When agricultural land is allowed to return to its natural forest state, CO_2 is removed from the air by growing trees. At present, scientists calculate that forests are growing more rapidly than they are being destroyed in the midlatitude regions of the

northern hemisphere. This plant growth helps to counteract the buildup of CO_2 produced by fossil fuel burning. But it may be outweighed by the tropical deforestation that is taking place in South America, Africa, and Asia. (See *Eye on the Environment • Exploitation of the Low-Latitude Rainforest Ecosystem,* Chapter 8.)

Laboratory experiments have shown that plants exposed to increased concentrations of CO_2 will grow faster and better. The faster they grow, the more CO_2 they can take in, which helps to reduce the amount in the atmosphere. However, scientists are unsure whether increased CO_2 will stimulate plant growth under natural conditions.

Another part of the cycle involves the oceans. The ocean's surface layer contains microscopic plant life that takes in carbon dioxide. The CO_2 in the ocean water initially comes from the atmosphere and is mixed into the ocean by surface waves. When these microscopic floating plants die, their bodies sink to the ocean bottom. There they decompose and release CO_2,

Increase in atmospheric carbon dioxide, observed in 1990 and predicted into the twenty-first century.

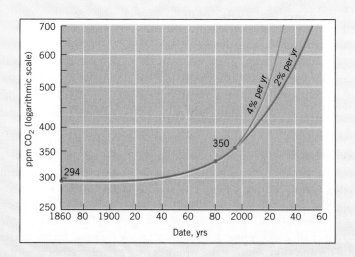

enriching waters near the ocean floor.

This CO_2 eventually returns to the surface through a system of global ocean current flows involving both bottom and surface currents (shown below). In this system, cold CO_2-rich bottom waters rise to the surface in the northernmost Pacific, where CO_2 is released. Meanwhile, warmer, CO_2-poor

The movement of deep and surface ocean CO_2 currents is like a giant conveyor belt, carrying cold, salty, CO_2-rich waters eastward along ocean floors. The returning warm surface current, which loses CO_2 and nutrients as they are used by tiny floating plants, flows westward. (After NASA.)

waters sink in the northernmost Atlantic. In fact, the ocean acts like a slow conveyor belt, moving CO_2 from the surface to ocean depths and releasing it again in a cycle lasting about 1500 years. At present, scientists estimate that ocean surface waters absorb more CO_2 than they release, owing to increased atmospheric levels of CO_2. Therefore, carbon dioxide may be accumulating in ocean depths. However, current studies of global climate computer simulations indicate that the oceans may not be as effective in removing excess CO_2 as scientists previously thought.

Although there is a great deal of uncertainty about the movements and buildup rate of excess CO_2 released to the atmosphere by fossil

fuel burning, one thing is certain. Without conversion to solar, nuclear, and hydroelectric power, fuel consumption will continue to release carbon dioxide, and it is only a matter of time before the earth will feel its impact.

Questions

1. Why has the atmospheric concentration of CO_2 increased in recent years?

2. How does plant life affect the level of atmospheric CO_2?

3. What is the role of the ocean in influencing atmospheric levels of CO_2?

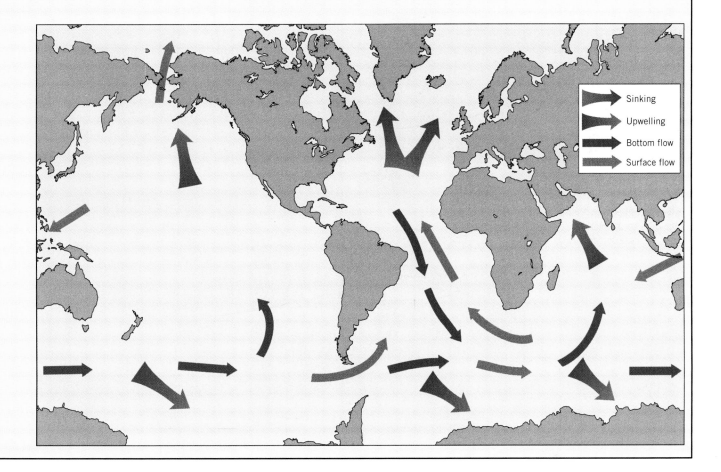

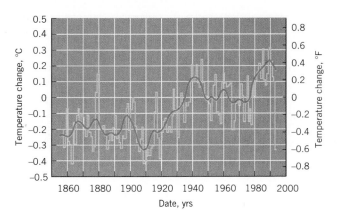

Figure 3.21 Mean annual surface temperature of the earth, 1856–1990. The vertical scale shows departures in degrees from a zero line of reference representing the average for the years 1951–1980. The bar graph shows the mean for each year. The smooth graph shows a running one-year average. Note the effect of Mount Pinatubo in 1992. (From Philip D. Jones and Tom M. L. Wigley, University of East Anglia/UK Meteorological Office. Used by permission.)

stratosphere (Figure 3.22). The layer of particles reduced the solar radiation reaching the earth's surface between 2 and 3 percent for the year or so following the blast. In response, global temperatures fell 0.5 to 0.7°C (0.9 to 1.3°F). Thus, volcanic activity may offset warming of the earth's surface from other causes.

Although the period of direct air temperature measurement does not extend past the middle of the last century, the record can be extended further back by using tree-ring analysis. The principle is simple—each

year, trees grow in diameter. This growth appears as a ring in climates where the seasons are distinct. If growing conditions are good, the ring is wide. If poor, the ring is narrow. For trees along the timber line in North America, the width indicates temperature—the trees grow better when temperatures are warmer. Since only one ring is formed each year, the date of each ring is easy to determine. Because the trees are quite old, the temperature record may be extended backward several centuries.

Figure 3.22 Eruption of Mt. Pinatubo, Philippine Islands, April, 1991. Volcanic eruptions like this can inject particles and gases into the stratosphere, influencing climate for several years afterwards.

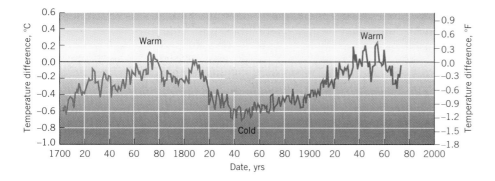

Figure 3.23 A reconstruction of the departures of northern hemisphere temperatures from the 1950–1965 mean, based on analyses of tree rings sampled along the northern tree limit of North America. (Courtesy of Gordon C. Jacoby of the Tree-Ring Laboratory of the Lamont-Doherty Geological Observatory of Columbia University.)

Figure 3.23 shows a reconstruction of northern hemisphere temperatures from 1700 to about 1980 using tree-ring analysis. Like Figure 3.21, it is expressed as a difference from a recent average (1950–1965). From 1880 to 1970, the temperatures reconstructed from tree-ring analysis seem to fit the observed temperatures (Figure 3.21) quite well. Analysis of an earlier period shows us another cycle of temperature increase and decline. The low point, around 1840, marks a cold event, during which European alpine glaciers became more active and advanced. Other evidence indicates that the two cycles in Figure 3.23 are part of a natural global cycle of temperature warming and cooling lasting about 150 to 200 years. These cycles have occurred regularly over the last thousand years. What causes these temperature cycles? Again, a number of theories have been offered, but no consensus has yet been reached as to the actual cause or causes.

Future Scenarios

What is the message of these past temperature records? Apparently, we cannot be certain right now that global temperatures are warming in response to a buildup of greenhouse gases. Other climatic factors, as yet undetermined, may be more important. Still, most scientists have little doubt that continued increases in greenhouse gases will eventually cause climatic warming. In 1988 an international group of scientists published three scenarios of rates of global warming in the next century. The most conservative scenario calls for warming at the rate of 0.06°C (0.1°F) per decade, with a 0.5°C (0.9°F) increase by 2100. A moderate scenario yields a rate of 0.3°C (0.5°F) per decade with a 3°C (5.4°F) rise by 2100. An extreme scenario uses a rate of 0.8°C (1.4°F) per decade to reach a rise of 5°C (9°F) at 2100. This last rate would mean a temperature change 10 to 100 times greater than what has been seen in the natural records of the last thousand years.

Why would a temperature rise cause concern? The problem is that many other changes may accompany a rise in temperature. For example, global patterns of weather systems could change. Changes mean precipitation would increase in some regions and decrease in others. Deserts would expand or contract, and favorable agricultural climates might shift from one nation to the next. Glaciers and ice sheets could melt enough to raise sea levels significantly, flooding coastal cities and islands. On the other hand, some scientists have theorized that increased CO_2 and warmer temperature levels could cause ice sheets to grow and expand! This could happen if snowfall increased at high latitudes because the warmer temperatures caused increased atmospheric moisture. Whatever the outcome, it seems wise to take steps to keep the buildup of greenhouse gases under control at least until we know more about the possible effects.

The world has become widely aware of the problem of the buildup in CO_2 and other greenhouse gases. At the Rio de Janeiro Earth Summit in 1992, nearly 150 nations signed a treaty limiting emissions of greenhouse gases. Under the terms of the treaty, emissions are to be reduced and held to 1990 levels. The large industrial nations will accomplish this objective by making more efficient use of fuels. Curbing production of chlorofluorocarbons, which is also required for ozone layer protection, will further reduce the annual increase in greenhouse gases. The ultimate solution is probably greater reliance on solar and geothermal energy sources, which produce power without releasing CO_2.

Looking ahead, we will study the climates of the earth in Chapter 7. Air temperatures are a very important part of climate, and in this chapter we have developed an understanding of air temperatures, their cycles, and their causes. The other key ingredient of climate is precipitation—the subject covered in the next chapter.

AIR TEMPERATURE IN REVIEW

The two major cycles of air temperature—daily and annual—are controlled by the cycles of insolation produced by the rotation and revolution of the earth. These cycles induce cycles of net radiation at the surface. When net radiation in the daily cycle is positive, air temperatures increase. When negative, air temperatures decrease. This principle applies for both daily and annual temperature cycles. Days are warm and nights are cool as net radiation goes from positive to negative. Summers are warm and winters are cold because average net radiation is high in the summer and low in the winter. Surface characteristics affect temperatures, too. Rural surfaces are generally moist and slow to heat, while urban surfaces are dry and absorb heat readily. This difference creates an urban heat island effect.

Air temperatures normally fall with altitude in the troposphere. At the tropopause, this decrease stops. In the stratosphere above, temperatures increase slightly with altitude. Air temperatures observed at mountain locations are lower with higher elevation, and day–night temperature differences increase with elevation.

Daily and annual air temperature cycles are influenced by maritime or continental location. Ocean temperatures vary less than land temperatures because water heats more slowly and can both mix and evaporate freely. Maritime locations that receive oceanic air therefore show smaller ranges of daily and annual temperature.

Global temperature patterns for January and July show the effects of latitude and maritime–continental location. Equatorial temperatures vary little from season to season. Poleward, temperatures decrease with latitude, and continental surfaces at high latitudes can become very cold in winter. Isotherms over continents swing widely north and south with the seasons, while isotherms over oceans move through a much smaller range of latitude.

Our planet's global temperature changes from year to year. Within the last few decades, global temperatures have been increasing. Some scientists attribute the increase to the human-induced buildup of carbon dioxide and other gases that enhance the greenhouse effect. Still others conclude that natural cycles, such as variations in the sun's output, explain the warming. However, most scientists agree that eventually global temperatures will rise significantly if we continue to release large quantities of greenhouse gases at present rates.

KEY TERMS

conduction	heat island	stratosphere
evaporation	lapse rate	temperature inversion
air temperature	environmental temperature lapse rate	isotherm
transpiration	troposphere	greenhouse gases

REVIEW QUESTIONS

1. How are mean daily temperature and mean monthly temperature determined?

2. Sketch graphs showing how insolation, net radiation, and temperature might vary from midnight to midnight during a 24-hour cycle at a midlatitude station such as Chicago.

3. How does the daily temperature cycle measured within a few centimeters or inches of the surface differ from the cycle at normal air temperature measurement height?

4. Compare the characteristics of urban and rural surfaces and describe how the differences affect urban and rural air temperatures. Include a discussion of the urban heat island.

5. What are the two layers of the lower atmosphere? How are they distinguished?

6. How and why are the temperature cycles of high mountain stations different from those of lower elevations?

7. Sketch a graph of air temperature with height showing a low-level temperature inversion. Where and when is such an inversion likely to occur?

8. Why do large water bodies heat and cool more slowly than land masses? What effect does this have on daily and annual temperature cycles for coastal and interior stations?

9. What three factors are most important in explaining the world pattern of isotherms? Explain how and why each factor is important, and what effect it has.

10. Turn to the January and July world temperature maps shown in Figure 3.18. Make six important observations about the patterns and explain why each occurs.

11. Turn to the world map of annual temperature range in Figure 3.20. What five important observations can you make about the annual temperature range patterns? Explain each.

12. Describe how global air temperatures have changed in the recent past. Identify some factors or processes that influence global air temperatures on this time scale.

IN-DEPTH QUESTIONS

1. Portland, Oregon, on the north Pacific coast, and Minneapolis, Minnesota, in the interior of the North American continent, are at about the same latitude. Sketch the annual temperature cycle you would expect for each location. How do they differ and why? Select one season, summer or winter, and sketch a daily temperature cycle for each location. Again, describe how they differ and why.

2. Many scientists have concluded that human activities are acting to raise global temperatures. What human processes are involved? How do they relate to natural processes? Are global temperatures increasing now? What other effects could be influencing global temperatures? What are the consequences of global warming?

Chapter 4

Atmospheric Moisture and Precipitation

The thick, humid air is perfectly still. Not a single leaf trembles in the oak shading the front yard. The songbirds and insects are strangely silent. The intense heat of the late July afternoon breaks, as clouds slip quietly in overhead. A low, distant rumble shatters the stillness. A thunderstorm is approaching! Rising from a lawn chair, you move to the wide front porch and watch the storm come in.

A rushing sound, low at first, slowly builds. Strong wind gusts rustle in the trees. Suddenly, the top of the oak tree twists violently, its upper branches jerking chaotically. A flood of air, cooling and refreshing, invades the porch.

The first large, heavy drops begin to fall. The rushing of the wind deepens as the first wave of heavy rainfall approaches. The sky explodes with light, and a loud crack of thunder startles you. The rain now drives intensely toward the ground, swirling onto the porch. A crackle, a flash, and then a deafening bang explodes. A huge oak branch crashes to the ground. You retreat from the porch for the safety of the house, hoping that the next lightning strike will be far, far away.

Not all precipitation events are this dramatic. Many involve nothing more than a brief shower or a rainy day. But have you ever wondered why it rains? Or how a powerful thunderstorm is formed? These are the questions answered in this chapter.

Precipitation is the fall of liquid or solid water from the atmosphere to reach the earth's surface. It is formed by condensation or deposition of water vapor as liquid droplets or solid crystals. Because warm air can hold much more water vapor than cold air, warm air will generally yield more precipitation.

What causes condensation or deposition? These processes occur when air is strongly cooled. The drops of water that form on a glass of an iced tea in summer are a familiar example of cooling and condensation. But what produces cooling in the free atmosphere? This subject forms the heart of our chapter.

Whenever air moves upward in the atmosphere, it is cooled. This is not simply because surrounding air temperatures tend to get lower as you go up. It is also because of a simple physical principle stating that when air expands, it cools.

And, because atmospheric pressure decreases with altitude, air moving upward is expanding—and so at the same time getting cooler.

What causes air to move upward? There are three major causes of upward movement of air. One occurs when winds move air up and over a mountain barrier. The second takes place when unequal heating at the ground surface creates a bubble of air that is warmer than the surrounding air. Since it is warmer, it is less dense and is buoyed upward—like a cork under water bobbing to the surface, or a hot air balloon ascending into the sky. This chapter focuses on these two causes of uplift. The third occurs when a mass of cooler, denser air slides under a mass of warmer, lighter air, lifting the warmer air aloft. This type of upward movement occurs in weather systems, which we will cover in Chapter 6.

Another important topic in this chapter is air pollution. How and where does air pollution arise? What forms does it take? How do local conditions concentrate pollution, and what are the results? Although air pollution has little effect on rainfall amounts, weather systems, or climate, it is important because of its influence on life at the surface of our planet.

Lightning storm, near Lava Beds National Monument, northeastern California.

Water exists in the air in the form of humidity, clouds, fog, and precipitation. Humidity is the dampness in the air that can be so uncomfortable on hot, muggy days. Clouds are visible as fleecy white or ominous gray masses in the sky overhead. Fog is a dense haze that surrounds you like a cold, damp blanket. And precipitation occurs as cool raindrops and delicate snowflakes that fall from the sky, providing our source of fresh water.

How are humidity, clouds, fog, and precipitation related? How does water vapor, an invisible gas, become transformed into a torrent of rain that falls to earth? This subject is the main focus of this chapter. But before we begin our study of atmospheric moisture and precipitation, we will briefly review the three states of water and the conversion of one state to another.

As shown in Figure 4.1, water can exist in three states—solid (ice), liquid (water), and gas (water vapor). A change of state from solid to liquid, liquid to gas, or solid to gas requires the input of heat energy. As we noted in Chapter 2, this energy is *latent heat*, which is drawn in from the surroundings. When the change goes the other way, from liquid to solid, gas to liquid, or gas to solid, this latent heat is released to the surroundings.

Figure 4.1 A schematic diagram of the three states of water. Arrows show the ways that any one state can change into either of the other two states. Heat energy is absorbed or released, depending on the direction of change.

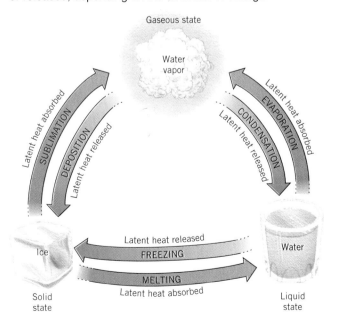

For each type of transition, there is a specific name, shown in the figure. Melting, freezing, evaporation, and condensation are all familiar terms. **Sublimation** describes the direct transition from solid to vapor. Perhaps you have noticed that old ice cubes in your refrigerator's freezer seem to shrink away from the sides of the ice cube tray and even get smaller over time. In this case, they are shrinking through sublimation—which is induced by the constant circulation of cold, dry air through the freezer. The ice cubes never melt, yet they lose bulk directly as vapor. **Deposition** is the reverse process, when water vapor crystallizes directly as ice. Frost formed on a cold winter night is an example.

WATER—THE GLOBAL PERSPECTIVE

Water plays several key roles on our planet. First, the oceans cover more than two-thirds of the planet's surface. Oceans act as a huge heat storage reservoir and redistribute heat from low to high latitudes by ocean currents. The oceans also act as a storage reservoir for dissolved compounds, ranging from salts to nutrients. Second, water falls on land in the form of rain or snow. As it runs off to the sea, water erodes rocks and soils and creates landscapes and landforms. This flow moves nutrients from one location to another, which also influences the distribution of plant and animal life. Third, water in the air moves huge quantities of heat from one place to another by absorbing surface heat in evaporation over warm oceans and releasing that latent heat in condensation or deposition over cooler regions. Like the flow of heat in ocean currents, this movement is generally poleward.

Let's turn first to the nature of the hydrosphere and the flows of water among ocean, land, and atmosphere.

The Hydrosphere and the Hydrologic Cycle

Recall from Chapter 1 that the hydrosphere includes water on the earth in all its forms. About 97.2 percent of the hydrosphere consists of ocean saltwater, as shown in Figure 4.2. The remaining 2.8 percent is fresh water. The next largest reservoir of fresh water is stored as ice in the world's ice sheets and mountain glaciers. This water accounts for 2.15 percent of total global water.

Fresh liquid water on land is a very small part of the earth's total store of water.

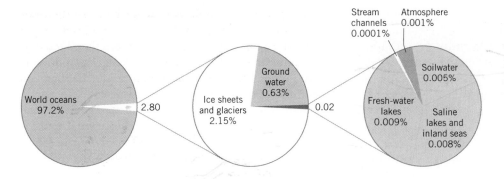

Figure 4.2 Volumes of global water in each reservoir of the hydrosphere. Nearly all the earth's water is contained in the world ocean. Fresh surface and soil water make up only a small fraction of the total volume of global water.

Fresh liquid water is found both on top of and beneath the earth's land surfaces. Water occupying openings in soil and rock is called *subsurface water.* Most of it is held in deep storage as *ground water,* at a level where plant roots cannot access it. Ground water makes up 0.62 percent of the hydrosphere, leaving 0.03 percent of the water remaining.

The right-hand portion of Figure 4.2 shows how this small remaining proportion of the earth's water is distributed. This proportion is important to us because it includes the water available for plants, animals, and human use. *Soil water,* which is held in the soil within reach of plant roots, comprises 0.005 percent of the global total. Water held in streams, lakes, marshes, and swamps is called *surface water.* Most of this surface water is about evenly divided between freshwater lakes and salty lakes. An extremely small proportion is held in streams and rivers as they flow toward the sea or inland lakes.

The hydrologic cycle describes the global flow of water among its reservoirs.

Note that the quantity of water held as vapor and cloud water droplets in the atmosphere is also very small—0.001 percent of the hydrosphere. Though small, this reservoir of water is of enormous importance. It provides the supply of precipitation that replenishes all freshwater stocks on land. And, as we will see in the next chapter, the flow of water vapor from warm tropical oceans to cooler regions provides a global flow of heat, in latent form, from low to high latitudes.

The movements of water among the great global reservoirs constitute the **hydrologic cycle** (Figure 4.3). The cycle begins with evaporation from water or land surfaces, in which water changes state from liquid to vapor and enters the atmosphere. Total evaporation is about six times greater over oceans than land, however. This is because the oceans cover most of the planet and because land surfaces are not always wet enough to yield much evaporated water. Once in the atmosphere, water vapor can condense or deposit to form precipitation, which falls to earth as rain or snow. Precipitation over the oceans is nearly four times greater than precipitation over land.

Upon reaching the land surface, precipitation has three fates. First, it can evaporate and return to the atmosphere as water vapor. Second, it can sink into soil and then into the surface rock layers below. As we will see in later chapters, this subsurface water emerges from below to feed rivers, lakes, and even ocean margins. Third, precipitation can run off the land, concentrating in streams and rivers that eventually carry it to the ocean or to a lake in a closed inland basin. This flow of water is known as *runoff.*

The Global Water Balance

Since our planet contains only a fixed amount of water, a global balance must be maintained among flows of water to and from the lands, oceans, and atmosphere. Let's examine this idea in more detail. For our analysis, we will assume that the volume of ocean waters and the overall volume of fresh water in surface and subsurface water remains constant from year to year. This is probably quite reasonable, unless climate is changing rapidly.

The global water balance is diagrammed in Figure 4.3. Let's consider the ocean first, shown on the right side of the diagram. There are three flows to and from the ocean: evaporation (out), precipitation (in), and runoff (in). For precipitation and runoff, water enters the ocean, while water leaves the ocean by evaporation. Since "in" and "out" must balance, we have

$$\text{Precipitation} + \text{Runoff} = \text{Evaporation} \quad \text{(ocean)}$$

Using the values in the figure, which are given as thousands of cubic kilometers, we find that $380 + 40 = 420$.

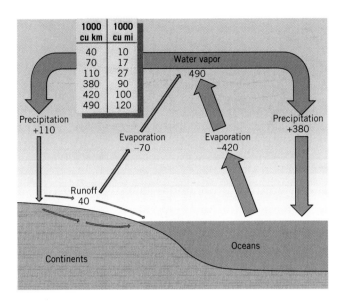

Figure 4.3 The global water balance. Figures give average annual water flows in and out of world land areas and world oceans. Values are given in thousands of cubic kilometers (cubic miles). Global precipitation equals global evaporation. (Based on data of John R. Mather).

Now let's turn to the continents, shown on the left side of the figure. Again, there are three flows—precipitation (in), evaporation (out), and runoff (out). This time, precipitation is entering (falling onto) the land, while evaporation and runoff are leaving. Thus,

$$\text{Precipitation} = \text{Evaporation} + \text{Runoff} \qquad \text{(land)}$$

Using the values in the figure, we observe that 110 = 70 + 40.

If we rearrange this last equation to put evaporation first, we have

$$\text{Evaporation} = \text{Precipitation} - \text{Runoff} \qquad \text{(land)}$$

compared to

$$\text{Evaporation} = \text{Precipitation} + \text{Runoff} \qquad \text{(ocean)}$$

Taking the land and oceans together and considering only the flow of water from the planet's surface to the atmosphere and the reverse, we can conclude that the runoff terms within the last two equations cancel out. This leaves evaporation to equal precipitation, as it must:

Total Evaporation	Total Precipitation
70 (land) + 420 (ocean) cubic kilometers	= 110 (land) + 380 (ocean) cubic kilometers

The point of this simple analysis is not to write equations, but to show that we have constructed a budget

that accounts for the flows of water between the reservoirs of the hydrologic cycle. Like a household budget, it must balance. If it does not, then there must be some flow that is not being considered. This is why scientists use budgets as tools to help understand natural processes, like the hydrologic cycle.

The global water balance describes the flows of water between ocean, atmosphere, and land.

For most of this chapter, we will be concerned with one aspect of the hydrologic cycle—the flow of water from the atmosphere to the surface in the form of precipitation. The examination of this aspect of the hydrologic cycle will involve descriptions of how the water vapor content of air is measured and how clouds and precipitation form.

HUMIDITY

The amount of water vapor present in the air varies widely from place to place and time to time. It ranges from almost nothing in the cold, dry air of arctic regions in winter to as much as 4 or 5 percent of a given volume of air in the warm wet regions near the equator. The general term **humidity** refers to the amount of water vapor present in the air.

Understanding humidity and how the moisture content of air is measured involves an important principle—namely, that the maximum quantity of moisture that can be held at any time in the air is dependent on temperature. Warm air can hold more water vapor than cold air—a lot more. Air at room temperature (20°C, 68°F) can hold about three times as much water vapor as air at freezing (0°C, 32°F).

The measure of humidity that we encounter every day is *relative humidity*. This measure compares the amount of water vapor present to the maximum amount that the air can hold at that temperature, expressed as a percentage. For example, if the air currently holds half the moisture possible at the present temperature, then the relative humidity is 50 percent. When the humidity is 100 percent, the air holds the maximum amount possible and is *saturated*.

A change in relative humidity of the atmosphere can happen in one of two ways. The first is by evaporation. If an exposed water surface or wet soil is present, additional water vapor can enter the air. This process is slow because the water vapor molecules must diffuse upward from the surface into the air layer above.

Relative humidity compares the amount of water held by air to the maximum amount that can be held at that temperature.

The second way is through a change of temperature. Even though no water vapor is added, a lowering of temperature results in a rise of relative humidity. Recall that the capacity of air to hold water vapor is dependent on temperature. When the air is cooled, this capacity is reduced. The existing amount of water vapor then represents a higher percentage of the total capacity.

Relative Humidity Through the Day

An example may help to illustrate these principles (Figure 4.4). At 10 A.M., the air temperature is 16°C (60°F), and the relative humidity is 50 percent. By 3 P.M., the air has been warmed by the sun to 32°C (90°F). The relative humidity has dropped to 20 percent, which is very dry air. The same amount of water vapor is present in the air, but the capacity of the air to hold water vapor has greatly increased. At night the air is chilled, and by 4 A.M. its temperature has fallen to 5°C (40°F). Now, with the same amount of vapor the relative humidity has risen to 100 percent, and the air is saturated.

Another way of describing the water vapor content of air is by its **dew-point temperature.** If air is slowly chilled, it eventually will reach saturation, with a relative humidity of 100 percent. At this temperature, the air holds the maximum amount of water vapor possible. If further cooling continues, condensation will begin and dew will form. The temperature at which saturation occurs is therefore known as the dew-point

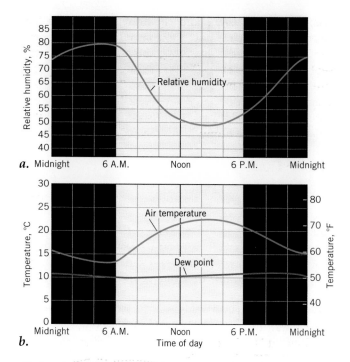

Figure 4.5 (*a*), Relative humidity, and (*b*), air temperature and dew-point temperature throughout the day at Washington, D.C. The curves show average values for the month of May. (Data from National Weather Service.)

temperature. Moist air will have a higher dew-point temperature than drier air.

The principle of relative humidity change caused by temperature change is further illustrated by a graph of these two properties throughout the day (Figure 4.5). Values shown are averages for May at Washington, D.C. As air temperature rises during the day, relative humidity falls, and vice versa. The dew-point temperature, which indicates the true moisture content, remains nearly constant.

How is humidity measured? The simplest instrument that measures humidity is called a *hygrometer*. It uses strands of human hair or special fibers that lengthen and shorten according to the relative humidity. The expansion or contraction in length activates a dial that reads relative humidity.

A more accurate method of measuring humidity can be devised by using the evaporation principle. Two thermometers are mounted together side by side in an instrument called a *sling psychrometer* (Figure 4.6). The wet-bulb thermometer bulb is covered with a sleeve of cotton fibers and is wetted. The dry-bulb thermometer remains dry. To operate the sling psychrometer, water is applied to the sleeve covering the wet bulb, and the two thermometers are then whirled in the open air. Evaporation cools the wet bulb. If the air is saturated, then evaporation cannot occur. In this

Figure 4.4 Relative humidity changes with temperature because the capacity of warm air to hold water vapor is greater than that of cold air. In this example, the amount of water vapor stays the same, and only the capacity changes.

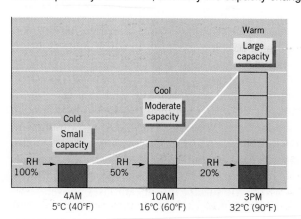

case, no cooling results, and the temperature of the two thermometers is the same. If the air is very dry, evaporation is strong, and the cooling effect is large. So, the wet-bulb thermometer will read a much cooler temperature than the dry-bulb thermometer. To determine the relative humidity, a special sliding scale is used (Figure 4.6). Wet-bulb and dry-bulb temperatures are set on the scale, and relative humidity is read off directly.

Specific Humidity

Relative humidity measures only the amount of water vapor present relative to the amount held at saturation. The actual quantity of water vapor held by the air is its *specific humidity*. This quantity is stated as the mass of water vapor contained in a given mass of air, and is expressed as grams of water vapor per kilogram of air (gm/kg).

We stated earlier that the amount of water vapor that air can hold depends on its temperature. Figure 4.7 shows this relationship. We see, for example, that at 20°C (68°F), the maximum amount of water vapor that the air can hold—that is, the maximum specific humidity—is about 15 gm/kg. At 30°C (86°F), it is almost doubled—about 26 gm/kg. For cold air, the values are quite small. At –10°C (14°F), the maximum is only about 2 gm/kg.

Specific humidity is often used to describe the moisture characteristics of a large mass of air. For example, extremely cold, dry air over arctic regions in winter

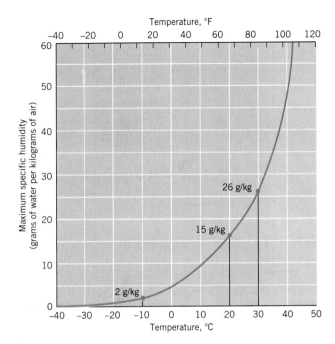

Figure 4.7 The maximum specific humidity of a mass of air increases sharply with rising temperature.

may have a specific humidity as low as 0.2 gm/kg. In comparison, the extremely warm, moist air of equatorial regions often holds as much as 18 gm/kg. The total natural range on a worldwide basis is very wide. In fact, the largest values of specific humidity observed are from 100 to 200 times as great as the smallest.

Specific humidity is a geographer's yardstick for a basic natural resource—water—that can be applied from equatorial to polar regions. It is a measure of the quantity of water that can be extracted from the atmosphere as precipitation. Cold, moist air can supply only a small quantity of rain or snow, but warm, moist air is capable of supplying large quantities.

Warm air can hold much more water vapor than cold air.

Figure 4.8 is a set of global profiles showing how specific humidity varies with latitude and how it relates to mean surface air temperature. Both humidity and temperature are measured at the same locations in standard thermometer shelters the world over, and also on ships at sea. Note that the horizontal axis of the graphs has an uneven scale of units—they get closer together toward both left and right edges, which represent the poles. In this case, the units are adjusted so that they reflect the amount of the earth's surface present at that latitude. That is, since there is much more surface area between 0° and 10° latitude than between 60° and

Figure 4.6 This standard sling psychrometer uses paired thermometers. The cloth-covered wet-bulb thermometer projects beyond the dry-bulb thermometer. The handle is used to swing the thermometers in the free air. Evaporation lowers the wet-bulb temperature. Thus, the dryer the air the greater the evaporation and the cooler the reading in relation to the dry-bulb thermometer. The sliding scale below enables rapid determination of relative humidity from wet and dry bulb readings.

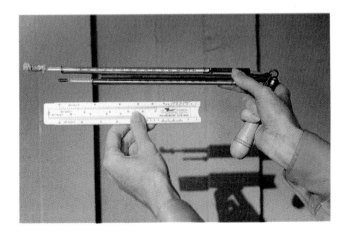

70°, the region between 0° and 10° is allocated more width on the graph.

Let's look first at specific humidity (*a*). The curve clearly shows the largest values for the equatorial zones, with values falling off rapidly toward both poles. This curve follows the pattern of insolation quite nicely (see Figure 2.7). More insolation is available at lower latitudes, on average, to evaporate water in oceans or on moist land surfaces. Therefore, specific humidity values are higher at low latitudes than at high latitudes.

The global profile of mean (average) surface air temperature (*b*), shows a similar shape to the specific humidity profile. We would expect this to be the case, since air temperature and maximum specific humidity vary together, as shown in Figure 4.7.

THE ADIABATIC PROCESS

Given that ample water vapor is present in a mass of air, how is that related to precipitation? In other words, how is the water vapor turned into liquid or solid particles that can fall to earth? The answer is by natural cooling of the air. Since the ability of air to hold water vapor is dependent on temperature, the air must give up water vapor if it is cooled to the dew point and below. Think about a moist sponge. To extract the water, you have to squeeze the sponge—that is, reduce its ability to hold water. Chilling the air is like squeezing the sponge.

So how is air chilled sufficiently to produce precipitation? One mechanism for chilling air is nighttime cooling. As we have seen, the ground surface can become quite cold on a clear night through loss of longwave radiation. Thus, still air near the surface can be cooled below the condensation point, producing dew or frost.

However, nighttime cooling is not sufficient to form precipitation. Only when large masses of air experience a steady drop in temperature below the dew point can precipitation occur in reasonable amounts. The primary mechanism for cooling large masses of air to produce precipitation is uplift. As we will see shortly, when air rises in the atmosphere, it is chilled. So precipitation will always be linked with the rise of moist air.

Dry Adiabatic Rate

An important principle of physics is that when a gas is allowed to expand, its temperature drops. Conversely, when a gas is compressed, its temperature increases. You have probably observed this latter effect yourself, if you have ever pumped up a bicycle tire using a hand pump. In this process, the pump gets hot because air

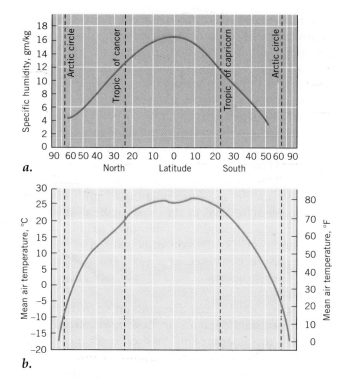

Figure 4.8   Pole-to-pole profiles of temperature (above) and specific humidity (below). The two are similar because the ability of air to hold water vapor (measured by specific humidity) is limited by temperature. (Data of J. von Hann, R. Süring, and J. Szava-Kovats as shown in Haurwitz and Austin, *Climatology*.)

inside the pump is being compressed and therefore heated. In the same way, a small jet of air escaping from a high-pressure hose feels cool. Perhaps you have flown on an airplane and noticed a small nozzle overhead that directs a stream of cool air toward you. In moving from high pressure in the air hose that feeds the nozzle to low pressure in the cabin, the air expands and cools. Physicists use the term **adiabatic process** to refer to heating or cooling that occurs solely as a result of pressure change.

The adiabatic principle states that compression warms a gas, while expansion cools it.

Given that air cools or heats when the pressure on it changes, how does that relate to uplift and precipitation? The missing link is simply that atmospheric pressure decreases with an increase in altitude. So, if a mass of air is uplifted to a level of lower atmospheric pressure, it expands and cools. And if a mass of air descends to a level of higher atmospheric pressure, it is compressed and warms (Figure 4.9). The **dry adiabatic lapse rate** describes this behavior. This rate has a value of 10°C per 1000 m (5.5°F per 1000 ft) of vertical rise. That is, if a mass of air is raised 1 km, its temperature will drop by 10°C. Or, in English units, if raised 1000 ft, its temperature will drop by 5.5°F. This is the "dry" rate because condensation does not occur.

Note that in the previous chapter we encountered the environmental temperature lapse rate. This is quite different from the dry adiabatic lapse rate. The dry adiabatic rate is always constant and is determined by physical laws. It applies to a mass of air in vertical motion. The temperature lapse rate is simply an expression of how the temperature of still air varies with altitude. This rate will vary from time to time and from one place to another, depending on the state of the atmosphere. No motion of air is implied for the temperature lapse rate.

Wet Adiabatic Rate

With the adiabatic process firmly in mind, let's examine the fate of a parcel of moist air that is moved upward in the atmosphere (Figure 4.10). We will assume that the parcel starts with a temperature of 20°C (68°F). As the parcel is moved upward, its temperature drops at the dry adiabatic rate, 10°C/1000 m (5.5°F/1000 ft). At 500 m (1600 ft), the temperature has dropped by 5°C (9°F) to 15°C (59°F). At 1000 m (3300 ft), the temperature has fallen to 10°C (50°F).

If the rising process continues, the air will be cooled to saturation and condensation will start to occur. This is shown on the figure as the *level of condensation*. If cooling continues, water droplets will form and we

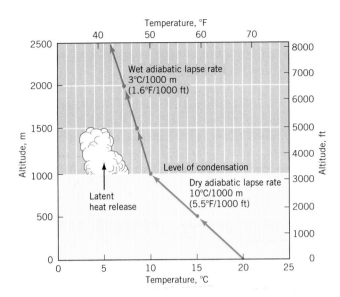

Figure 4.10 Adiabatic decrease of temperature in a rising parcel of air leads to condensation of water vapor into water droplets and the formation of a cloud. (Copyright © by A. N. Strahler.)

have a cloud. If the parcel of saturated air continues to rise, however, a new principle comes into effect—latent heat release. That is, when condensation occurs, latent heat is released and warms the uplifted air. In other words, two effects are occurring at once. First, the uplifted air is being cooled by the reduction in atmospheric pressure. Second, it is being warmed by the release of latent heat from condensation.

Figure 4.9 A schematic diagram of adiabatic cooling and heating that accompanies the rising and sinking of a mass of air. When air is forced to rise, it expands and its temperature decreases. When air is forced to descend, its temperature increases. (A. N. Strahler.)

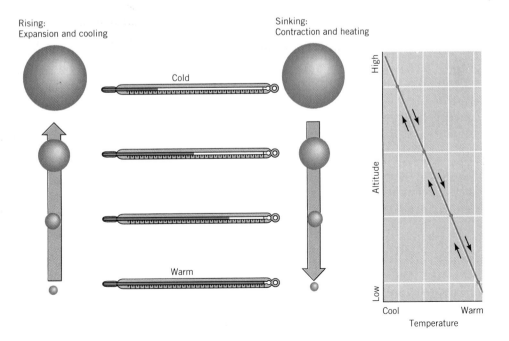

> **Rising air cools less rapidly when condensation is occurring, owing to the release of latent heat.**

Which effect is stronger? As it turns out, the cooling effect is the stronger, so the air will continue to cool as it is uplifted. But because of the release of latent heat, the cooling will occur at a lesser rate. This cooling rate is called the **wet adiabatic lapse rate,** and ranges between 3 and 6°C per 1000 m (1.6 and 2.7°F per 1000 ft). It is variable because it depends on the temperature of the air and its moisture content. For very warm air with a high moisture content, such as you might find over the equatorial oceans, the lower cooling rate values will apply. This is because such moist air releases lots of latent heat. In Figure 4.10, the cooling rate of 3°C/1000 m (1.6°F/1000 ft) is used. For cooler and somewhat drier air, such as you might find in midlatitude continental regions in the spring or fall, the latent heat released is not so great and a higher cooling rate applies.

CLOUDS

A **cloud** is made up of water droplets or ice particles suspended in air. These particles have a diameter in the range of 20 to 50 μm (0.0008 to 0.002 in.). Recall from Chapter 2 that μm denotes the *micrometer,* or one-millionth of a meter. Each cloud particle is formed on a tiny center of solid matter, called a *condensation nucleus.* This nucleus has a diameter in the range 0.1 to 1 μm (0.000004 to 0.00004 in.).

An important source of condensation nuclei is the surface of the sea. When winds create waves, droplets of spray from the crests of the waves are carried rapidly upward in turbulent air. Evaporation of sea water droplets leaves a tiny residue of crystalline salt suspended in the air. This aerosol strongly attracts water molecules. Even very clean and clear air contains enough condensation nuclei for the formation of clouds.

> **Cloud particles are tiny water droplets and ice crystals that form around condensation nuclei.**

In our everyday life at the earth's surface, liquid water turns to ice when the surrounding temperature falls to the freezing point, 0°C (32°F), or below. However, the water in tiny cloud particles can remain in the liquid state at temperatures far below freezing.

Such water is described as *supercooled.* Clouds consist entirely of water droplets at temperatures down to about –12°C (10°F). As cloud temperatures grow colder, a mix of water droplets and ice crystals occur. The coldest clouds, with temperatures below –40°C (–40°F), occur at altitudes of 6 to 12 km (20,000 to 40,000 ft) and are formed entirely of ice particles.

Cloud Forms

Clouds come in many shapes and sizes—from the small, white puffy clouds often seen in summer to the dark layers that produce a good, old-fashioned rainy day. Meteorologists classify clouds into four families, arranged by height: high, middle, and low clouds, and clouds with vertical development. These are shown in Figures 4.11 and 4.12. Some individual types with their names are also shown.

Clouds are grouped into two major classes on the basis of form—stratiform, or layered clouds, and cumuliform, or globular clouds. *Stratiform* clouds are blanket-like and cover large areas. A common type is *stratus,* which covers the entire sky. Stratus clouds are formed when large air layers are forced to rise gradually. This can happen when one air layer overrides another. As the rising layer is cooled, condensation occurs over a large area, and a blanket-like cloud forms. If the layer is quite moist and rising continues, dense, thick stratiform clouds result that can produce abundant rain or snow.

> **Stratiform clouds form horizontal layers, while cumuliform clouds are globular in shape.**

Cumuliform clouds are globular masses of cloud that are associated with small to large parcels of rising air. The air parcels rise because they are warmer than the surrounding air. Like bubbles in a fluid, they move upward. Of course, as they are buoyed upward, they are cooled by the adiabatic process. When condensation occurs, a cloud is formed. The most common cloud of this type is the *cumulus* cloud (Figure 4.12). Sometimes the upward movement yields dense, tall clouds that produce thunderstorms. This form of cloud is the *cumulonimbus.* ("Nimbus" is the latin word for rain cloud, or storm.)

Fog

Fog is simply a cloud layer at or very close to the surface. In our industrialized world, fog is a major environmental hazard. Dense fog on high-speed highways

Classification of clouds according to height and form

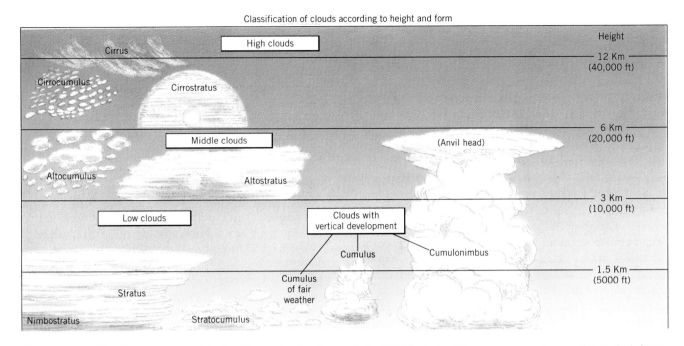

Figure 4.11   Clouds are grouped into families on the basis of height. Individual cloud types are named according to their form.

Figure 4.12   (Left) Rolls of altocumulus clouds, also known as a "mackerel sky." (Right) High cirrus clouds lie above puffy cumulus clouds in this scene from Holland.

can cause chain-reaction accidents, sometimes involving dozens of vehicles and causing injury or death to their occupants. At airports, landing delays and shutdowns because of fog bring economic losses to airlines and can inconvenience thousands of travelers in a single day. For centuries, fog at sea has been a navigational hazard, increasing the danger of ship collisions and groundings. In addition, polluted fogs, like those of London in the early part of this century, can injure urban dwellers' lungs and take a heavy toll in lives.

One type of fog, known as a *radiation fog*, is formed at night when temperature of the air layer at the ground level falls below the dew point. This kind of fog is associated with a low-level temperature inversion (Figure 3.10). Another fog type, an *advection fog*, results when a warm, moist air layer moves over a cold or surface. As the air layer loses heat to the surface, its temperature drops below the dew point, and condensation sets in. Advection fog commonly occurs over oceans where warm and cold currents occur side by side. When warm, moist air above the warm current moves over the cold current, condensation occurs. Fogs of the Grand Banks off Newfoundland are formed in this way, because here the cold Labrador current comes in contact with warm waters of the Gulf Stream.

Sea fog is frequently found along the California coast. It forms when warm, moist air moves eastward across the cold waters of the California current (Figure 4.13). Similar fogs are found on continental west coasts in the tropical latitude zones where cool currents parallel the shoreline.

PRECIPITATION

With a firm understanding of the adiabatic process and how it produces cooling, condensation or deposition, and clouds, we can now turn to the final topic of our chapter—precipitation. Precipitation can form in two ways. In the first, cloud droplets collide and coalesce into larger and larger raindrops that can fall as rain. In the second, ice crystals form and grow in a cloud that contains a mixture of both ice crystals and water droplets.

In the first process, saturated air is rapidly uplifted, and cooling forces additional condensation. Cloud particles grow by added condensation and attain a diameter of 50 to 100 μm (0.002 to 0.004 in.). In collisions with one another, the particles grow into droplets of about 500 μm in diameter (about 0.02 in.). This is the size of water droplets in drizzle. Further collisions increase drop size and yield *rain*. Average raindrops have diameters of about 1000 to 2000 μm (0.04 to 0.1 in.), but they can reach a maximum diameter of about 7000 μm (0.25 in.). Above this value they become unstable and break into smaller drops while falling. This type of precipitation formation occurs in warm clouds typical of the equatorial and tropical zones.

Snow is produced by the second process, and forms in clouds that are a mixture of ice crystals and supercooled water droplets. When an ice crystal collides with a droplet of supercooled water, it induces freezing of the droplet. The ice crystals further coalesce to form snow particles, which can become heavy enough to fall

Figure 4.13 Coastal fog under the Bay Bridge, San Francisco.

Figure 4.14 These individual snow crystals, greatly magnified, were selected for their beauty. Most snow particles are lumps of ice without such a delicate structure.

from the cloud. Some ice crystals grow directly by deposition. Snowflakes formed entirely by deposition can have intricate crystal structures (Figure 4.14), but most particles of snow are simply tiny lumps of ice.

In warm clouds experiencing strong condensation, water droplets grow by condensation, then collide and coalesce to form raindrops.

When the underlying air layer is below freezing, snow reaches the ground as a solid form of precipitation. Otherwise, it melts and arrives as rain. A reverse process, the fall of raindrops through a deep, cold air layer, results in the freezing of rain and produces pellets or grains of ice. These are commonly referred to in North America as *sleet*. (Among the British, sleet refers to a mixture of snow and rain.)

Perhaps you have experienced an *ice storm*. This occurs when the ground is frozen and the lowest air layer is also below freezing. Rain falling through the layer is chilled and freezes onto ground surfaces as a clear glaze. Ice storms cause great damage, especially to telephone and power lines and to tree limbs pulled down by the weight of the ice. In addition, roads and sidewalks are made extremely hazardous by the slippery glaze.

Actually, the ice storm is more accurately named an "icing" storm, since it is not ice that is falling but supercooled rain. *Hail*, another form of precipitation, consists of large pellets or spheres of ice. The formation of hail will be explained in our discussion of the thunderstorm.

Precipitation is measured in units of depth of fall per unit of time—for example, centimeters or inches per hour or per day. A centimeter (inch) of rainfall would cover the ground to a depth of 1 cm (1 in.), if the water did not run off or sink into the soil. Rainfall is measured with a *rain gauge*. The simplest rain gauge is a straight-sided, flat-bottomed pan, which

Eye on the Environment • Acid Deposition and Its Effects

"What? Acid rain? You mean now we're even poisoning the rain?" You've heard about acid rain killing fish and poisoning trees. Acid rain is part of the phenomenon of **acid deposition,** which includes not only acid rain, but also dry acidic dust particles. Acid rain simply consists of raindrops that have been acidified by air pollutants. Dry acidic particles are dust particles that are acidic in nature. They fall to earth and coat the surface in a thin dust layer. When wetted by rain or fog, they acidify the water on leaves and soils. In winter, acid particles can mix with snow as it accumulates. In spring, when the snow melts, a surge of acid water is then released to soils and streams.

Where does this type of pollution come from? Acid deposition is produced by the release of sulfur dioxide (SO_2) and nitric oxide (NO_2) into the air. This occurs when fossil fuels are burned, as gasoline and diesel fuels for transportation and as coal and oil for electric and industrial power generation. Once in the air, SO_2 and NO_2 readily combine with oxygen and water in the presence of sunlight and dust particles to form sulfuric and nitric acid aerosols. These aero-sols can further serve as conden-sation nuclei and acidify the tiny water droplets created. When these coalesce in precipitation, acid raindrops result. Sulfuric and nitric acids can also be formed on dust particles, creating dry acid particles. These can be as damaging to plants, soils, and aquatic life as acid rain.

What are the effects of acid deposition? A primary effect is acidification of lakes and streams, injuring aquatic plants and animals and perhaps eventually sterilizing a lake, pond, or stream. Another problem is soil damage. Through a series of chemical effects, acid deposition can cause soils to lose nutrients, affecting the natural ecosystem within a region.

To measure acidity, scientists use the pH scale. Values on this scale range from 1 to 14. High values indicate alkalinity, and low values indicate acidity. Pure water has a pH of 7. A value of 1 is very acidic—like battery acid. At this pH, acid will eat through clothes and burn the skin. Normally, rainwater is slightly acid, showing pH values of 5 to 6. In the 1960s European water chemists observed that the pH of rain in northwestern Europe had dropped to values as low as 3 in some samples, similar to household vinegar. Because pH numbers are on a logarithmic scale, these values mean that rain in these samples was 100 to 1000 times more acid than normal.

In North America, scientists studying the chemical quality of rainwater have reported that since 1975 rainwater over a large area of the northeastern United States has had an average pH of about 4. Values less than pH 4 occur at times over many heavily industrialized U.S. cities, including Boston, New York, Philadelphia, Birmingham, Chicago, Los Angeles, and San Francisco. Values between pH 4 and 5 have been observed near smaller cities such as Tucson, Arizona; Helena, Montana; and Duluth, Minnesota—localities we do not usually associate with heavy air pollution.

At the left is a map of the United States and Canada showing the average acidity of rainwater. Compare this with the figure on the right, showing emissions of sulfur dioxide and nitrogen oxides by states. There is a strong relationship, especially between SO_2 and the average acidity of rainwater.

An important factor in the level of impact of acid deposition on the environment is the ability of the soil and surface water to absorb and neutralize acid. This factor ranges widely. In dry climates, acids can be readily neutralized because surface waters are normally somewhat alkaline. Areas where soil water is naturally somewhat acidic are most

Acidity of rain water for the United States and Canada, averaged over the year 1982. Values are pH units. The northeastern United States and southeastern Canada are most strongly affected. Other recent years show essentially similar values. (From NOAA Air Resources Laboratory.)

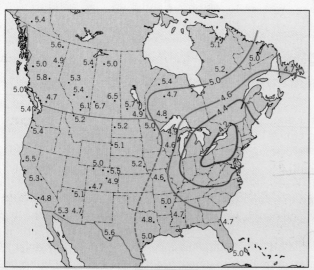

sensitive to acid deposition. Such areas are associated with moist climates generally and include the eastern United States, high-mountain regions of the western states, and the Pacific Northwest.

Acid deposition in Europe and North America has severely impacted some ecosystems. In Norway, acidification of stream water has virtually eliminated salmon runs by inhibiting salmon egg development. At present, the acidity of Norwegian streams has leveled off as air pollution levels have stabilized. However, salmon kills continue to occur, especially when rainstorms carry large amounts of sulfate into streams.

The increased fish mortality observed in Canadian lakes has also been attributed to acidification. In 1980 Canada's Department of

Environment reported that 140 Ontario lakes had no fish and that thousands of other lakes of that province were threatened with a similar fate. In 1990 American scientists estimated that 14 percent of Adirondack lakes were heavily acidic, along with 12 to 14 percent of the streams in the Mid-Atlantic states.

Forests, too, have been damaged by acid deposition. In western Germany, the impact has been especially severe in the Harz Mountains and the Black Forest. In 1983 it was reported that about one-third of West Germany's forests showed visible damage. In the eastern United States, pine and spruce trees have experienced damage in recent years. In the early 1990s Europe continued to experience heavy losses in timber, resulting

from fallout of sulfur dioxide in combination with nitrogen oxide emissions from vehicles, industry, and farm wastes. Fallout of sulfur oxides was most heavily concentrated in Germany, Czechoslovakia, Poland, and Hungary.

In 1990 the National Acid Precipitation Assessment Program (NAPAP) issued a report after its ten-year study of the causes and effects of acid precipitation in the United States. Emissions of sulfur dioxide and volatile organic compounds had declined 30 percent since 1970. Furthermore, nitrogen oxides had also declined over the period by 8 to 16 percent. These effects were due primarily to improved industrial emission controls. In 1993 a report by the U.S. Geological Survey confirmed that the acidity of rainfall for many stations in the northeast and northern midwest had lessened since 1980. However, streams and lakes in these regions have not shown many signs of recovery.

Although the situation is improving, acid deposition is still a very important problem in many parts of the world—especially Eastern Europe and the states of the former Soviet Union. There, air pollution controls have been virtually nonexistent for decades. Reducing pollution levels and cleaning up polluted areas will be a major task for these nations as they enter the next century.

Emissions of sulfur dioxide and nitrogen oxides by states for the year 1980. Sulfur dioxide releases are especially heavy in the states that burn Appalachian coal. Releases of nitrogen oxides are high in Texas, due to oil processing, and in California, due to auto emissions. Since 1980, releases of SO_2 and NO_x have decreased somewhat, but the pattern has stayed about the same. (Office of Technology Assessment, U.S. Congress, U.S. Government Printing Office, Washington, D.C.)

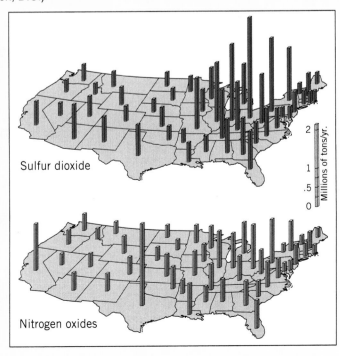

Questions

1. What is acid deposition? Where does the acid come from? In what parts of the United States is rainfall most acid? Why?

2. What are the effects of acid deposition? Why is it harmful?

Figure 4.15 This rain gauge consists of two clear plastic cylinders, here partly filled with red-tinted water. The inner cylinder receives rainwater from the funnel top. When it fills, the water overflows into the larger, outer cylinder.

is set outside before a rainfall event. (An empty coffee can does nicely.) After the rain, the depth of water in the pan is measured.

A very small amount of rainfall, such as 2 mm (0.1 in.), makes too thin a layer to be accurately measured in a flat pan. To avoid this difficulty, a typical rain gauge is constructed from a narrow cylinder with a funnel at the top (Figure 4.15). The funnel gathers rain from a wider area than the mouth of the tube, so the cylinder fills more quickly. The water level gives the amount of precipitation, which is read on a graduated scale.

Precipitation is measured with a rain gauge.

Snowfall is measured by melting a sample column of snow and reducing it to an equivalent in rainfall. In this way rainfall and snowfall records may be combined in a single record of precipitation. Ordinarily, a 10 cm (or 10 in.) layer of snow is assumed to be equivalent to

1 cm (or 1 in.) of rainfall, but this ratio may range from 30 to 1 in very loose snow to 2 to 1 in old, partly melted snow.

Precipitation Processes

So far, we have seen how air that is moving upward will be chilled by the adiabatic process to the saturation point and then to condensation, and how eventually precipitation will form. However, one key piece of the precipitation puzzle has been missing—what causes air to move upward? Air can move upward in three ways. First, it can be forced upward as a through-flowing wind. Consider the case of a mass of air moving up and over a mountain range, for example. If the mountain range is high enough and the air is moist enough, then precipitation will occur. This is called **orographic precipitation,** and we'll describe it further in the following section.

The three types of precipitation are orographic, convectional, and cyclonic.

A second way for air to be forced upward is through convection. In this process, a parcel of air is heated, perhaps by a patch of warm ground, so that it is warmer and therefore less dense than the air around it. Like a bubble, it rises. If the air is quite moist and condensation sets in, then the release of latent heat of condensation will ensure that the bubble of air remains warmer than the surrounding air as it rises. This type of precipitation is **convectional precipitation,** and it will be described after orographic precipitation.

A third way for air to be forced upward is through the movement of air masses. As we will see in the next chapter, air masses are large bodies of air, each with a set of uniform temperature and moisture properties. Air masses move normally from west to east in midlatitudes, and one can overtake another. When this happens, one of them will be forced aloft. Since this overtaking usually occurs in a common type of storm called a cyclone, this type of precipitation is known as *cyclonic precipitation.* We will return to this subject in Chapter 6.

Orographic Precipitation

In orographic precipitation, through-flowing winds move moist air up and over a mountain barrier. (The term *orographic* means "related to mountains.") Figure 4.16 shows how this process works. Moist air arrives at the coast after passing over a large ocean surface (1). As the air rises on the windward side of the range, it is cooled at the dry adiabatic rate. When cooling is suffi-

Figure 4.16 The forced ascent of a warm, moist oceanic air mass over a mountain barrier produces precipitation and a rainshadow desert. As the air moves up the mountain barrier, it loses moisture through precipitation. As it descends the far slope, it is warmed.

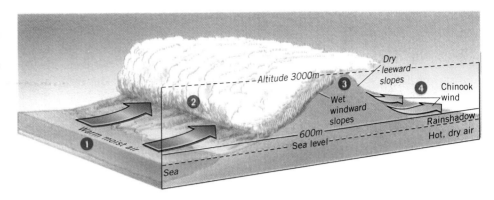

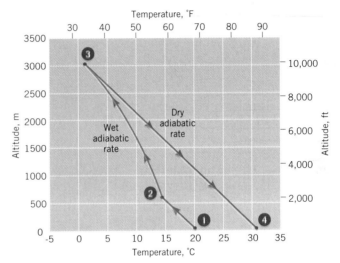

cient, condensation and/or deposition sets in (2). Cooling now proceeds at the wet adiabatic rate. Eventually, precipitation begins. As the air continues up the slope, precipitation continues to fall.

After passing over the mountain summit, the air begins to descend down the leeward slopes of the range (3). However, as it descends, it is compressed. This means that it is warmed, according to the adiabatic principle. Since cooling is no longer occurring, cloud droplets and ice crystals evaporate or sublimate. The air clears rapidly. As it continues to descend, it warms further. At the base of the mountain on the far side (4), the air is now hot—and dry, since its moisture has been removed.

In orographic precipitation, moist air is forced up and over a mountain barrier, producing cooling, condensation, and precipitation.

The net result of the trip over the mountain is that the warm, moist air has become hot and dry. It was warmed by the latent heat released to the surrounding air as water vapor formed water droplets and ice crys-

tals in the precipitation process. Its water was shed to the mountain slopes below, on the uphill journey. This effect creates a *rainshadow* on the far side of the mountain—a belt of dry climate that extends on down the leeward slope and beyond. Several of the earth's great deserts are of this type.

California's rainfall patterns provide an excellent example of orographic precipitation and the rainshadow effect. A map of California (Figure 4.17) shows mean annual precipitation. It uses lines of equal precipitation, called *isohyets.* Focus now on central California and follow the arrow on the map. Prevailing westerly winds bring moist air in from the Pacific Ocean, first over the Coast Ranges of central and northern California, and then, after rain is deposited on the Coast Ranges, the air descends into the broad Central Valley. Here, precipitation is low, averaging less than 25 cm (10 in.) per year.

Next the air continues up and over the great Sierra Nevada, whose summits rise to 4200 m (14,000 ft) above sea level. Heavy precipitation, largely in the form of winter snow, falls on the western slopes of these ranges and nourishes rich forests. Passing down the steep eastern face of the Sierra Nevada, the air descends quickly into the Owens Valley, at about

1000 m (3300 ft) elevation. The adiabatic heating warms and dries the air, creating a rainshadow desert. In the Owens Valley, annual precipitation is less than 10 cm (4 in.). In this way, the orographic effect on air moving across the mountains of California produces a part of America's great interior desert zone, extending from eastern California and across Nevada.

Convectional Precipitation

The second process of inducing uplift in a parcel of air is *convection,* the exchange of heat by vertical movement. In this process, strong updrafts occur within *convection cells.* Air rises in a convection cell because it is warmer, and therefore less dense, than the surrounding air.

The convection process starts when a surface is heated unequally. For example, consider an agricultural field surrounded by a forest. Since the field surface consists largely of bare soil and only a low layer of vegetation, it will be warmer under steady sunshine than the adjacent forest. This means that the air above the field will eventually grow warmer than the air above the forest.

Now, the density of air depends on its temperature—warm air is less dense than cooler air. Perhaps you have seen a hot-air balloon, which uses this principle. The balloon is open at the bottom, and in the basket below a large gas burner forces heated air into the balloon. Because the heated air is less dense than the surrounding air, the balloon rises. The same principle will cause a bubble of air to form over the field, rise, and break free from the surface. Figure 4.18 diagrams this process.

If a bubble of air is warmer than the surrounding air, it will be less dense and rise upward.

As the bubble of air rises, it is cooled adiabatically and its temperature will decrease as it rises. However, we know that the temperature of the surrounding air will normally decrease with altitude as well. Nonetheless, as long as the bubble is still warmer than the surrounding air, it will be less dense and will therefore continue to rise.

If the bubble remains warmer than the surrounding air and uplift continues, adiabatic cooling chills the bubble below the dew point. Condensation occurs, and the rising air column becomes a puffy cumulus cloud. The flat base shows the level at which condensation begins. The bulging "cauliflower" top of the cloud is

the top of the rising warm-air column, pushing into higher levels of the atmosphere. Normally, the small cumulus cloud will encounter winds aloft that mix it into the local air. After drifting some distance downwind, the cloud evaporates.

Figure 4.17 The effect of mountain ranges on precipitation is strong in the state of California, because of the prevailing flow of moist oceanic air from west to east. Centers of high precipitation coincide with the western slopes of mountain ranges, including the coast ranges and Sierra Nevada. To the east, in their rainshadows, lie desert regions.

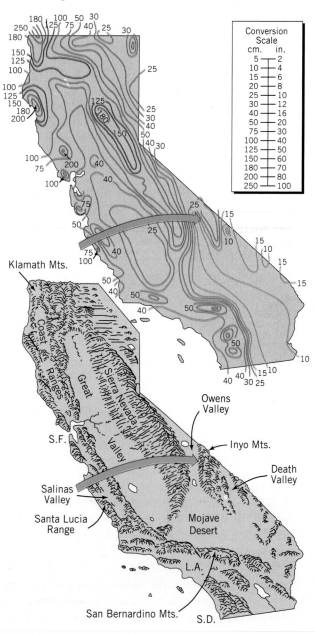

Figure 4.18 Rise of a bubble of heated air to form a cumulus cloud.

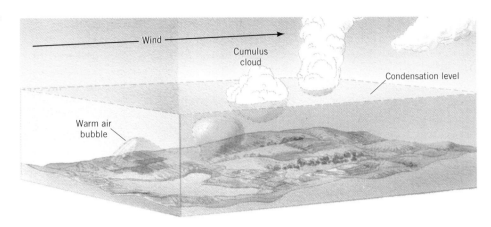

Unstable Air

Sometimes, however, convection continues strongly, and the cloud develops into a dense cumulonimbus mass, or thunderstorm, from which heavy rain will fall. Two conditions encourage the development of thunderstorms—air that is very warm and moist, and a steep environmental temperature lapse rate. By "steep," we mean that temperature decreases quite rapidly with altitude. Air with these characteristics is referred to as **unstable air.**

> **Unstable air—warm, moist, and heated at the surface—can produce abundant convectional precipitation.**

For the first factor, recall that there are two adiabatic rates—dry and wet. The wet rate is about half the dry rate. While condensation is occurring, the lesser wet rates apply. Thus, air in which condensation is occurring cools less rapidly with uplift. Furthermore, the wet rate is smallest for warm, moist air. This means that the temperature decrease experienced by warm, moist, rising, condensing air is quite small. And, since the temperature decrease is small, the rising air is more likely to stay warmer than the surrounding air. Thus, uplift continues.

With regard to the second factor—a steep environmental temperature lapse rate—keep in mind that this is the temperature of the surrounding, still air. Therefore, the still air gets colder quite rapidly as you go up. This means that the condensing air in the rising bubble will be more likely to stay warmer than the surrounding air. Again, uplift continues.

A simple example may make convection in unstable air clearer. Figure 4.19 shows how temperature changes with altitude for a parcel moving upward by convection in unstable air. The surrounding air is at a temperature of 29°C (84°F) at ground level and has a lapse rate of 12°C/1000 m (6.6°F/1000 ft). A parcel of air is heated by 1°C (1.8°F) to 30°C (86°F), and it begins to rise. At first, it cools at the dry adiabatic rate. At 400 m (1300 ft), the parcel is at 26°C (79°F), while the surrounding air is at 24°C (75°F). Since it is still warmer than the surrounding air, uplift continues.

As condensation occurs, the wet adiabatic rate applies, here using the value 3°C/1000 m (1.6°F/1000 ft). Now the parcel cools more slowly as it rises. At 900 m (3000 ft), the parcel is 23°C (73°F), while the surrounding air is 18°C (64°F). Note that the difference in temperature between the rising parcel and surrounding air is actually increasing with altitude! This means that the parcel will be pushed ever more strongly upward, forcing even more condensation. And we may safely predict that precipitation will form.

The key to the convectional precipitation process is latent heat. When water vapor condenses into cloud droplets or ice particles, it releases latent heat to the rising air parcel. By keeping the parcel warmer than the surrounding air, this latent heat fuels the convection process, driving the parcel ever higher. When the parcel reaches a high altitude, most of its water will have condensed. As adiabatic cooling continues, less latent heat will be released. As a result, the convection cell will weaken. Eventually, uplift will stop, since the energy source, latent heat, is gone. The cell dies and dissipates into the surrounding air.

> **Latent heat release in condensation fuels the convectional precipitation process.**

Unstable air is typical of summer air masses in the central and southeastern United States. As we will see later, summer weather patterns sweep warm, humid air from the Gulf of Mexico over the continent. The

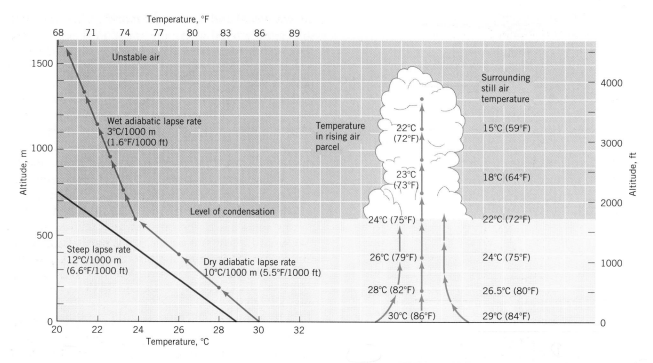

Figure 4.19 Convection in unstable air. When the air is unstable, a parcel of air that is heated sufficiently to rise will continue to rise to great heights.

intense summer insolation, over a period of days, strongly heats the air layer near the ground, producing a steep lapse rate. Thus, both of the conditions that create unstable air are present. As a result, thunderstorms are very common in these regions during the summer.

Unstable air is also likely to be found in the vast, warm and humid regions of the equatorial and tropical zones. In these regions, convective showers and thundershowers are common. At low latitudes, much of the orographic rainfall is actually in the form of heavy showers and thundershowers produced by convection. In this case, the forced ascent of unstable air up a mountain slope easily produces rapid condensation, which then triggers the convection mechanism.

Thunderstorms

The thunderstorm is an awesome event (Figure 4.20). Intense rain and violent winds, coupled with lightning and thunder, renew our respect for nature's power. What is the anatomy of a thunderstorm?

A **thunderstorm** is an intense local storm associated with a tall, dense cumulonimbus cloud in which there are very strong updrafts of air. A single thunderstorm consists of several individual convection cells. A single convection cell is diagrammed in Figure 4.21. Air rises within the cell as a succession of bubblelike air parcels. As each bubble rises, air in its wake is brought in from the surrounding region. Intense adiabatic cooling

Figure 4.20 This majestic cumulonimbus cloud marks a thunderstorm, moving from left to right. Lightning strikes lead the storm. A dark plume of falling rain follows, on the left.

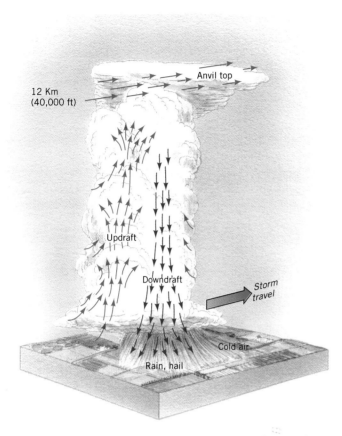

Figure 4.21 Anatomy of a thunderstorm cell. Successive bubbles of moist condensing air push upward in the cell. Their upward movement creates a corresponding downdraft, expelling rain, hail, and cool air from the storm as it moves forward.

within the bubbles produces precipitation. It can be in the form of water at the lower levels, mixed water and snow at intermediate levels, and snow at high levels where cloud temperatures are coldest.

As the rising air parcels reach high levels, which may be 6 to 12 km (20,000 to 40,000 ft) or even higher, the rising rate slows. Strong winds typically present at such high altitudes drag the cloud top downwind, giving the thunderstorm cloud its distinctive shape—resembling an old-fashioned blacksmith's anvil.

Thunderstorms are formed of convection cells in which rising bubbles of air are chilled to yield heavy precipitation.

Ice particles falling from the cloud top act as nuclei for freezing and deposition at lower levels. Large ice crystals form and begin to sink rapidly. Melting, they coalesce into large, falling droplets. The rapid fall of raindrops adjacent to the rising air bubbles pulls the

air downward and feeds a downdraft. This downdraft strikes the ground where precipitation is heaviest, creating strong local winds. Wind gusts can sometimes be violent enough to topple trees and raise the roofs of weak buildings.

In addition to powerful wind gusts and heavy rains, thunderstorms can produce hail. Hailstones (Figure 4.22) are formed by the accumulation of ice layers on an ice pellet suspended in the strong updrafts of the thunderstorm. In this process, they can reach diameters of 3 to 5 cm (1 to 2 in.). When they become too heavy for the updraft to support, they fall to earth.

Annual losses from crop destruction caused by hailstorms amount to several hundred million dollars. Damage to wheat and corn crops is particularly severe in the Great Plains, running through Nebraska, Kansas, Missouri, Oklahoma, and northern Texas (Figure 4.23). Figure 4.24 shows a hailstorm in progress.

Another effect of convection cell activity is to generate lightning. This occurs when updrafts and downdrafts cause positive and negative static charges to accumulate within different regions of the cloud. Lightning is a great electric arc—a series of gigantic sparks—passing between differently charged parts of the cloud mass or between cloud and ground (see Figure 4.20). During a lightning discharge, a current

Figure 4.22 Close up view of hailstones.

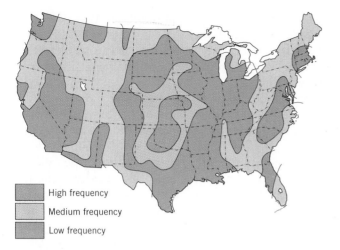

High frequency

Medium frequency

Low frequency

Figure 4.23 Map of frequency of severe hailstorms in the 48 contiguous states of the United States. A severe hailstorm is defined as a local convective storm producing hailstones equal to or greater than 1.9 cm (0.75 in.) in diameter. (From Richard H. Skaggs, *Proc. Assoc. American Geographers*, vol. 6, Figure 2. Used by permission.)

of as much as 60,000 to 100,000 amperes may develop. This current heats the air intensely, which makes it expand very rapidly—much like an explosion. This expansion sends out sound waves, which we recognize as a thunderclap. Most lightning discharges occur within the cloud, but a significant proportion strike land. In the United States, lightning causes a yearly average of about 150 human deaths and property damage of hundreds of millions of dollars, including loss by structural and forest fires set by lightning.

AIR POLLUTION

"Smog alert! All persons should refrain from vigorous outdoor activity today." Perhaps you've heard a warning like this on the radio, especially if you live in a metropolitan area with a smog problem. If so, you've had some first-hand experience with air pollution. Here, we provide a short introduction to the topic.

In preceding chapters, we have discussed two kinds of substances in air—aerosols and gases. Aerosols are small bits of matter in the air, so small that they float freely with normal air movements. Gases are molecular compounds that are mixed together to form the main body of air. To these two categories, we add a third category, *particulates*—larger, heavier particles that sooner or later fall back to earth.

Figure 4.24 A hailstorm in progress in Santa Cruz County, Arizona. Although the hail is falling from a cloud directly overhead, the low afternoon sun still illuminates the scene at an angle. The marble-sized hailstones are accumulating on the ground. (Copyright © John Hoffman/Bruce Coleman, Inc.)

Figure 4.25 Air pollution emissions in the United States in percentage by weight. *Left:* pollutants emitted. *Right:* sources of emission. (Data from National Air Pollution Control Administration, HEW.)

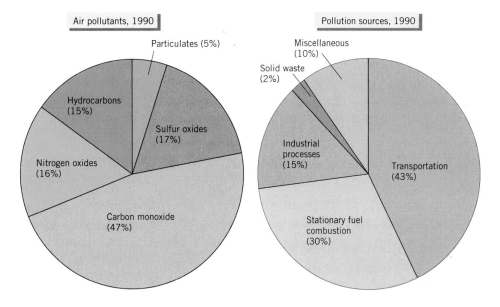

An *air pollutant* is an unwanted substance injected into the atmosphere from the earth's surface by either natural or human activities. Air pollutants come as aerosols, gases, and particulates. Pollutants generated by human activity can arise in two ways. The first is through the day-to-day activities of large numbers of people, for example, in driving automobiles within urban centers. The second is through industrial activities, such as fossil fuel combustion or the smelting of mineral ores to produce metals.

Figure 4.25 shows the relative proportions of the most common air pollutants and their sources. The gases are carbon monoxide (CO), sulfur oxides, and nitrogen oxides. The nitrogen oxides can be NO, NO_2, or NO_3; this mixture is usually referred to as NO_x. Similarly, the sulfur oxides SO_2 and SO_3 are referred to as SO_x. Hydrocarbon compounds occur both as gases and as aerosols. The remaining pollutants are in the form of particulates.

By weight, fossil fuel burning accounts for most emissions of air pollutants.

Combustion of fossil fuel, in transportation or as stationary fuel combustion, is the most important source of all these pollutants. As shown in the right half of the figure, it accounts for 73 percent of the emissions. Exhausts from gasoline and diesel engines contribute most of the carbon monoxide, half the hydrocarbons, and about a third of the nitrogen oxides.

Stationary sources of fuel combustion include power plants that generate electricity and various industrial plants that also burn fossil fuels. They contribute most

Figure 4.26 This satellite image shows the devastating effects of pollutants emitted by a smelter near Wawa, Ontario, Canada. Healthy vegetation is shown in red to pink tones. The white streak from lower right to upper left is barren of vegetation.

Figure 4.27 Smog in the city. New York's World Trade Center projects skyward through a brownish-pink layer of smog.

of the sulfur oxides because they burn coal or lower grade fuel oils. These fuels are richer in sulfur than gasoline and diesel fuel. Stationary sources also supply most of the particulate matter. Some of it is fly ash—coarse soot particles emitted from smokestacks of power plants. These particles settle out quite quickly within close range of the source. Smaller and finer carbon particles, however, can remain suspended almost indefinitely and become aerosols. Industrial processes, such as the smelting of ore, are also important sources of air pollutants (Figure 4.26). A small contribution to air pollution is made by the burning of solid wastes.

Smog and Haze

When aerosols and gaseous pollutants are present in considerable density over an urban area, the resultant mixture is known as **smog** (Figure 4.27). This term was coined by combining the words "smoke" and "fog." Typically, smog allows hazy sunlight to reach the ground, but it may also hide aircraft flying overhead from view. Smog irritates the eyes and throat, and it can corrode structures over long periods of time.

Modern urban smog has three main toxic ingredients: nitrogen oxides, hydrocarbons, and ozone. Nitrogen oxides and hydrocarbons are largely automobile pollutants, although significant amounts can be released by industrial power plants and petroleum proc-essing and storage facilities. Ozone is not normally produced directly by pollution sources in the city but forms through a photochemical reaction in the air. In this reaction, nitrogen oxides react with hydrocarbons in the presence of sunlight to form ozone. Ozone in urban smog can harm plant tissues and eventually kill sensitive plants. It is also harmful to human lung tissue, and it aggravates bronchitis, emphysema, and asthma. This is the same compound that absorbs ultraviolet radiation in the ozone layer within the stratosphere. So, ozone in the stratosphere is desirable, but if close to earth, it is undesirable indeed. Photochemical reactions can also produce other toxic compounds in smog.

Haze is a condition of the atmosphere in which aerosols obscure distant objects. Haze builds up naturally in stationary air as a result of human and natural activity. When the air is humid and abundant water vapor is available, water films grow on suspended nuclei. This creates particles large enough to obscure and scatter light, reducing visibility. Natural haze aerosols include soil dust, salt crystals from the sea surface, hydrocarbon compounds from plants, plant pollen, and smoke from forest and grass fires. Some regions are noted for naturally occurring haze—the Great Smoky Mountains are an example.

Fallout and Washout

Pollutants generated by a combustion process are contained within hot exhaust air emerging from a factory smokestack or an auto tailpipe. Since hot air rises, the pollutants are at first carried aloft by convection. However, the larger particulates soon settle under gravity and return to the surface as *fallout*. Particles too small to settle out are later swept down to earth by precipitation in a process called *washout*. Through a combination of fallout and washout, the atmosphere tends to be cleaned of pollutants. Although a balance between input and output of pollutants is achieved in the long run, the quantities stored in the air at a given time fluctuate widely.

Fallout and washout move pollutants from the air to the surface. While this action cleans the air, it also deposits the pollutants on soils and vegetation. Because SO_x and NO_x can readily be converted to sulfuric and nitric acids, the pollutant particles, dry or wet, can be quite acidic. (See Eye on the Environment: Acid Deposition and Its Effects, in this chapter.)

Pollutants are also eliminated from the air over their source areas by wind. Strong, through-flowing winds will disperse pollutants into large volumes of cleaner air in the downwind direction. Strong winds can quickly sweep away most pollutants from an urban area, but, during periods when winds are light or absent, the concentrations can rise to high values.

Inversion and Smog

The concentration of pollutants over a source area rises to its highest levels when vertical mixing (convection) of the air is inhibited. This happens in an inversion—a condition in which the temperature of the air increases with altitude.

Why does the inversion inhibit mixing? Recall that a heated air parcel, perhaps emerging from a smokestack or chimney, will rise as long as it is warmer than the surrounding air. Recall, too, that as it rises, it is cooled according to the adiabatic principle. In an inversion, however, the surrounding air gets warmer, not colder, with altitude. So, the parcel will cool as it rises, while the surrounding air becomes warmer. Thus, the parcel will quickly arrive at the temperature of the surrounding air and uplift will stop. Under these conditions, heated air will move only a short distance upward. Pollutants in the air parcel will disperse at low levels, keeping concentrations near the ground high.

Inversions suppress convectional mixing, sealing pollutants into an air layer close to the ground.

Two types of inversions are important in causing high air pollutant concentrations—low-level and high-level inversions. When a *low-level temperature inversion*

develops over an urban area with many air pollution sources, pollutants are trapped under the "inversion lid." Heavy smog or highly toxic fog can develop. An example is the tragedy that occurred in Donora, Pennsylvania, in late October 1948, when a persistent low-level inversion developed. The city occupies a valley floor hemmed in by steeply rising valley walls. The walls prevented the free mixing of the lower air layer with that of the surrounding region (Figure 4.28) and helped keep the inversion intact. Industrial smoke and gases from factories poured into the inversion layer for five days, increasing the pollution level. A poisonous fog formed and began to take its toll. Twenty persons died, and several thousand persons were stricken before a change in weather patterns dispersed the smog layer.

Another type of inversion is responsible for the smog problem experienced in the Los Angeles basin and other California coastal regions, ranging north to San Francisco and south to San Diego. Here special climatic conditions produce prolonged inversions and smog accumulations (Figure 4.29). Off the California coast is a persistent fair-weather system, which is especially strong in the summer. This system produces a layer of hot, dry air at upper elevations. However, a cold current of upwelling ocean bottom water runs along the coast, just offshore. Moist ocean air at the surface moves across this cool current and is chilled, creating a cool, marine air layer.

Now, the Los Angeles basin is a low, sloping plain lying between the Pacific Ocean and a massive mountain barrier on the north and east sides. Weak winds from the south and southwest move the cool, marine air inland over the basin. Further landward movement is blocked by the mountain barrier. Since there is a warm layer above this cool marine air layer, the result is an inversion. Pollutants accumulate in the cool air layer and produce smog. The upper limit of the smog stands out sharply in contrast to the clear air above it, filling the basin like a lake and extending into valleys in the bordering mountains (Figure 4.30). Since this type of

Figure 4.28 A low-level inversion held air pollutants close to the ground, inducing a poison fog accumulation at Donora, Pennsylvania, in October 1948.

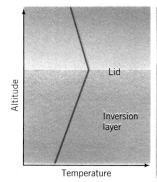

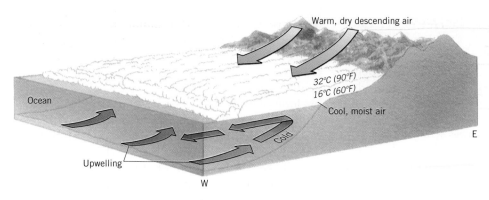

Figure 4.29 A layer of warm, dry descending air from a persistent fair-weather system rides over a cool, moist marine air layer at the surface to create a persistent temperature inversion.

inversion persists to a higher level in the atmosphere, we can refer to it as a *high-level temperature inversion.*

An actual temperature inversion, in which temperature close to the surface increases with altitude, is not essential for building a high concentration of pollutants above a city. All that is required is light or calm winds and stable air. *Stable air* has a temperature profile that decreases with altitude, but at a slow rate. Some convectional mixing occurs in stable air, but convectional precipitation is inhibited. At certain times of the year, slow-moving masses of dry, stable air occupy the central and eastern portions of the North American continent. Under these conditions, a broad *pollution dome* can form over a city or region, and air quality will suffer (Figure 4.31). When there is a regional wind, the pollution from a large city will be carried downwind to form a *pollution plume.*

Climatic Effects of Urban Air Pollution

Urban air pollution reduces visibility and illumination. Specifically, a smog layer can cut illumination by 10 percent in summer and 20 percent in winter. Ultraviolet radiation is absorbed by ozone in smog. At

Figure 4.30 A dense layer of smog and marine haze lying over the Los Angeles Basin. The view is from a point over the San Gabriel Mountains, looking southwest.

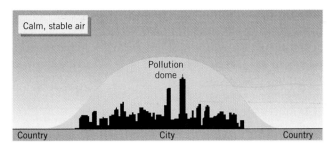

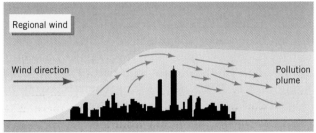

Figure 4.31 If calm, stable air overlies a major city, a pollution dome can form. When a wind is present, pollutants are carried away as a pollution plume.

times, ultraviolet radiation is completely prevented from reaching the ground. While this reduces the risk of human skin cancer, it may also permit increased viral and bacterial activity at ground level.

Winter fogs are much more frequent over cities than over the surrounding countryside. The fog is enhanced by the abundance of urban aerosols and particulates. Coastal airports, such as those of New York City, Newark, and Boston, suffer from an increased frequency of fog resulting from urban air pollution. Cities also show an increase in cloudiness and precipitation

as compared to the surrounding countryside. This increase results from intensified convection generated when the lower air is heated by human activities.

Air Pollution Control

The United States has some of the strictest laws in the world limiting air pollution. As a result, many strategies have been developed to reduce emissions. Sometimes these strategies involve trapping and processing pollutants after they are generated. The "smog controls" on automobiles are an example. Sometimes the use of alternative, nonpolluting technology is in order—for example, substituting solar, wind, or geothermal power for coal burning to generate electricity.

In any event, both as individuals and as a society we can do much to preserve the quality of the air we breathe. Reducing fossil fuel consumption is an example. It will be a considerable challenge to our society and others to preserve and increase global air quality in the face of an expanding human population and increasing human resource consumption.

Precipitation has been the major focus of this chapter. Keep in mind that no matter what the cause of precipitation, latent heat is always released. Because a significant amount of oceanic evaporation results in precipitation over land, there is a very significant flow of latent heat from oceans to land. And, as we will see in the following chapter, global air circulation patterns move this latent heat poleward, helping to warm the continents and creating the distinctive pattern of climates that differentiates our planet.

ATMOSPHERIC MOISTURE AND PRECIPITATION IN REVIEW

Water moves freely between ocean, atmosphere, and land in the hydrologic cycle. The global water balance describes these flows. The fresh water in the atmosphere and on land in lakes, streams, rivers, and ground water is only a very small portion of the total water in the hydrosphere.

Humidity describes the amount of water vapor present in air. The ability of air to hold water vapor depends on temperature. Warm air can hold much more water vapor than cold air. Relative humidity measures water vapor in the air as the percentage of the maximum amount of water vapor that can be held, at the given air temperature. Specific humidity simply measures grams of water vapor per kilogram of air.

The adiabatic principle states that when a gas is compressed, it warms, and when a gas expands, it cools. When an air parcel moves upward in the atmosphere, it encounters a lower pressure and so expands and cools. The dry adiabatic rate describes the rate of cooling with altitude. If the air is cooled below the dew point, condensation or deposition occurs, and latent heat is released. This heat reduces the rate of cooling with altitude, which is described as the wet adiabatic rate if condensation or deposition is occurring.

Clouds are composed of droplets of water or crystals of ice that form on condensation nuclei. Clouds typically occur in layers, as stratiform clouds, or in globular masses, as cumuliform clouds. Fog occurs when a cloud forms at ground level. Precipitation from clouds occurs as rain, hail, snow, and sleet. When supercold rain falls on a surface below freezing, it produces an ice storm.

There are three types of precipitation—orographic, convectional, and cyclonic. In orographic precipitation, air moves up and over a mountain barrier. As it moves up, it is cooled adiabatically and rain forms. As it descends the far side of the mountain, it is warmed. In convectional precipitation, unequal heating of the surface causes an air parcel to become warmer and less dense than the surrounding air. Because it is less dense, it rises. As it moves upward, it cools, and condensation with precipitation may occur. Under conditions of unstable air, thunderstorms can form, yielding hail and lightning. Cyclonic precipitation is described in Chapter 6.

Air pollution is defined as unwanted gases, aerosols, and particulates injected into the air by human and natural activity. Polluting gases are generated largely by fuel combustion. Aerosols and particulates are also released by fossil fuel burning. Smog, a common form of air pollution, contains nitrogen oxides, hydrocarbons, and ozone. Inversions can trap smog and other pollutant mixtures in a layer close to the ground, creating unhealthy air.

KEY TERMS

sublimation	adiabatic process	orographic precipitation
deposition	dry adiabatic lapse rate	convectional precipitation
hydrologic cycle	wet adiabatic lapse rate	unstable air
humidity	cloud	thunderstorm
dew-point temperature	acid deposition	smog

REVIEW QUESTIONS

1. What is the hydrosphere? Where is water found on our planet? In what amounts? How does water move in the hydrologic cycle?

2. Define relative and specific humidity. How is relative humidity measured? Sketch a graph showing relative humidity and temperature through a 24-hour cycle.

3. Use the terms saturation, dew point, and condensation properly in describing what happens when an air parcel of moist air is chilled.

4. What is the adiabatic process? Why is it important?

5. Distinguish between dry and wet adiabatic lapse rates. In a parcel of air moving upward in the atmosphere, when do they apply? Why is the wet adiabatic lapse rate less than the dry adiabatic rate? Why is the wet adiabatic rate variable in amount?

6. How are clouds classified? Name four cloud families, two broad types of cloud forms, and three specific cloud types.

7. What is fog? Explain how radiation fog and advection fog form.

8. How is precipitation formed? Describe the process for warm and cold clouds.

9. Describe the orographic precipitation process. What is a rainshadow? Provide an example of the rainshadow effect.

10. What is unstable air? What are its characteristics?

11. Describe the convectional precipitation process. What is the energy source that powers this source of precipitation? Explain.

12. Sketch a convection cell within a thunderstorm. Show rising bubbles of air, updraft, downdraft, precipitation, and other features.

13. What are the most abundant air pollutants, and what are their sources?

14. What is smog? What important pollutant forms within smog, and how does this happen?

15. Distinguish low-level and high-level inversions. How are they formed? What is their effect on air pollution? Give an example of a pollution situation of each type.

IN-DEPTH QUESTIONS

1. Compose an essay on water in the atmosphere. What part of the global supply of water is atmospheric? Why is it important? What is its global role? How does the capacity of air to hold water vapor vary? How is the moisture content of air measured? Clouds and fog visibly demonstrate the presence of atmospheric water. What are they? How do they form?

2. Compare and contrast orographic and convectional precipitation. Begin with a discussion of the adiabatic process and the generation of precipitation within clouds. Then compare the two processes, paying special attention to the conditions that create uplift. Can convectional precipitation occur in an orographic situation? Under what conditions?

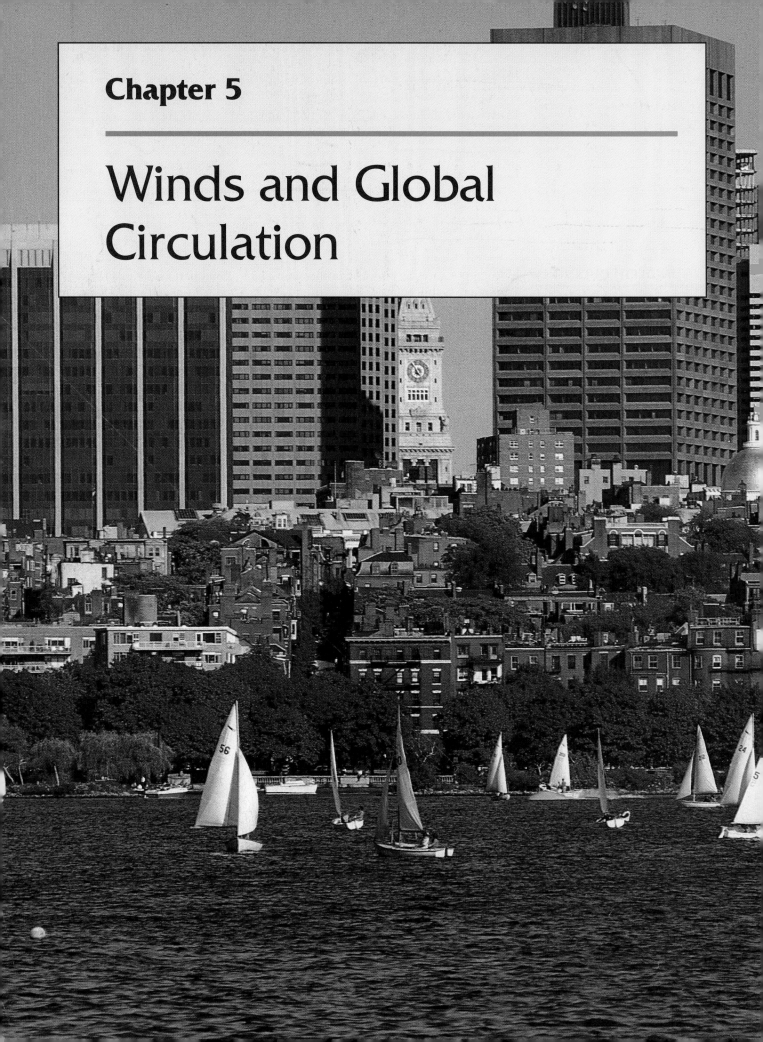

Chapter 5

Winds and Global Circulation

Coming about! You push the tiller hard to the starboard side of the boat, and the small craft turns easily into the wind. The sail goes slack and begins to swing across the boat. You duck under the boom that holds the bottom of the sail taut, stepping neatly to your right. The sail now catches the wind from the starboard side and begins to fill. You pull in the line attached to the sail, and, steering with the tiller, you correct the rudder for your new course. The sail tightens, and the boat leans away from the wind, gathering speed. You quickly move to the boat's edge and, sitting on the rim, lean out over the water. Your weight is held in check by the strong pull of the sail on the line in your right hand. The boat swings upright. With the tiller in your left hand guiding the boat, you are skimming across the water again, thanks to the cool, southeast breeze.

What a marvelous setting for a sail! With the soaring skyscrapers of Boston on the south and the leafy Cambridge campus of MIT on the north, the mouth of the Charles River is the magnificent jewel in the center of the famed necklace of parks, green spaces, and beaches that frames the map of the city. Tree-lined paths flank the river on both banks, with bikers, runners, skaters, and walkers out in force on this mild June afternoon. Unfortunately, you're not the only one out for a sail. There are perhaps a dozen other boats sharing the water with you, and since it's still early in the afternoon, the fleet is sure to increase. Keeping track of them requires attention.

Sailing on the Charles River in downtown Boston is always something of a challenge. Not only are there other sailors, but also the wind can be a bit fickle, to say the least. Today, the tall buildings stand as obstacles to the southeast breeze, creating currents and eddies in the downstream airflow as it crosses the river mouth. At any moment you can become becalmed, or worse—the breeze can suddenly shift direction, swinging your sail and the force of the wind over to the side of the boat carrying your weight. Without a fast scramble to the other side to balance the force, you can wind up in the water!

The river bank is fast approaching, and it's time to turn. Instead of heading close-hauled into the wind, you'll be sailing free, with the wind at an angle behind your boat. You let out the sail and ease the tiller to starboard. So far, so good. But wait, there's another boat about to sail into your bow! Turn left, fast! You kick the tiller over hard, and the boat swings cleanly to port. Oh, oh. The wind's shifting, just at the wrong moment. You're going to jibe! Duck! The sail comes flying across the boat, and the line attached to the boom jerks tight in your right hand. You'd better move fast, now, you're on the wrong side of the boat. Scrambling to the other side of the boat, you drop the tiller! A major mistake. The boat spins and tilts violently in the strong breeze. Water enters the boat as the rim tips below the waterline. Now the boom is in the water! You put all your weight on the opposite rim of the boat, trying to right the boat against the forces aligned to tip it over. Completing the spin, the boat comes to rest pointed into the wind, the unrestrained sail flapping violently. You are crouched in the bottom of the boat, breathless, under the wriggling boom. Ankle-deep water covers your feet. Humbled by the power of the wind, you think, "Welcome to the Charles!"

Wind is the subject of this chapter. Did you ever wonder why there is wind? Winds are caused by unequal heating of the earth's surface. As you know from Chapter 4, heating causes air to expand and become less dense. When this occurs, cooler, denser, surrounding air presses inward to buoy up the lighter, heated air, and a wind along the surface is created. This principle directly explains many types of local winds, such as the sea breeze you might experience on an afternoon at the beach. It is also the cause of global wind motions, when we consider that the equatorial and tropical regions are heated more intensely by the sun than are the mid- and higher latitudes. In response to this difference in heating, vast tongues of warm air move poleward, and huge pools of cool air shift equatorward, influencing our day-to-day weather strongly. For global wind motions, however, wind direction is strongly influenced by the earth's rotation—a factor that is also a subject of this chapter.

Sailboats on the Charles River, Boston.

The air around us is always in motion. From a gentle breeze on a summer's day to a cold winter wind, we experience moving air much of the time. Why does the air move? What are the forces that cause winds to blow? Why do winds blow more often in some directions than others? What is the pattern of wind flow in the upper atmosphere, and how does it affect our weather? How does the global wind pattern produce ocean currents? These are some of the questions we will answer in this chapter.

But first we need to examine atmospheric pressure and how wind develops in response to pressure differences. These are the subjects for the next few pages.

ATMOSPHERIC PRESSURE

Did you know that you are living at the bottom of a vast ocean? Yes, that's right! However, it's an ocean of air—the earth's atmosphere. Like the water in the ocean, the air in the atmosphere is constantly pressing on the solid or liquid surface beneath it. This pressure is exerted in all directions. For example, try taking a deep breath. What happens? As you expand your chest volume, air is forced into your lungs by atmospheric pressure. When you contract your chest volume and exhale, you push the air out against atmospheric pressure.

The pressure of air at sea level is about 1 kilogram per square centimeter (1 kg/sq cm), or about 15 pounds per square inch (Figure 5.1). Pressure exists because air has mass and is constantly being pulled toward the earth by the force of gravity.

Measuring Atmospheric Pressure

Perhaps you know that atmospheric pressure is measured by a **barometer.** But how does a barometer work? The principle is simple, and you have experienced it often. Visualize sipping a soda through a straw. By lowering your jaw and moving your tongue, you expand the volume of your mouth. This creates a partial vacuum, and air pressure forces the soda up through the straw and into your mouth. A *mercury barometer* works on the same principle (Figure 5.2)—that atmospheric pressure will force a liquid up into a tube when a vacuum is present. To construct the barometer, a glass tube about 1 m (3 ft) long, sealed at one end, is completely filled with mercury. The open end is temporarily held closed. Then the tube is inverted, and the end is immersed into a dish of mercury. When the opening of the tube is uncovered, the mercury in the tube falls some distance but then remains fixed at a level about

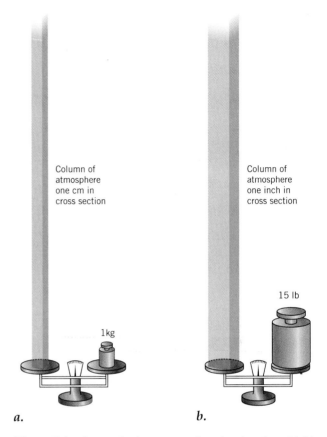

Figure 5.1 Atmospheric pressure imagined as the weight of a column of air. (*a*) metric system. The weight of a column of air 1 cm on a side is balanced by the weight of a 1 kg mass. (*b*) English system. The weight of a column of air 1 in. on a side is balanced by a weight of 15 pounds.

76 cm (30 in.) above the surface of the mercury in the dish.

What is happening here? At first, mercury flows out of the tube, under the force of gravity, and into the dish. Since there is no air in the top of the tube, a vacuum is formed. At the same time, atmospheric pressure pushes on the surface of the mercury in the dish, pressing the mercury upward into the tube. Since there is no air in the tube, there is no air pressure to push back against it. The pull of gravity on the mercury column then balances the pressure of the air, and the mercury level stays at a stable height. If air pressure changes, so will the height of the mercury column. So, this device will measure air pressure directly—but as a height in centimeters or inches rather than a pressure. The mercury barometer has become the standard instrument for weather measurements because of its high accuracy.

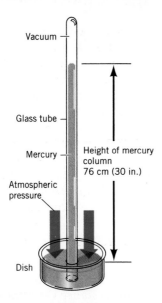

Figure 5.2 A simple diagram of a mercury barometer. Atmospheric pressure pushes the mercury upward into the tube, balancing the pressure exerted by the weight of the mercury column. As atmospheric pressure changes, the level of mercury in the tube changes. At sea level, the average height of the mercury level is about 76 cm (30 in.).

A more common type of barometer is the *aneroid barometer* (Figure 5.3). It uses a sealed canister of air, much like a small, flat can you might see in a supermarket. The canister has a slight partial vacuum, tensing the lid, or diaphragm, against a spring. The diaphragm flexes as air pressure changes, and through a mechanical linkage this fine movement drives a hand indicator that moves along a scale.

The widespread use of the mercury barometer has led to the common measurement of air pressure in heights of mercury, as centimeters or inches. On this scale, standard sea-level pressure is 76 cm Hg (29.92 in. Hg). The letters Hg are the chemical symbol for mercury. The metric unit of pressure is the *bar*. One bar is, coincidentally, very close to the value of normal atmospheric pressure at sea level. In meteorology, the **millibar** (mb), one-thousandth of a bar, is the unit most commonly used. In millibars, standard sea-level pressure is 1013.2 mb.

In the mercury barometer, air pressure balances the pull of gravity of a column of mercury about 76 cm (30 in.) high.

Atmospheric pressures at a location vary from day to day by only a small percentage. On a cold, clear winter day, the barometric pressure might be as high as 1030 mb (30.4 in. Hg), while in the center of a storm, pressure might drop to 980 mb (28.9 in. Hg) or lower. These changes, however small, are associated with traveling weather systems that we will discuss later in this chapter.

How Air Pressure Changes with Altitude

Figure 5.4 shows how atmospheric pressure decreases with altitude. From the earth's surface, the decrease of pressure with altitude is quite rapid initially. At higher altitudes, the decrease is much slower. Because atmospheric pressure decreases rapidly with altitude near the surface, a small change in elevation will often produce a significant change in air pressure. For example, you may have noticed that your ears sometimes "pop" during a fast elevator ride. For your eardrums to feel comfortable, the pressure inside your ear must be close to that outside. If the outside pressure changes abruptly, the popping sensation is produced. You may also have noticed your ears popping on an airplane ride. Aircraft cabins are pressurized to about 800 mb (24 in. Hg), which corresponds to an elevation of about 1800 m (6000 ft).

The decrease of pressure with elevation also affects human physiology. Perhaps you've felt out of breath, or tired more easily, walking along a high mountain trail. With decreased air pressure, oxygen moves into lung tissues more slowly. This typically produces shortness of breath. Many persons feel the effects of high

Figure 5.3 Diagram of an aneroid barometer. As air pressure varies, the diaphragm moves up and down. This motion is transmitted mechanically to move an indicator hand along a scale.

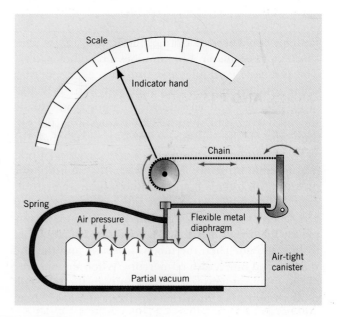

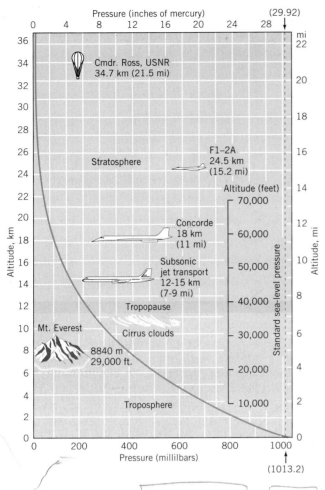

Figure 5.4 Atmospheric pressure decreases with increasing altitude above the earth's surface.

elevation as mountain sickness and experience weakness, headache, nosebleed, or nausea—usually at altitudes of 3000 m (10,000 ft) or higher. However, most people are able to adjust to the reduced air pressure after several days.

WINDS AND PRESSURE GRADIENTS

From a gentle, cooling breeze on a warm summer afternoon to the howling gusts of a major winter storm, wind is a familiar phenomenon. **Wind** is defined as air motion with respect to the earth's surface. Wind movement is dominantly horizontal. Air motions that are dominantly vertical are not called winds, but rather are known by other terms, such as updrafts or downdrafts.

Wind is caused by differences in atmospheric pressure from place to place. Air tends to move from high to low pressure until the air pressures are equal. This follows the simple physical principle that any flowing fluid (such as air) subjected to gravity will move until the pressure at every level is uniform.

Figure 5.5 illustrates this principle. It is a weather map showing two locations, taken here as Wichita and Columbus. At Wichita, the pressure is high (H), and the barometer (adjusted to sea-level pressure) reads 1028 mb (30.4 in. Hg). At Columbus, the barometer is low (L) and reads 997 mb (29.5 in. Hg). Thus, there is a **pressure gradient** between Wichita and Columbus. The lines drawn on the map around Wichita and Columbus are lines of equal pressure, called **isobars.** As you move between Wichita and Columbus, you cross these isobars, encountering pressures as a result of the drop from 1025 to 1000 mb (30.2 to 29.6 in. Hg).

Pressure gradients within the atmosphere cause winds.

Because atmospheric pressure is unequal at Wichita and Columbus, a *pressure gradient force* will push air from Wichita toward Columbus. The greater the pressure difference between the two locations, the greater this force will be. This pressure gradient will produce a wind.

What causes a pressure gradient to occur between two locations? If fluids flow naturally to equalize pressures at every level, how does a pressure gradient develop? Why does it persist? The answer is that pressure gradients develop because of unequal heating of the earth's surface. This unequal heating produces air of different temperatures. Recall from Chapter 4 that air parcels of different temperature will have different densities. That is, warm air is less dense than cool air. Thus, a column of warmer, less dense air over a location will weigh less, producing low pressure. Conversely, a column of colder, denser air will weigh more, producing high pressure.

Figure 5.5 Isobars and a pressure gradient. High pressure is centered at Wichita, and low pressure is centered at Columbus.

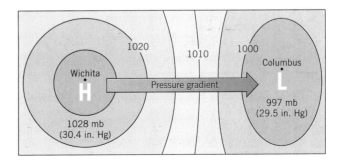

Sea and Land Breezes

If you've ever vacationed at the beach, you may have noticed that in the afternoons a *sea breeze* often sets in. This wind brings cool air off the water, dropping temperatures and refreshing beachgoers and residents living close to the beach. Late at night, a *land breeze* may develop. This wind moves cooler air, which is chilled over land by nighttime radiant cooling, toward the water. Sea and land breezes are simple examples of how heating and cooling produce a pressure gradient that causes a wind. Figure 5.6 shows them in more detail.

On a clear day, the sun heats the land surface rapidly, and a shallow air layer near the ground is strongly warmed (Figure 5.6*a*). The warm air expands. Because the warm air is less dense than the cool air off-shore, it weighs less and exerts less pressure. Thus, low pressure is formed over the coastal belt. Pressure remains higher over the water. This means that there is a pressure gradient from ocean to land, and a wind is set in motion. The sea breeze can become very strong in a shallow air layer close to shore.

Figure 5.7 A combination cup anemometer and wind vane. Wind speed and compass direction are displayed on the meter below. Also shown are a barometer and maximum-minimum thermometer. Normally, the wind vane and anemometer will be mounted outside, with a cable from the instruments leading to the meter, which is located indoors. (Courtesy of Taylor Instrument Company and Ward's Natural Science Establishment, Rochester, N.Y.).

Figure 5.6 Sea breeze and land breeze alternate in direction from day to night. They are caused by heating and cooling of land, while the ocean remains at an even temperature. (After R. G. Barry and R. J. Chorley, *Atmosphere, Weather and Climate,* 6th ed., Routledge. Used by permission.)

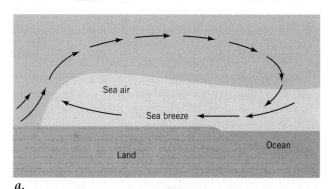

a.

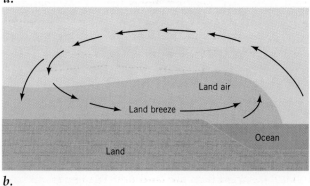

b.

Because the air over the land is warmer and less dense, it will also tend to rise. Over the ocean, the air moving landward will be replaced by air moving downward from above, causing a sinking motion. The rising over land and the sinking over ocean will create a return flow aloft that is opposite in direction to the sea breeze. We refer to this as a convection loop, in which unequal heating causes a circuit of moving air.

At night, the land surface cools rapidly if skies are clear. A cool air layer of higher pressure develops (Figure 5.6*b*). Now the pressure gradient is from land to ocean, setting up a land breeze. Like the sea breeze, the convection loop is completed by a wind aloft moving in the opposite direction.

Measurement of Winds

To describe winds fully, we must measure both wind speed and wind direction (Figure 5.7). Wind direction can be determined by a *wind vane,* the most common of the weather instruments. The design of the wind vane keeps the vane facing into the wind. Wind direction is given as the direction from which the wind is coming. So, an east wind is coming from the east, moving to the west (Figure 5.8).

Wind speed is measured by an *anemometer.* The most common type is the cup anemometer, which consists of three funnel-shaped cups mounted on the ends of

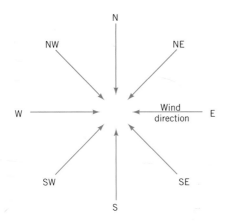

Figure 5.8 Winds are designated according to the compass point from which the wind comes. An east wind comes from the east, but the air is moving westward.

spokes of a horizontal wheel. The cups rotate with a speed proportional to that of the wind. One version of the anemometer turns a small electric generator. The more rapidly the wheel rotates, the more current is generated. The current is measured by a meter, which

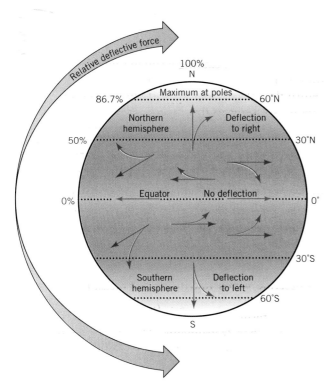

Figure 5.9 The Coriolis effect acts to deflect the paths of winds or ocean currents to the right in the northern hemisphere and to the left in the southern hemisphere as viewed from the starting point.

is calibrated in units of wind speed. Common units are meters per second or miles per hour.

The Coriolis Effect and Winds

The pressure gradient force tends to move air from high pressure to low pressure. For sea and land breezes, which are local in nature, this push produces a wind motion in about the same direction as the pressure gradient. For larger wind systems, however, the direction of air motion will be somewhat different. The difference is due to the earth's rotation, through the Coriolis effect. What is this effect, and how does it come about?

Through the **Coriolis effect** an object in motion on the earth's surface always appears to be deflected to the right (or left) of its course. The object will move as though a force were pulling it sidewards. The apparent deflection is to the right in the northern hemisphere and to the left in the southern hemisphere. The effect is strongest near the poles and decreases to no apparent deflection at the equator. Figure 5.9 diagrams the effect. Note that the apparent force does not depend on direction—the deflection occurs whether the object is moving toward the north, south, east or west, except along the equator.

> **The Coriolis effect causes the apparent motion of winds to be deflected away from the direction of the pressure gradient.**

The Coriolis effect is a result of the earth's rotation. Imagine that a rocket is launched from the North Pole toward New York, following the 74° W longitude meridian (Figure 5.10). However, as it travels toward New York, the earth is rotating beneath its trajectory, moving from west to east. The rocket's path will apparently curve to the right, with respect to the earth underneath. Using reasonable values for calculation, the rocket might well fall to earth near Chicago!

What would happen if the rocket were launched from New York heading along the 74° W meridian toward the pole? At the latitude of New York (about 40° N), a point on the earth's surface is moving eastward at about 1300 km/hr (800 mi/hr). Although the rocket is aimed properly along the meridian, its motion will have an eastward component of 1300 km/hr. However, as the rocket moves northward, it passes over land that is moving less swiftly eastward. At 60° N latitude (point A in Figure 5.10), the eastward velocity of the earth's surface is only about 800 km/hr (500 mi/hr). So, the rocket will be moving more quickly eastward than the land underneath. It will pass

Figure 5.10 The Coriolis effect. The path of a rocket launched from the North Pole toward New York will be apparently deflected to the right by the earth's rotation and might land near Chicago. If launched from New York toward the North Pole, its path will be apparently deflected to the right as well.

well east of point A, and its path will apparently be deflected to the right.

Cyclones and Anticyclones

If the Coriolis effect deflects air moving from high to low pressure, how does that affect the wind pattern? A parcel of air in motion near the surface is subjected to three influences. First is the pressure gradient that propels the parcel toward low pressure. Second is the Coriolis effect that turns the parcel to the side. Third is friction with the ground surface. The result of all these influences is that air will move toward low pressure but at an angle to the pressure gradient.

Figure 5.11 shows wind patterns around surface high- and low-pressure centers in northern and southern hemispheres. When low pressure is at the center, the pressure gradient is straight inward (left side of figure). When high pressure is at the center, the gradient is straight outward (right side). But when air actually moves in response to these pressure gradients, it will move at an angle across the gradient.

For low-pressure centers (left side), the wind will be deflected into an inward-moving spiral pattern, or *inspiral*. In the northern hemisphere (upper portion), the inspiral will be counterclockwise, since the Coriolis deflection is to the right. In the southern hemisphere

Figure 5.11 Surface winds spiral inward toward the center of a cyclone, but outward and away from the center of an anticyclone. Because the Coriolis effect deflects moving air to the right in the northern hemisphere and to the left in the southern hemisphere, the direction of inspiraling and outspiraling reverses from one hemisphere to the other.

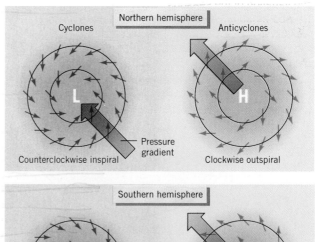

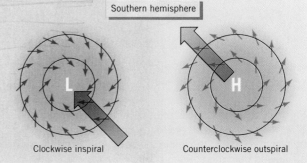

(lower portion), the inspiral will be clockwise, since the Coriolis deflection is to the left. Meteorologists use the term **cyclone** to refer to a center of low pressure with inspiraling air.

For high-pressure centers, the situation is reversed. Air moving outward forms an *outspiral,* clockwise in the northern hemisphere and counterclockwise in the southern hemisphere. This outspiral is referred to as an **anticyclone** (Figure 5.11).

In a cyclone, winds converge spiraling inward and upward. In an anticyclone, winds diverge, spiraling outward and downward.

Recall that in the case of the sea and land breezes, air moves upward at the site of low pressure and downward at the site of high pressure. The same is true of cyclones and anticyclones. In the cyclone, air spiraling inward converges, causing a rising motion at the center. And what happens when air is forced aloft? It is cooled according to the adiabatic principle, and condensation and precipitation can occur. This is why

cyclones, or low-pressure centers, are associated with cloudy or rainy weather.

In the anticyclone, air spiraling outward diverges, causing a sinking motion at the center. When air descends, we know from Chapter 4 that it is warmed, again by the adiabatic principle, and condensation cannot occur. So, anticyclones, or high-pressure centers, are associated with clear weather.

Cyclones and anticyclones are typically large features of the atmosphere—perhaps a thousand kilometers (600 mi) across, or more. For example, a fair weather system, which is an anticyclone, may stretch from the Rockies to the Appalachians. Cyclones and anticyclones can remain more or less in one location, or they can move, sometimes rapidly, to create weather disturbances.

Surface Winds on an Ideal Earth

Consider wind and pressure patterns on an ideal earth—one without a complicated pattern of land and water and no seasonal changes. Pressure and winds for such an ideal earth are shown in Figure 5.12. The

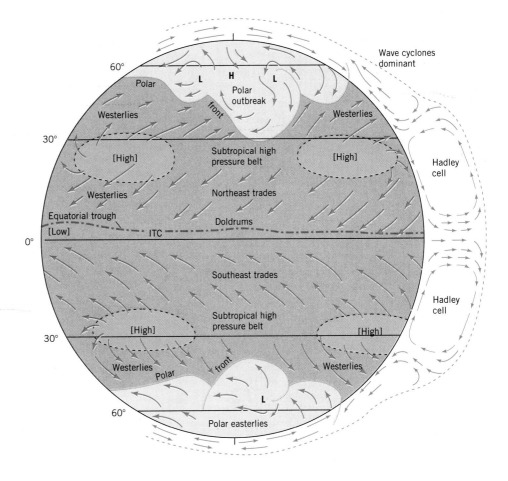

Figure 5.12 This schematic diagram of global surface winds and pressures shows the features of an ideal earth, without the disrupting effect of oceans and continents and the variation of the seasons. Surface winds are shown on the disk of the earth, while the cross section at the right shows winds aloft.

surface wind patterns are shown on the globe's surface. On the right, a cross section through the atmosphere shows the wind patterns aloft.

The most important features for understanding the pattern at low latitudes are shown on the cross section to the right, labeled "Hadley cells." Recall from our study of land and sea breezes that, when air is heated, low pressure forms and air rises. Since insolation is strongest when the sun is directly overhead, the equator will be heated more strongly than other places on this featureless earth. The result will be two convection loops, the **Hadley cells,** which form in the northern and southern hemispheres. In each Hadley cell, air rises over the equator, flows poleward, and descends at about a 30° latitude. (The Hadley cell is named for George Hadley, who first proposed its existence in 1735.)

In the Hadley cell convection loop, air rises at the intertropical convergence zone and descends in the subtropical high-pressure cells.

Since air is heated and rises at the equator, a zone of low pressure will result. This zone is referred to as the *equatorial trough.* As shown by the wind arrows, air moves toward the equatorial trough, where it converges and moves aloft as part of the Hadley cell circulation. Convergence occurs in a narrow zone, referred to as the **intertropical convergence zone (ITC).** Because the air is generally rising, winds in the ITC are light and variable. In earlier centuries mariners on sailing vessels referred to this region as the *doldrums,* where they were sometimes becalmed for days at a time.

On the poleward side of the Hadley cell circulation, air descends and surface pressures are high. This produces two **subtropical high-pressure belts,** each centered at about 30° latitude. Within the belts, two, three, or four very large and stable anticyclones are formed. Air is descending, and winds are weak at the center of these anticyclones. Calm prevails as much as one-quarter of the time. Because of the high frequency of calms, mariners named this belt the *horse latitudes.* The name is said to have originated in colonial times from the experiences of New England traders carrying cargoes of horses to the West Indies. When their ships were stranded for long periods of time, the freshwater supplies ran low and the horses were thrown overboard.

Winds around the subtropical high-pressure centers are outspiraling and move toward equatorial as well as middle latitudes. The winds moving equatorward are the strong and dependable *trade winds.* North of the equator, these are from the northeast and are referred to as the *northeast trades.* To the south of the equator, they are from the southeast and are the *southeast trades.* Poleward of the subtropical anticyclones, air spiraling outward produces southwesterly winds in the northern hemisphere and northwesterly winds in the southern hemisphere.

Between about 30° and 60° latitude, the pressure and wind pattern becomes more complex. This latitudinal belt is a zone of conflict between air bodies with different characteristics. Masses of cool, dry air move into the region, heading eastward and equatorward. These form *polar outbreaks.* The border of the polar outbreak is known as the *polar front.* (We'll have much more to say about air masses and the polar front in later sections.) As a result of this activity, pressures and winds can be quite variable in this region from day to day and week to week. On average, however, winds are more often from the west, so the region is said to have *prevailing westerlies.*

At the poles, the air is intensely cold. As a result, high pressure occurs. Outspiraling of winds around a polar anticyclone should create surface winds from a generally easterly direction, known specifically as *polar easterlies.* As we will see in discussing actual pressure and wind patterns, this situation exists only in the south polar region. In the north polar region, winds tend to have an eastward component, but there is too much variation in direction to consider polar easterlies as dominant winds there.

GLOBAL WIND AND PRESSURE PATTERNS

So far, we've discussed the wind pattern for a seasonless, featureless earth. Let's turn now to actual global wind and pressure patterns. Here, we will use the global maps of wind and pressure for January and July, shown in Figures 5.13 and 5.17. The pressures and winds shown are averages over many years for all daily observations in either January or July. They are corrected for the elevation of the recording station, so that pressures are shown for sea level. Average barometric pressure is 1013 mb (30.0 in. Hg). Values greater than that are "high" (red lines), while those lower are "low" (green lines). The maps make use of both Mercator and polar projections. As we discuss the

Figure 5.13 Mean monthly atmospheric pressure and prevailing surface winds for January and July. Pressure units are millibars reduced to sea level. Many of the wind arrows are inferred from isobars. See also Figure 5.17. (Data compiled by John E. Oliver.)

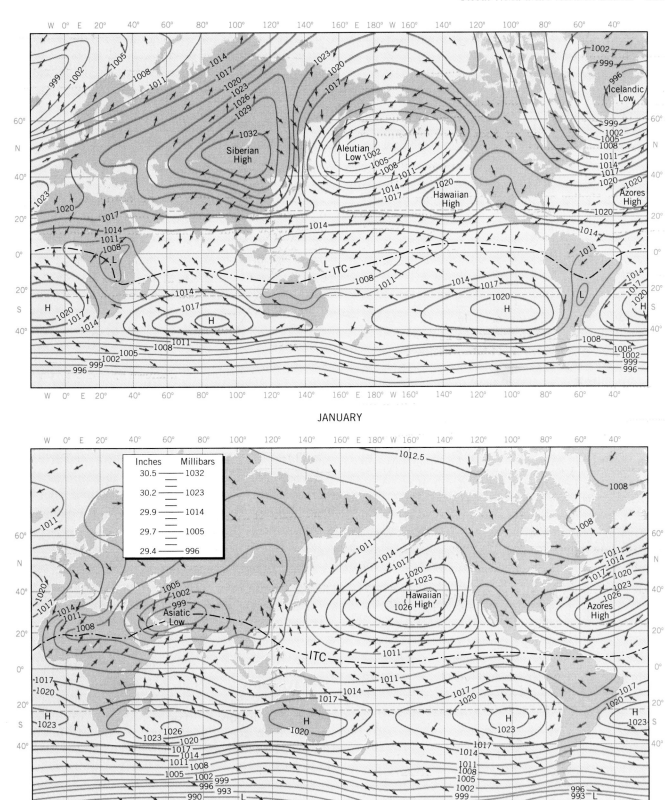

JANUARY

JULY

various features of global pressure and winds, be sure to examine both sets of maps.

Subtropical High-Pressure Belts

The largest and most prominent features of the maps are the subtropical high-pressure belts, created by the Hadley cell circulation. The high-pressure belt in the southern hemisphere follows best the pattern of Figure 5.12. It has three large high-pressure cells, each developed over oceans, that persist year round. A fourth, weaker high-pressure cell forms over Australia in July, as the continent cools during the southern hemisphere winter.

In the northern hemisphere, the situation is somewhat different. The subtropical high-pressure belt shows two large anticyclones centered over oceans—the Hawaiian High in the Pacific and the Azores High in the Atlantic. From January to July, these intensify and move northward. From their positions off the east and west coasts of the United States, they have a dominant influence on our summer weather.

Figure 5.14 is a schematic map of two large high-pressure cells bounded by continents, showing the features of circulation around the cells. As we noted earlier, the outspiraling circulation produces the trade winds in the tropical and equatorial zones, and the westerlies in the subtropical zones and poleward. In

the high-pressure cells, air on the east side subsides more intensely. This means that winds spiraling outward on the eastern side are drier. On the west side, subsidence is less strong. Also, these winds travel long distances across warm, tropical ocean surfaces before reaching land, and so are warm and moist.

When the Hawaiian and Azores Highs intensify and move northward in the summer, our east and west coasts feel their effects. On the west coast, dry subsiding air from the Hawaiian High dominates, so fair weather and rainless conditions prevail. On the east coast, warm, moist air from the Azores High flows across the continent from the southeast, producing generally hot, humid weather for the central and eastern United States. In winter, these two anticyclones weaken and move to the south—leaving our weather to the mercy of colder winds and air masses from the north and west.

The ITC and the Monsoon Circulation

Recall from Chapter 2 that insolation is most intense when the sun is directly overhead. Remember, too, that the latitude at which the sun is directly overhead changes with the seasons, migrating between the tropics of cancer and capricorn. Since the Hadley cell circulation is driven by this heating, we can expect that the elements of the Hadley cell circulation—ITC and subtropical high-pressure belts—will also shift with the seasons. We have already noted this shift for the Hawaiian and Azores Highs, which intensify and move northward as July approaches.

In general, the ITC also shifts with the seasons, as we might expect. The shift is moderate in the western hemisphere. The ITC moves a few degrees north from January to July over the oceans. In South America, the ITC lies across the Amazon in January and swings northward by about 20° to the northern part of the continent. But what happens in Africa and Asia? By comparing the two Mercator maps of Figure 5.13, you can see that there is a huge shift indeed! In January, the ITC runs south across eastern Africa and crosses the Indian Ocean to northern Australia at a latitude of about 15° S. In July, it swings north across Africa along the southern side of the Sahara, and then it rises further north to lie along the south rim of the Himalayas, in India, at a latitude of about 25° N. This is a shift of about 40 degrees of latitude!

Why does such a large shift occur in Asia? To answer this question, look at the pressure and wind pattern over Asia. In January, an intense high-pressure system, the Siberian High, is found there. Recall from Chapter 3 that, in winter, temperatures in northern Asia are very cold, so this high-pressure is to be expected. In

Figure 5.14 Over the oceans, surface winds spiral outward from the subtropical high pressure cells, feeding the trades and the westerlies. On the eastern side of the cells, air subsides more strongly, producing dry winds. On the western sides, subsidence is not as strong, and a long passage over oceans brings warm, moist air to the continents.

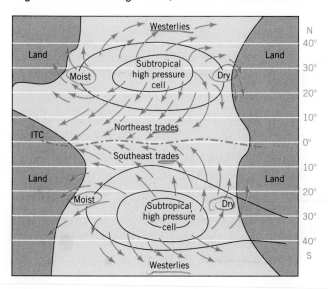

5800 ft per Mile

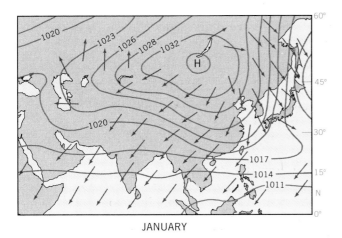

JANUARY

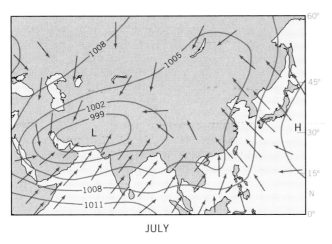

JULY

Figure 5.15 The Asiatic monsoon winds alternate in direction from January to July, responding to reversals of barometric pressure over the large continent.

monsoon, dry conditions prevail. In the summer, warm, humid air from the Indian Ocean and the southwestern Pacific moves northward and northwestward into Asia, passing over India, Indochina, and China. This airflow is known as the *summer monsoon* and is accompanied by heavy rainfall in southeastern Asia.

North America does not have the remarkable extremes of monsoon winds experienced in Asia. Even so, in summer there is a prevailing tendency for warm, moist air originating in the Gulf of Mexico to move northward across the central and eastern part of the United States. This Gulf air sometimes moves westward across the southwestern deserts of New Mexico and Arizona, occasionally reaching the California coast. This airflow is locally referred to as a "monsoon." In winter, the airflow across North America generally reverses, and dry, continental air from Canada moves south and eastward.

Wind and Pressure Features of Higher Latitudes

The northern and southern hemispheres are quite different in their geography. As you can easily see from Figure 5.16, the northern hemisphere has two large continental masses, separated by oceans. An ocean is also at the pole. In the southern hemisphere, we find a large ocean with a cold, glacier-covered continent at the center. These differing land—water patterns strongly influence the development of high- and low-pressure centers with the seasons.

Let's examine the northern hemisphere first (Figure 5.17). Recall from Chapter 3 that continents will be cold in winter and warm in summer, as compared to oceans at the same latitude. But we know from our analysis earlier in the chapter that cold air will be associated with high pressure and warm air with low

July, this high-pressure center is absent, replaced instead by a low centered over the Middle Eastern desert region. The Asiatic low is produced by the intense summer heating of the landscape.

> **The Asian monsoon wind pattern consists of a cool, dry air flow from the northeast during the low-sun season, and a warm, moist airflow from the southwest during the high-sun season.**

The movement of the ITC and the change in the pressure pattern with the seasons creates a reversing wind pattern in Asia known as the **monsoon** (Figure 5.15). In the winter, there is a strong outflow of dry, continental air from the north across China, Southeast Asia, India, and the Middle East. During this *winter*

Figure 5.16 Northern and southern hemispheres—a study in contrasts in the distribution of lands and oceans. One pole bears a deep ocean and the other a great continental land mass.

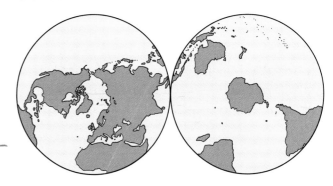

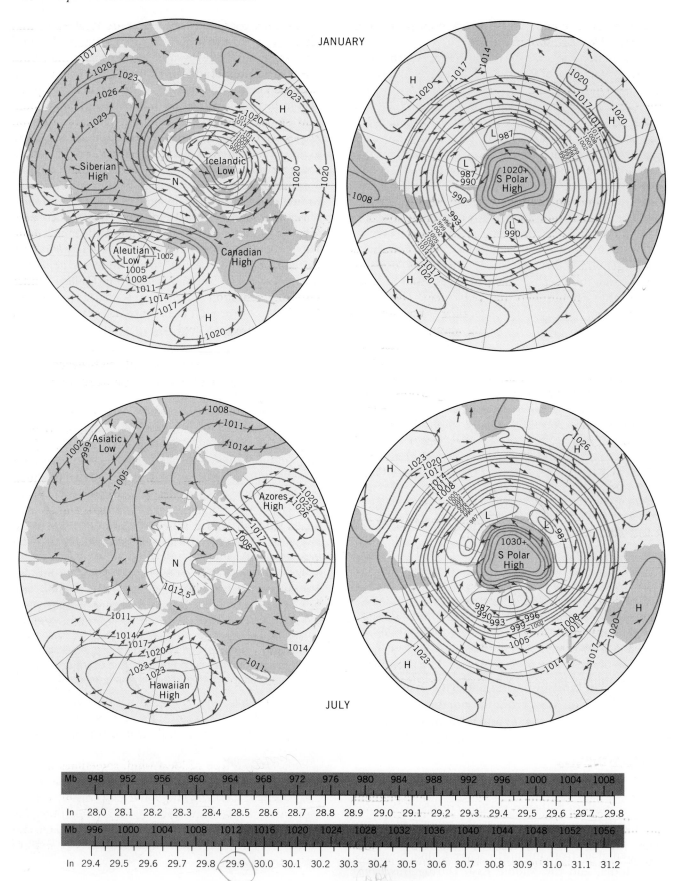

Figure 5.17 Mean monthly atmospheric pressure and prevailing surface winds for January and July, northern and southern hemispheres. Pressure units are millibars reduced to sea level. See also Figure 5.13. (Data compiled by John E. Oliver.)

pressure. Thus, continents will show high pressure in winter and low pressure in summer.

This pattern is very clear in the northern hemisphere polar maps of Figure 5.17. In winter, the strong Siberian High is found in Asia, and a weaker cousin, the Canadian High, is found in North America. From these high-pressure centers, air spirals outward, bringing cold air to the south. Over the oceans, two large centers of low pressure are striking—the Icelandic Low and the Aleutian Low. These two low-pressure centers are not actually large stable features that we would expect to find on every daily world weather map in January. Rather, they are regions of average low pressure where winter storm systems are spawned.

In summer, the pattern reverses. The continents show generally low pressure, while high pressure builds over the oceans. This pattern is easily seen on the northern hemisphere July polar map. The Asiatic Low is strong and intense. Inspiraling winds, forming part of the monsoon cycle, bring warm, moist Indian Ocean air over India and Southeast Asia. A lesser low forms over the deserts of the southwestern United States and northwestern Mexico. The two subtropical highs, the Hawaiian Highs and the Azores Highs, strengthen and dominate the Atlantic and Pacific Ocean regions. Outspiraling winds, following the pattern of Figure 5.14, keep the west coasts of North America and Europe warm and dry, and the east coasts of North America and Asia warm and moist (see Figure 5.13 also).

The higher latitudes of the southern hemisphere present a polar continent surrounded by a large ocean. Since Antarctica is covered by a glacial ice sheet and is cold at all times, a permanent anticyclone, the South Polar High, is centered there. Easterly winds spiral outward from this high. Surrounding this high-pressure center is a band of deep low pressure, with strong, inward-spiraling westerly winds. As early mariners sailed southward, they encountered this band, in which wind strength intensifies toward the pole. Because of the strong prevailing westerlies, they named these southern latitudes the "roaring forties," "flying fifties," and "screaming sixties."

LOCAL WINDS

Not every wind is produced by global pressure gradients in large weather systems. In some locations, *local winds* are important. These are often influenced by local terrain. We have already discussed one class of local winds—sea and land breezes—earlier in this chapter.

Mountain winds and *valley winds* are local winds that alternate in direction in a manner similar to the land and sea breezes. During the day, mountain hill slopes are heated intensely by the sun, causing air to rise. To take the place of the rising air, a current moves up valleys from the plains below—upward over rising mountain slopes, toward the summits. At night, the hill slopes are chilled by radiation. The cooler, denser hill slope air then moves valleyward, down the hill slopes, to the plain below. These winds are responding to local pressure gradients, set up by heating or cooling of the local air.

Another group of local winds are known as *drainage winds*, in which cold, dense air flows under the influence of gravity from higher to lower regions. In a typical situation, cold, dense air accumulates in winter over a high plateau or high interior valley. Under favorable conditions, some of this cold air spills over low divides or through passes, flowing out on adjacent lowlands as a strong, cold wind.

Local winds include land and sea breezes, mountain and valley winds, drainage winds, and Santa Ana winds.

Drainage winds occur in many mountainous regions of the world and go by various local names. The *mistral* of the Rhône valley in southern France is a well-known example—it is a cold, dry local wind. On the ice sheets of Greenland and Antarctica, powerful drainage winds move down the gradient of the ice surface and are funneled through coastal valleys. Picking up loose snow, these winds produce blizzardlike conditions that last for days at a time.

Another type of local wind occurs when the outward flow of dry air from an anticyclone is combined with the local effects of mountainous terrain. An example is the Santa Ana—a hot, dry easterly wind that sometimes blows from the interior desert region of southern California across coastal mountain ranges to reach the Pacific coast. This wind is funneled through local mountain gaps or canyon floors where it gains great force. At times the Santa Ana wind carries large amounts of dust. Because this wind is dry, hot, and strong, it can easily fan wildfires in brush or forest out of control (Figure 5.18).

Still another type of local wind, sometimes bearing the name *chinook*, results when strong regional winds pass over a mountain range and descend on the lee side. As we know from studying orographic precipitation in Chapter 4, the descending air is heated and dried. A chinook wind can sublimate snow or dry out soils very rapidly.

Figure 5.18   Strong downslope winds of hot, dry air fan this brush fire in Oakland, California.

WINDS ALOFT

The surface wind systems we have examined describe airflows at or near the surface. How does air move at the higher levels of the troposphere? Just as we saw for air near the surface, air aloft will move in response to pressure gradients and will be influenced by the Coriolis effect.

Pressure gradients aloft are generated by large, slow-moving, high- and low-pressure systems. Figure 5.19 shows a typical pattern of high and low pressure and winds that might be found at an altitude of 5 to 7 km (16,000 to 23,000 ft). Notice that the wind arrows are shown as following the isobars, not cutting across them as they do at the surface. The difference is that a moving parcel of air at high altitude moves freely, without the drag of surface friction. As a result, air at upper levels in the atmosphere moves smoothly around centers of high and low pressure, following a curving path. This type of wind is called the *geostrophic wind*.

Global Circulation at Upper Levels

Figure 5.20 sketches the general pattern of airflows at higher levels in the troposphere. The pattern has four major features—equatorial easterlies, a tropical high-pressure belt, upper-air westerlies, and a polar low.

The *upper-air westerlies* blow in a complete circuit about the earth, from about 25° lat. almost to the poles. They comprise the first major tropospheric circulation system. At high latitudes these westerlies form a huge circumpolar spiral, circling a great polar low-pressure center. Toward lower latitudes, atmospheric pressure rises steadily, forming a *tropical high-pressure belt* at 15°–20° N and S lat. This is the high-altitude part of the surface subtropical high-pressure belt, but it is shifted somewhat equatorward. Between the high-pressure ridges is a zone of weak low pressure in which the winds are easterly. These winds comprise the second major circulation system of the troposphere. They are called the *equatorial easterlies*.

The general pattern of winds aloft is a band of easterly winds, flanked by tropical high-pressure belts, and a spiral of westerly winds to poleward.

So, the overall picture of upper-air wind patterns is really quite simple—a band of easterly winds in the equatorial zone, belts of high pressure near the tropics of cancer and capricorn, and westerly winds, with some variation in direction, spiraling around polar lows.

Rossby Waves

The smooth westward flow of the upper-air westerlies frequently forms undulations, called **Rossby waves.** Figure 5.21 shows how these waves develop and grow in the northern hemisphere. The waves arise in a zone of contact between cold, polar air and warm, tropical air.

For a period of several days or weeks, winds may flow smoothly from west to east. Then, an undulation devel-

Figure 5.19 At high levels above the earth's surface, the wind blows parallel to the isobars. The northern hemisphere is shown here.

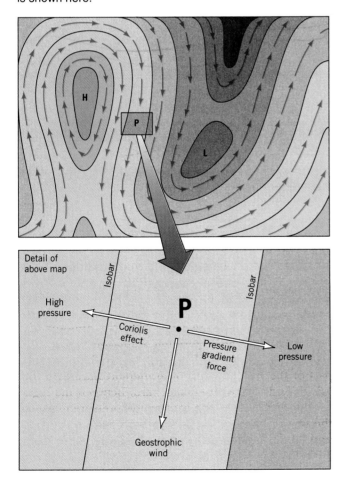

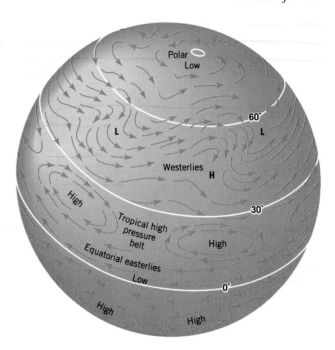

Figure 5.20 A generalized plan of global winds high in the troposphere. Easterly winds flow in the band in between the subtropical high-pressure belts. Poleward of these belts, winds are westerly but often sweep to the north or south around centers of high and low pressure aloft. (Copyright © by A. N. Strahler.)

ops, and warm air pushes poleward while a tongue of cold air is brought to the south. Eventually, the tongue is pinched off, leaving a pool of cold air at a latitude far south of its normal location. This cold pool may persist for some days or weeks.

Jet Streams

Jet streams are important features of upper-air circulation. They are narrow zones at a high altitude in which wind streams reach great speeds. Along a jet stream, pulselike movements of air follow broadly curving tracks (Figure 5.22). The greatest wind speeds occur in the center of the stream, with velocities decreasing away from the center. There are three kinds of jet streams. Two are westerly wind streams. The third is a weaker jet with easterly winds that develops in Asia as part of the summer monsoon circulation. These are shown in Figure 5.23.

The most poleward type of jet stream is located along the polar front—the fluctuating boundary between cold polar air and warmer tropical air (Figure 5.12). It is designated the *polar-front jet stream* (or simply, the "polar jet"). Located generally between 35° and 65° latitude, it is present in both hemispheres. The polar jet is typically found at altitudes of 10 to 12 km (30,000 to 40,000 ft), and wind speeds in the jet

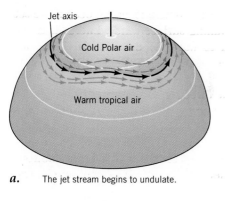

a. The jet stream begins to undulate.

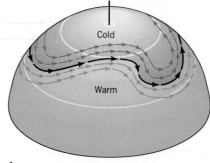

b. Rossby waves begin to form.

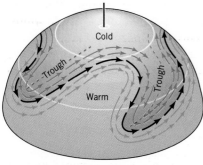

c. Waves are strongly developed. The
cold air occupies troughs of low pressure.

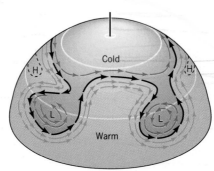

d. When the waves are pinched off,
they form cyclones of cold air.

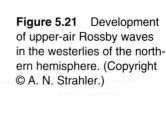

Figure 5.21 Development of upper-air Rossby waves in the westerlies of the northern hemisphere. (Copyright © A. N. Strahler.)

range from 350 to 450 km/hr (200 to 250 mi/hr). In the northern hemisphere, traveling aircraft often use the polar jet to increase ground speed when flying eastward. In the westward direction, flight paths are chosen to avoid the strong winds of the polar jet.

A second type of jet stream forms in the subtropical latitude zone—the *subtropical jet stream*. It occupies a position at the tropopause just poleward of the Hadley cell in each hemisphere (Figure 5.24). Here, westerly

Figure 5.22 The polar jet stream is shown on this map by lines of equal wind speed. (National Weather Service.)

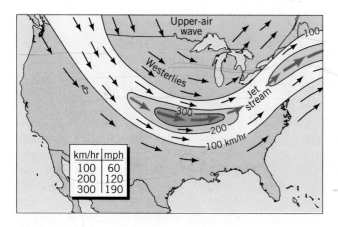

wind speeds reach maximum values of 345 to 385 km/hr (215 to 240 mph). Cloud bands of the subtropical jet stream are pictured in Figure 5.24.

A third type of jet stream is found at even lower latitudes. Known as the *tropical easterly jet stream*, it runs from east to west—opposite in direction to that of the polar-front and subtropical jet streams. The tropical easterly jet occurs only in the summer season and is limited to a northern hemisphere location over Southeast Asia, India, and Africa.

OCEAN SURFACE CURRENTS

Just as there is a circulation pattern to the atmosphere, so there is a circulation pattern to the oceans. An *ocean current* is any persistent, dominantly horizontal flow of ocean water. Because ocean currents move warm water poleward and cold waters toward the equator, they are important regulators of air temperatures. Warm currents keep winter temperatures in the British Isles from falling much below freezing in winter. Cold currents keep weather on the California coast cool, even in the height of summer. Current systems act to exchange heat between low and high latitudes and are essential in sustaining the global energy balance.

Nearly all large-surface currents of the oceans are set in motion by prevailing surface winds. Energy is transferred from wind to water by the friction of the air blowing over the water surface. Because of the Coriolis effect, the actual direction of water drift is deflected about 45° from the direction of the driving wind.

Figure 5.25 presents a world map that shows ocean currents for the month of January. The primary features are large circular movements, called **gyres,** that are centered at latitudes of 20° to 30°. These motions track the movements of air around the subtropical high-pressure cells.

In the equatorial region, ocean currents flow westward, pushed by northeast and southeast trade winds. As these equatorial currents approach land, they are turned poleward along the west sides of the oceans, forming warm currents paralleling the coast. Examples are the Gulf Stream of eastern North America and the Kuroshio Current of Japan. In the Pacific, a thin equatorial countercurrent flows eastward. It separates the northern and southern hemisphere equatorial westward flows.

In the zone of westerly winds, a slow eastward motion of water is named the *west-wind drift*. As west-wind drift waters approach the western sides of the continents, they are deflected equatorward along the coast. The equatorward flows are cool currents. They are often accompanied by *upwelling* along continental margins. In this process, colder water from greater depths rises to the surface. Examples of cool currents with

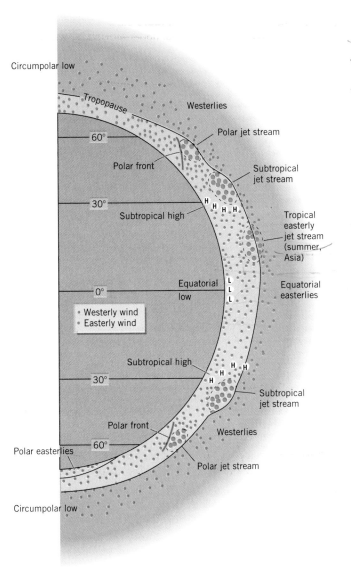

Figure 5.23   A schematic diagram of wind directions and jet streams along an average meridian from pole to pole. The four polar and subtropical jets are westerly in direction, in contrast to the single tropical easterly jet. (Copyright © A. N. Strahler.)

Figure 5.24   A strong subtropical jet stream is marked in this space photo by a narrow band of cirrus clouds, generated by turbulence in the air stream. This jet stream is moving from west to east at an altitude of about 12 km (40,000 ft). The cloud band lies at about 25°N latitude. In this view, astronauts aimed their camera toward the southeast, taking in the Nile River valley and the Red Sea. At the left is the tip of the Sinai Peninsula. (NASA.)

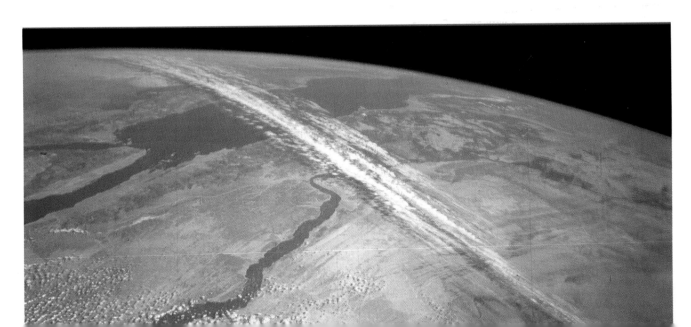

Eye on the Environment • El Niño

At intervals of about three to eight years, a remarkable disturbance of ocean and atmosphere occurs. It begins in the eastern Pacific Ocean and spreads its effects widely over the globe. This disturbance lasts for more than a year, bringing droughts, heavy rainfalls, severe spells of heat and cold, or a high incidence of cyclonic storms to various parts of the Pacific and its eastern coasts. This phenomenon is called *El Niño*. The expression comes from Peruvian fishermen, who refer to the *Corriente del Niño*, or the "Current of the Christ Child," in describing an invasion of warm surface water that occurs once every few years around Christmas time and greatly depletes their catch of fish. El Niño occurs at irregular intervals and with varying degrees of intensity. Notable El

Niño events occurred in 1891, 1925, 1940–1941, 1965, 1972–1973, 1982–1983, and 1991–1992.

Normally, the cool Humboldt (Peru) Current flows northward off the South American coast, and then at about the equator it turns westward across the Pacific as the South Equatorial Current (see Figure 5.25). The Humboldt Current is characterized by upwelling of cold, deep water, bringing with it nutrients that serve as food for plankton. Fish feed on these high concentrations of plankton. The anchoveta, a small fish used commercially to produce fish meal for animal feed, thrives in great numbers and is harvested by the Peruvian fishermen. With the onset of El Niño, upwelling ceases, the cool water is replaced by warm water from the west, and the plankton and their

anchoveta predators disappear. Vast numbers of birds that feed on the anchoveta die of starvation. Only since the event of 1982–1983 has the full story of El Niño been uncovered, although certain far-reaching weather phenomena that accompany El Niño have long been linked with it.

Cessation of the Humboldt Current upwelling is linked to a major change in barometric pressures across the entire stretch of the equatorial zone as far west as southeastern Asia. The figure to the left shows a section of the world map of pressure and winds for January, for the Pacific region near the equator. Normally, low pressure prevails over northern Australia, the East Indies, and New Guinea, where the largest and warmest body of ocean water can be found. Abundant rainfall normally occurs in this area during this season, which is the high-sun period in the southern hemisphere. During an El Niño event, this low-pressure system is replaced by weak high pressure, and drought ensues. In contrast, in the equatorial zone of the eastern Pacific, pressures become lower than normal, strengthening the equatorial trough. Rainfall is abundant in this new low-pressure region.

Surface winds and currents also change with this change in pressure. During normal conditions, the strong, prevailing trade winds blow steadily westward, causing very warm ocean water to move to the western Pacific and to "pile up" near the western equatorial low. This westward motion causes the normal upwelling along the South American coast, as bottom water is carried up to replace the water

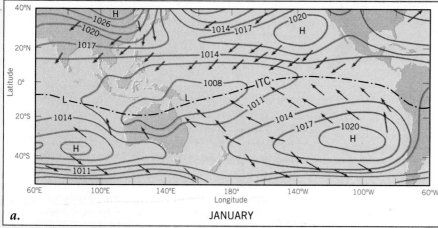

a. JANUARY

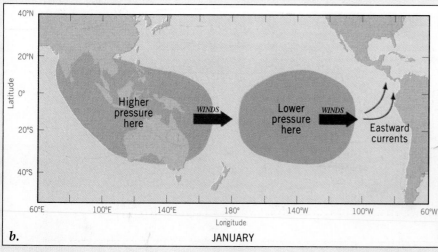

b. JANUARY

(*a*) January average surface barometric pressure and winds of the Pacific equatorial zone. (*b*) Changes during an El Niño event.

dragged to the west. With the change in atmospheric pressure, the easterly trade winds die. A weak westerly wind flow sometimes occurs, completely reversing the normal wind direction. Without the pressure of the trade winds to hold them back, warm waters surge eastward. Sea-surface temperatures and actual sea levels rise off the tropical western coasts of the Americas.

The major change in sea-surface temperatures that accompanies an El Niño can also shift rainfall patterns dramatically in large regions. Among the regions that receive more rainfall at some time during a typical El Niño event are the southeastern and interior northwestern United States, southeastern South America, and the southern tip of India. Drought regions include the western Pacific, northern South America, southeast Africa, and northern India.

In the El Niño of 1982–1983, torrential rainfalls occurred in the Andean highlands of Peru, Bolivia, and Colombia causing devastating floods. The warm ocean surface layer spread far northward along the Central American and North American coast, reaching the latitude of Oregon.

Associated with the mature phase of the 1982–1983 El Niño was an important change in the pattern of the northern hemisphere subtropical jet stream, as shown in the figure below. Notice that the El Niño jet swept across Mexico and the Caribbean and continued strong across the Atlantic. It then entered North Africa and continued in strength across the Arabian Peninsula and northern India.

The far southward position of the subtropical jet stream over North America allowed frequent and intense invasions of cyclonic storms. These dumped heavy rains on coastal areas and heavy snows over the western mountain ranges, as well as heavy rains over the Gulf Coast and Florida. Precipitation in January and February 1983 exceeded 200 percent of normal in these locations. Another strong El Niño event occurred in 1991–1992, with similar effects in North America. Severe droughts and famine occurred in South Africa, Sudan, and Somalia.

As research on the global changes in winds and currents linked to the El Niño has continued, scientists have recently recognized a condition called *La Niña* (the girl child). This event presents a situation roughly opposite to El Niño. During a La Niña period, sea-surface temperatures in the central and eastern Pacific Ocean fall to lower than average levels. Why does this occur? It happens because the South Pacific subtropical high becomes very strongly developed during the high-sun season. The result is abnormally strong southeast trade winds. The force of these winds drags a more than normal amount of warm surface water westward, bringing cooler water to the surface off western continental coasts. La Niña conditions were recognized during 1988 and may have been related to a drought experienced in some parts of North America in the summer of 1988. In some manner not yet well understood, La Niña conditions give rise to the onset of the following El Niño, which in turn gives rise to the next La Niña, perpetuating an unending succession of these events.

Questions

1. Compare the normal pattern of wind, pressure, and ocean currents in the equatorial Pacific with the pattern during an El Niño event.

2. What are some of the weather changes reported for the 1982–1983 El Niño?

3. What is La Niña, and how does it compare with the normal pattern?

(blue) The typical prevailing subtropical jet stream pattern for December through February. (red) Pattern during the El Niño of 1982–1983. (Data of E. M. Rasmusson, 1984.)

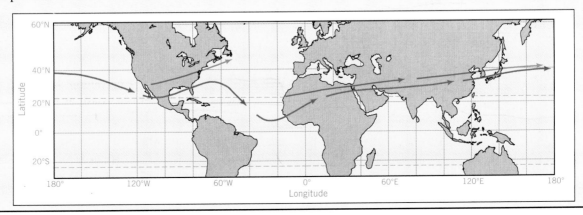

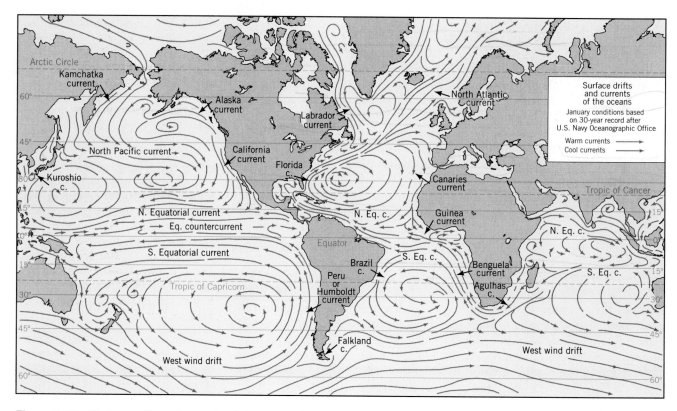

Figure 5.25 Surface drifts and currents of the oceans in January. (Based on data from U.S. Navy Oceanographic Office. Redrawn and revised by A. N. Strahler.)

upwelling are the Humboldt (or Peru) Current, off the coast of Chile and Peru, the Benguela Current, off the coast of southern Africa, and the California Current.

The strong west winds of southern high latitudes produce an Antarctic circumpolar current of cold water. Some of this flow branches equatorward along the west coasts of South America and Africa, adding to the Humboldt and Benguela currents.

Global surface currents are dominated by huge circular gyres centered near the subtropical high-pressure cells.

West-wind drift water also moves poleward to join arctic and antarctic circulations. In the northeastern Atlantic Ocean, the west-wind drift forms a relatively warm current. This is the North Atlantic Drift, which spreads around the British Isles, into the North Sea, and along the Norwegian coast. The Russian port of Murmansk, on the arctic circle, remains ice-free year round because of this warm drift current.

In the northern hemisphere, where the polar sea is largely landlocked, cold water flows equatorward along

the east sides of continents. Two examples are the Kamchatka Current, which flows southward along the Asian coast across from Alaska, and the Labrador Current, which flows between Labrador and Greenland to reach the coasts of Newfoundland, Nova Scotia, and New England.

Figure 5.26 shows a satellite image of ocean temperature along the east coast of North America for a week in April. The Gulf Stream stands out as a tongue of red and yellow (warm) color, moving along the coast and heading off from North Carolina in a northeasterly direction. Cooler water from the Labrador Current, in green and blue tones, hugs the northern Atlantic coast. This current heads south and then turns to follow the Gulf Stream to the northeast. Instead of mixing, the two flows remain quite distinct. The boundary between them shows a wavelike flow, much like Rossby waves in the atmosphere. Warm and cold bodies of water are cut off to float freely, forming warm-core and cold-core rings.

The global circulation of winds and currents paves the way for our next subject—weather systems. Recall from Chapter 4 that when warm, moist air rises, precipitation can occur. This happens in the centers of

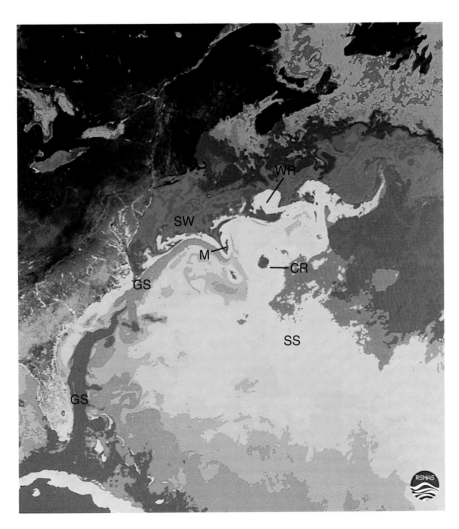

Figure 5.26 A satellite image in false colors showing the Gulf Stream (GS) and its interactions with cold water of the Continental Slope (SW), brought down by the Labrador Current, and the warm water of the Sargasso Sea (SS). Cold water appears in green and blue tones; warm water in red and yellow tones. Other features include a meander (M), a warm-core ring (WR), and a cold-core ring (CR).
The image was made from data collected by the NOAA-7 orbiting satellite. Data of 35 satellite passes during a week in April were processed together to remove the effects of clouds on single images. (NASA. Courtesy of Otis B. Brown, Robert Evans, and M. Carle, University of Miami, Rosenstiel School of Marine and Atmospheric Science, Florida.)

cyclones, where air converges. Although cyclones and anticyclones are generally large features, they move from day to day and are steered by the global pattern of winds. Thus, your knowledge of global winds and pressures will help you to understand how weather systems and storms develop and migrate. Your knowledge will also be very useful for the study of climate, which we take up in Chapter 7.

WIND AND GLOBAL CIRCULATION IN REVIEW

The term atmospheric pressure describes the weight of air pressing on a unit of surface area. Atmospheric pressure decreases rapidly as altitude increases. Wind occurs when air moves with respect to the earth's surface. Air motion is produced by pressure gradients that form between one region and another. A pressure gradient is formed when air in one location is heated to a temperature that is warmer than another. Because warm air is less dense than cold air, local low pressure is associated with warmer air and local high pressure with cooler air. The sea breeze is an example of a wind caused by a pressure gradient that is produced by a difference in surface heating. A convection loop is formed as warm air rises over land and sinks over the nearby water surface.

At a global scale, atmospheric circulation is strongly influenced by the earth's rotation. The Coriolis effect deflects winds and ocean currents, producing circular or spiraling flow paths around cyclones (centers of low pressure) and anticyclones (centers of high pressure).

Because the equatorial and tropical regions are heated more intensely than the higher latitudes, two convection loops develop—the Hadley cells. These loops drive the northeast and southeast trade winds, the convergence and lifting of air at the intertropical convergence zone (ITC), and the sinking and divergence of air in the subtropical high-pressure cells. The monsoon circulation of Asia responds to a reversal of atmospheric pressure over the continent with the seasons. A winter monsoon flow of cool, dry air from the northeast alternates with a summer monsoon flow of warm, moist air from the southwest.

In the midlatitudes and poleward, westerly winds prevail. In winter, continents develop high pressure, and intense oceanic low-pressure centers are found off the Aleutian Islands and near Iceland in the northern hemisphere. In the summer, the continents develop low pressure as oceanic subtropical high-pressure cells intensify and move poleward.

Local winds are generated by local pressure gradients. Sea and land breezes, as well as mountain and valley winds, are examples of winds caused by local surface heating. Other local winds include drainage winds, Santa Ana, and chinook winds.

Winds aloft follow a simple basic pattern—easterlies between two belts of tropical high pressure, and westerlies to the north and south of these belts. Rossby waves develop in the westerlies, bringing cold, polar air equatorward and warmer air poleward. The polar and subtropical jet streams are concentrated westerly wind streams with high wind speeds.

Ocean surface currents are dominated by huge gyres that are driven by the global surface wind pattern. Equatorial currents move warm water westward and then poleward along the east coasts of continents. Return flows bring cold water equatorward along the west coasts of continents.

KEY TERMS

barometer	Coriolis effect	subtropical high-pressure belts
millibar	cyclone	monsoon
wind	anticyclone	Rossby waves
pressure gradient	Hadley cell	jet stream
isobars	intertropical convergence zone (ITC)	gyres

REVIEW QUESTIONS

1. Explain atmospheric pressure. Why does it occur? How is atmospheric pressure measured, and in what units? What is the normal value of atmospheric pressure at sea level? How does atmospheric pressure change with altitude?

2. Describe land and sea breezes. How do they illustrate the concepts of pressure gradient and convection loop?

3. What is the Coriolis effect, and why is it important? What produces it? How does it influence the motion of wind and ocean currents in the northern hemisphere? In the southern hemisphere?

4. Define cyclone and anticyclone. How does air move within each? What is the direction of circulation of each in the northern and southern hemispheres? What type of weather is associated with each and why?

5. Sketch an ideal earth (without seasons or ocean—continent features) and its global wind system. Label the following on your sketch: doldrums, equatorial trough, Hadley cell, ITC, northeast trades, polar easterlies, polar front, polar outbreak, southeast trades, subtropical high-pressure belts, and westerlies.

6. What is the Asian monsoon? Describe the features of this circulation in summer and winter. How is the ITC involved? How is the monsoon circulation related to the high- and low-pressure centers that develop seasonally in Asia?

7. Compare the winter and summer patterns of high and low pressure that develop in the northern hemisphere with those that develop in the southern hemisphere.

8. List four types of local winds and describe three of them.

9. Describe the basic pattern of global atmospheric circulation at upper levels.

10. What are Rossby waves? How are they formed?

11. Identify five jet streams. Where do they occur? In which direction do they flow?

12. What is the general pattern of ocean surface current circulation? How is it related to global wind patterns? To air patterns?

IN-DEPTH QUESTIONS

1. An airline pilot is planning a nonstop flight from Los Angeles to Sydney, Australia. What general wind conditions can the pilot expect to find in the upper atmosphere as the airplane travels? What jet streams will be encountered? Will they slow or speed the aircraft on its way?

2. You are planning to take a round-the-world cruise, leaving New York in October. Your vessel's route will take you through the Mediterranean Sea to Cairo, Egypt, in early December. Then you will pass through the Suez Canal and Red Sea to the Indian Ocean, calling at Bombay, India, in January. From Bombay, you will sail to Djakarta, Indonesia, and then go directly to Perth, Australia, arriving in March. Rounding the southern coast of Australia, your next port of call is Auckland, New Zealand, which you will reach in April. From Auckland, you head directly to San Francisco, your final destination, arriving in June. Describe the general wind and weather conditions you will experience on each leg of your journey.

Weather Systems

Our day-to-day weather may occasionally be stormy, bringing the inconvenience of rain or snow to our daily activities. But rare storms, with high winds, high coastal ocean waters, and torrential rains or blizzards of snow can cause great destruction and loss of life. What is it like to experience a severe storm?

On August 23, 1992, Hurricane Andrew approached the coast of south Florida, near Miami. Following the path of the storm was veteran weather watcher and photographer, Warren Faidley, who decided to ride out the storm with two friends, Steve and Mike. For their shelter, they selected a sturdy parking garage, built of concrete and steel, in a coastal community south of Miami. Here is part of Faidley's account of the terrifying night of the storm.

At sunset, my friends and I gathered at the garage. Like sentries awaiting an advancing army, we took turns scanning the Miami skyline from our borrowed fortress, searching for early signs of Andrew. Bright flashes from downed power lines on the horizon marked Andrew's imminent arrival. . . .

The electrical explosions across the city grew more intense, yet the fury of the wind swallowed up all sound save the car alarms whining inside the garage. Eventually even the alarms were overwhelmed by the whistling wind.

Around 4 A.M. all hell broke loose. I have experienced severe storms with winds in excess of 75 m.p.h., but these gusts were blowing at well over 140 m.p.h. The anemometer and later damage surveys indicated even stronger gusts. By now only an occasional thud—some building collapsing or losing a roof—would punctuate the noise.

Windows from the surrounding buildings imploded, scattering glass everywhere. I looked down at my arm and saw blood, not knowing when I'd been cut. Putting on my reading glasses to protect my eyes, I made a mental note: next time, bring goggles.

My senses were on red alert, a self-preservation mode I adopt only in the most extreme danger. Everything I see, hear, feel—even taste—goes by so quickly that my mind simply slows it all down to preserve rational thought. The alternative is sheer panic.

The infamous "hurricane wail" people talk about is in many ways unreal—"the scream of the devil," Steve and I agreed. The piercing sound alone is wicked enough, but mixed with breaking glass and crashing debris, it rattled me inside. I will never forget that sound.*

Hurricanes, or tropical cyclones, as they are termed by meteorologists, can be the most deadly of storms, claiming tens of thousands of lives in a single event. They are but one of a number of types of weather systems that are the subject of this chapter.

---

* From *Weatherwise,* vol. 45, no. 6, Copyright © Warren Faidley. Used by permission.

Hurricane Andrew approaching the Louisiana coast on August 25, 1992, as depicted in a computer-enhanced satellite image.

What are cold fronts and warm fronts? You may have heard these terms while watching the weather on television or listening to weather reports on the radio. You probably know that cold fronts generally bring cold weather and that warm fronts generally bring warm weather. But did you know that cold and warm fronts are often organized together as part of the same weather system? Cold fronts and warm fronts are found together in *wave cyclones*—low-pressure centers in the midlatitudes that move eastward and provide much of North America's precipitation. Wave cyclones are just one type of **weather system**—an organized state of the atmosphere associated with a characteristic weather pattern. Other weather systems we will encounter in this chapter include traveling anticyclones, tornadoes, tropical cyclones (including hurricanes), easterly waves, and weak equatorial lows.

TRAVELING CYCLONES AND ANTICYCLONES

What's the difference between a cyclone and an anticyclone? If the sun is shining and skies are blue, it's an anticyclone. If the day is overcast and gray, it's a cyclone. Most types of cyclones and anticyclones move slowly across the earth's surface bringing changes in the weather as they move. These are referred to as *traveling cyclones* and *traveling anticyclones*.

In cyclones, low pressure at the center draws air inward and carries it aloft. And when moist air rises, condensation or deposition can occur. Thus, cyclones normally have dense cloud layers, often releasing rain or snow. This is **cyclonic precipitation.** Many cyclones are weak and pass overhead with little more than a period of cloud cover and light precipitation. However, when pressure gradients are steep, strong winds and heavy rain or snow can accompany the cyclone. In the second case, the disturbance is called a **cyclonic storm.**

In middle and higher latitudes, traveling cyclones and anticyclones bring changing weather systems.

In anticyclones, high pressure pushes air outward, causing air to descend from above to take its place. Since descending air warms adiabatically, condensation does not occur and skies are fair, except for occasional puffy cumulus clouds. Toward the center of an anticyclone, the pressure gradient is weak, and winds are light and variable. Traveling anticyclones are found in the middle latitudes. They are typically associated with ridges or domes of cool, dry air that move equatorward and eastward.

Traveling cyclones fall into three types. First is the wave cyclone of midlatitude, arctic, and antarctic zones. This type of cyclone ranges in intensity from a weak disturbance to a powerful storm. Second is the tropical cyclone of tropical and subtropical zones. This type of cyclone ranges in intensity from a mild disturbance to the highly destructive hurricane, or typhoon. A third type is the tornado, a small, intense cyclone of enormously powerful winds. The tornado is much, much smaller in size than other cyclones, and it is related to strong convectional activity.

Air Masses

An **air mass** is a large body of air with fairly uniform temperature and moisture characteristics. It can be several thousand kilometers or miles across, and extend upward to the top of the troposphere. A given air mass is characterized by a distinctive combination of surface temperature, environmental temperature lapse rate, and surface-specific humidity. Air masses range widely in temperature, from searing hot to icy cold, as well as in moisture content.

Air masses acquire their characteristics in *source regions.* In a source region, air moves slowly or stagnates, which allows the air to acquire temperature and moisture characteristics from the surface. For example, an air mass with warm temperatures and a high water vapor content develops over a warm equatorial ocean. Over a large tropical desert, slowly subsiding air forms a hot air mass with low relative humidities. A very cold air mass with a low water vapor content is generated over cold, snow-covered land surfaces in the arctic zone in winter.

Air masses move from one region to another under the influence of barometric pressure gradients and are sometimes pushed or blocked by high-level jet stream winds. When an air mass moves to a new area, its properties will begin to change, because it is influenced by the new surface environment. For example, the air mass may lose heat or take up water vapor.

Air masses are classified on the basis of the latitudinal position and the nature of the underlying surface of their source regions. Latitudinal position primarily determines surface temperature and environmental temperature lapse rate of the air mass, while the nature of the underlying surface—continent or ocean—usually determines the moisture content. For latitudinal position, five types of air masses are distinguished as shown in the accompanying table.

Air Mass	Symbol	Source Region
Arctic	A	Arctic ocean and fringing lands
Antarctic	AA	Antarctica
Polar	P	Continents and oceans, lat. 50–60° N and S
Tropical	T	Continents and oceans, lat. 20–35° N and S
Equatorial	E	Oceans close to equator

For the type of underlying surface, two subdivisions are used:

Air Mass	Symbol	Source Region
Maritime	m	Oceans
Continental	c	Continents

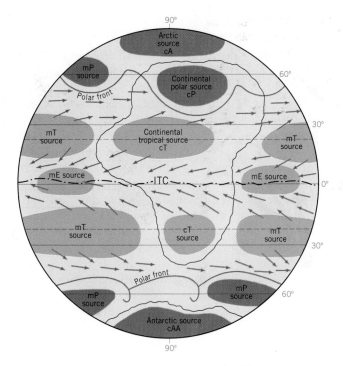

Figure 6.1 A schematic diagram of global air masses and source regions. An idealized continent, producing continental (c) air masses, is shown at the center. It is surrounded by oceans, producing maritime air masses (m). Tropical (T) and equatorial (E) source regions provide warm or hot air masses, while polar (P), arctic (A), and antarctic (AA) source regions provide colder air masses with lower specific humidities.

Combining these two types of labels produces a list of six important types of air masses, shown in Table 6.1. The table also gives some typical values of surface temperature and specific humidity, although these can vary widely, depending on season. Air mass temperatures can range from −46°C (−50°F) for continental arctic (cA) air masses to 27°C (80°F) for maritime equatorial air masses (mE). Specific humidities show a very high range—from 0.1 gm/kg for the cA air mass to as much as 19 gm/kg for the mE air mass. In other words, maritime equatorial air can hold about 200 times as much moisture as continental arctic air.

Figure 6.1 shows schematically the global distribution of source regions of these air masses. An idealized continent is shown in the center of the figure surrounded by ocean. Note from the figure that the polar air masses (mP, cP) originate in the subarctic latitude

zone, not in the polar latitude zone. Recall that in Chapter 2 we defined "polar" as a region containing one of the poles. However, meteorologists use the word "polar" to describe air masses from the subarctic and subantarctic zones, and we will follow their usage when referring to air masses.

Table 6.1 Properties of typical air masses

Air Mass	Symbol	Source Region	Properties	Temperature °C	(°F)	Specific Humidity (gm/kg)
Maritime equatorial	mE	Warm oceans in the equatorial zone	Warm, very moist	27°	(80°)	19
Maritime tropical	mT	Warm oceans in the tropical zone	Warm, moist	24°	(75°)	17
Continental tropical	cT	Subtropical deserts	Warm, dry	24°	(75°)	11
Maritime polar	mP	Midlatitude oceans	Cool, moist (winter)	4°	(39°)	4.4
Continental polar	cP	Northern continental interiors	Cold, dry (winter)	−11°	(12°)	1.4
Continental-arctic (and continental antarctic)	cA (cAA)	Regions near north and south poles	Very cold, very dry (winter)	−46°	(−50°)	0.1

Figure 6.2 A cold front, along which a cold air mass lifts a warm air mass aloft. The upward motion sets off a line of thunderstorms. The frontal boundary is actually much less steep than shown in this schematic drawing. (Drawn by A.N. Strahler.)

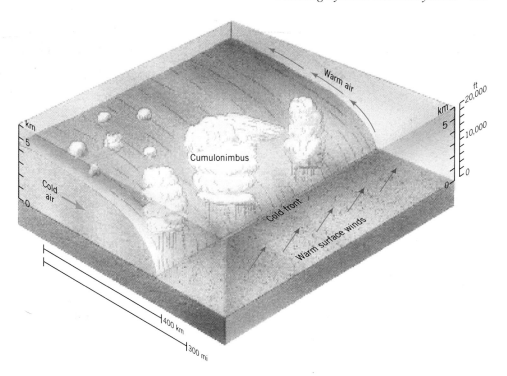

The maritime tropical air mass (mT) and maritime equatorial air mass (mE) originate over warm oceans in the tropical and equatorial zones. They are quite similar in temperature and water vapor content. With very high values of specific humidity, both are capable of very heavy yields of precipitation. The continental tropical air mass (cT) has its source region over subtropical deserts of the continents. Although it may have a substantial water vapor content, it tends to be stable and has low relative humidity when highly heated during the daytime.

The maritime polar air mass (mP) originates over midlatitude oceans. Although the quantity of water vapor it holds is not as large as maritime tropical air masses, the mP air mass can still yield heavy precipitation. Much of this precipitation is orographic and occurs over mountain ranges on the western coasts of continents. The continental polar air mass (cP) originates over North America and Eurasia in the subarctic zone. It has low specific humidity and is very cold in winter. Last is the continental arctic (and continental antarctic) air mass type (cA, cAA), which is extremely cold and holds almost no water vapor.

Cold, Warm, and Occluded Fronts

A given air mass usually has a sharply defined boundary between itself and a neighboring air mass. This boundary is termed a **front.** We saw an example of a front in the contact between polar and tropical air

masses, shown in Figures 5.12, 5.21, and 5.23. This feature is the *polar front,* and it is located below the axis of the jet stream in the upper-air waves. If one air mass is advancing on another, the surface of contact between the two will be at a relatively low angle. If the two air masses are stationary with respect to each other, the frontal contact will be nearly vertical.

In an occluded front, a cold front overtakes a warm front, lifting a pool of warm, moist air upward.

Figure 6.2 shows the structure of a front along which a cold air mass invades a zone occupied by a warm air mass. A front of this type is called a **cold front.** Because the colder air mass is denser, it remains in contact with the ground. As it moves forward, it forces the warmer air mass to rise above it. If the warm air is unstable, severe thunderstorms may develop. Thunderstorms near a cold front often form a long line of massive clouds stretching for tens of kilometers (Figure 6.3).

Figure 6.4 diagrams a **warm front** in which warm air moves into a region of colder air. Here, again, the cold air mass remains in contact with the ground because it is denser. The warm air mass is forced to rise on a long ramp over the cold air below. The rising motion causes stratus clouds to form, and precipitation often follows. If the warm air is stable, the precipitation will be steady (Figure 6.4). If the warm air is unstable, convection cells can develop, producing cumulonimbus clouds with heavy showers or thunderstorms (not shown in the figure).

Cold fronts normally move along the ground at a faster rate than warm fronts. Thus, when both types are in the same neighborhood, the cold front overtakes the warm front. The result is an **occluded front,** diagrammed in Figure 6.5. The colder air of the fast-moving cold front remains next to the ground, forcing both the warm air and the less cold air to rise over it. The warm air mass is lifted completely free of the ground.

Wave Cyclones

In middle and high latitudes, the dominant form of weather disturbance is the **wave cyclone.** The wave cyclone is a large inspiral of air that repeatedly forms, intensifies, and dissolves along the polar front. Figure 6.6 shows a situation favorable to the formation of a wave cyclone. Two large anticyclones are in contact on the polar front. One contains a cold, dry polar air mass, and the other a warm, moist maritime air mass.

Airflow converges from opposite directions on the two sides of the front, setting up an unstable situation. The wave cyclone will begin to form between the two high-pressure cells in a zone of lower pressure referred to as a low-pressure trough.

How does a wave cyclone form, grow, and eventually dissolve? Figure 6.7 shows the life history of a wave cyclone. In Block A (early stage), the polar-front region shows a wave beginning to form. Cold air is turned in a southerly direction and warm air in a northerly direction, so that each advances on the other. As these frontal motions develop, precipitation will form.

In Block B (open stage), the wave disturbance along the polar front has deepened and intensified. Cold air actively pushes southward along a cold front, and warm air actively moves northeastward along a warm front. The zones of precipitation along the two fronts are now strongly developed but the zone is wider along the warm front than along the cold front.

In Block C (occluded stage), the cold front has overtaken the warm front, producing an occluded front. The warm air mass at the center of the inspiral is forced off the ground, intensifying precipitation.

Figure 6.3   A line of cumulus clouds marks the advance of a cold front, moving from left to right. The cold air pushes warmer, moister air aloft, triggering cloud formation.

Figure 6.4 A warm front. Warm air advances toward cold air and rides up and over the cold air. A notch of cloud is cut away to show rain falling from the dense stratus cloud layer. (Drawn by A.N. Strahler.)

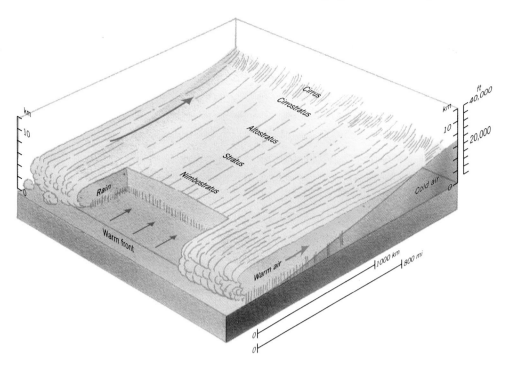

Eventually, the polar front is reestablished (Block D, dissolving stage), but a pool of warm, moist air remains aloft. As the moisture content of the pool is reduced, precipitation dies out, and the clouds gradually dissolve.

Keep in mind that a wave cyclone is quite a large feature—a thousand kilometers (600 mi) or more across. Also, the cyclone normally moves eastward as it develops, propelled by prevailing westerlies aloft. Therefore, Blocks A–D are like three-dimensional snapshots taken at intervals along an eastbound track.

You may recall from Chapter 4 that we recognized three types of precipitation—orographic, convectional, and cyclonic. With your knowledge of how precipitation occurs in cold fronts, warm fronts, and occluded fronts, you now understand cyclonic precipitation as

Figure 6.5 An occluded front. A warm front is overtaken by a cold front. The warm air is pushed aloft, and it no longer contacts the ground. Abrupt lifting by the denser cold air produces precipitation. (Drawn by A.N. Strahler.)

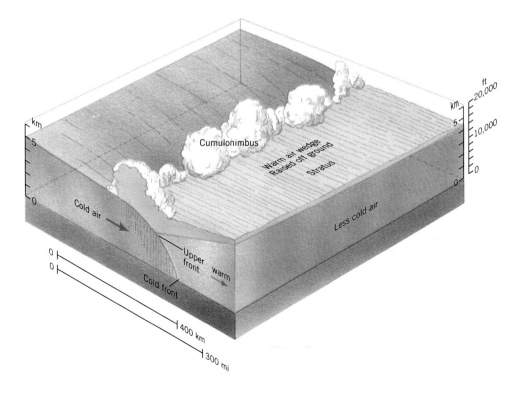

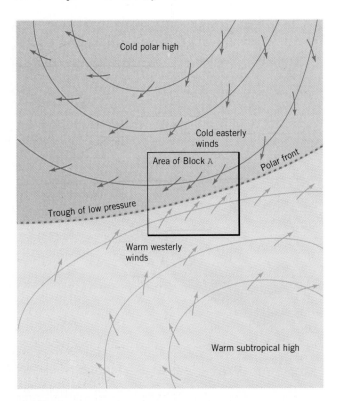

Figure 6.6 Conditions for formation of a wave cyclone. Warm subtropical air and cold polar air are in contact on the polar front. The shaded area is shown in Block A of the following figure as the early stage of development of a wave cyclone.

the precipitation associated with air mass movements, fronts, and cyclonic storms.

Weather Changes Within a Wave Cyclone

How does weather change as a wave cyclone passes through a region? Figure 6.8 shows two simplified weather maps of the eastern United States depicting

conditions on successive days. The structure of the storm is defined by the isobars, labeled in millibars. The three kinds of fronts are shown by special line symbols. Areas of precipitation are shown in color.

Map *a* shows the cyclone in an open stage, similar to Figure 6.7, Block B. The isobars show that the cyclone is a low-pressure center with inspiraling winds. The cold front is pushing south and east, supported by a flow of cold, dry continental polar air from the northwest filling in behind it. Note that the wind direction changes abruptly as the cold front passes. There is also a sharp drop in temperature behind the cold front as cP air fills in. The warm front is moving north and somewhat east, with warm, moist maritime tropical air following. The precipitation pattern includes a broad zone near the warm front and the central area of the cyclone. A thin band of precipitation extends down the length of the cold front. Cloudiness generally prevails over much of the cyclone.

A cross section through Map *a* along the line A–A′ shows how the fronts and clouds are related. Along the warm front is a broad layer of stratus clouds. These take the form of a wedge with a thin leading edge of cirrus. (See Figures 4.11 and 4.12 for cloud types.) Westward, this wedge thickens to altostratus, then to stratus, and finally to nimbostratus with steady rain. Within the sector of warm air, the sky may partially clear with scattered cumulus. Along the cold front are cumulonimbus clouds associated with thunderstorms. These yield heavy rains but only along a narrow belt.

The second weather map, Map *b*, shows conditions 24 hours later. The cyclone has moved rapidly northeastward, its track shown by the dashed line. The center has moved about 1600 km (1000 mi) in 24 hours—a speed of just over 65 km (40 mi) per hour. The cold front has overtaken the warm front, forming an occluded front in the central part of the disturbance. A high-pressure area, or tongue of cold polar air, has moved in to the area west and south of the

Figure 6.7 Development of a wave cyclone. In Block A, a cyclonic motion begins at a point along the polar front. Cold and warm fronts intensify in Block B. In Block C, the cold front overtakes the warm front, producing an occluded front in the center of the cyclone. Later, the polar front is reestablished with a mass of warm air isolated aloft (Block D). (Drawn by A.N. Strahler.)

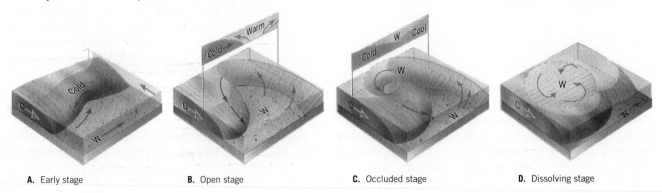

A. Early stage   **B.** Open stage   **C.** Occluded stage   **D.** Dissolving stage

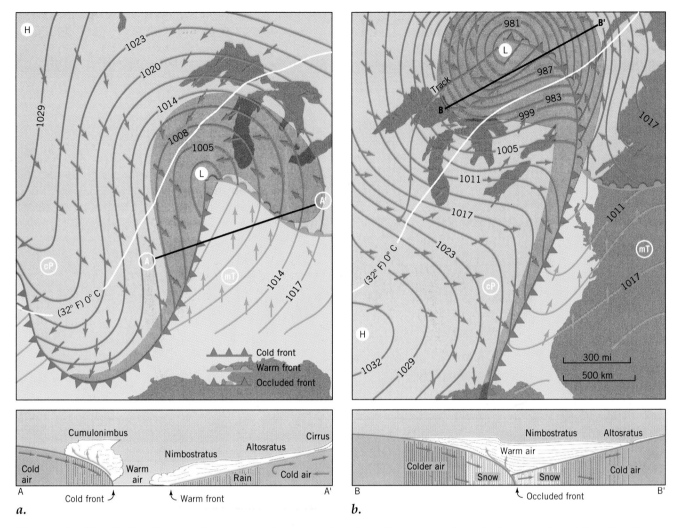

Figure 6.8 Simplified surface weather maps and cross sections through a wave cyclone. In the open stage (*a*), cold and warm fronts pivot around the center of the cyclone. In the occluded stage (*b*), the cold front has overtaken the warm front, and a large pool of warm, moist air has been forced aloft.

cyclone, and the cold front has pushed far south and east. Within the cold air tongue, the skies are clear. A cross section below the map shows conditions along the line B–B', cutting through the occluded part of the storm. Notice that the warm air mass is lifted well off the ground and yields heavy precipitation.

Cyclone Tracks and Cyclone Families

Wave cyclones tend to form in certain areas and travel common paths until they dissolve. Figure 6.9 is a world map showing common cyclone paths. Wave cyclone tracks are shown as dashed lines, with tropical cyclones shown as solid lines. (We will discuss tropical cyclones in detail shortly.) The western coast of North America commonly receives wave cyclones arising in the North Pacific Ocean. Wave cyclones also originate over land,

shown by the tracks starting in Alaska, the Pacific Northwest, the south central United States, and along the Gulf coast. Most of these tracks converge toward the northeast and pass out into the North Atlantic, where they tend to concentrate in the region of the Icelandic Low.

In the northern hemisphere, wave cyclones are heavily concentrated in the neighborhood of the Aleutian and Icelandic Lows. These cyclones commonly form in a succession, traveling as a chain across the North Atlantic and North Pacific oceans. Figure 6.10, a world weather map, shows several such cyclone families. Each cyclone moves northeastward, deepening in low pressure and occluding to form an upper-air low. For this reason, intense cyclones arriving at the western coasts of North America and Europe are usually occluded.

In the southern hemisphere, storm tracks are more nearly along a single lane, following the parallels of

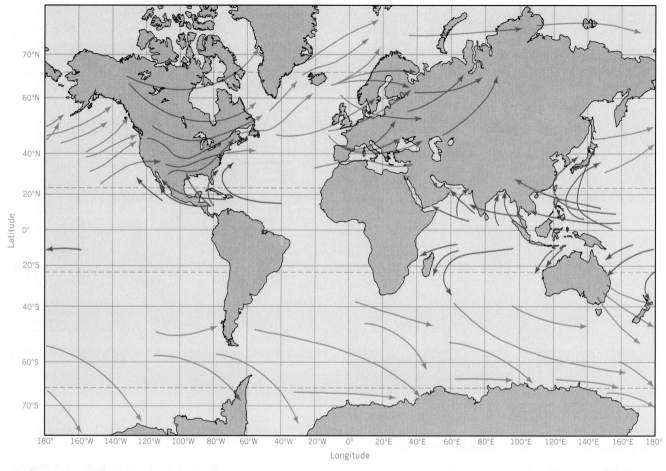

Figure 6.9 Typical paths of tropical cyclones (solid lines) and wave cyclones of middle latitudes (dashed lines). (Based on data of S. Pettersen, B. Haurwitz and N. M. Austin, J. Namias, M. J. Rubin, and J-H. Chang.)

latitude. Three such cyclones are shown in Figure 6.10. This track is more uniform because of the uniform pattern of ocean surface circling the globe at these latitudes. Only the southern tip of South America projects southward to break the monotonous expanse of ocean.

The Tornado

A **tornado** is a small but intense cyclonic vortex in which air spirals at tremendous speed. It is associated with thunderstorms spawned by fronts in midlatitudes of North America. Tornadoes also occur inside tropical cyclones (hurricanes).

> **Tornado winds are the highest of all storm winds and can cause great local damage.**

The tornado appears as a dark funnel cloud hanging from the base of a dense cumulonimbus cloud (Figure 6.11). At its lower end, the funnel may be 100 to 450 m (300 to 1500 ft) in diameter. The base of the funnel

appears dark because of the density of condensing moisture, dust, and debris swept up by the wind. Wind speeds in a tornado exceed speeds known in any other storm. Estimates of wind speed run as high as 400 km (250 mi) per hour. As the tornado moves across the country, the funnel writhes and twists. Where it touches the ground, it can cause the complete destruction of almost anything in its path.

Tornadoes occur as parts of cumulonimbus clouds traveling in advance of a cold front. They seem to originate where turbulence is greatest. They are most common in the spring and summer, but can occur in any month. Where a cold front of maritime polar air lifts warm, moist maritime tropical air, conditions are most favorable for tornadoes. As shown in Figure 6.12, they occur in greatest numbers in the central and southeastern states and are rare over mountainous and forested regions. They are almost unknown west of the Rocky Mountains and are relatively less frequent on the eastern seaboard. Tornadoes are a typically American phenomenon, since they are most frequent and violent in the United States. They also occur in Australia in

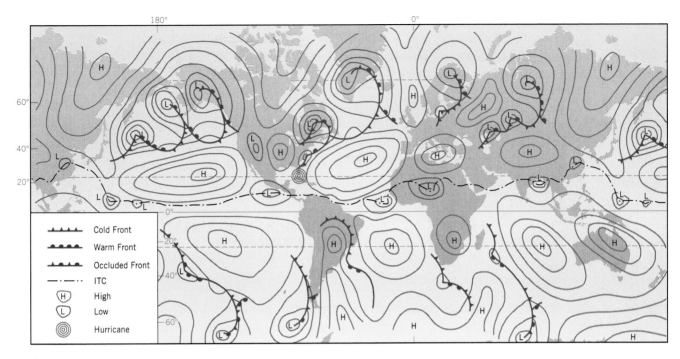

Figure 6.10 A daily weather map of the world for a given day during July or August might look like this map, which is a composite of typical weather conditions. (After M. A. Garbell.)

Figure 6.11 This tornado touched down near Clearwater, Kansas, on May 16, 1991. Surrounding the funnel is a cloud of dust and debris carried into the air by the violent winds.

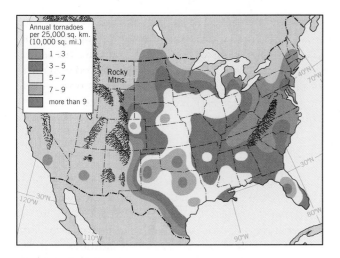

Figure 6.12 Frequency of occurrence of observed tornadoes in the 48 contiguous United States. The data span a 30-year record, 1960–1989. (Courtesy of Edward W. Ferguson, National Severe Storms Forecast Center, National Weather Service.)

substantial numbers and are occasionally reported from other midlatitude locations.

Devastation from a tornado is often complete within the narrow limits of its path (Figure 6.13). Only the strongest buildings constructed of concrete and steel can resist major structural damage from the extremely violent winds. The National Weather Service maintains a tornado forecasting and warning system. Whenever weather conditions favor tornado development, the danger area is alerted, and systems for observing and reporting a tornado are set in readiness.

TROPICAL AND EQUATORIAL WEATHER SYSTEMS

So far, the weather systems we have discussed are those that are important in the midlatitudes and poleward. Weather systems of the tropical and equatorial zones show some basic differences from those of the midlati-

Figure 6.13 A swath of destruction left by a tornado that swept through Plainfield, Illinois, in August of 1990. Undamaged houses delimit the tornado path.

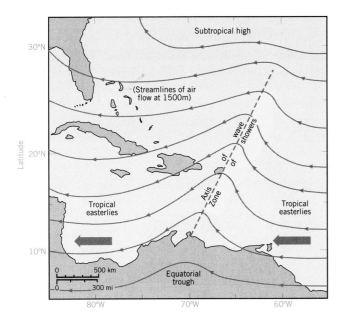

Figure 6.14 An easterly wave passing over the West Indies. (Data from H. Riehl, *Tropical Meteorology*, McGraw-Hill, New York.)

tudes. The Coriolis effect is weak close to the equator, and there is a lack of strong contrast between air masses. So, clearly defined fronts and large, intense wave cyclones are missing.

On the other hand, there is intense convectional activity because of the high moisture content of the maritime air masses at these low latitudes. In other words, evaporation from tropical oceans produces an enormous reservoir of energy in the form of latent heat that can be released to the atmosphere.

Easterly Waves and Weak Equatorial Lows

One of the simplest forms of weather disturbance is an *easterly wave,* a slowly moving trough of low pressure within the belt of tropical easterlies (trades). These waves occur in latitudes 5° to 30° N and S over oceans, but not over the equator itself. Figure 6.14 is a simplified upper-air map of an easterly wave showing wind patterns, the axis of the wave, isobars, and the zone of showers. At the surface, a zone of weak low pressure underlies the axis of the wave. The wave travels westward at a rate of 300 to 500 km (200 to 300 mi) per day. Surface airflow converges on the eastern, or rear, side of the wave axis. This convergence causes the moist air to be lifted, producing scattered showers and thunderstorms. The rainy period may last a day or two as the wave passes over.

Another related disturbance is the *weak equatorial low,* a disturbance that forms near the center of the equatorial trough. Moist equatorial air masses converge on the center of the low, causing rainfall from

many individual convectional storms. Several such weak lows are shown on the world weather map (Figure 6.10), lying along the ITC. Because the map is for a day in July or August, the ITC is shifted well north of the equator. At this season, the rainy monsoon is in progress in Southeast Asia. It is marked by a weak equatorial low in northern India.

Tropical Cyclones

The most powerful and destructive type of cyclonic storms is the **tropical cyclone,** which is known as the *hurricane* in the Atlantic Ocean or the *typhoon* in the Pacific and Indian oceans. This type of storm develops over oceans in 8° to 15° N and S latitudes, but not closer to the equator. It originates as a weak low, which deepens and intensifies, growing into a deep, circular low. High sea-surface temperatures, over 27°C (80°F), are required for tropical cyclones to form. Once formed, the storm moves westward through the trade-wind belt, often intensifying as it travels. It can then curve northwest, north, and northeast, steered by westerly winds aloft. The storm can penetrate well into the midlatitudes, as many residents of the southern and eastern coasts of the United States who have experienced a hurricane can attest.

Figure 6.15 A simplified weather map of a hurricane passing over the western tip of Cuba. Daily locations, beginning on September 3, are shown as circled numerals. The recurring path will take the storm over Florida. Shaded areas show dense rain clouds as seen in a satellite image.

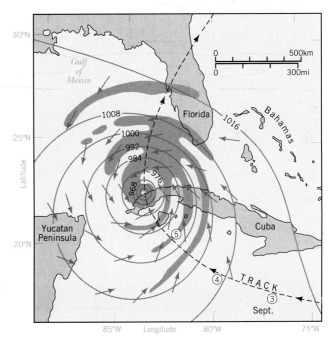

Eye on the Environment • Cloud Cover, Precipitation, and Global Warming

Earlier, we examined climatic change from several different perspectives. Let's now focus on how global climate might be influenced by an increase in clouds and precipitation.

Recall that global temperatures have been rising over the last 20 years. As part of this global warming, satellite data seem to have detected a rise in temperature of the global ocean surface of about 1°C (2°F) over the past decade. Any rise in sea-surface temperature increases the rate of evaporation. And the increase in evaporation will raise the average atmospheric content of water vapor. What effect will this have on climate?

Water has several roles in global climate. First, in its vapor state it is one of the greenhouse gases. That is, it absorbs and emits longwave radiation, thus enhancing the warming effect of the atmosphere above the earth's surface. In fact, water is more important than CO_2

in creating the greenhouse effect. Thus, one result of an increase in global water vapor in the atmosphere should be warming.

Second, water vapor can condense or deposit, forming clouds. Will more clouds increase or decrease global temperatures? That question is still being debated by the scientists who study and model the atmosphere. It seems to depend on where and at what level clouds form. As large, white objects in the atmosphere, clouds reflect solar energy back to space. Thus, an increase in cloud cover should increase the earth's albedo, causing the earth to cool.

On the other hand, clouds have a warming effect by absorbing longwave radiation from the ground and returning that emission as counterradiation. This is a greenhouse-type effect, which is much stronger when the water is present as cloud droplets or ice particles rather than vapor. Therefore, clouds should cre-

ate a warming effect.

Which effect, warming or cooling, will dominate? The answer appears to depend on the altitude of the clouds. If the clouds are low in the troposphere, solar reflection dominates, and they tend to cool temperatures. If clouds are high, greenhouse blanketing dominates, and they tend to increase temperatures. The best information at present is that clouds in general act to cool global climate. Therefore, more clouds would presumably cause global cooling. However, this conclusion is far from certain at this time. If additional cloud production at high levels were the main result of a warmer ocean, the effect could well be warming.

What about precipitation? This is the third role of water in global climate. With more water vapor and more clouds in the air, more precipitation should result. But where? Because of the dynamic nature of our climate system, it is not possible

---

The tropical cyclone is an almost circular storm center of extremely low pressure. Winds spiral inward at high speed, accompanied by very heavy rainfall (Figure 6.15). The storm gains its energy through the release of latent heat as the intense precipitation forms. The storm's diameter may be 150 to 500 km (100 to 300 mi). Wind speeds range from 120 to 200 km (75 to 125 mi) per hour and sometimes much higher. Barometric pressure in the storm center commonly falls to 950 mb (28.1 in. Hg) or lower.

A characteristic feature of the tropical cyclone is its central eye, in which clear skies and calm winds prevail (Figure 6.16 and chapter opener photo). The eye is a cloud-free vortex produced by the intense spiraling of the storm. In the eye, air descends from high altitudes and is adiabatically warmed. As the eye passes over a site, calm prevails, and the sky clears. Passage of the eye may take about half an hour, after which the storm strikes with renewed ferocity, but with winds in the opposite direction.

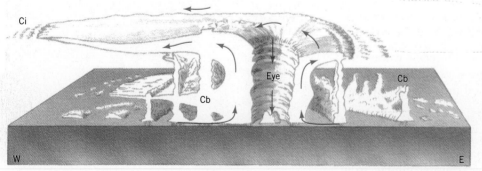

Figure 6.16 Schematic diagram of a hurricane. Cumulonimbus clouds (Cb) in concentric rings rise through dense stratiform clouds. Cirrus clouds (Ci) fringe out ahead of the storm. Width of diagram represents about 1000 km (600 mi). (Redrawn from NOAA, National Weather Service.)

at present to predict accurately where precipitation will increase.

Think however, about what might happen if precipitation increases in arctic and subarctic zones. In this case, more of the earth's surface could be covered by snow. And since snow is a good reflector of solar energy, this could increase the earth's albedo, thus tending to reduce global temperatures.

Which of these effects will dominate? Scientists are unsure. As time goes by, however, our understanding of global climate and the ability to predict its changes is certain to increase.

Question

How does water, as vapor, clouds, and precipitation, influence global climate? How will water in these forms act to enhance or retard climatic warming?

This satellite image shows the cloud patterns of earth for a typical day.

Figure 6.9, presented earlier, shows typical paths of wave cyclones and tropical cyclones. As you can see, tropical cyclones always form over oceans. In the western hemisphere, hurricanes originate in the Atlantic off the west coast of Africa, in the Caribbean Sea, or off the west coast of Mexico. Curiously, tropical cyclones do not form in the South Atlantic or southeast Pacific regions, so South America is never threatened by these severe storms. In the Indian Ocean, typhoons originate both north and south of the equator, moving north and east to strike India, Pakistan, and Bangladesh, as well as south and west to strike the eastern coasts of Africa and Madagascar. Typhoons of the western Pacific also form both north and south of the equator, moving into northern Australia, Southeast Asia, China, and Japan.

Tracks of tropical cyclones of the North Atlantic are shown in detail in Figure 6.17. Most of the storms originate at 10° to 20° N latitude, travel westward and northwestward through the trades, and then turn northeast at about 30° to 35° N latitude into the zone of the westerlies. Here their intensity lessens, especially

Figure 6.17 Tracks of some typical hurricanes occurring during August. The storms arise in warm tropical waters and move northwest. On entering the region of prevailing westerlies, the storms change direction and move toward the Northeast.

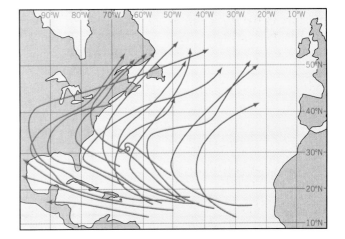

if they move over land. In the trade-wind belt, the cyclones travel 10 to 20 km (6 to 12 mi) per hour. In the zone of the westerlies, their speed is more variable.

Tropical cyclones occur only during certain seasons. For hurricanes of the North Atlantic, the season runs from May through November, with maximum frequency in late summer or early autumn. In the southern hemisphere, the season is roughly the opposite. These periods follow the annual migrations of the ITC to the north and south with the seasons, and correspond to when ocean temperatures are warmest.

Impacts of Tropical Cyclones

Tropical cyclones can be tremendously destructive storms. Islands and coasts feel the full force of the high winds and flooding as tropical cyclones move onshore. For example, Hurricane Andrew, striking the Florida coast near Miami, was the most devastating storm ever to occur in the United States, claiming as much as $30 billion in property damage and 43 lives (Figure 6.18).

The most serious effect of tropical cyclones is coastal destruction by storm waves and very high tides. Since the atmospheric pressure at the center of the cyclone is so low, sea level rises toward the center of the storm. High winds create a damaging surf and push water toward the coast, raising sea level even higher. Waves attack the shore at points far inland of the normal tidal range. Low pressure, winds, and the underwater shape of a bay floor can combine to produce a sudden rise of water level, known as a **storm surge**. During surges ships are lifted bodily and carried inland to become stranded.

If high tide accompanies the storm, the limits reached by inundation are even higher. Enormous death tolls can be created by this flooding. At Galveston, Texas, in 1900, a sudden storm surge generated by a severe hurricane flooded the low coastal city and drowned about 6000 persons. At the mouth of the Hooghly River on the Bay of Bengal, 300,000 persons died in 1737 as a result of flooding by a 12 m (40 ft) typhoon-generated storm surge. Low-lying coral atolls of the western Pacific may be entirely swept over by wind-driven sea water, washing away palm trees and houses and drowning the inhabitants.

Also important is the large amount of rainfall produced by tropical cyclones. For some coastal regions, these storms provide much of the summer rainfall. Although this rainfall is a valuable water resource, it can also produce freshwater flooding, raising rivers and streams out of their banks. On steep slopes, soil saturation and high winds can topple trees and produce disastrous mudslides and landslides.

Figure 6.18 The violent winds and high waters of Hurricane Andrew, striking the Florida coast on August 23–24, 1992, swept the boats in this Palm Beach marina into a huge pile.

POLEWARD TRANSPORT OF HEAT AND MOISTURE

Recall from Chapter 2 that a vast global heat flow extends from the equatorial and tropical regions, where the sun's energy is concentrated and net radiation is positive, to the middle and upper latitudes, where net radiation is negative. This heat flow was referred to as *poleward heat transport*. With our knowledge of the global circulation patterns of the atmosphere and oceans, and our knowledge of precipitation processes, we can now examine this flow of heat in more detail.

Atmospheric Heat and Moisture Transport

Figure 6.19 is a schematic diagram of global heat and moisture flow within the atmosphere. In the tropical and equatorial zones, the Hadley cell circulation acts like a heat pump to contain and concentrate heat between the two subtropical high-pressure belts. Descending air in the belts is warmed, and much of it

Figure 6.19 Global atmospheric transport of heat and moisture.

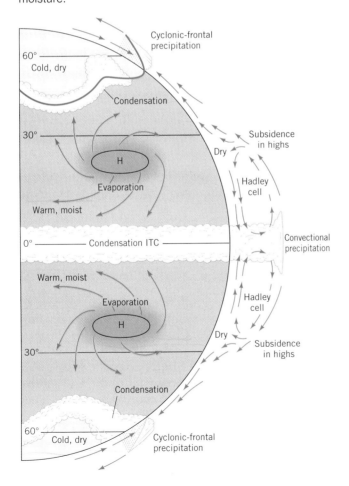

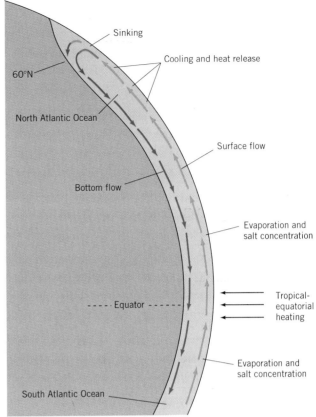

Figure 6.20 Warm surface waters flow into the North Atlantic, cool, and sink to the deep Atlantic basin, creating a circulation that warms westerly winds moving onto the European continent.

heads toward the equator. Although dry at first, it absorbs moisture from the ocean surface, evaporated by solar energy striking the water. This adds to the latent heat content of the air.

As the air converges and rises in the intertropical convergence zone (ITC), the latent heat is released. This heat adds to the sensible heat content of the air. Much of the added heat content is retained as the air completes the cycle and subsides again in the subtropical high-pressure belts. As a result, the air is even warmer when it reaches the surface. This warmer air has a twofold effect. (1) It produces a broad band around the middle of the earth where warm temperatures and moist air prevail. (2) It also moves water from the tropical regions, where evaporation takes place, to the equatorial zones, where it falls as precipitation.

The Hadley cell pump leaks rather badly, however, and some of its heat also flows poleward. This occurs when some diverging air from the subtropical high-pressure belts heads poleward into the midlatitudes. This air also acquires latent heat by picking up moisture as it moves over warm oceans. Through the Rossby wave mechanism, occasional lobes of cold, dry polar air plunge toward the equator, while tongues of warm, moist air flow toward the poles. These exchanges provide a heat flow that warms the mid- and higher latitudes well beyond the capabilities of the sun. Water is transported in this cycle, too, from warm subtropical oceans where it is evaporated, to midlatitude, arctic, antarctic, and even polar zones where it condenses or deposits and falls as precipitation.

Oceanic Heat Transport

Just as atmospheric circulation plays a role in moving heat from one region of the globe to another, so does oceanic circulation. Although no Hadley cells are present in the ocean, there is an overall circulation pattern involving surface and deep waters. (See Eye on the Environment: The Greenhouse Effect and Global Temperature Change, Chapter 3.) This circulation imports heat into the North Atlantic, making the climate of Europe a good deal warmer than you might expect from its latitude. This very large circulation pattern is superimposed on the ocean surface currents shown in Figure 5.25, and it is more like a slow drifting of huge volumes of water that contain more rapidly moving currents on their surfaces.

Heat and moisture are moved poleward from equatorial and tropical regions by global winds and ocean currents.

Figure 6.20 diagrams the features of this flow. Surface water moves northward from the South Atlantic toward the North Atlantic. Exposed to intense insolation in the tropical zones, the water is warmed. Evaporation also concentrates salt in the surface layer, making the water slightly more dense. As it travels northward, it also cools, losing heat and warming the atmosphere. Finally, the water is chilled almost to its freezing point. Because of its cold temperature and saltiness, the North Atlantic water is dense enough to sink to the bottom of the Atlantic basin. This cold, Atlantic bottom water then heads slowly southward.

This circulation pattern acts like a heat pump, moving warm surface waters northward, where they lose heat to the overlying air. Since wind patterns move air eastward at north latitudes, this heat ultimately warms Europe. The amount of heat released is quite large. In fact, a recent calculation shows that it accounts for about 35 percent of the total solar energy received by the Atlantic Ocean north of 40° latitude! This type of circulation does not occur in the Pacific or Indian oceans.

With a knowledge of the global circulation patterns of the atmosphere and oceans, as well as an understanding of weather systems and how they produce precipitation, the stage is set for global climate—which is the topic of Chapter 7. As we will see, the annual cycles of temperature and precipitation that most regions experience are quite predictable, given the changes in wind patterns, air mass flows, and weather systems that occur with the seasons. The result will be a description of the world's climates that grows easily and naturally from the principles you have mastered in your study of physical geography thus far.

WEATHER SYSTEMS IN REVIEW

A weather system is an organized state of the atmosphere associated with a characteristic weather pattern. Weather systems include wave cyclones, traveling anticyclones, tornadoes, easterly waves, weak equatorial lows, and tropical cyclones.

Air masses are distinguished by the latitudinal location and type of surface of their source regions. The boundaries between air masses are termed fronts. Wave cyclones typically form in the midlatitudes at the boundary between cool, dry air masses and warm, moist air masses. In the wave cyclone, a vast inspiraling motion produces cold and warm fronts, and eventually an occluded front. Precipitation normally occurs with each type of front. Tornadoes are very small, intense cyclones that occur as a part of thunderstorm activity. Their high winds can be very destructive.

Tropical weather systems include easterly waves and weak equatorial lows. Easterly waves occur when a weak low-pressure trough develops in the easterly wind circulation of the tropical zones, producing convergence, uplift, and shower activity. Weak equatorial lows occur near the intertropical convergence zone. They are convergence zones where convectional precipitation is frequent.

Tropical cyclones can be the most powerful of all storms. They develop over very warm tropical oceans and can intensify to become vast inspiraling systems of very high winds with very low central pressures. As they move onto land, they can bring heavy surf and storm surges of very high waters. Tropical cyclones have caused great death and destruction in coastal regions.

Global air and ocean circulation provides the mechanism for poleward heat transport by which excess heat moves from the equatorial and tropical regions toward the poles. In the atmosphere, the heat is carried primarily in the movement of warm, moist air poleward, which releases its latent heat when precipitation occurs. In the oceans, a global circulation moves warm surface water northward through the Atlantic Ocean. Heated in the equatorial and tropical regions, the surface water loses its heat to the air in the North Atlantic and sinks to the bottom. These heat flows help make northern and southern climates warmer than we might expect based on solar heating alone.

KEY TERMS

weather system	front	wave cyclone
cyclonic precipitation	cold front	tornado
cyclonic storm	warm front	tropical cyclone
air mass	occluded front	storm surge

REVIEW QUESTIONS

1. Define air mass. What two features are used to classify air masses?

2. Compare the characteristics and source regions for mP and cT air mass types.

3. Identify three types of fronts. Draw a cross section through each, showing the air masses involved, the contacts between them, and the direction of air mass motion.

4. What is a wave cyclone? How is it formed? Sketch two weather maps, showing a wave cyclone in open and occluded stages. Include isobars on your sketch. Identify the center of the cyclone as a low. Lightly shade areas where precipitation is likely to occur.

5. Describe a tornado. Where and under what conditions do tornadoes typically occur?

6. Identify three weather systems that bring rain in equatorial and tropical regions. Describe each system briefly.

7. Describe the structure of a tropical cyclone. What conditions are necessary for the development of a tropical cyclone? Give a typical path for the movement of a tropical cyclone in the northern hemisphere.

8. Why are tropical cyclones so dangerous?

9. How does the global circulation of the atmosphere and oceans provide polar heat transport?

IN-DEPTH QUESTIONS

1. Compare and contrast midlatitude and tropical weather systems. Be sure to include the following terms or concepts in your discussion: air mass, convectional precipitation, cyclonic precipitation, easterly wave, polar front, stable air, traveling anticyclone, tropical cyclone, unstable air, wave cyclone, and weak equatorial low.

2. Prepare a description of the annual weather patterns that are experienced through the year at your location. Refer to the general temperature and precipitation pattern as well as the types of weather systems that occur in each season.

Chapter 7

Global Climates

London, Rome, Cairo, Delhi, Singapore, Tokyo, and Honolulu! An around-the-world tour! It was hard to believe that you were the winner of the sweepstakes drawing sponsored by the credit card company, but now that you've picked up your ticket, it's finally hitting home. Although you'll only spend a few days in each of these cosmopolitan cities, it will be a real thrill to experience them first-hand.

But what kind of weather should you expect? Will it be hot or cold, sunny or rainy? Should you pack your overcoat and sweaters? Or will your linen suit be more appropriate? Will you need an umbrella in any of these cities? You'll be leaving in January, so you know what to expect for your trip to the airport, but what happens after that? Fortunately, you still have your college physical geography textbook to refer to. That evening, you brush up on the climates you will encounter.

"Let's see, London. Oh yes, that's a marine west-coast climate type. Lots of occluded wave cyclones in January. Hmm. I'd better have my raincoat and umbrella. Wool would be good. Perhaps that lightweight gray sweater will fit in my bag.

Then on to Rome. That's the Mediterranean climate. Well, winter is the wet season for the Mediterranean climate, but the rainfall is so much lighter than in London that I'm likely to get away with just an umbrella. Still, I'll need some kind of light coat, since it won't be very warm. Looks like I'll need the raincoat for Rome, too.

"Cairo—now, that's the dry subtropical climate. Not much chance of rain in that climate, and the hottest temperatures aren't till later in the year, when the sun moves to the tropic of cancer and is nearly overhead. Better have something light and cool. And I'll need my swimsuit for lounging by the hotel pool. Then there's Delhi. It's dry tropical there as well. Hmm. With the winter monsoon bringing cold, continental air down the slope of the Himalayas, it should be pretty clear, with warm days and cool nights.

"Then on from Delhi to Singapore. Well, Singapore is right on the equator, so that must be the wet equatorial climate. I'd better make sure my umbrella isn't buried too deeply in my suitcase. Those convectional showers don't last long, but they can be pretty intense. It'll certainly be hot and steamy. Now, Tokyo is in the moist continental climate, at about 35° N latitude. That's the same as New York. In January? Looks like I'll need those wool clothes, again. And I'd better keep that umbrella handy.

"Now, Honolulu! That's the wet equatorial climate, the same as Singapore. But Hawaii's at about 20° N, so the ocean will be cooler and the daily temperature should be just perfect! What a way to end the trip!"

Who said learning physical geography wouldn't come in handy some day? After finishing this chapter, you'll be set to pack your clothes for any destination, any time. The world's global climates are easy to master. Your earlier studies of the principles governing air temperatures, precipitation, global winds and circulation patterns, and weather systems will all combine to help you understand and predict the world's climates as they occur over the globe.

Sand dunes and date palms, Sahara Desert.

What do we mean by climate? In its most general sense, **climate** is the average weather of a region. To describe the average weather of a region, we could use many of the measures describing the state of the atmosphere that we have already encountered. These could include daily net radiation, barometric pressure, wind speed and direction, cloud cover and type, presence of fog, precipitation type and intensity, incidence of cyclones and anticyclones, frequency of frontal passages, and other such weather information. However, these observations are not made regularly at most weather stations around the globe. To study climate on a worldwide basis, we must turn to the two simple measurements that are made daily at every weather station—temperature and precipitation.

Annual cycles of mean monthly temperature and precipitation are used to determine the climate type.

Temperature and precipitation are usually related to the natural vegetation of a region—forest or grassland, shrubland or tundra. The natural vegetation cover is often a distinctive feature of a climatic region and typically influences the human use of the area. Temperature and precipitation are also important factors in cultivation of crop plants—a necessary process for human survival. And, the development of soils, as well as the types of processes that shape landforms, is partly dependent on temperature and precipitation. For these reasons, we will find that climates defined on the basis of temperature and precipitation also help set apart many features of the environment, not just climate alone. So, the material covered in this chapter will be useful in our study of vegetation in Chapter 8, soils in Chapter 9, and landforms especially in Chapters 13 and 15.

In Chapter 3, we examined the global pattern of air temperatures in detail (Figure 3.15). Recall that two major factors influence the annual cycle of air temperature experienced at a location.

- *Latitude.* The annual cycle of temperature depends on latitude. Near the equator, temperatures are warmer and the annual range is low. Toward the poles, temperatures are colder and the annual range is greater. These effects are produced by the changes in the annual cycle of insolation that occur with the seasons.
- *Coastal-continental location.* Coastal regions show a smaller annual variation in temperature, while the

variation is larger for continental regions. This effect occurs because ocean surface temperatures vary less with the seasons than temperatures of land surfaces.

Air temperature also has an important effect on precipitation. Recall this principle from Chapter 4:

- *Warm air can hold more moisture than cold air.* This means that colder regions generally have lower precipitation than warmer regions. Also, precipitation will tend to be greater during the warmer months of the temperature cycle.

Keeping these three key ideas in mind as you read this chapter will help make the climates easy to understand and explain.

GLOBAL PRECIPITATION

Figure 7.1 is a map of mean annual precipitation over the world. This map uses **isohyets**—lines drawn through all points having the same annual precipitation. For regions where all or most of the precipitation is rain, we use the word "rainfall." For regions where snow is a significant part of the annual total, we use the word "precipitation." Let's look at moist and dry regions in more detail.

Rainfall is generally high in a broad belt around the equator, including large regions of South America, Africa, and Southeast Asia. This is largely convectional precipitation, which occurs in the warm, moist, unstable air that converges in the intertropical zone. Further from the equator, easterly waves and tropical cyclones also generate precipitation. Rainfall is especially high in Southeast Asia. Recall that this is produced by the monsoon system, which brings warm, moist air from the Indian Ocean across the subcontinent during the rainy season.

Orographic precipitation occurs along mountainous coasts in the belt of the trade winds. In these regions, southeast or northeast trades bring warm, moist air ashore, where it rises and releases precipitation as it crosses the coastal mountains. The east coast of Central America, the northeast coast of South America, the Philippines, Malaysia, and the eastern coast of Indochina are examples.

Beyond the equatorial and tropical regions, the world map shows high precipitation in a few narrow bands in the midlatitude region. These are all located along the western edges of continents. The Pacific Northwest coast, including parts of British Columbia and the Alaska panhandle, provides the best example.

Orographic precipitation occurs on mountainous coasts exposed to moist trade winds or frequent cyclonic storms.

Bands of high coastal precipitation are also found in southern South America and in northwestern Europe. The bands occur because large numbers of cyclonic storms develop over oceans to the west and pass across these coasts, headed eastward. In North and South America, mountain belts cause the rainfall to be intensified by the orographic effect.

Let's now look at the dry regions of the globe. Dry regions occur along the central coast of western South America (Chile), along the western coast of southern Africa, and in the American Southwest. These locations correspond well with the eastern sides of subtropical high-pressure cells, where subsiding air is generally clear, dry, and stable (see Figure 5.13).

The largest dry region consists of a broad band ranging from the Sahara Desert of North Africa, through the Middle East, and well into Central Asia. The western part, including the Sahara and other Middle Eastern deserts, owes its dryness to descending air in the subtropical high-pressure belt. Farther to the east, in Central Asia, conditions are dry largely because the region is far from oceans. Moist oceanic air loses its water vapor in precipitation as it passes over land before reaching this region. This inland dryness effect also occurs in the western American high plains. Here moist Pacific air is dried out as it moves eastward over the mountain ranges and high country of the West, producing a dry interior region.

Figure 7.1 World precipitation. Isohyets are labeled with cm (in.).

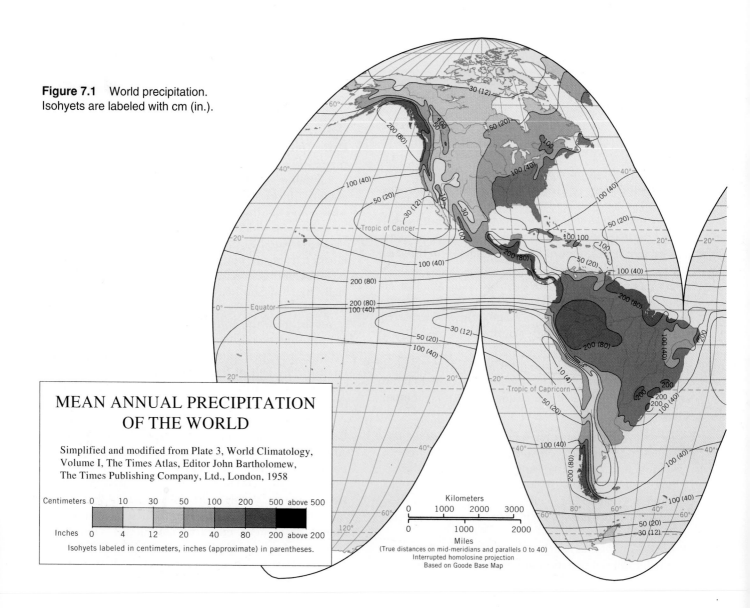

MEAN ANNUAL PRECIPITATION OF THE WORLD

Simplified and modified from Plate 3, World Climatology, Volume I, The Times Atlas, Editor John Bartholomew, The Times Publishing Company, Ltd., London, 1958

Centimeters 0 10 30 50 100 200 500 above 500

Inches 0 4 12 20 40 80 200 above 200

Isohyets labeled in centimeters, inches (approximate) in parentheses.

The high-latitude regions are also areas of low precipitation. Examples are the Arctic islands of Canada, northern Greenland, and eastern Siberia, which are cold and show low precipitation nearly year round.

CLIMATE CLASSIFICATION

Mean monthly air temperature and precipitation values can describe the climate of a weather station and its nearby region quite accurately. To study climates from a global viewpoint, climatologists classify these values into a set of climate types. This requires developing a set of rules to use in examining monthly temperature and precipitation values. By applying the rules, the climatologist can determine the climate to which a station belongs. This textbook recognizes 13 distinctive climate types that are designed to be understood and explained by air mass movements and frontal zones. They are shown on our global climate map, Figure 7.2. The types follow quite naturally from the global temperature and precipitation processes described in prior chapters.

Recall from Chapter 6 that air masses are classified according to the general latitude of their source regions and their surface type—land or ocean—within that region (see Figure 6.1). The latitude determines the temperature of the air mass, which can also depend on the season. The kind of surface, land or ocean, controls the moisture content. Marine air masses will be more moist and yield more precipitation, while continental air masses will be drier and yield less precipitation. Since the air mass characteristics control the two most important climate variables—

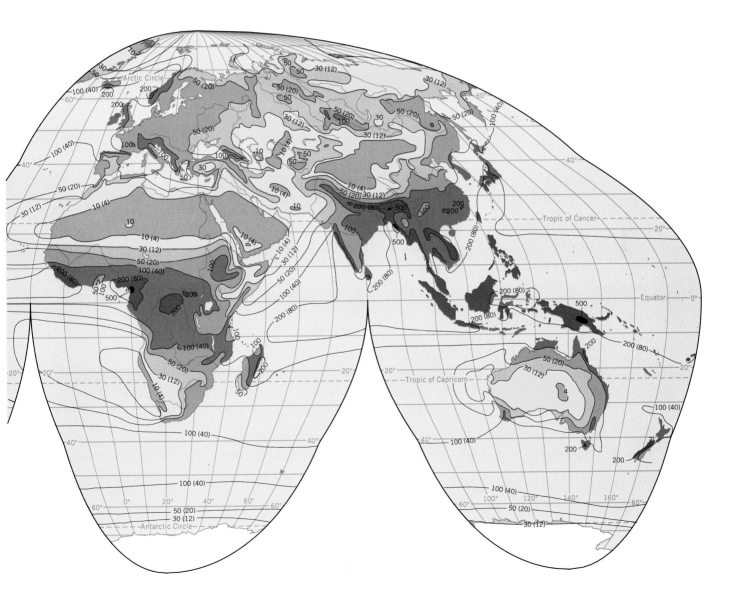

Figure 7.2 Climates of the world. Compiled from station data by A. N. Strahler.

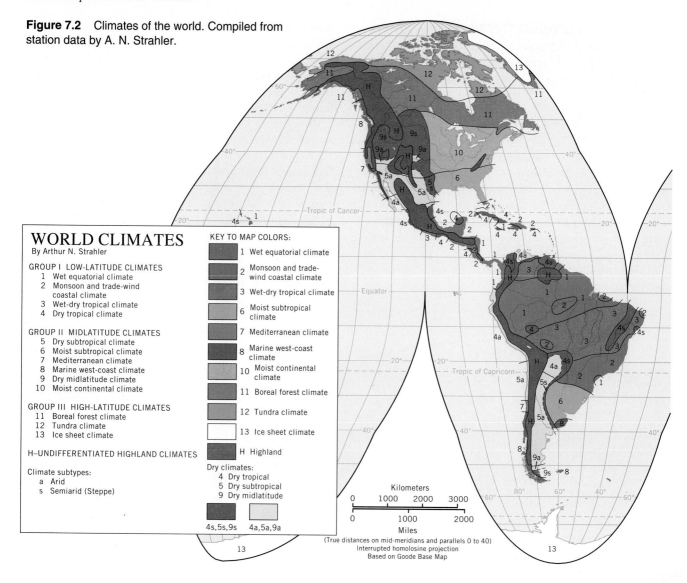

WORLD CLIMATES
By Arthur N. Strahler

GROUP I LOW-LATITUDE CLIMATES
1 Wet equatorial climate
2 Monsoon and trade-wind coastal climate
3 Wet-dry tropical climate
4 Dry tropical climate

GROUP II MIDLATITUDE CLIMATES
5 Dry subtropical climate
6 Moist subtropical climate
7 Mediterranean climate
8 Marine west-coast climate
9 Dry midlatitude climate
10 Moist continental climate

GROUP III HIGH-LATITUDE CLIMATES
11 Boreal forest climate
12 Tundra climate
13 Ice sheet climate

H—UNDIFFERENTIATED HIGHLAND CLIMATES

Climate subtypes:
a Arid
s Semiarid (Steppe)

KEY TO MAP COLORS:
1 Wet equatorial climate
2 Monsoon and trade-wind coastal climate
3 Wet-dry tropical climate
6 Moist subtropical climate
7 Mediterranean climate
8 Marine west-coast climate
10 Moist continental climate
11 Boreal forest climate
12 Tundra climate
13 Ice sheet climate
H Highland

Dry climates:
4 Dry tropical
5 Dry subtropical
9 Dry midlatitude

4s,5s,9s   4a,5a,9a

Kilometers
0   1000   2000   3000

Miles
0   1000   2000

(True distances on mid-meridians and parallels 0 to 40)
Interrupted homolosine projection
Based on Goode Base Map

temperature and precipitation—we can explain climates using air masses as a guide.

We also know from Chapter 6 that frontal zones are regions in which air masses are in contact. And, when unlike air masses are in contact, cyclonic precipitation is likely to develop. Since the position of frontal zones changes with the seasons, these seasonal movements of frontal zones influence the annual cycles of temperature and precipitation. For example, the polar-front zone lies generally across the middle latitudes of the United States in winter, but it retreats northward to Canada during the summer.

Figure 7.3 shows our schematic diagram of air mass source regions. We have subdivided this diagram into global bands that contain three broad groups of climates: low-latitude (Group I), midlatitude (Group II), and high-latitude (Group III).

The region of low-latitude climates (Group I) is dominated by the source regions of continental tropi-

cal, maritime tropical, and maritime equatorial air masses. These source regions are related to the three most obvious atmospheric features that occur within their latitude band—the two subtropical high-pressure belts and the equatorial trough at the intertropical convergence zone (ITC). Air of polar origin occasionally invades regions of low-latitude climates.

The region of midlatitude climates (Group II) lies in the *polar front zone*—a zone of intense interaction between unlike air masses. Here tropical air masses moving poleward and polar air masses moving equatorward are in conflict. Wave cyclones are normal features of the polar front, and this zone may contain as many as a dozen wave cyclones around the globe.

The region of high-latitude climates (Group III) is dominated by polar and arctic (including antarctic) air masses. In the arctic belt of the 60th to 70th parallels, continental polar air masses meet arctic air masses along an *arctic-front zone*, creating a series of eastward-

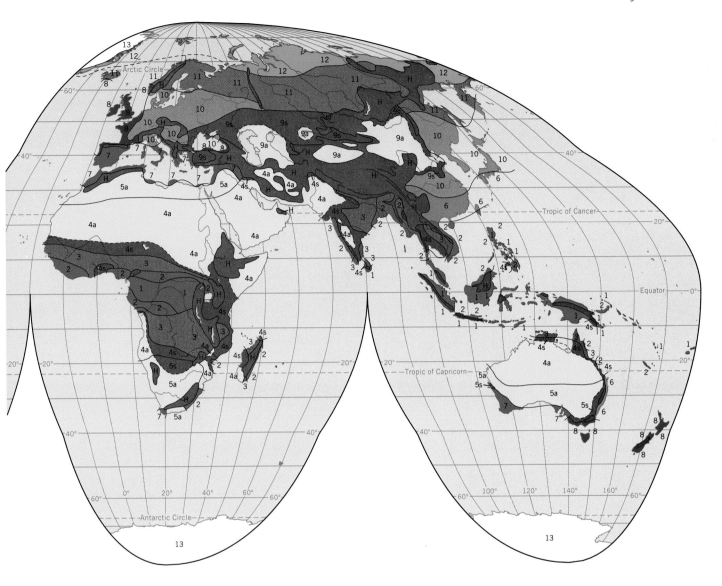

moving wave cyclones. In the southern hemisphere, there are no source regions in the subantarctic belt for continental polar air—just a great single oceanic source region for maritime polar (mP) air masses. The pole-centered continent of Antarctica provides a single great source of the extremely cold, dry antarctic airmass (cAA). These two airmasses interact along the *antarctic front zone*.

Within each latitudinal climate group are a number of climate types (or simply, climates)—four low-latitude climates (Group I), six midlatitude climates (Group II), and three high-latitude climates (Group III)—for a total of 13 climate types. In this textbook, the climates are numbered for ease in identification on maps and diagrams. We will refer to each climate by name, because the names describe the general nature of the climate and also suggest its global location. For convenience, we include the climate number next to its name in the text.

The 13 climate types can be understood and explained by air mass movements and frontal zones.

The world map of climates, Figure 7.2, shows the actual distribution of climate types on the continents. This map, based on data collected at a large number of observing stations, is simplified because the climate boundaries are uncertain in many areas where observing stations are thinly distributed.

In presenting the climates, we will make use of a pictorial device called a **climograph.** It shows the annual cycles of monthly mean air temperature and monthly mean precipitation for a location, along with some other useful information. Figure 7.4 is an example of a climograph. The data plotted are for Kayes, a station in the African country of Mali. At the top of the climograph, the mean monthly temperature is plotted as a

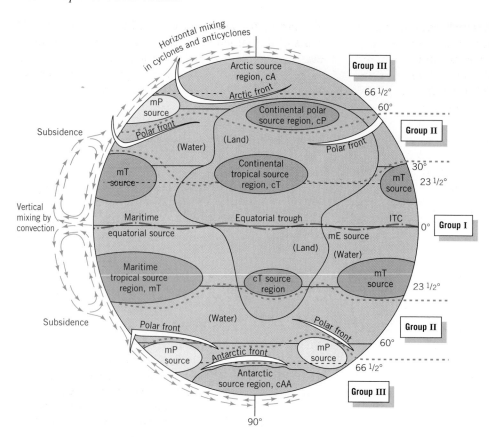

Figure 7.3 Using the map of air mass source regions, we can identify five global bands associated with three major climate groups. Within each group is a set of distinctive climates with unique characteristics. The characteristics are explained by the movements of air masses and frontal zones.

line graph. At the bottom, the mean monthly precipitation is shown as a bar graph. The annual range in temperature and the total annual precipitation are stated on every climograph as well. Most climographs also display dominant weather features, which are shown using picture symbols. For Kayes, the two dominant features are the subtropical high and equatorial trough (ITC). Many climographs also include a small graph of the sun's declination in order to help show when solstices and equinoxes occur.

LOW-LATITUDE CLIMATES

The low-latitude climates lie for the most part between the tropics of cancer and capricorn. In terms of world latitude zones, the low-latitude climates occupy all the equatorial zone (10° N to 10° S), most of the tropical zone (10–15° N and S), and part of the subtropical zone. In terms of prevailing pressure and wind systems,

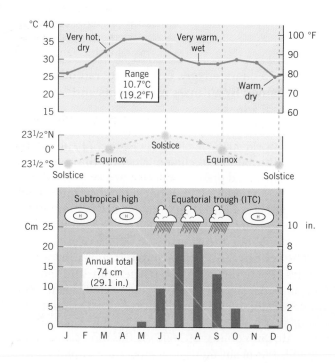

Figure 7.4 Climograph for Kayes, Mali, lat. 14° N. This station is located in western Africa, just south of the Sahara Desert. Here, the climate is nearly rainless for seven months of the year. During this period, dry subsiding air from subtropical high pressure cells dominates. Rainfall occurs as the equatorial trough moves northward, reaching the vicinity of Kayes in June or July. Temperatures are hottest in April and May, before the rainy season begins.

the region of low-latitude climates includes the equatorial trough of the intertropical convergence zone (ITC), the belt of tropical easterlies (northeast and southeast trades), and large portions of the oceanic subtropical high-pressure belt (Figure 7.2).

Figure 7.5 shows climographs for the four low-latitude climates. These climates range from extremely moist—the wet equatorial climate ①—to extremely dry—the dry tropical climate ④. They also vary strongly in the seasonality of their rainfall. In the wet equatorial climate ①, rainfall is abundant all year round. But in the wet-dry tropical climate ③, rainfall is abundant for only part of the year. During the remainder of the year, little or no rain falls. The seasonal temperature cycle also varies among these climates. In the wet equatorial climate ①, temperatures are nearly uniform throughout the year. In the dry tropical climate ④, there is a strong annual temperature cycle. Table 7.1 summarizes some of the characteristics of these climates.

The Wet Equatorial Climate ① (Köppen: *Af*)

The wet equatorial climate is a climate of the intertropical convergence zone (ITC), which is nearby for most of the year. The climate is dominated by warm, moist maritime equatorial (mE) and maritime tropical (mT)

Table 7.1 Low-latitude Climates

Climate	Temperature	Precipitation	Explanation
Wet equatorial ①	Uniform temperatures, mean near 27°C (80°F).	Abundant rainfall, all months, from mT and mE air masses. Annual total may exceed 250 cm (100 in.).	The ITC dominates this climate, with abundant convectional precipitation generated by convergence in weak equatorial lows. Rainfall is heaviest when the ITC is nearby.
Monsoon and trade-wind coastal ②	Temperatures show an annual cycle, with warmest temperatures in the high-sun season.	Abundant rainfall but with a strong seasonal pattern.	Trade-wind coastal: Rainfall from mE and mT air masses is heavy when the ITC is nearby, lighter when the ITC moves to the opposite hemisphere. Asian monsoon coasts: dry air flowing southwest in low-sun season alternates with moist oceanic air flowing northeast, producing seasonal rainfall pattern on west coasts.
Wet-dry tropical ③	Marked temperature cycle, with hottest temperatures before the rainy season.	Wet high-sun season alternates with dry low-sun season.	Subtropical high pressure moves into this climate in the low-sun season, bringing very dry conditions. In the high-sun season, the ITC approaches and rainfall occurs. Asiatic monsoon climate: alternation of dry continental air in low-sun season with moist oceanic air in high-sun season brings a strong pattern of dry and wet seasons.
Dry tropical ④	Strong temperature cycle, with intense hot temperatures during high-sun season.	Low precipitation. Sometimes rainfall occurs when the ITC is near.	This climate is dominated by subtropical high pressure, which provides clear, stable air for much or all of year. Insolation is intense during high-sun period.

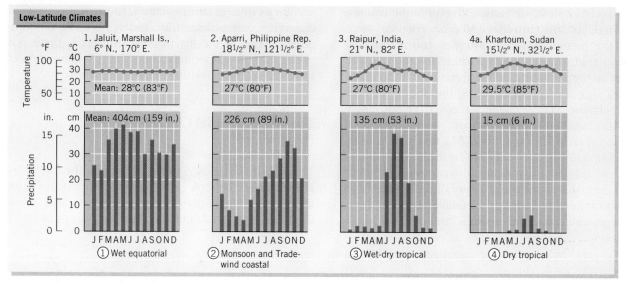

Figure 7.5 Climographs for the four low-latitude climates.

air masses that yield heavy convectional rainfall. Rainfall is plentiful in all months, and the annual total often exceeds 250 cm (100 in.). However, there is usually a seasonal pattern to the rainfall, so that rainfall is greater during some part of the year. This period of heavier rainfall occurs when the ITC migrates into the region. Remarkably uniform temperatures prevail throughout the year. Both mean monthly and mean annual temperatures are typically close to 27°C (80°F).

Figure 7.6 shows the world distribution of the wet equatorial climate. This climate is found in the latitude range 10° N to 10° S. Its major regions of occurrence include the Amazon lowland of South America, the Congo basin of equatorial Africa, and the East Indies, from Sumatra to New Guinea.

Figure 7.7 is a climograph for Iquitos, Peru (located in Figure 7.6), a typical wet equatorial station located close to the equator in the broad, low basin of the

Figure 7.6 World map of wet equatorial ① and monsoon and trade-wind coastal climate ②. (Based on Goode Base Map.)

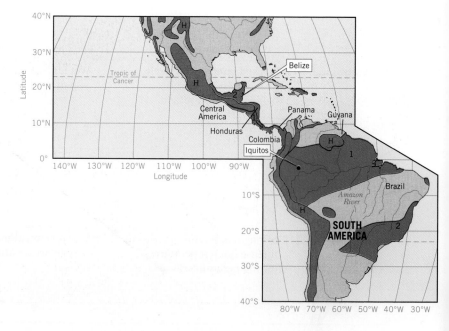

upper Amazon River. Notice the very small annual range in temperature and the very large annual rainfall total. The monthly air temperatures are extremely uniform in the wet equatorial climate. Typically, mean monthly air temperature will range between 26° and 29°C (79° and 84°F) for stations at low elevation in the equatorial zone.

The Monsoon and Trade-Wind Coastal Climate ② (Köppen: *Af, Am*)

Like the wet equatorial climate ①, the monsoon and trade-wind coastal climate ② has abundant rainfall. But unlike the wet equatorial climate, the rainfall of the monsoon and trade-wind coastal climate always shows a strong seasonal pattern. This seasonal pattern is due to the migration of the intertropical convergence zone. In the high-sun season ("summer," depending on the hemisphere), the ITC is near and monthly rainfall is greater. In the low-sun season, when the ITC has migrated to the other hemisphere, subtropical high pressure dominates and monthly rainfall is less.

Figure 7.6 shows the global distribution of the monsoon and trade-wind coastal climate ②. The climate occurs over latitudes from 5° to 25° N and S.

As the name of the monsoon and trade-wind coastal climate ② suggests, two somewhat different situations produce it. On trade-wind coasts, rainfall is produced by moisture-laden maritime tropical (mT) and maritime equatorial (mE) air masses. These are moved onshore onto narrow coastal zones by trade winds or by monsoon circulation patterns. As the warm, moist air passes over coastal hills and mountains, the orographic effect touches off convectional shower activity.

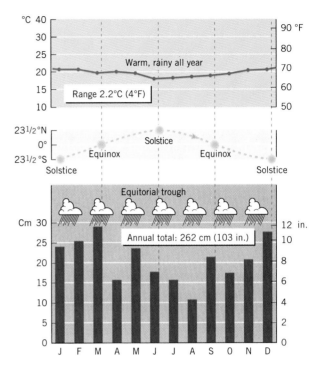

Figure 7.7   Wet equatorial climate ①. Iquitos, Peru, lat. 3°S, is located in the upper Amazon lowland, close to the equator. Temperatures differ little from month to month, and there is abundant rainfall throughout the year.

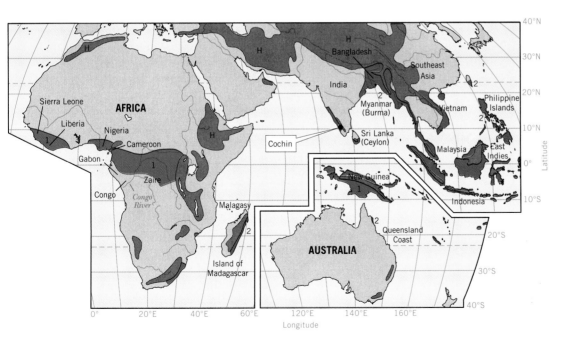

Shower activity is also intensified by easterly waves, which are more frequent when the ITC is nearby. The east coasts of land masses experience this trade-wind effect, because the trade winds blow from east to west. Trade-wind coasts are found along the east sides of Central and South America, the Caribbean Islands, Madagascar (Malagasy), Indochina, the Philippines, and northeast Australia (see Figure 7.6).

The coastal precipitation effect also applies to the summer monsoon of Asia, when the monsoon circulation brings mT air onshore. However, the onshore monsoon winds blow from southwest to northeast, so it is the western coasts of land masses that are exposed to this moist airflow. Western India and Myanmar (formerly Burma) are examples. Moist air also penetrates well inland in Bangladesh, providing the very heavy monsoon rains for which the region is well known.

In central and western Africa, and southern Brazil, the monsoon pattern shifts the intertropical convergence zone over 20° of latitude, or more (see Figure 5.13). Here, heavy rainfall occurs in the high-sun season, when the ITC is nearby. Drier conditions prevail in the low-sun season, when the ITC is far away.

Temperatures in the monsoon and trade-wind coastal climate ②, though warm throughout the year,

also show an annual cycle. Warmest temperatures occur in the high-sun season, just before arrival of the ITC brings clouds and rain. Minimum temperatures occur at the time of low sun.

Figure 7.8 is a climograph for the city of Belize in Belize (Figure 7.6). This Central American east coast city at lat. 17° N is exposed to the tropical easterly trade winds. Rainfall is abundant from June through November, when the ITC is in this latitude zone. Easterly waves are common in this season, and an occasional tropical cyclone strikes the coast, bringing torrential rainfall. Following the December solstice, rainfall is greatly reduced, with minimum values in March and April. At this time, the ITC lies farthest away and the climate is dominated by the subtropical high pressure. Air temperatures show an annual range of 5° C (9° F) with maximum in the high-sun months.

The Asiatic monsoon shows a similar pattern, but there is an extreme peak of rainfall during the high-sun period and a well-developed dry season with two or three months of only small rainfall amounts. The climograph for Cochin, India, provides an example (Figure 7.9). Located at lat. 10° N on the west coast of lower peninsular India (see Figure 7.6), Cochin receives the warm, moist southwest winds of the summer monsoon. In this season, monthly rainfall is extreme in both June and July. A strongly pronounced season of low rainfall occurs at time of low sun— December through March. Air temperatures show only a very weak annual cycle, cooling a bit during the rains, but the annual range is small at this low latitude.

Figure 7.8 Trade-wind coastal climate ②. This climograph for Belize, a Central American east coast city at lat. 17° N, shows a marked season of low rainfall following the period of low sun. For the remainder of the year, precipitation is high, produced by warm, moist northeast trade winds.

The Low-Latitude Rainforest Environment

Our first two climates—wet equatorial ① and monsoon and trade-wind coastal ②—are very uniform in temperature and have very high annual rainfall. These factors create a special environment—the *low-latitude rainforest environment* (Figure 7.10). Streams flow abundantly throughout most of the year, and river channels are lined with dense forest vegetation. These form the transportation arteries of the rainforest, shared by natives and traders alike.

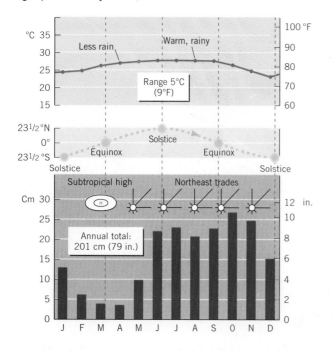

> **Taken together, the wet equatorial ① and monsoon and trade-wind coastal ② climates comprise the low-latitude rainforest environment.**

Low-latitude rainforests, unlike higher latitude forests, possess a great diversity of plant and animal species. The low-latitude rainforest features as many as

Special Supplement • The Köppen Climate System

Air temperature and precipitation data have formed the basis for several climate classifications. One of the most important of these is the Köppen climate system, devised in 1918 by Dr. Vladimir Köppen of the University of Graz in Austria. For several decades, this system, with various later revisions, was the most widely used climate classification among geographers. Köppen was both a climatologist and plant geographer, so that his main interest lay in finding climate boundaries that coincided approximately with boundaries between major vegetation types.

Under the Köppen system, each climate is defined according to assigned values of temperature and precipitation, computed in terms of annual or monthly values. Any given station can be assigned to its particular climate group and subgroup solely on the basis of the records of temperature and precipitation at that place.

The Köppen system features a shorthand code of letters designating major climate groups, sub-groups within the major groups, and further subdivisions to distinguish particular seasonal characteristics of temperature and precipitation. Five major climate groups are designated by capital letters as follows:

A *Tropical rainy climates*

Average temperature of every month is above 18°C (64.4°F). These climates have no winter season. Annual rainfall is large and exceeds annual evaporation.

B *Dry climates*

Evaporation exceeds precipitation on the average throughout the year. There is no water surplus; hence, no permanent streams originate in B climate zones.

C *Mild, humid (mesothermal) climates*

The coldest month has an average temperature of under 18°C (64.4°F), but above –3°C (26.6°F); at least one month has an average temperature above 10°C (50°F). The *C* climates thus have both a summer and a winter.

D *Snowy-forest (microthermal) climates*

The coldest month has an average temperature of under –3°C (26.6°F). The average temperature of the warmest month is above 10°C (50°F). (Forest is not generally found where the warmest month is colder than 10°C (50°F).)

E *Polar climates*

The average temperature of the warmest month is below 10°C (50°F). These climates have no true summer.

Note that four of these five groups (*A, C, D,* and *E*) are defined by temperature averages, whereas

Figure S7.1 Highly generalized world map of major climate regions according to the Köppen classification. Highland areas are in black. (Based on Goode Base Map.)

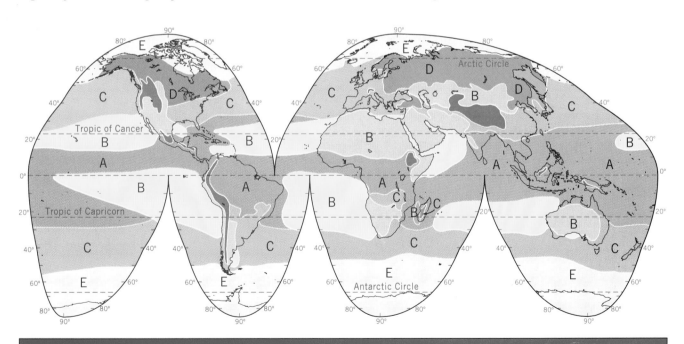

one (*B*) is defined by precipitation-to-evaporation ratios. Groups *A, C,* and *D* have sufficient heat and precipitation for the growth of forest and woodland vegetation. Figure S7.1 shows the boundaries of the five major climate groups, and Figure S7.2 is a world map of Köppen climates.

Subgroups within the five major groups are designated by a second letter according to the following code.

S *Semiarid (steppe)*

W *Arid (desert)*

(The capital letters *S* and *W* are applied only to the dry *B* climates.)

f *Moist, adequate precipitation in all months, no dry season. This modifier is applied to **A, C,** and **D** groups.*

w *Dry season in the winter of the respective hemisphere (low-sun season).*

s *Dry season in the summer of the respective hemisphere (high-sun season).*

m *Rainforest climate, despite short, dry season in monsoon type of pre-cipitation cycle. Applies only to **A** climates.*

From combinations of the two letter groups, 12 distinct climates emerge:

Af *Tropical rainforest climate*

The rainfall of the driest month is 6 cm (2.4 in.) or more.

Am *Monsoon variety of Af*

The rainfall of the driest month is less than 6 cm (2.4 in.). The dry season is strongly developed.

Figure S7.2 World map of climates according to the Köppen-Geiger-Pohl system.

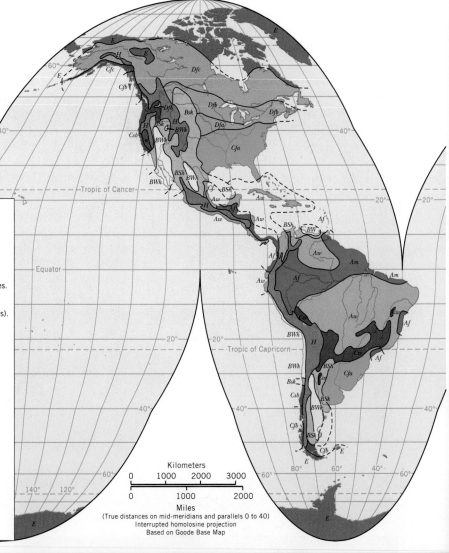

KÖPPEN-GEIGER SYSTEM OF CLIMATE CLASSIFICATION

After R. Geiger and W. Pohl (1953)

Key to letter code designating climate regions:

FIRST LETTER

A C D Sufficient heat and precipitation for growth of high-trunked trees.
A Tropical climates. All monthly mean temperatures over 64.4°F (18°C).
B Dry climates. Boundaries determined by formula using mean annual temperature and mean annual precipitation (see graphs).
C Warm temperature climates. Mean temperature of coldest month: 64.4°F (18°C) down to 26.6°F (-3°C).
D Snow climates. Warmest month mean over 50°F (10°C) Coldest month mean under 26.6°F (-3°C).
E Ice climates. Warmest month mean under 50°F (10°C).

SECOND LETTER

S Steppe climate. ⎫ Boundaries determined by formulas (See graphs).
W Desert climate. ⎭
f Sufficient precipitation in all months.
m Rainforest despite a dry season (i.e., monsoon cycle).
s Dry season in summer of the respective hemisphere.
w Dry season in winter of the respective hemisphere.

THIRD LETTER

a Warmest month mean over 71.6°F (22°C).
b Warmest month mean under 71.6°F (22°C). At least 4 months have means over 50°F (10°C).
c Fewer than 4 months with means over 50°F (10°C).
d Same as c, but coldest month mean under -36.4°F (-38°C).
h Dry and hot. Mean annual temperature over 64.4°F (18°C).
k Dry and cold. Mean annual temperature under 64.4°F (18°C).

H Highland climates.

Kilometers
0 1000 2000 3000

0 1000 2000
Miles

(True distances on mid-meridians and parallels 0 to 40)
Interrupted homolosine projection
Based on Goode Base Map

Aw *Tropical savanna climate*

At least one month has rainfall less than 6 cm (2.4 in.). The dry season is strongly developed.

Figure S7.3 shows the boundaries between *Af, Am,* and *Aw* climates as determined by both annual rainfall and rainfall of the driest month.

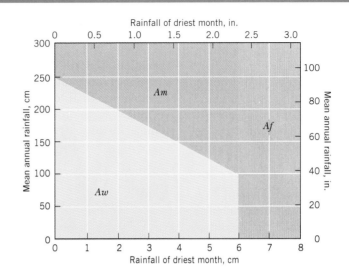

Figure S7.3   Boundaries of the A climates.

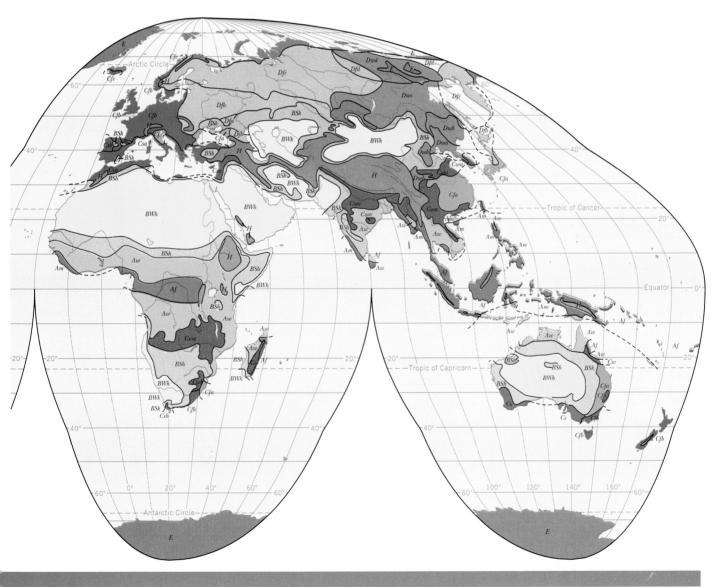

BS *Steppe climate*

A semiarid climate characterized by grasslands, it occupies an intermediate position between the desert climate (*BW*) and the more humid climates of the *A, C,* and *D* groups. Boundaries are determined by formulas given in Figure S7.4.

BW *Desert climate*

Desert has an arid climate with annual precipitation of usually less than 40 cm (15 in.). The boundary with the adjacent steppe climate (*BS*) is determined by formulas given in Figure S7.4.

Cf *Mild humid climate with no dry season*

Precipitation of the driest month averages more than 3 cm (1.2 in.).

Cw *Mild humid climate with a dry winter*

The wettest month of summer has at least 10 times the precipitation of the driest month of winter. (Alternative definition: 70 percent or more of the mean annual precipitation falls in the warmer six months.)

Cs *Mild humid climate with a dry summer*

Precipitation of the driest month of summer is less than 3 cm (1.2 in.). Precipitation is at least three times as much as the driest month of summer. (Alternative definition: 70 percent or more of the mean annual precipitation falls in the six months of winter.)

Df *Snowy-forest climate with a moist winter*

No dry season.

Dw *Snowy-forest climate with a dry winter*

ET *Tundra climate*

The mean temperature of the warmest month is above 0°C (32°F) but below 10°C (50°F).

EF *Perpetual frost climate*

In this ice sheet climate, the mean monthly temperatures of all months are below 0°C (32°F).

To denote further variations in climate, Köppen added a third letter to the code group. The meanings are as follows:

a With hot summer; warmest month is over 22°C (71.6°F); *C* and *D* climates.

b With warm summer; warmest month is below 22°C (71.6°F); *C* and *D* climates.

c With cool, short summer; less than four months are over 10°C (50°F); *C* and *D* climates.

d With very cold winter; coldest month is below −38°C (−36.4°F); *D* climates only.

h Dry-hot; mean annual temperature is over 18°C (64.4°F); *B* climates only.

k Dry-cold; mean annual temperature is under 18°C (64.4°F); *B* climates only.

As an example of a complete Köppen climate code, *BWk* refers to a cool desert climate, and *Dfc* refers to a cold, snowy-forest climate with cool, short summer.

Figure S7.4 Boundaries of the B climates. Upper figures: metric system. Lower figures: English system.

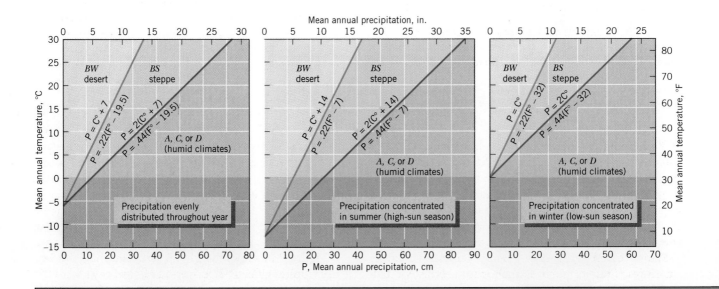

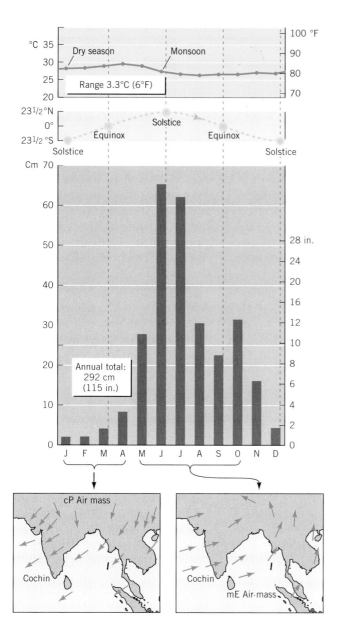

Figure 7.9 Monsoon coastal climate ②. Cochin, India, on a windward coast at lat. 10° N, shows an extreme peak of rainfall during the rainy monsoon, contrasting with a short dry season at time of low sun.

Figure 7.10 Rainforest of the western Amazon lowland, near Manaus, Brazil. The river is a tributary of the Amazon. Note the many different types of trees of varying shapes.

sawood, is an important export. Quinine, cocaine, and other drugs come from the bark and leaves of tropical plants. Cocoa is derived from the seed kernel of the cacao plant. Natural rubber is made from the sap of the rubber tree, which originated in South America.

The Wet-Dry Tropical Climate ③ (Köppen: *Aw, Cwa*)

In the monsoon and trade-wind coastal climate ②, we noted that the movements of the intertropical convergence zone into and away from the climate region produce a seasonal cycle of rainfall and temperature. As we move farther poleward, this cycle becomes stronger, and the monsoon and trade-wind coastal climate ② grades into the tropical wet-dry climate ③.

The tropical wet-dry climate ③ is distinguished by a very dry season low sun that alternates with a very wet season at high sun. During the low-sun season, when the equatorial trough is far away, dry continental tropical (cT) air masses prevail. In the high-sun season, when the ITC is nearby, moist maritime tropical (mT) and maritime equatorial (mE) air masses dominate. Cooler temperatures accompany the dry season, but give way to a very hot period before the rains begin.

Figure 7.11 shows the global distribution of the wet-dry tropical climate. It is found at latitudes of 5° to 20°

3000 different tree species in an area of only a few square kilometers, whereas the midlatitude forest possesses fewer than one-tenth that number. The number of types of animals found in the rainforest is also very large. A 16 km² (6 mi²) area in Panama near the Canal, for example, contains about 20,000 species of insects, whereas all of France has only a few hundred.

Many forest products have economic value. Rainforest lumber, such as mahogany, ebony, or bal-

Eye on the Environment • Drought and Land Degradation in the African Sahel

The wet-dry tropical climate ③ is subject to years of devastating drought as well as to years of abnormally high rainfall that can result in severe floods. Climate records show that two or three successive years of abnormally low rainfall (a drought) typically alternate with several successive years of average or higher than average rainfall. This variability is a permanent feature of the wet-dry tropical ③ climate, and the plants and animals inhabiting this region have adjusted to the natural variability in rainfall, with one exception: the human species.

The wet-dry tropical climate ③ of North Africa, including the adjacent semiarid southern belt of the dry tropical climate ④ to the north, provides a lesson on human impact on a delicate ecological system.

Countries of this perilous belt, called the *Sahel*, or *Sahelian zone*, are shown in the figure below. From 1968 through 1974, all these countries were struck by a severe drought. Both nomadic cattle herders and grain farmers share this zone. During the drought, grain crops failed and foraging cattle could find no food to eat. In the worst stages of the Sahel drought, nomads were forced to sell their remaining cattle. Because the cattle were their sole means of subsistence, the nomads soon starved. Some 5 million cattle perished, and it has been estimated that 100,000 people died of starvation and disease in 1973 alone.

The Sahelian drought of 1968–1974 was associated with a special phenomenon, which at that time was called *desertification*—the permanent transformation of the land surface by human activities to resemble a desert, largely through the destruction of grasses, shrubs, and trees. That term has now been largely abandoned in favor of *land degradation*. Also visible are the effects of accelerated soil erosion, such as gullying of slopes and accumulations of sediment in stream channels. The removal of soil by wind is also intensified.

Land degradation in the African Sahel is the direct result of greatly increased numbers of humans and their cattle. European managers of African colonies contributed to population increases by supplying food in time of famine, by reducing the death rate from disease, and by developing ground-water supplies for crop irrigation and livestock. With each succeeding drought, the increased population made a heavier impact on the vegetation and soil. Ultimately, land degradation became too sustained and too severe to permit recovery of the plant cover during the rainy season or over a period of wetter years. As long as the population in the area remains high, the land degradation is permanent.

Periodic droughts throughout past decades are well documented in the Sahel, as they are in other world regions of wet-dry tropical climate. To the right is a graph showing the percentage of departures from the long-term mean of each year's rainfall in the western Sahel from 1901 through 1990. Since about 1950, the durations of peri-

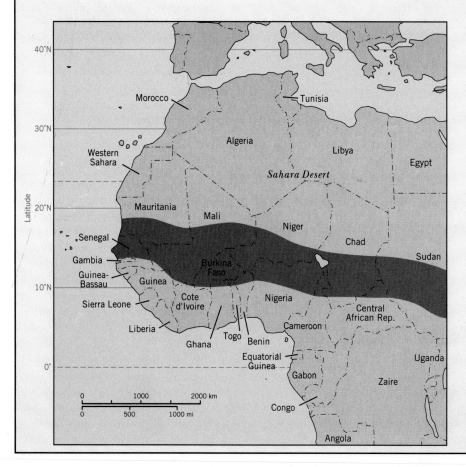

The Sahel, or Sahelian zone, shown in color, lies south of the great Sahara Desert of North Africa.

At the height of the Sahelian drought, vast numbers of cattle had perished and even the goats were hard pressed to survive. Trampling of the dry ground prepared the region for devastating soil erosion in the rains that eventually ended the drought.

ods of continuous departures both above and below the mean seem to have increased substantially. The period of sustained high-rainfall years in the 1950s contrasts sharply with the severe drought years of 1971–1973 and 1976–1977. Although rainfall improved somewhat through the rest of the decade, severe drought returned in 1982 and continued through 1984. During this period, the effects were

particularly severe in Ethiopia. The period of continuous deficiencies spanned 20 years and was broken in 1988 by near-average rainfall. The Sudan experienced exceptionally heavy rains in 1988 that caused disastrous flooding of the Nile River. A drought year followed in 1990.

Going much further back in the record, we can point to earlier periods of rainfall deficiency and excess—1820–1840, below normal; 1870–1895, above normal; 1895–1920, below normal. Evidence of these periods is preserved in the record of the changing shorelines of Lake Chad, which lies within the Sahelian zone. The climatic record shows that long drought periods occurred in the past, well before human populations had reached today's high levels.

The effect of an accelerated global warming in the coming decades on the climate of the Sahelian zone is difficult to predict. The Sahelian zone may experience greater swings between drought and surplus precipitation. At the same time, the location of the Sahel, along with the desert zone, may shift northward, as a result of intensification of the Hadley cell circulation. However, these changes are highly speculative. At this time, there is no scientific consensus on how climatic change will affect the Sahel.

Questions

1. What is meant by the term *land degradation?* Provide an example of a region in which land degradation has occurred.
2. Examine the Sahel rainfall graph carefully. Compare the pattern of rainfall fluctuations for the periods 1901–1950 and 1951–present. How do they differ?

Rainfall fluctuations for stations in the western Sahel, 1901–1991, expressed as a percent departure from the long-term mean. (Courtesy of Sharon E. Nicholson, Department of Meteorology, Florida State University, Tallahassee.)

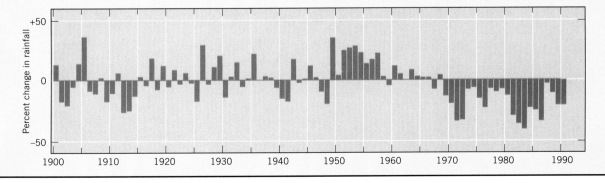

N and S in Africa and the Americas, and at 10° to 30° N in Asia. In Africa and South America, the climate occupies broad bands poleward of the wet equatorial and monsoon and trade-wind coastal climates. Because these regions are further from the ITC, less rainfall is triggered by the ITC during the rainy season, and subtropical high pressure can dominate more strongly during the low-sun season. In central India and Indochina, the regions of tropical wet-dry climate are somewhat protected by mountain barriers from the warm, moist mE and mT airflows provided by trade and monsoon winds. These barriers create a rain-shadow effect, so that less rainfall occurs during the rainy season and the dry season is still drier.

Figure 7.12 is a climograph for Timbo, Guinea, at lat. 10° N in West Africa (Figure 7.11). Here the rainy season begins just after the March equinox and reaches a peak in August, about two months following June solstice. At this time, the ITC has migrated to its most northerly position, and moist mE air masses flow into the region from the ocean lying to the south. Monthly rainfall then decreases as the low-sun season arrives and the ITC moves to the south. Three months—December through February—are practically rainless. During this season, subtropical high pressure dominates the climate, and stable, subsiding continental tropical (cT) air pervades the region. The temperature cycle is closely linked to both the solar cycle and the precipitation pattern. In February and March, insolation increases, and air temperature rises sharply. A brief hot season occurs. As soon as the rains set in, the effect of cloud cover and evaporation of rain causes

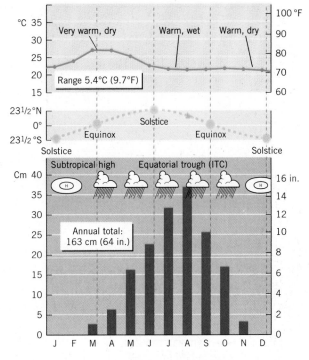

Figure 7.12   Wet-dry tropical climate ③. Timbo, Guinea, at lat. 10° N, is in West Africa. A long wet season at time of high sun alternates with an almost rainless dry season at time of low sun.

the temperature to decline. By July, temperatures have resumed an even level.

The native vegetation of the wet-dry tropical climate ③ must survive alternating seasons of very dry and very

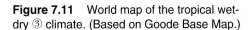

Figure 7.11   World map of the tropical wet-dry ③ climate. (Based on Goode Base Map.)

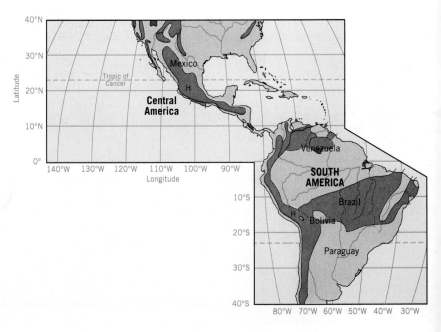

Figure 7.13 Savanna woodland of the Serengeti Plains, Tanzania, East Africa. Acacia trees with flattened crowns remain green, although the coarse grasses have turned to straw in the dry season.

wet weather. Most plants enter a dormant phase during the dry period, then burst forth into leaf and bloom with the coming of the rains. For this reason, the native plant cover can be described as *rain-green vegetation*. A dominant type of rain-green vegetation is the *savanna*—a grassy plain with scattered trees (Figure 7.13).

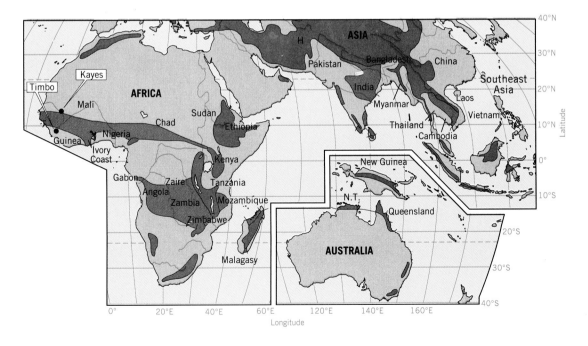

The Dry Tropical Climate ④
(Köppen: *BWh, BSh*)

The dry tropical climate ④ is found in the center and east sides of subtropical high-pressure cells. Here, air strongly subsides, warming adiabatically and inhibiting condensation. Rainfall is very rare and occurs only when unusual weather conditions move moist air into the region. Since skies are almost always clear, the sun heats the surface intensely, keeping air temperatures high. During the high-sun period, heat is extreme. During the low-sun period, temperatures are cooler. Given the dry air and lack of cloud cover, the daily temperature range is very large.

The driest areas of the dry tropical climate ④ are near the tropics of cancer and capricorn. As we travel from the tropics toward the equator, we find that somewhat more rain falls. Continuing in this direction, we encounter a short rainy season at the time of year when the ITC is near, and the climate grades into the wet-dry tropical ③ type.

Figure 7.14 shows the global distribution of the dry tropical climate ④. Nearly all of the dry tropical climate ④ areas lie in the latitude range 15° to 25° N and S. The largest region is the Sahara-Saudi Arabia-Iran-Thar desert belt of North Africa and southern Asia. This vast desert expanse includes some of the driest regions on earth. Another large region of dry tropical climate ④ is the desert of central Australia. The west coast of South America, including portions of Ecuador, Peru, and Chile, also exhibits the dry tropical climate ④. However, temperatures are moderated by the cool marine air layer that blankets the coast.

Figure 7.15 is a climograph for a dry tropical station in the heart of the North African desert. Wadi Halfa, Sudan (Figure 7.14), lies at lat. 22° N, almost on the tropic of cancer. The temperature record shows a strong annual cycle with a very hot period at the time of high sun, when three consecutive months average 32°C (90°F). Daytime maximum air temperatures are frequently between 43° and 48°C (110° to 120°F) in the warmer months. There is a comparatively cool season at the time of low sun, but the coolest month averages a mild 16°C (60°F) and freezing temperatures are rarely recorded. No rainfall bars are shown on the climograph because precipitation averages less than 0.25 cm (0.1 in.) in all months. Over a 39-year period, the maximum rainfall recorded in a 24-hour period at Wadi Halfa was only 0.75 cm (0.3 in.).

The world's dry climates consist in large part of extremely dry areas—the true *arid deserts*. In addition, there are broad zones at the margins of these very dry areas that are best described as *semiarid*. These zones have a short wet season that supports the growth of grasses on which animals (both wild and domestic) graze. Geographers also call these semiarid regions

Figure 7.14 World map of the dry tropical ④, dry subtropical ⑤, and dry midlatitude ⑨ climates. (Based on Goode Base Map.)

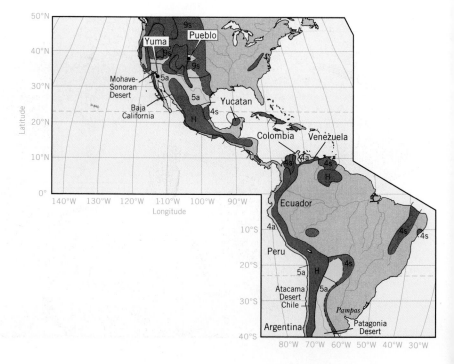

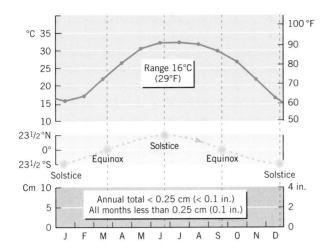

Figure 7.15   Dry tropical climate ④, dry desert. Wadi Halfa is a city on the Nile River in the Sudan at lat. 22° N, close to the Egyptian border. Too little rain falls to be shown on the graph. Air temperatures are very high during the high-sun months.

steppes. Nomadic tribes and their herds of animals visit these areas during and after the brief moist period. In Figures 7.2 (world climate map) and 7.14 (world map of dry climates), the two subdivisions of dry climates are distinguished with the letters *a* (arid) and *s* (semiarid).

An example of the semiarid dry tropical climate ④ is that of Kayes, Mali, which we presented earlier as a sample climograph (Figure 7.4). Located in the Sahel region of Africa, which is equatorward of the Sahara, this station has a distinct rainy season that occurs when the intertropical convergence moves north in the high-sun season. This precipitation pattern shows the semiarid subtype as a transition between the arid dry tropical climate ④ and the wet-dry tropical climate ③.

Much of the arid desert landscape consists of barren areas of drifting sand or sterile salt flats. However, in semiarid regions, thorny trees and shrubs are often abundant, since the climate includes a small amount of regular rainfall. Figure 7.16 presents a scene from the semiarid part of the Kalahari Desert in southwest Africa (Figure 7.14) that shows this type of vegetation in the dry season.

MIDLATITUDE CLIMATES

The midlatitude climates almost fully occupy the land areas of the midlatitude zone and a large proportion of the subtropical latitude zone. Along the western fringe of Europe, they extend into the subarctic latitude zone as well, reaching to the 60th parallel. Unlike the low-latitude climates, which are about equally distributed between northern and southern hemispheres,

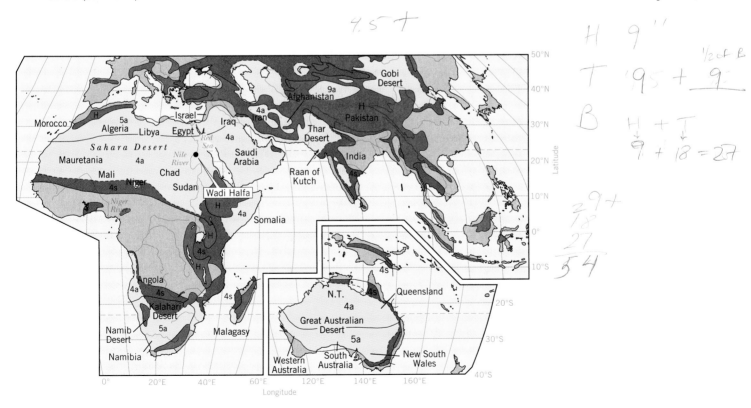

Figure 7.16 This strange-looking baobab tree is a common inhabitant of the thorntree semidesert of Botswana, in the Kalahari region of southern Africa.

nearly all of the midlatitude climate area is in the northern hemisphere. In the southern hemisphere, the land area poleward of the 40th parallel is so small that the climates are dominated by a great southern ocean and do not develop the continental characteristics of their counterparts in the northern hemisphere.

The midlatitude climates of the northern hemisphere lie in a broad zone of intense interaction between two groups of very unlike air masses (Figure 7.3). From the subtropical zone, tongues of maritime tropical (mT) air masses enter the midlatitude zone. There, they meet and conflict with tongues of maritime polar (mP) and continental polar (cP) air masses along the discontinuous and shifting polar-front zone.

In terms of prevailing pressure and wind systems, the midlatitude climates include the poleward halves of the great subtropical high-pressure systems and much of the belt of prevailing westerly winds. As a result, weather systems, such as traveling cyclones and their fronts, characteristically move from west to east. This dominant global eastward airflow influences the distribution of climates from west to east across the North American and Eurasian continents.

The six midlatitude climate types range from two that are very dry to three that are extremely moist (see Figure 7.17). The midlatitude climates span the range from those with strong wet and dry seasons—the Mediterranean climate ⑦—to those with precipitation that is more or less uniformly distributed through the year. Temperature cycles are also quite varied. Low annual ranges are seen along the windward west coasts. In contrast, annual ranges are large in the continental interiors. Table 7.2 summarizes the important features of these climates.

The midlatitude climates lie in a zone of frontal interaction between tropical air masses, derived from subtropical high-pressure cells, and polar air masses, derived from source regions at high latitudes.

Figure 7.17 Climographs for the six midlatitude climates.

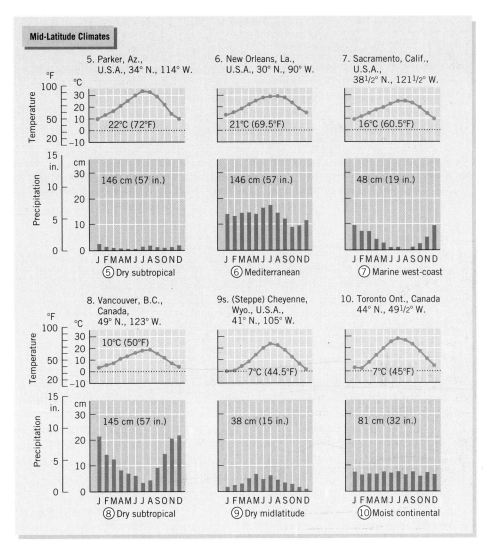

Table 7.2 Midlatitude Climates

Climate	Temperature	Precipitation	Explanation
Dry subtropical ⑤	Distinct cool or cold season at low-sun period.	Precipitation is low in nearly all months.	This climate lies poleward of the subtropical high-pressure cells and is dominated by dry cT air most of the year. Rainfall occurs when moist mT air reaches the region, either in summer monsoon flows or in winter frontal movements.
Moist subtropical ⑥	Temperatures show strong annual cycle, but with no winter month below freezing.	Abundant rainfall, cyclonic in winter and convectional in summer. Humidity generally high.	The flow of mT air from the west sides of subtropical high-pressure cells provides moist air most of the year. cP air may reach this region during winter.
Mediterranean ⑦	Temperature range is moderate, with warm to hot summers and mild winters.	Unusual pattern of wet winter and dry summer. Overall, drier when nearer to subtropical high pressure.	The poleward migration of subtropical high-pressure cells moves clear, stable air cT into this region in the summer, In winter, cyclonic storms and polar frontal precipitation reach the area.

(continued on following page)

Table 7.2 Midlatitude Climates*(continued)*

Climate	Temperature	Precipitation	Explanation
Marine west-coast ⑧	Temperature cycle is moderated by marine influence.	Abundant precipitation but with a winter maximum.	Moist mP air, moving inland from the ocean to the west, dominates this climate most of the year. In the summer, subtropical high pressure reaches these regions, reducing precipitation.
Dry midlatitude ⑨	Strong temperature cycle with large annual range. Summers warm to hot, winters cold to very cold.	Precipitation is low in all months but usually shows a summer maximum.	This climate is dry because of its interior location, far from mP source regions. In winter, cP dominates. In summer, a local dry continental air mass develops.
Moist continental ⑩	Summers warm, winters cold with three months below freezing. Very large annual temperature range.	Ample precipitation, with a summer maximum.	This climate lies in the polar-front zone. In winter, cP air dominates, while mT invades frequently in summer. Precipitation is abundant, cyclonic in winter, and convectional in summer.

The Dry Subtropical Climate ⑤
(Köppen: *BWh, BWk, BSh, BSk*)

The dry subtropical climate ⑤ is simply a poleward extension of the dry tropical climate ④, caused by somewhat similar air mass patterns. A point of difference is in the annual temperature range, which is

Figure 7.18 Dry subtropical climate ⑤. Yuma, Arizona, lat. 33° N, has a strong seasonal temperature cycle. Compare with Wadi Halfa (Figure 7.15).

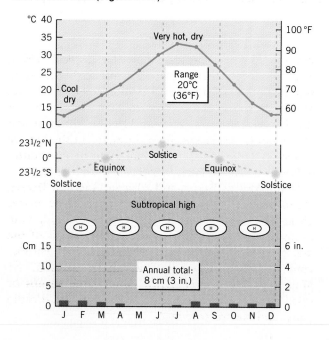

greater for the dry subtropical climate ⑤. The lower latitude portions have a distinct cool season, and the higher latitude portions, a cold season. The cold season, which occurs at a time of low sun, is due in part to invasions of cold continental polar (cP) air masses from higher latitudes. Precipitation occurring in the low-sun season is produced by midlatitude cyclones that occasionally move into the subtropical zone. As in the dry tropical climate ④, both arid and semiarid subtypes are recognized.

Figure 7.14 shows the global distribution of the dry subtropical climate ⑤. A broad band of this climate type is found in North Africa, connecting with the Near East. Southern Africa and southern Australia also contain regions of dry subtropical climate ⑤ poleward of the dry tropical climate ④. In South America, a band of dry subtropical climate ⑤ occupies Patagonia, a region east of the Andes in Argentina. In North America, the Mojave and Sonoran deserts of the American Southwest and northwest Mexico are of the dry subtropical type.

Figure 7.18 is a climograph for Yuma, Arizona, a city within the arid subtype of the dry subtropical climate ⑤, close to the Mexican border at lat. 33° N. The pattern of monthly temperatures shows a strong seasonal cycle with a dry, hot summer. A cold season brings monthly means as low as 13°C (55°F). Freezing temperatures—0°C (32°F) and below—can be expected at night in December and January. The annual range is 20°C (36°F). Precipitation, which totals about 8 cm (3 in.), is small in all months, but it has peaks in late winter

Figure 7.19   A Mojave desert landscape. The strange-looking Joshua tree, shown in the foreground, is abundant at higher elevations in the Mojave desert.

and late summer. The August maximum is caused by the invasion of maritime tropical (mT) air masses, which bring thunderstorms to the region. Higher rainfalls from December through March are produced by midlatitude wave cyclones following a southerly path. Two months, May and June, are nearly rainless.

The Subtropical Desert Environment

The environment of the dry subtropical climate ⑤ is similar to that of the dry tropical climate ④, in that both are very dry. The boundary between these two climate types is gradational. But if we were to travel northward in the subtropical climate zone of North America, arriving at about 34° N in the interior Mojave Desert of southeastern California, we would encounter environmental features significantly different from

those of the low-latitude deserts of tropical Africa, Arabia, and northern Australia. Although the great summer heat of the low-elevation regions of the Mojave Desert is comparable to that experienced in the Sahara Desert, the low sun brings a winter season that is not found in the tropical deserts. In the Mojave Desert, cyclonic precipitation can occur in any month, although it is more likely in the cool low-sun months.

In the Mojave Desert and adjacent Sonoran Desert, plants are often large and numerous, in some places giving the appearance of an open woodland. One example is the occurrence of forestlike stands of the tall, cylindrical saguaro cactus. Another is the woodland of Joshua trees found in higher parts of the Mojave Desert (Figure 7.19). Other large shrubs or small trees include the prickly pear cactus, the ocotillo plant, and the creosote bush.

The deserts of the dry subtropical climate ⑤ are marked by a distinctive cover of desert plants and an assemblage of animals adapted to drought.

Both plants and animals of the Mojave and Sonoran deserts show adaptations to the dry environment. Many annual plants remain dormant as seeds during long dry periods, then spring to life, flower, and bloom very quickly when rain falls. Often invertebrate animals adopt the same life pattern. For example, the tiny brine shrimp of the North American desert may wait many years in dormancy until normally dry lakebeds fill with water, an event that occurs perhaps three or four times per century. The shrimp then emerge and complete their life cycles before the lake evaporates.

The Moist Subtropical Climate ⑥ (Köppen: *Cfa*)

Recall that circulation around the subtropical high-pressure cells provides a flow of warm, moist air onto the eastern side of continents. This flow of maritime tropical (MT) air dominates the moist subtropical climate ⑥. Summer in this climate sees abundant rainfall, much of it convectional. Occasional tropical cyclones further enhance summer precipitation. Summer temperatures are warm, with persistent high humidity.

Winter precipitation in the moist subtropical climate ⑥ is also plentiful, produced in midlatitude cyclones. Invasions of continental polar (cP) air masses are frequent in winter, bringing spells of subfreezing weather. No winter month has a mean temperature below 0°C (32°F). In Southeast Asia, this climate is characterized by a strong monsoon effect, with summer rainfall much increased above winter rainfall.

Figure 7.20 presents a global map of the moist subtropical climate ⑥. It is found on the eastern sides of continents in the latitude range 20° to 35° N and S. In South America, it includes parts of Uruguay, Brazil, and Argentina. In Australia, it consists of a narrow band between the eastern coastline and the eastern interior ranges. Southern China, Taiwan, and southernmost Japan are regions of the moist subtropical climate ⑥ in Asia. In the United States, the moist subtropical climate ⑥ covers most of the Southeast, from the Carolinas to east Texas.

Figure 7.20 World map of the moist subtropical climate ⑥. (Based on Goode Base Map).

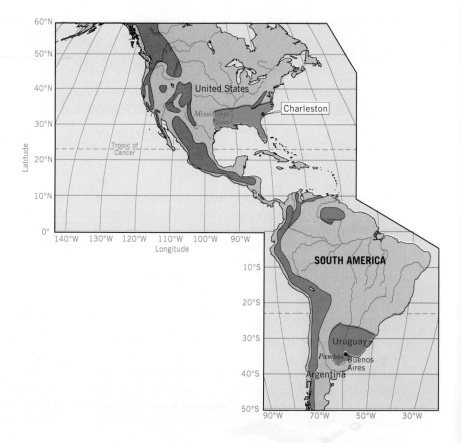

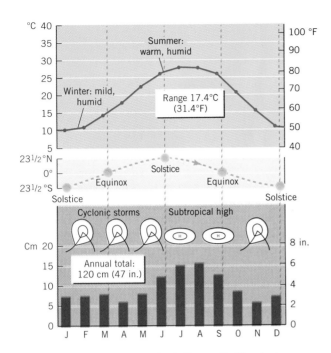

Figure 7.21 Moist subtropical climate ⑥. Charleston, South Carolina, lat. 33° N, has a mild winter and a warm summer. There is ample precipitation in all months but a definite summer maximum.

Figure 7.21 is a climograph for Charleston, South Carolina (Figure 7.21), located on the eastern seaboard at lat. 33° N. In this region, a marked summer maximum of precipitation is typical. Total annual rainfall is abundant—120 cm (47 in.)—and ample precipitation falls in every month. The annual temperature cycle is strongly developed, with a large annual range of 17°C (31°F). Winters are mild, with the January mean temperature well above the freezing mark.

With its abundant rainfall in all seasons, the dominant vegetation type of the moist subtropical climate ⑥ is forest. However, large areas of forest within this climate region have been replaced with agricultural croplands. The warm temperatures favor crop growth, but the heavy rainfall can wash nutrients from soils, requiring fertilization for crop growth. In Asia, staples such as rice and tea are commonly planted.

The Mediterranean Climate ⑦
(Köppen: *Csa, Csb*)

The Mediterranean climate ⑦ is unique among the climate types because its annual precipitation cycle has a wet winter and a very dry summer. The reason for this precipitation cycle lies in the poleward movement of

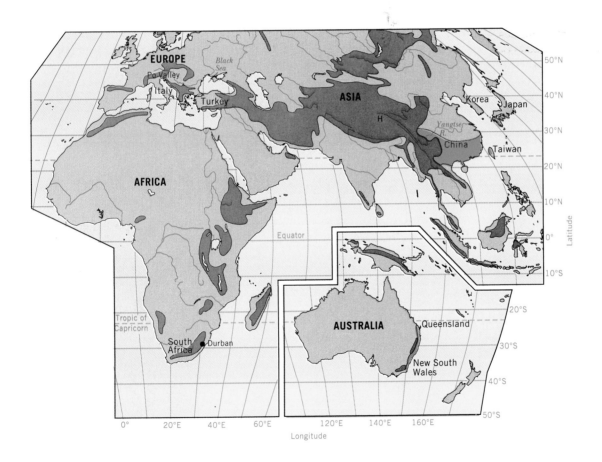

the subtropical high-pressure cells during the summer season. The Mediterranean climate ⑦ is located along the west coasts of continents, just poleward of the dry, eastern side of the subtropical high-pressure cells (Figures 5.13, 5.14). When the subtropical high-pressure cells move poleward in summer, they enter the region of this climate. Dry continental tropical (cT) air then dominates, producing the dry summer season. In winter, the moist mP air mass invades with cyclonic storms and generates ample rainfall.

In terms of total annual rainfall, the Mediterranean climate ⑦ spans a wide range from arid to humid, depending on location. Generally, the closer an area is to the tropics, the stronger the influence of subtropical high pressure will be, and thus the drier the climate. The temperature range is moderate, with warm to hot summers and mild winters. Coastal zones between lat. 30° and 35° N and S, such as southern California, show a smaller annual range, with very mild winters.

A global map of the Mediterranean ⑦ climate is shown in Figure 7.22. It is found in the latitude range 30° to 45° N and S. In the southern hemisphere, it occurs along the coast of Chile, in the Cape Town region of South Africa, and along the southern and western coasts of Australia. In North America, it is found in central and southern California. In Europe, this climate type surrounds the Mediterranean Sea, which gives the climate its distinctive name.

Figure 7.23 is a climograph for Monterey, California, a Pacific coastal city at lat. 36° N. The annual temperature cycle is very weak. The small annual range and cool summer reflect the strong control of the cold California current and its cool, marine air layer. Rainfall drops to nearly zero for four consecutive summer months, but rises to substantial amounts in the rainy winter season. In California's Central Valley, the dry summers persist, but the annual temperature range is greater.

The vegetation of the Mediterranean environment is adapted to the pattern of wet winters and dry summers. Many plant species rely on thick leaves and bark to keep from losing scarce water in the dry season. Coastal California's chaparral, a type of tough, sometimes-spiny shrub cover, is an example (Figure 7.24). The hot, dry summers can make the brush extremely flammable, potentially leading to destructive fires.

Figure 7.22   World map of the Mediterranean ⑦ and marine west coast ⑧ climates. (Based on Goode Base Map.)

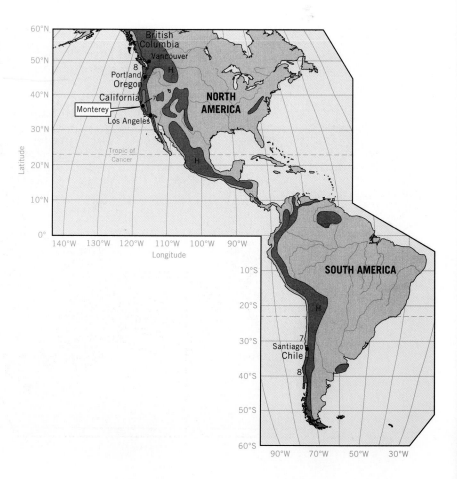

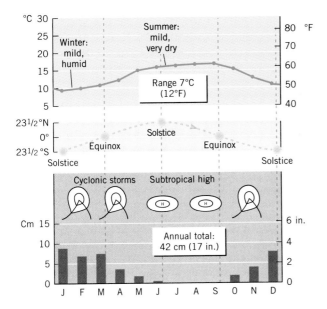

Figure 7.23 Mediterranean climate ⑦. Monterey, California, lat. 36° N, has a very weak annual temperature cycle because of its closeness to the Pacific Ocean. The summer is very dry.

The Marine West-Coast Climate ⑧
(Köppen: *Cfb, Cfc*)

The marine west-coast climate ⑧ occupies midlatitude west coasts. These locations receive the prevailing westerlies from over a large ocean and experience frequent cyclonic storms involving cool, moist mP air masses. Where the coast is mountainous, the orographic effect causes a very large annual precipitation. In this moist climate, precipitation is plentiful in all months, but there is often a distinct winter maximum. In summer, subtropical high pressure extends poleward into the region, reducing rainfall. The annual temperature range is comparatively small for midlatitudes. The marine influence keeps winter temperatures mild, as compared with inland locations at equivalent latitudes.

The global map of the marine west-coast climate ⑧ (Figure 7.22) shows the areas in which this climate occurs. In North America, the climate occupies the western coast from Oregon to northern British Columbia. In Western Europe, the British Isles, Portugal, and much of France fall into the marine west-coast climate. New Zealand and the southern tip of

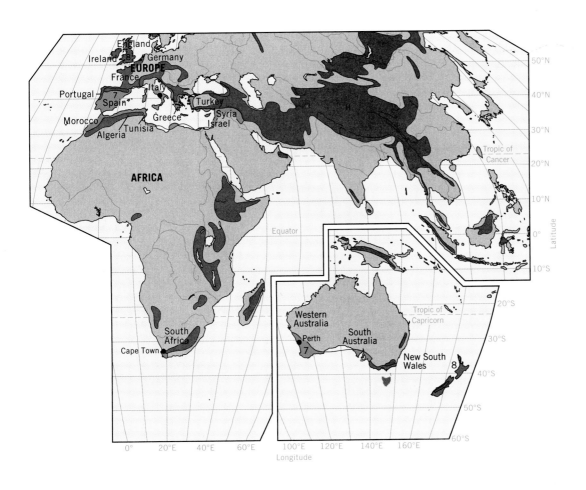

Australia, as well as the island of Tasmania, are marine west-coast climate regions found in the southern hemisphere, as is the Chilean coast south of 35° S. The general latitude range of this climate is 35° to 60° N and S.

Figure 7.25 is a climograph for Vancouver, British Columbia, just north of the U.S.-Canadian border. The annual precipitation is very great, and most of it falls during the winter months. Notice the greatly reduced rainfall in the summer months. The temperature cycle shows a remarkably small range for this latitude. Even the winter months have averages above the freezing mark.

With abundant rainfall and mild temperatures, the marine west-coast climate ⑧ in North America is associated with dense, evergreen forests of fir, cedar, hemlock, and spruce (Figure 7.26). These forests dominate the mountainous terrain of the region. In Western Europe, extensive forests disappeared many centuries ago, yielding to cultivation for crops and dairy farming. The remaining small plots of forest remain under intense management, providing lumber for construction and wood for fuel.

The Dry Midlatitude Climate ⑨ (Köppen: *BWk, BSk*)

The dry midlatitude climate ⑨ is limited almost exclusively to interior regions of North America and Eurasia, where it lies within the rainshadow of mountain ranges on the west or south. Maritime air masses are effectively blocked out much of the time, so that the conti-

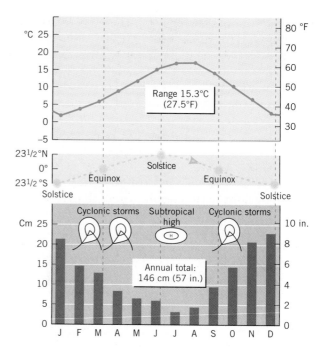

Figure 7.25   Marine west coast climate ⑧. Vancouver, British Columbia, lat. 49° N, has a large annual total precipitation but with greatly reduced amounts in the summer. The annual temperature range is small and winters are very mild for this latitude.

nental polar (cP) air mass dominates the climate in winter. In summer, a dry continental air mass of local origin is dominant. Summer rainfall is mostly convectional and is caused by occasional invasions of

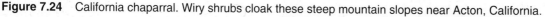

Figure 7.24   California chaparral. Wiry shrubs cloak these steep mountain slopes near Acton, California.

maritime air masses. The annual temperature cycle is strongly developed, with a large annual range. Summers are warm to hot, but winters are cold to very cold.

The largest expanse of the dry midlatitude climate ⑨ is in Eurasia, stretching from the southern republics of the former Soviet Union to the Gobi Desert and northern China (Figure 7.14). In the central portions of this region lie true deserts of the arid climate subtype, with very low precipitation. Extensive areas of highlands occur here as well. In North America, the dry western interior regions, including the Great Basin, Columbia Plateau, and the Great Plains, are of the semiarid subtype. A small area of dry midlatitude climate ⑨ is found in southern Patagonia, near the tip of South America. The latitude range of this climate is 35° to 55° N.

Figure 7.27 is a climograph for Pueblo, Colorado, a semiarid station located at lat. 38° N, just east of the Rocky Mountains. Total annual precipitation is 31 cm (12 in.). Most of this precipitation is in the form of convectional summer rainfall, which occurs when moist maritime tropical (mT) air masses invade from the south and produce thunderstorms. In winter, snowfall is light and yields only small monthly

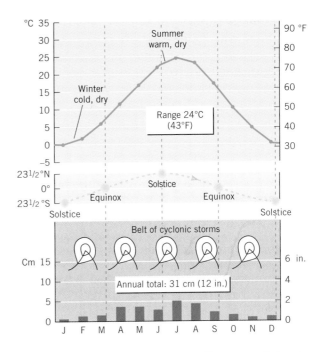

Figure 7.27 Dry midlatitude climate ⑨. Pueblo, Colorado, lat. 38° N, shows a marked maximum of rainfall in the summer months. Note also the cold, dry winter season.

Figure 7.26 Needleleaf forest of Sitka spruce and hemlock in the Hoh rainforest, Olympic National Park, Washington. Ferns and mosses provide a lush ground cover in this very wet environment.

precipitation averages. The temperature cycle has a large annual range, with warm summers and cold winters. January, the coldest winter month, has a mean temperature just below freezing.

The low precipitation and cold winters of this semi-arid climate produce a steppe landscape dominated by hardy perennial short grasses. In North America, this cover is termed *short-grass prairie*, rather than steppe. Wheat is a major crop of the semiarid, dry midlatitude steppelands, but wheat harvests are at the mercy of rainfall variations from year to year. With good spring rains, there is a good crop, but if spring rains fail, so does the wheat crop.

The Moist Continental Climate ⑩
(Köppen: *Dfa, Dfb, Dwa, Dwb*)

The moist continental climate ⑩ is located in central and eastern parts of North America and Eurasia in the midlatitudes. This climate lies in the polar-front zone—the battleground of polar and tropical air masses. Seasonal temperature contrasts are strong, and day-to-day weather is highly variable. Ample precipitation throughout the year is increased in summer by invading maritime tropical (mT) air masses. Cold winters are dominated by continental polar (cP) and continental arctic (cA) air masses from subarctic source regions.

In eastern Asia—China, Korea, and Japan—the seasonal precipitation pattern shows more summer rainfall and a drier winter than in North America. This is an effect of the monsoon circulation, which moves moist maritime tropical (mT) air across the eastern side of the continent in summer and dry continental polar southward through the region in winter. In Europe, the moist continental climate ⑩ lies in a higher latitude belt (45° to 60° N) and receives precipitation from mP air masses coming from the North Atlantic.

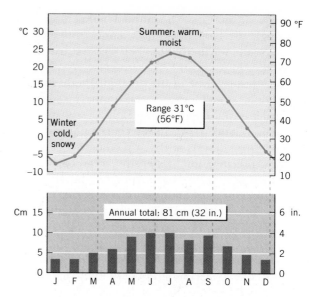

Figure 7.29 Moist continental climate ⑩. Madison, Wisconsin, lat. 43° N, has cold winters and warm summers, making the annual temperature range very large.

Figure 7.28 shows the global locations of the moist continental climate ⑩. It is restricted to the northern hemisphere, occurring in latitudes 30° to 55° N in North America and Asia, and in latitudes 45° to 60° N in Europe. In Asia, it is found in northern China, Korea, and Japan. Most of Central and Eastern Europe has a moist continental climate, as does most of the eastern half of the United States from Tennessee to the north, as well as the southernmost strip of eastern Canada.

Madison, Wisconsin, lat. 43° N, in the American Midwest (Figure 7.29), provides an example of the moist continental climate ⑩ (Figure 7.30). The annual temperature range is very large. Summers are warm, but winters are cold, with three consecutive monthly

Figure 7.28 World map of the moist continental climate ⑩ (Based on Goode Base Map.)

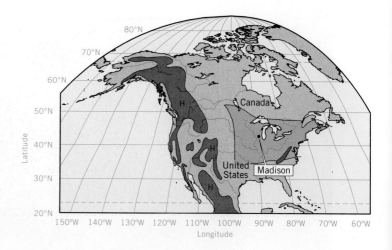

Figure 7.30   The Mosel River in its winding, entrenched meandering gorge through the Rhineland-Pfalz province of western Germany. On the steep, undercut valley wall at the right are vineyards and forested land.

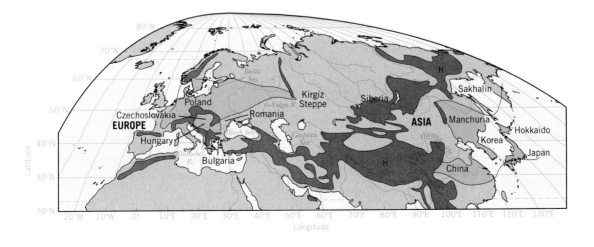

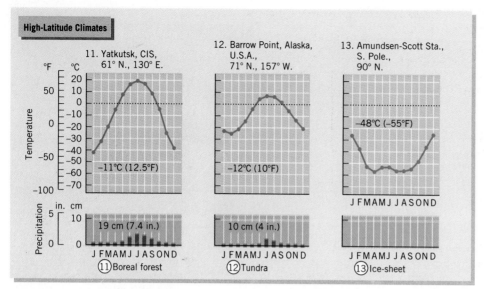

Figure 7.31 Climographs of the three high-latitude climates.

means well below freezing. Precipitation is ample in all months, and the annual total is large. There is a summer maximum of precipitation when the maritime tropical (mT) air mass invades, and thunderstorms are formed along moving cold fronts and squall lines. Much of the winter precipitation is in the form of snow, which remains on the ground for long periods.

Throughout most of the moist continental climate ⑩, forests are the dominant natural vegetation cover. However, where the climate grades into drier climates, such as the dry midlatitude climate, tall, dense grasses may be the natural cover. The *tall-grass prairie* of Illinois, Iowa, and Nebraska is an example. These lands are now the heart of the American corn belt. In Europe, lands of the moist continental climate have been in cultivation for many centuries, providing a multitude of crops. Forests are planted and managed for timber yield. The result is a landscape that shows human influence everywhere (Figure 7.30).

HIGH-LATITUDE CLIMATES

By and large, the high-latitude climates are climates of the northern hemisphere (Figure 7.31). They occupy the northern subarctic and arctic latitude zones, but also extend southward into the midlatitude zone as far south as about the 47th parallel in eastern North America and eastern Asia. One of these, the ice sheet climate ⑬, is present in both hemispheres in the polar zones. Table 7.3 provides an overview of the three high-latitude climates.

The high-latitude climates coincide closely with the belt of prevailing westerly winds that circles each pole. In the northern hemisphere, this circulation sweeps maritime polar (mP) air masses, formed over the northern oceans, into conflict with continental polar (cP) and continental arctic (cA) air masses on the continents. Rossby waves form in the westerly flow, bringing lobes of warmer, moister air poleward into the

Figure 7.32 World map of the boreal forest climate ⑪. (Based on Goode Base Map.)

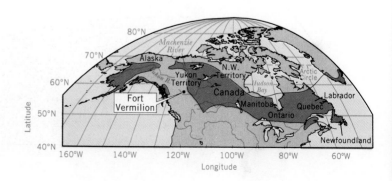

Table 7.3   High-latitude Climates

Climate	Temperature	Precipitation	Explanation
Boreal forest ⑪	Short, cool summers and long, bitterly cold winters. <u>Greatest annual range of all climates.</u>	Annual precipitation small, falling mostly in summer months.	This climate falls in the source region for cold, dry, stable cP air masses. Traveling cyclones, more frequent in summer, bring precipitation from mP air.
Tundra ⑫	No true summer but a short mild season. Otherwise, cold temperatures.	Annual precipitation small, falling mostly during mild season.	The coastal arctic fringes occupied by this climate are dominated by cP, mP, and cA air masses. The maritime influence keeps winter temperatures from falling to the extreme lows of interiors.
Ice sheet ⑬	All months below freezing, with lowest global temperatures on earth experienced during Antarctic winter.	Very low precipitation, but snow accumulates since temperatures are always below freezing.	Ice sheets are the <u>source</u> regions of <u>cA and cAA</u> air masses. Temperatures are intensely cold.

region in exchange for colder, drier air moved equatorward. The result of these processes is frequent <u>wave cyclones</u>, produced along a discontinuous and constantly fluctuating arctic-front zone. In summer, tongues of maritime tropical air masses (mT) reach the subarctic latitudes to interact with polar air masses and yield important amounts of precipitation.

The Boreal Forest Climate ⑪
(Köppen: *Dfc, Dfd, Dwc, Dwd*)

The boreal forest climate is a continental climate with long, bitterly cold winters and short, cool summers. It occupies the source region for cP air masses, which are cold, dry, and stable in the winter. Invasions of the very cold cA air mass are common. The annual range of temperature is greater than that of any other climate and is greatest in Siberia. Precipitation increases substantially in summer, when maritime air masses penetrate the continent with traveling cyclones, but the total annual precipitation is small. Although much of the boreal forest climate is moist, large areas in western Canada and Siberia have low annual precipitation and are therefore cold and dry.

The global extent of the boreal forest climate ⑪ is presented in Figure 7.32. In North America, it stretches from central and western Alaska, across the Yukon and Northwest Territories to Labrador on the Atlantic coast. In Europe and Asia, it reaches from the Scandinavian Peninsula eastward across all of Siberia to the Pacific. In latitude, this climate type ranges from 50° to 70° N.

Figure 7.33 is a climograph for Fort Vermilion, Alberta, at lat. 58° N. The very great annual temperature range shown here is typical for North America. Monthly mean air temperatures are below freezing for

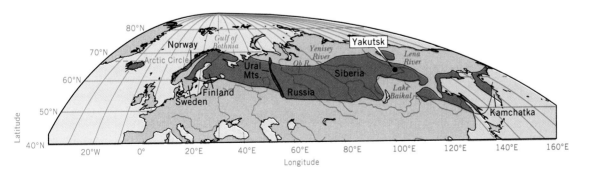

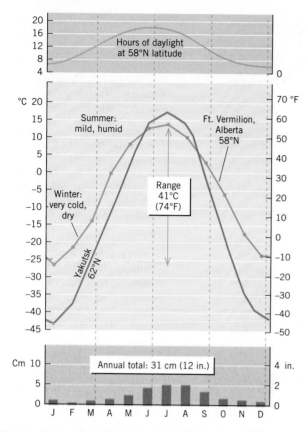

Figure 7.33 Boreal forest climate ⑪. Extreme winter cold and a very great annual range in temperature characterizes the climate of Fort Vermilion, Alberta. The temperature range of Yakutsk, Siberia, is even greater.

seven consecutive months. The summers are short and cool. Precipitation shows a marked annual cycle with a summer maximum, but the total annual precipitation is small. Although precipitation in winter is small, a snow cover remains over solidly frozen ground through the entire winter. On the same climograph, temperature data are shown for Yakutsk, a Siberian city at lat. 62° N. in the Russian Republic. The enormous annual range is evident, as well as the extremely low means in winter months. January reaches a mean of about –42°C (–45°F). Except for the ice sheet interiors of Antarctica and Greenland, this region is the coldest on earth. Precipitation is not shown for Yakutsk, but the annual total is very small.

The Boreal Forest Environment

Land surface features of much of the region of boreal forest climate were shaped beneath the great ice sheets of the last ice age. Severe erosion by the moving ice exposed hard bedrock over vast areas and created numerous shallow rock basins. Bouldery rock debris mantles the rock surface in many places. Many of the

shallower rock basins have been filled by organic bog materials. The dominant upland vegetation of this climate region is boreal forest, consisting of needle-leaf trees.

Although the growing season in the boreal forest climate ⑪ is short, crop farming is still possible. It is largely limited to lands surrounding the Baltic Sea, bordering Finland and Sweden. Crops grown in this area include barley, oats, rye, and wheat. Along with dairying, these crops primarily supply food for subsistence. The needleleaf trees of the boreal forest provide pulpwood and construction lumber.

**The Tundra Climate ⑫
(Köppen: *ET*)**

The tundra climate ⑫ occupies arctic coastal fringes and is dominated by polar (cP, mP) and arctic (cA) air masses. Winters are long and severe. A moderating influence of the nearby ocean water prevents winter temperatures from falling to the extreme lows found in the continental interior. There is a very short mild season, which many climatologists do not recognize as a true summer.

The world map of the tundra climate ⑫ (Figure 7.34) shows the tundra ringing the Arctic Ocean and extending across the island region of northern Canada. It includes the Alaskan north slope, the Hudson Bay region, and the Greenland coast in North America. In Eurasia, this climate type occupies the northernmost fringe of the Scandinavian Peninsula and Siberian coast. The Antarctic Peninsula (not shown in Figure 7.34) belongs to the tundra climate ⑫. The latitude range for this climate is 60° to 75° N and S., except for the northern coast of Greenland, where tundra occurs at latitudes greater than 80° N.

Figure 7.35 is a climograph for Upernivik, located on the west coast of Greenland at lat. 73° N. A short mild period, with above-freezing temperatures, is equivalent to a summer season in lower latitudes. The long winter is very cold, but the annual temperature range is not as large as that for the boreal forest climate to the south. Total annual precipitation is small. Increased precipitation beginning in July is explained by the melting of the sea-ice cover and a warming of ocean water temperatures. This increases the moisture content of the local air mass, allowing more precipitation.

The Tundra Environment

The term **tundra** describes both an environmental region and a major class of vegetation (Figure 7.36). (An equivalent climatic environment—called alpine tundra—prevails in many global locations in high

Figure 7.34 World map of the tundra climate ⑫.

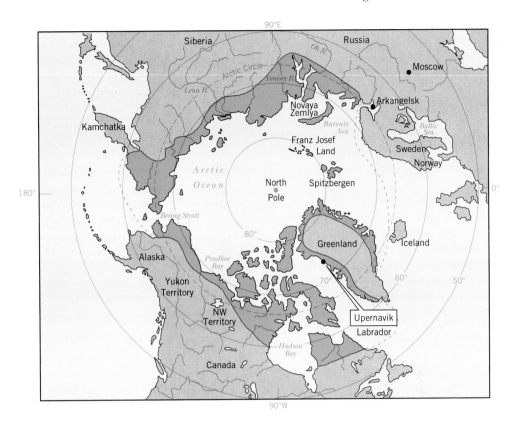

mountains above the timberline.) Trees exist in the tundra only as small, shrublike plants. They are stunted because of the seasonal damage to roots by freeze and thaw of the soil layer and to branches exposed to the abrading action of wind-driven snow. Vegetation of the treeless tundra consists of grasses, sedges, and lichens, along with shrubs of willow. Peat bogs are numerous. Because soil water is solidly and permanently frozen not far below the surface, the summer thaw brings a condition of water saturation to the soil.

At high latitudes, perennially frozen ground, or permafrost, occurs. Human disturbances can cause permafrost to thaw.

Because of the cold temperatures experienced in the tundra and northern boreal forest climate zones, the ground is typically frozen to great depth. This perennially frozen ground, or **permafrost,** prevails over the tundra region and a wide bordering area of boreal forest climate. Normally, a top layer of the ground will thaw each year during the mild season. This active layer of seasonal thaw is from 0.6 to 4 m (2 to 14 ft) thick, depending on latitude and the nature of the ground.

The permafrost environment is in a stable, delicate equilibrium that can be easily disturbed. When human

activities remove the cover of moss or peat and living plants that insulates the permafrost, the summer thaw is extended to a greater depth. Ice wedges or ice masses beneath the surface melt, causing the soil sur-

Figure 7.35 Tundra climate ⑫. Upernavik, Greenland, lat. 73° N, shows a smaller annual range than Fort Vermilion (Figure 7.33).

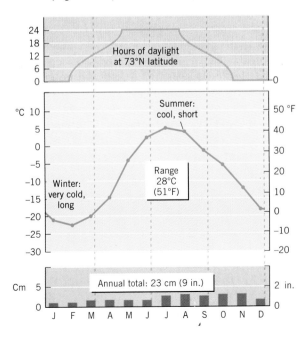

face to sink. Meltwater mixes with the fine soil to form mud, which is then eroded and transported by water in streams. This activity is called thermal erosion and is reversed only with great difficulty.

The Ice Sheet Climate ⑬
(Köppen: *EF*)

The ice sheet climate coincides with the source regions of arctic (A) and antarctic (AA) air masses, situated on the vast, high ice sheets of Greenland and Antarctica and over polar sea ice of the Arctic Ocean. Mean annual temperature is much lower than that of any other climate, with no monthly mean above freezing. Strong temperature inversions develop over the ice sheets. In Antarctica and Greenland, the high surface altitude of the ice sheets intensifies the cold. Strong cyclones with blizzard winds are frequent. Precipita-

tion, almost all occurring as snow, is very low, but accumulates because of the continuous cold. The latitude range for this climate is 65° to 90° N and S.

Figure 7.37 shows temperature graphs for several representative ice sheet stations. The graph for Eismitte, a research station on the Greenland ice cap, shows the northern hemisphere temperature cycle, whereas the other four examples are all from Antarctica. Temperatures in the interior of Antarctica have proved to be far lower than those at any other place on earth. A Russian meteorological station at Vostok, located about 1300 km (800 mi) from the south pole at an altitude of about 3500 m (11,400 ft), may be the world's coldest spot. Here a low of –88.3°C (–127°F) was observed in 1958. At the pole itself (Amundsen-Scott Station), July, August, and September of 1957 had averages of about –60°C (–76°F). Temperatures are considerably higher, month for month, at Little America in Antarctica

Figure 7.36 Caribou migration across the arctic tundra of northern Alaska.

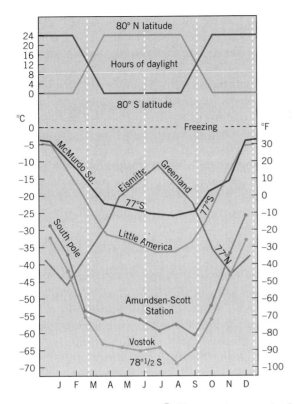

Figure 7.37 Ice-sheet climate ⑬. Temperature graphs for five ice sheet stations.

because it is located close to the Ross Sea and is at a low altitude.

The climates of the earth are remarkably diverse, ranging from the hot, humid wet equatorial climate ① at the equator to the bitterly cold and dry ice sheet climate ⑬ at the poles. Between these extremes are the other climates, each with distinctive features—such as dry summers or dry winters, or uniform or widely varying temperatures. Yet, these distinctive features can be explained using the principles controlling air temperatures (Chapter 3), atmospheric moisture and precipitation (Chapter 4), global circulation (Chapter 5) and weather systems (Chapter 6). In this way, the study of climate in this chapter acts to focus our existing knowledge on the problem of how the world's regions are differentiated by the annual patterns of weather they experience.

Climate exerts strong controls on vegetation and soils, especially at the global level. These are the subjects of the following chapters.

GLOBAL CLIMATES IN REVIEW

Climate classification makes use of the annual cycles of temperature and precipitation as they are observed at many locations around the world. The world precipitation map shows rainfall concentrated in a broad belt around the equator. Mountain regions also show high precipitation, especially where they are swept by moist winds off nearby oceans. The driest regions occur where air continually subsides in subtropical high pressure cells. Continental interiors are also dry. Precipitation generally decreases toward the poles.

Low-latitude climates are controlled by the characteristics and annual movements of the ITC and the subtropical highs. In the wet equatorial climate ①, the ITC is always nearby, so rainfall is abundant throughout the year. The monsoon and trade-wind coastal climate ② shows a dry period when the ITC has migrated toward the opposite tropic. In the wet-dry tropical climate ③, this dry period is more pronounced. The dry tropical climate ④ is dominated by subtropical high pressure and is often nearly rainless. The annual cycle of temperature is nearly uniform in the wet equatorial climate ①, but becomes more pronounced as dry conditions come to dominate the climate type.

Midlatitude climates are more varied. The dry subtropical ⑤ and dry midlatitude ⑨ climates extend the dry tropical ④ climate poleward into continental interiors, where they are far from oceanic sources of moist air. Both the Mediterranean ⑦ and marine west-coast ⑧ climates show a summer precipitation minimum, produced by the poleward movement of subtropical high pressure at that time of year. The moist subtropical ⑥ and moist continental ⑩ climates fall on the east sides of continents. They can receive moist tropical air or dry polar air at any time of year, but the moist subtropical climate ⑥ receives warm moist air more frequently and is located further equatorward. Midlatitude climate temperature cycles show strong seasonal effects. Temperature range increases in a poleward direction and as the climate becomes drier. The west-coast climates have a lesser range because of their maritime location.

High-latitude climates have low precipitation since air temperatures are low. The precipitation generally occurs during the short warm period. The boreal forest climate ⑪ occupies the cP source region in continental interiors and experiences a great annual range of temperatures. The tundra climate ⑫ is strongly influenced by nearby ocean and so experiences a lesser range. The ice sheet climate ⑬ is the coldest on earth.

KEY TERMS

climate
isohyet
climograph
wet equatorial climate ①
monsoon and trade-wind
   coastal climate ②
wet-dry tropical climate ③

dry tropical climate ④
steppes
dry subtropical climate ⑤
moist subtropical climate ⑥
Mediterranean climate ⑦
marine west-coast climate ⑧
dry midlatitude climate ⑨

moist continental
   climate ⑩
boreal forest climate ⑪
tundra climate ⑫
tundra
permafrost
ice sheet climate ⑬

REVIEW QUESTIONS

1. Why are annual cycles of temperature and precipitation useful measures of climate?

2. Identify two moist regions and two dry regions of the world, and explain why each region is either moist or dry.

3. Why is the annual temperature cycle of the wet equatorial climate ① so uniform?

4. Sketch the temperature and rainfall cycles for a typical station in the monsoon and trade-wind coastal climate ②. What factors contribute to the seasonality in the two cycles?

5. The wet-dry tropical climate ③ has two distinct seasons. What factors produce the dry season? The wet season?

6. Why are the dry tropical ④, dry subtropical ⑤, and dry midlatitude ⑨ climates dry? How do they differ in temperature and precipitation cycles?

7. Both the moist subtropical ⑥ and moist continental ⑩ climates are found on eastern sides of continents in the midlatitudes. What are the major factors that determine their temperature and precipitation cycles? How do these two climates differ?

8. Both the Mediterranean ⑦ and marine west-coast ⑧ climates are found on the west coasts of continents. Why do they experience more precipitation in winter than in summer? How do the two climates differ?

9. The subtropical high pressure cells influence several climate types in the low and midlatitudes. Identify the climates and describe the effects of the subtropical high-pressure cells on them.

10. Both the boreal forest ⑪ and tundra ⑫ climates are climates of the northern regions, but the tundra is found fringing the Arctic Ocean and the boreal forest is located further inland. Compare these two climates from the viewpoint of coastal-continental effects.

11. What is the coldest climate on earth? How is the annual temperature cycle of this climate related to the cycle of insolation?

IN-DEPTH QUESTIONS

1. The intertropical convergence zone (ITC) moves north and south with the seasons. Describe how this movement affects the four low-latitude climates.

2. Suppose South America were turned over. That is, imagine that the continent was cut out and flipped over end-for-end so that the southern tip was at about 10° N latitude, and the northern end (Venezuela) was positioned at about 55° S. The Andean chain will still be on the west side, but the shape of the land mass will now be quite different. Sketch this continent and draw possible climate boundaries, using your knowledge of global air circulation patterns, frontal zones, and air mass movements.

Chapter 8

Global Vegetation

All aboard! The conductor's cry resounds through Union Station, Los Angeles, and at precisely 11:50 A.M., your train begins to move. At last your journey is underway! You're headed for Chicago on a three-day train ride that will take you through some of the most breathtaking scenery in America.

An hour later, you are relaxing in the comfortable armchair of your roomette, gazing out the large picture window. The Mojave Desert flashes by. The strange forms of Joshua trees contrast against the gray and tan colors of the desert earth. The coach rushes across a dry lakebed, its white salt sparkling in the sun. Sharp peaks appear in the background from time to time, their angular foothills terminating the broad plains of grayish creosote bush that sweep toward the mountain fronts.

Early that evening, the train pulls out of Las Vegas, headed up toward the high desert of the Colorado Plateau. As the sun slowly sets, the reds and browns of the rocks and soils intensify. Soon the desert sky darkens to a blue-black. Stars appear, and finally the moon. Sagebrush carpets the landscape, looking soft and inviting in the half light. A dark stubble of pines cloaks the mountain ranges that appear from time to time in the middle distance. Reluctantly, you draw the shade and turn in for the night.

"More coffee?" The steward expertly refills your breakfast coffee cup, swaying with the motion of the dining car. During the night, the train has slowly gained elevation, passing through Salt Lake City, then crossing the crest of the Wasatch Plateau into the upper reaches of the Colorado basin. The landscape is more rugged now.

Twisted crowns of piñon pines and junipers dot the hills. Gray-green sagebrush blankets the lowlands, with green, willowy saltbush lining the dry creek beds.

By early afternoon, you are well into your crossing of the Rocky Mountains. The track follows the narrow canyon of the upper Colorado. The river twists and turns, splashing and foaming through rapids as its water courses toward the low desert. Steep slopes enclose the rushing train, the north-facing ones on the right clothed in spruce and fir, the south-facing ones on the left in pines and junipers. Here and there the valley widens, providing glimpses of white-capped Rocky Mountain summits with evergreen forests below.

Soon you are in the high country, bordering Rocky Mountain National Park. Ragged, windswept peaks jut skyward, spread with a green and white quilt of snowfields, conifer forests, and aspen groves. Your coach sweeps past fields of alpine wildflowers, their blues, yellows, and whites merging in a kaleidoscope of colors. Suddenly the train plunges into the Moffat Tunnel for the long passage under the continental divide. Emerging on the east side, you descend steeply toward the high plains. As the dinner hour approaches, the skyscrapers of Denver are silhouetted against the darkening sky.

After dinner, you sit alone in your darkened roomette, gazing out the window. The stars fill the sky, glistening in its vast blue-blackness. Soon the moon appears on the far distant horizon. It rises higher, washing the landscape with a silvery twilight and dimming the stars. The featureless prairie of short-grass tufts sweeps into the distance, its moonlit surface like a table spread with faded lace. Soon, the rocking motion of the train lulls you into a contented sleep.

Noises in the station at Omaha awaken you early in the morning. As you raise the window shade a few minutes later, the reddish glow of early morning light glints off the broad Missouri River. Relaxing after breakfast, you watch Iowa roll by. Like the landscape of yesterday evening, it is flat to the far horizon, breaking only as the train crosses small creeks and rivers draining into the Missouri. Rectangular fields stretch as far as the eye can see, the maturing corn and soybean crops forming a dense, green blanket. Trees mark the landscape only as windbreaks between fields and along the watercourses.

As the train rolls eastward, the density of settlement increases. Towns are more frequent, especially after you enter Illinois. By late afternoon, you reach the Chicago area, and precisely at 4:15 P.M., your train pulls into the station. Regretfully, you hoist your suitcase and step off the train. Your trip is over, but the magnificent landscapes and scenic vistas of the American West are sights you'll never forget.

Vegetation is the most obvious element of the landscape in many regions, and this chapter is devoted to a global study of vegetation. As we will see, vegetation is often strongly related to climate, so the knowledge of climate you obtained in Chapter 7 will be very useful as you study the world's vegetation.

The Mojave Desert in bloom. Anza Borrego State Park, California.

Most of the earth's land surface has some sort of plant cover. Except for ice sheet surfaces and the most barren of deserts, where a plant cover is absent, the vegetation is a visible and obvious part of the landscape. So, vegetation is an important subject for geographers to study. Plants are also important because humans depend on them for food, medicines, fuel, clothing, shelter, and many other life essentials. Human use of plant resources has been a persistent theme in the writings of geographers.

NATURAL VEGETATION

Over the last few thousand years, human societies have come to dominate much of the land area of our planet. In many regions, humans have changed the natural vegetation—sometimes drastically, other times subtly. What do we mean by natural vegetation? **Natural vegetation** is a plant cover that develops with little or no human interference. It is subject to natural forces of modification and destruction, such as storms or fires. Natural vegetation can still be seen over vast areas of the wet equatorial climate, although the rainforests there are being rapidly cleared. Much of the arctic tundra and the boreal forest of the subarctic zones is in a natural state.

In contrast to natural vegetation is *human-influenced vegetation*, which is modified by human activities. Much of the land surface in midlatitudes is totally under human control, through intensive agriculture, grazing, or urbanization. You can drive across an entire state, such as Ohio or Iowa, without seeing a single remnant of the type of plant cover that existed before European settlers arrived.

Most areas of the earth's land surface are influenced in some way by human activity.

Some areas of natural vegetation appear to be untouched but are actually dominated by human activity in a subtle manner. For example, most national parks and national forests have been protected from fire for many decades. When lightning starts a forest fire, the firefighters put out the flames as fast as possible. However, periodic burning is part of the natural cycle in many regions. One vital function of fire is to release nutrients that are stored in plant tissues. When the vegetation burns, the ashes containing the nutrients remain. These enrich the soil for the next cycle of vegetation cover. In recent years, some managers of parks and forests have stopped suppressing wildfires, allowing the return of more natural periodic burning.

Our species has influenced vegetation in yet another way—by moving plant species from their original habitats to foreign lands and foreign environments. The eucalyptus tree is a striking example. From Australia, various species of eucalyptus have been transplanted to such far-off lands as California, North Africa, and India. Sometimes exported plants thrive like weeds, forcing out natural species and becoming a major nuisance. Few of the grasses that clothe the coastal ranges of California are native species, yet a casual observer might think that these represent native vegetation.

Even so, all plants have limited tolerance to the environmental conditions of soil water, heat and cold, and soil nutrients. Consequently, the structure and outward appearance of the plant cover conforms to basic environmental controls, and each vegetation type stays within a characteristic geographical region—whether forest, grassland, or desert.

STRUCTURE AND LIFE-FORM OF PLANTS

Plants come in many types, shapes, and sizes. Botanists recognize and classify plants by species. However, the plant geographer is less concerned with individual species and more concerned with the plant cover as a whole. In describing the plant cover, plant geographers refer to the **life-form** of the plant—its physical structure, size, and shape. Although the life-forms go by common names and are well understood by almost everyone, we will review them to establish a uniform set of meanings.

Both trees and shrubs are erect, woody plants (Figure 8.1). They are *perennial*, meaning that their woody tissues endure from year to year. Most have life spans of many years. *Trees* are large, woody perennial plants having a single upright main trunk, often with few branches in the lower part, but branching in the upper part to form a crown. *Shrubs* are woody perennial plants having several stems branching from a base near the soil surface, so as to place the mass of foliage close to ground level.

Lianas are also woody plants, but they take the form of vines supported on trees and shrubs. Lianas include not only the tall, heavy vines of the wet equatorial and tropical rainforests (see Figure 8.9), but also some woody vines of midlatitude forests. Poison ivy and the tree-climbing form of poison oak are familiar American examples of lianas.

Herbs comprise a major class of plant life-forms. They lack woody stems and so are usually small, tender plants. They occur in a wide range of shapes and leaf types. Some are *annuals,* living only for a single season—while others are perennials—living for multiple seasons. Some herbs are broad-leaved, and others are narrow-leaved, such as grasses. Herbs as a class share few characteristics in common except that they usually form a low layer as compared with shrubs and trees.

Forest is a vegetation structure in which trees grow close together. Crowns are in contact, so that the foliage largely shades the ground. Many forests in moist climates show at least three layers of life-forms (Figure 8.1). Tree crowns form the uppermost layer, shrubs an intermediate layer, and herbs a lower layer. There is sometimes a fourth, lowermost layer that consists of mosses and related very small plants. In **woodland,** crowns of trees are mostly separated by open areas, usually having a low herb or shrub layer.

Lichens are another life-form seen in a layer close to the ground. They are plant forms in which algae and fungi live together to form a single plant structure. In some alpine and arctic environments, lichens grow in profusion and dominate the vegetation (Figure 8.2).

PLANTS AND ENVIRONMENT

A primary objective of this chapter is to describe the earth's vegetation cover as part of the earth's physical geography. But before we begin this task, we will examine some concepts from *plant ecology*—the study of the interrelationships among plants and their environment. These will be helpful in understanding how vegetation is related to climate and soils.

Plant Habitats

As we travel through a hilly, wooded area, it is easy to see that the vegetation is strongly influenced by landform and soil. (As we will see in Chapter 10, landform refers to the configuration of the land surface, including features such as hills, valleys, ridges, or cliffs.) Vegetation on an upland—relatively high ground with thick soil through which water drains easily—is quite different from that on an adjacent valley floor, where soils are wet much of the time. Vegetation is also strikingly different in form on rocky ridges and on steep cliffs, where water drains away rapidly and soil is thin or largely absent.

The total vegetation cover is actually a collection of small patches with different conditions of slope, water drainage, and soil type. Such subdivisions of the plant environment are described as *habitats*. Figure 8.3 presents an example taken from a Canadian forest. Here, there are six distinctive habitats: upland, bog, bottomland, ridge, cliff, and active sand dune. Each habitat supports a different type of vegetation cover.

Plants and Water Need

The two most important environmental factors influencing plant growth are water availability and temperature. Green plants give off large quantities of water to the atmosphere through **transpiration.** In this process, water from the soil is absorbed by root tissues and moves through the plant's stems to the leaves, where it evaporates. This water flow carries nutrients to the leaves and also helps keep the leaves cool. Evaporation at the leaf is controlled by specialized leaf pores, which provide openings in the outer layer of cells. When soil water is depleted, the pores are closed and evaporation is greatly reduced.

As you know from the study of climate in Chapter 7, many regions of the world have climates that are dry or have a significant dry season. Many species of plants have developed adaptations to help survive dry conditions. Plants that are adapted to drought conditions are termed **xerophytes.** The word "xerophyte" comes from the Greek roots *xero-,* meaning "dry," and *phyton,* meaning "plant." Because they are highly tolerant of drought, xerophytes can survive in habitats that dry quickly following rapid drainage of precipitation (for example, sand dunes, beaches, and bare rock surfaces). The adaptations of desert plants to dry conditions make them xerophytes as well.

> **Moisture and temperature are key climatic factors that influence the vegetation cover within a region.**

In some xerophytes, water loss is reduced by a thick layer of wax or waxlike material on leaves and stems. The wax helps to seal water vapor inside the leaf or stem. Still others adapt to a desert environment by greatly reducing their leaf area or by bearing no leaves at all. Needlelike leaves, or spines in place of leaves, are also adaptations of plants to conserve water. In cactus plants, the foliage leaf is not present, and transpiration is limited to thickened, water-filled stems that store water for use during long, dry periods.

Adaptations of plants to water-scarce environments also include improved abilities to obtain and store water. Roots may extend deeply to reach soil moisture far from the surface. In cases where the roots reach to the ground water zone, a steady supply of water is assured. Plants drawing from ground water may be

Figure 8.1   This schematic diagram shows the layers of a beech–maple–hemlock forest. The vertical dimensions of the lower layers are greatly exaggerated. (After P. Dansereau.)

found along dry stream channels and valley floors in desert regions. In these environments, ground water is usually near the surface. Other desert plants produce a widespread, but shallow, root system. This enables them to absorb water from short desert downpours that saturate only the uppermost soil layer.

Another adaptation to extreme aridity is a very short life cycle. Many small desert plants will germinate from seed, then leaf out, bear flowers, and produce seed in the few weeks immediately following a heavy rainshower. In this way, they complete their life cycle when soil moisture is available, and survive the dry period as seeds that require no moisture.

Certain climates, such as the wet-dry tropical climate ③ and the moist continental climate ⑩, have a yearly cycle with one season in which water is unavailable to plants because of lack of precipitation or because the soil water is frozen. This season alternates with one in which there is abundant water. Many plants respond to this pattern by dropping their leaves at the close of the

moist season and becoming dormant during the dry season. When water is again available, they leaf out and grow at a rapid rate. Trees and shrubs that shed their leaves seasonally are termed *deciduous*. In contrast are *evergreen* plants, which retain most of their leaves in a green state through one or more years.

The Mediterranean climate ⑦ also has a strong seasonal wet-dry alternation, with dry summers and wet winters. Plants in this climate are often xerophytic and characteristically have hard, thick, leathery leaves. An example is the live oak, which holds most of its leaves through the dry season (Figure 8.4). Such hard-leaved evergreen trees and woody shrubs are called *sclerophylls*. (The prefix *scler-* is from the Greek root for "hard" and is combined with the Greek word for leaf, *phyllon*.) Plants that hold their leaves through a dry or cold season have the advantage of being able to resume photosynthesis immediately when growing conditions become favorable, whereas the deciduous plants must grow a new set of leaves.

Figure 8.2   Reindeer moss, a white variety of lichen, seen here on rocky tundra of Alaska.

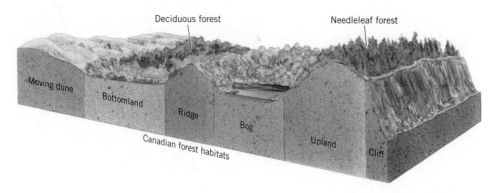

Deciduous forest   Needleleaf forest

Moving dune
Bottomland
Ridge
Bog
Upland
Cliff
Canadian forest habitats

Figure 8.3 Habitats within the Canadian forest. (After P. Dansereau.)

Plants and Temperature

Temperature is the second important climatic factor that influences global vegetation. It acts directly on plants by influencing the rates at which physiological processes take place in plant tissues. In general, each plant species has an optimum temperature associated with each of its functions, such as photosynthesis, flowering, fruiting, or seed germination. There are also limiting lower and upper temperatures for these individual functions and for the total survival of the plant itself.

Temperature can also act indirectly on plants. For example, higher air temperatures increase the water vapor holding capacity of the air. This means that liquid water will evaporate more readily if air temperature is increased. Therefore, plants lose more water through increased transpiration at higher temperatures.

In general, the colder the climate, the fewer the species that are capable of surviving. A large number of tropical plant species cannot survive below-freezing temperatures for more than a few hours. In the severely cold arctic and alpine environments of high latitudes and high altitudes, only a few plant species are found. This principle explains why a forest in the equatorial zone has many species of trees, whereas a forest of the subarctic zone will be dominated by only a few.

Freezing damages plant tissues largely by causing ice crystals to form inside the cells. These crystals physically disrupt the internal structure of the cell, damaging the fine cell structure. Cold-tolerant species can expel excess water from cells to spaces between cells, where freezing does no damage.

Plant geographers recognize that there is a critical level of climatic stress beyond which a plant species cannot survive. Where this critical level occurs, a geographical boundary will exist that marks the limit of the distribution of the species. Such a boundary is sometimes referred to as a *climatic frontier*. Although the frontier is determined by a complex of climatic elements, it is sometimes possible to single out one climate element related to soil water or temperature that coincides with the plant's frontier.

Figure 8.4 This California live oak is an example of a sclerophyll—a plant with thick, leathery leaves that is adapted to an environment with a very dry season.

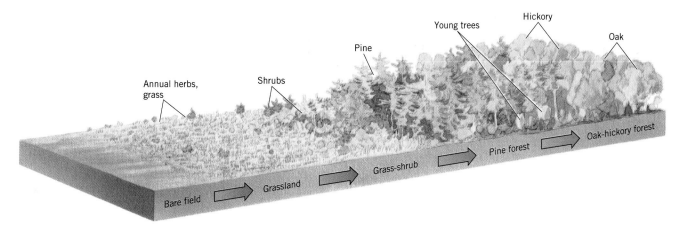

Figure 8.5   Old-field succession in the southeastern United States, following abandonment of cultivated fields. This is a pictorial graph of continuously changing plant composition spanning about 150 years.

Ecological Succession and Human Impact on Vegetation

Although climatic influences are important in determining the general type of vegetation cover within a region, local vegetation patterns often depend on the developmental stage of the vegetation cover. For example, a country drive in the Southeast reveals patches of vegetation in many stages of development—from open fallow fields, to grassy shrublands, to forests. As we will see shortly, these are stages in a process by which abandoned fields become mature forests. On a longer time scale, clear lakes gradually fill in with sediment from the rivers that drain into them and become marshes. The marshes may eventually dry out as drainage improves and be replaced by a tall, mature forest. The development process, in which plant communities succeed one another on the way to a stable endpoint, make up an *ecological succession.*

> **Vegetation on new or cleared land undergoes a series of stages of development, called a succession, until it reaches a stable vegetation type, called the climax.**

In general, succession leads to formation of the most complex community of organisms possible in an area, given its physical controlling factors of climate, soil, and water. The stable community that is the endpoint of succession is the *climax.* Succession may begin on a newly constructed mineral deposit, such as a sand dune or river bar. It may also occur on a previously vegetated area that has been recently disturbed by agents such as fire, flood, and windstorms, or by human activity.

An important case of succession occurs on farmlands that are intensively cultivated for decades and then abandoned to the natural sequence of vegetation development. An example of this "old-field" succession comes from the southeastern United States, where the moist subtropical climate ⑥ provides ample soil water much of the year and the growing season is long. Figure 8.5 shows the succession, starting at the left with the bare field.

The first plants to take hold in the hostile environment are called *pioneers.* They are annual herbs most persons would call weeds. When they die, these plants supply organic matter to the soil. They also begin to shade the soil and reduce the extremes of soil temperature. Next, grasses and shrubs move in, and these occupy the old field during the first two decades or so. Pine seedlings now enter the habitat and, as these grow, a shade cover develops. This greatly changes the climate near the ground.

In the next half century, as the pine forest matures, broad-leaved deciduous trees begin to displace the pines. By the middle of the second century, a mature forest of oak and hickory has dominated the habitat. This climax forest has a lower layer of shrubs and small trees, a basal herb layer, and a substantial accumulation of organic matter on the ground in various stages of decomposition. In this environment, soil water tends to be conserved, and the thermal environment is protected from extremes of heat or cold.

Although the oak–hickory climax forest is generally stable in composition and structure, serious natural upsets to the successional sequence may occur through fires and severe storms. Insect hordes and disease epidemics may also radically alter the climax forest. Such disturbances will be followed by an appropriate succession, with stages leading to the return of the oak–hickory forest.

Disturbances may also result from human activity. Cutting and clearing of forest trees to reclaim the land for agriculture removes the climax forest. Or a disease introduced from a foreign continent may cause the extinction of a particular plant species, changing the climax. An example is the chestnut blight, which eliminated the American chestnut from the forests of the northeastern United States. Imported insects may also wipe out most of the mature individuals of a plant species if there is no native predator available to combat the invasion. These are just a few of the ways in which humans influence natural vegetation and the succession process.

TERRESTRIAL ECOSYSTEMS—THE BIOMES

What is an ecosystem? If ecology is the study of the relationships between organisms and their environment, then the **ecosystem** can be taken as the total assemblage of components that form those interrela-

tionships. It includes the living organisms—plants and animals, bacteria and viruses—as well as the nonliving matter that organisms depend on. Nonliving matter includes water and plant nutrients from the soil, and oxygen and carbon dioxide from the air. Also part of the ecosystem is energy absorbed or released by organisms, such as the sunlight that is absorbed in photosynthesis.

Geographers view ecosystems as natural resource systems. Food, fiber, fuel, and structural material are products of ecosystems. These are manufactured by organisms using energy that is derived from the sun. Humans harvest that energy by using ecosystem products. The products and productivity of ecosystems depend to a large degree on climate. Where temperature and rainfall cycles permit, ecosystems provide a rich bounty for human use. Where temperature or rainfall cycles restrict ecosystems, human activities can

Figure 8.6 Natural vegetation of the world.

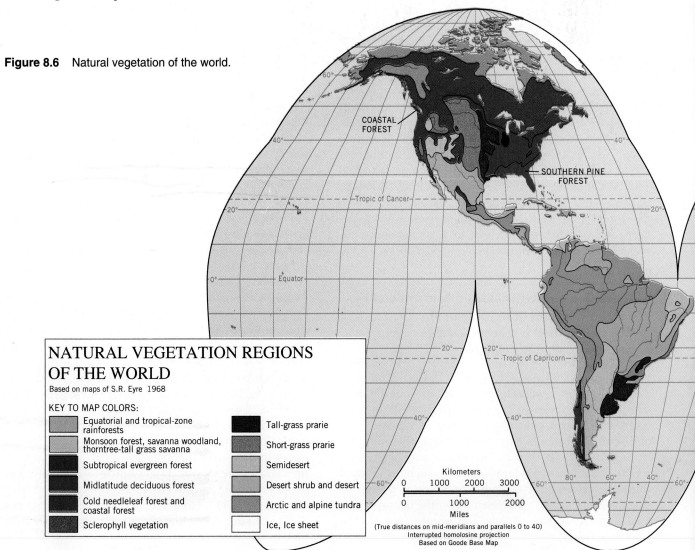

NATURAL VEGETATION REGIONS OF THE WORLD
Based on maps of S.R. Eyre 1968

KEY TO MAP COLORS:

Equatorial and tropical-zone rainforests

Monsoon forest, savanna woodland, thorntree-tall grass savanna

Subtropical evergreen forest

Midlatitude deciduous forest

Cold needleleaf forest and coastal forest

Sclerophyll vegetation

Tall-grass prarie

Short-grass prarie

Semidesert

Desert shrub and desert

Arctic and alpine tundra

Ice, Ice sheet

Kilometers
0   1000   2000   3000

0   1000   2000
Miles

(True distances on mid-meridians and parallels 0 to 40)
Interrupted homolosine projection
Based on Goode Base Map

also be limited. Of course, humans, too, are part of the ecosystem. A persistent theme of human geography is the study of how human societies function within ecosystems, utilizing ecosystem resources and modifying ecosystems for human benefit.

Ecosystems fall into two major groups—aquatic and terrestrial. *Aquatic ecosystems* include marine environments and the freshwater environments of the lands. Marine ecosystems include the open ocean, coastal estuaries, and coral reefs. Freshwater ecosystems include lakes, ponds, streams, marshes, and bogs. Our survey of physical geography will not include these aquatic environments. Instead, we will focus on the **terrestrial ecosystems,** which are dominated by land plants spread widely over the upland surfaces of the continents. The terrestrial ecosystems are directly influenced by climate and interact with the soil and, in

this way, are closely woven into the fabric of physical geography.

Within terrestrial ecosystems, the largest recognizable subdivision is the **biome.** Although the biome includes the total assemblage of plant and animal life interacting within the life layer, the green plants dominate the biome physically because of their enormous biomass, as compared with that of other organisms. Plant geographers concentrate on the characteristic life-form of the green plants within the biome. These life-forms are principally trees, shrubs, lianas, and herbs, but other life-forms are important in certain biomes. Biomes are recognized and mapped using their climax vegetation type. Areas intensively modified by agriculture and urbanization are assigned to a biome according to the climax vegetation assumed to have once been present.

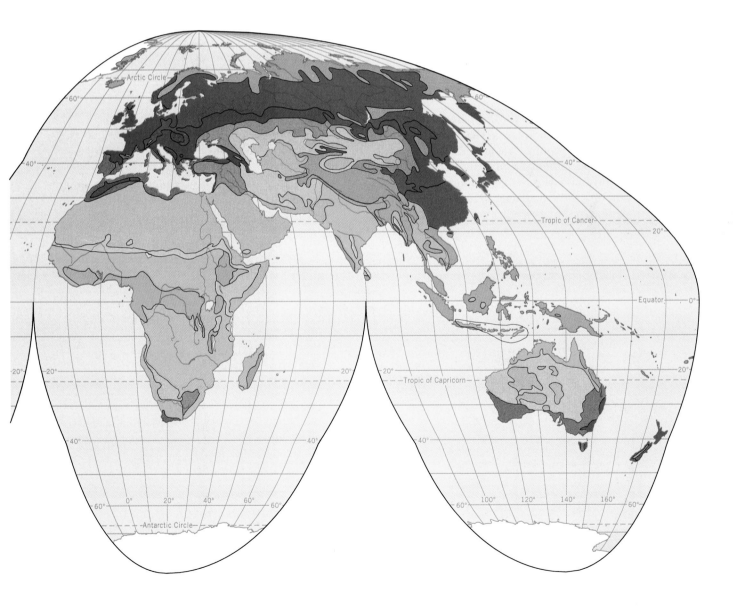

Figure 8.7 An aerial view of low-latitude rainforest. Middle Mazaruni River near Bartica, Guyana.

There are five principal biomes. The **forest biome** is dominated by trees, which form a closed or nearly closed canopy. Forest requires an abundance of soil water, so forests are found in the moist climates. Temperatures must also be suitable, requiring at least a warm season, if not warm temperatures the year round. The **savanna biome** is transitional between forest and grassland. It exhibits an open cover of trees with grasses and herbs underneath. The **grassland biome** develops in regions with moderate shortages of soil water. The semiarid regions of the dry tropical, dry subtropical, and dry midlatitude climates are the home of the grassland biome. Temperatures must also provide adequate warmth during the growing season. The **desert biome** includes organisms that can survive a moderate to severe water shortage for most, if not all, of the year. Temperatures can range from very hot to cool. Plants are often xerophytes, showing adaptations to the dry environment. The **tundra biome** is limited by cold temperatures. Only small plants that can grow quickly when temperatures warm above freezing in the warmest month or two can survive.

Ecologists recognize five principal biome types: forest, grassland, savanna, desert, and tundra.

Biogeographers break the biomes down further into smaller vegetation units, called *formation classes*, using the life-form of the plants. For example, at least four and perhaps as many as six kinds of forests are easily distinguished within the forest biome. At least three kinds of grasslands are easily recognizable. Deserts, too, span a wide range in terms of the abundance and life-form of plants. The formation classes described in the remaining portion of this chapter are major, widespread types that are clearly associated with specific

Figure 8.8 This diagram shows the typical structure of equatorial rainforest. (After J. S. Beard, *The Natural Vegetation of Trinidad*, Clarendon Press, Oxford.)

climate types. Figure 8.6 is a generalized world map of the formation classes. It simplifies the very complex patterns of natural vegetation to show large uniform regions in which a given formation class might be expected to occur.

Forest Biome

Low-Latitude Rainforest

Low-latitude rainforest, found in the equatorial and tropical latitude zones, consists of tall, closely set trees. Crowns form a continuous canopy of foliage and provide dense shade for the ground and lower layers (Figure 8.7). The trees are characteristically smooth-barked and unbranched in the lower two-thirds. Tree leaves are large and evergreen—thus, the equatorial rainforest is often described as "broadleaf evergreen forest."

Crowns of the trees of the low-latitude rainforest tend to form two or three layers (Figure 8.8). The highest layer consists of scattered "emergent" crowns that protrude from the closed canopy below, often rising to 40 m (130 ft). Some emergent species develop

wide buttress roots, which aid in their physical support (Figure 8.9). Below the layer of emergents is a second, continuous layer, which is 15 to 30 m (50 to 100 ft) high. A third, lower layer consists of small, slender trees 5 to 15 m (15 to 50 ft) high with narrow crowns.

Typical of the low-latitude rainforest are thick, woody lianas supported by the trunks and branches of trees. Some are slender, like ropes. Others reach thicknesses of 20 cm (8 in.). They climb high into the trees to the upper canopy, where light is available, and develop numerous branches of their own. *Epiphytes* ("air plants") are also common in low-latitude rainforest. These plants attach themselves to the trunk, branches, or foliage of trees and lianas. Their "host" is used solely as a means of physical support. Epiphytes include plants of many different types—ferns, orchids, mosses, and lichens (Figure 8.10).

Large numbers of species of trees coexist in the low-latitude rainforest. In equatorial regions of rainforest, as many as 3000 species may be found in a few square kilometers. Individuals of a given species are often widely separated. Many species of plants and animals in this very diverse ecosystem still have not been identified or named by biologists.

Figure 8.9 Buttress roots at the base of a large rainforest tree. From Daintree National Park, Queensland, Australia.

Figure 8.10 In this photo from the El Yunque rainforest, Carribean National Forest, Puerto Rico, red-flowering epiphytes adorn the trunks of sierra palms.

Equatorial and tropical rainforests are not "jungles" of impenetrable plant thickets. Rather, the floor of the low-latitude rainforest is usually so densely shaded that plant foliage is sparse close to the ground. This gives the forest an open aspect, making it easy to travel within the forest's interior. The ground surface is covered only by a thin litter of leaves. Dead plant matter rapidly decomposes because the warm temperatures and abundant moisture promote its breakdown by bacteria. Nutrients released by decay are quickly absorbed by roots. As a result, the soil is low in organic matter.

Equatorial and tropical rainforests are very diverse, containing large numbers of plant and animal species.

Low-latitude rainforest develops in a climate that is continuously warm, frost-free, and has abundant precipitation in all months of the year (or, at most, only one or two months are dry). These conditions occur in the wet equatorial climate ① and the monsoon and trade-wind littoral climate ②. In the absence of a cold or dry season, plant growth goes on continuously throughout the year. In this uniform environment, some plant species grow new leaves continuously, shedding old ones continuously as well. Still other plant species shed their leaves according to their own seasons, responding to the slight changes in day length that occur with the seasons.

World distribution of the low-latitude rainforest is shown in Figure 8.11. A large area of rainforest lies astride the equator. In South America, this equatorial rainforest includes the Amazon lowland. In Africa, the *equatorial rainforest* is found in the Congo lowland and in a coastal zone extending westward from Nigeria to Guinea. In Indonesia, the island of Sumatra, on the west, bounds a region of equatorial rainforest that stretches eastward to the islands of the western Pacific.

The low-latitude rainforest extends poleward through the tropical zone (lat. 10° to 25° N and S) along monsoon and trade-wind coasts. The monsoon and trade-wind coastal climate in which this *tropical-zone rainforest* thrives has a short dry season. However, the dry season is not intense enough to deplete soil water.

In the northern hemisphere, trade-wind coasts that have rainforests are found in the Philippine Island a also along the eastern coasts of Central America and the West Indies. These highlands receive abundant orographic rainfall in the belt of the trade winds. A good example of tropical-zone rainforest is the rainforest of the eastern mountains of Puerto Rico. In Southeast Asia, tropical-zone rainforest is extensive in monsoon coastal zones and highlands that have heavy rainfall and a very short dry season. These occur in Vietnam and Laos, southeastern China, and on the western coasts of India and Myanmar. In the southern hemisphere, belts of tropical-zone rainforest extend down the eastern Brazilian coast, the Madagascar coast, and the coast of northeastern Australia.

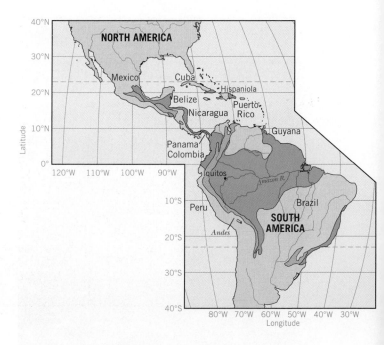

Figure 8.11   World map of low-latitude rainforest, showing equatorial and tropical rainforest types. (Data source same as Figure 8.6. Based on Goode Base Map.)

Figure 8.12 Monsoon woodland in the Bandipur Wild Animal Sanctuary in the Nilgiri hills of southern India. The scene was taken in the rainy season, with trees in full leaf.

Within the regions of low-latitude rainforest are many islandlike highland regions where climate is cooler and rainfall is increased by the orographic effect. Here, rainforest extends upward on the rising mountain slopes. Between 1000 and 2000 m (3000 and 6000 ft), the rainforest gradually changes in structure and becomes *montane forest* (Figure 8.11). The canopy of montane forest is more open, and tree heights are lower, than in the rainforest. With increasing elevation, the forest canopy height becomes even lower. Tree ferns and bamboos are numerous, and epiphytes are particularly abundant. Mist and fog are persistent much of the time, giving this high-elevation montane forest the name *cloud forest*.

Monsoon Forest

Monsoon forest of the tropical latitude zone differs from tropical rainforest because it is deciduous, with most of the trees of the monsoon forest shedding their leaves during the dry season. Shedding of leaves results from the stress of a long dry season that occurs at the time

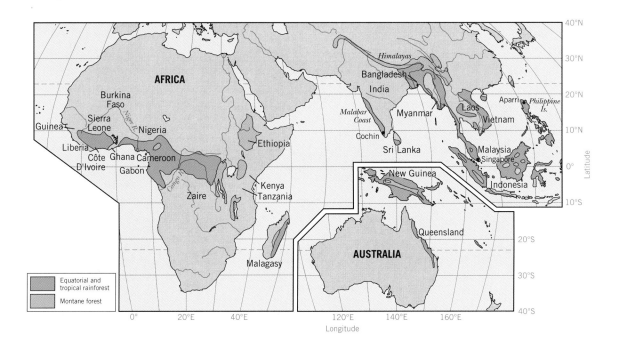

of low sun and cooler temperatures. In the dry season, the forest resembles the deciduous forests of the mid-latitudes during their leafless winter season.

Monsoon forest is typically open. It grades into woodland, with open areas occupied by shrubs and grasses (Figure 8.12). Because of its open nature, light easily reaches the lower layers of the monsoon forest. As a result, these lower layers are better developed than in the rainforest. Tree heights are also lower. Typically, many tree species are present—as many as 30 to 40 species in a small tract—although the rainforest has many more. Tree trunks are massive, often with thick, rough bark. Branching starts at a comparatively low level and produces large, round crowns.

Figure 8.13 is a world map of the monsoon forest and closely related formation classes. Monsoon forest develops in the wet-dry tropical climate ③ in which a long rainy season alternates with a dry, rather cool season. These conditions, though most strongly developed in the Asiatic monsoon climate, are not limited to that area. The typical regions of monsoon forest are in Myanmar, Thailand, and Cambodia. In the monsoon forest of southern Asia, the teakwood tree was once abundant and was widely exported to the Western world to make furniture, paneling, and decking. Now this great tree is logged out, and the Indian elephant, once trained to carry out this logging work, is unemployed. Large areas of monsoon forest also occur in south central Africa and in Central and South America, bordering the equatorial and tropical rainforests.

Subtropical Evergreen Forest

Subtropical evergreen forest is generally found in regions of moist subtropical climate ⑥, where winters are mild and there is ample rainfall throughout the year. This forest occurs in two forms: broadleaf and needleleaf. The *subtropical broadleaf evergreen forest* differs from the low-latitude rainforests, which are also broadleaf evergreen types, in having relatively few species of trees. Trees are not as tall as in the low-latitude rainforests. Their leaves tend to be smaller and more leathery, and the leaf canopy less dense. The subtropical broadleaf evergreen forest often has a well-developed lower layer of vegetation. Depending on the location, this layer may include tree ferns, small palms, bamboos, shrubs, and herbaceous plants. Lianas and epiphytes are abundant.

> **Subtropical evergreen forest occurs in broadleaf and needleleaf types. Most of the broadleaf type has been lost to centuries of cultivation.**

Figure 8.14 is a map of the subtropical evergreen forests of the northern hemisphere. Here subtropical evergreen forest consists of broad-leaved trees such as evergreen oaks and trees of the laurel and magnolia families. The name "laurel forest" is applied to these forests, which are associated with the moist subtropical climate ⑥ in the southeastern United States, southern

Figure 8.13 World map of monsoon forest and related types—savanna woodland and thorntree-tall grass savanna. (Data source same as Figure 8.6. Based on Goode Base Map.)

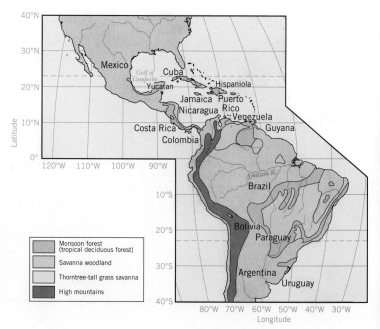

forest type is associated with the moist continental climate ⑩. Recall from Chapter 7 that this climate receives adequate precipitation in all months, normally with a summer maximum. There is a strong annual temperature cycle with a cold winter season and a warm summer.

Common trees of the deciduous forest of eastern North America, southeastern Europe, and eastern Asia are oak, beech, birch, hickory, walnut, maple, elm, and ash. A few needleleaf trees are often present as well—hemlock, for example. Where the deciduous forests have been cleared by lumbering, pines readily develop as second-growth forest.

In Western Europe, the midlatitude deciduous forest is associated with the marine west-coast climate ⑧. Here the dominant trees are mostly oak and ash, with beech in cooler and moister areas. In Asia, the midlatitude deciduous forest occurs as a belt between the boreal forest to the north and steppelands to the south. A small area of deciduous forest is found in Patagonia, near the southern tip of South America.

Needleleaf Forest

Needleleaf forest refers to a forest composed largely of straight-trunked, cone-shaped trees with relatively short branches and small, narrow, needlelike leaves. These trees are conifers. Most are evergreen, retaining their needles for several years before shedding them. When the needleleaf forest is dense, it provides continuous and deep shade to the ground. Lower layers of vegetation are sparse or absent, except for a thick carpet of mosses that may occur. Species are few—in fact, large tracts of needleleaf forest consist almost entirely of only one or two species.

Figure 8.17 Deciduous forest of the Catskill Mountains, New York. Some needleleaf trees—pine and hemlock—are also present.

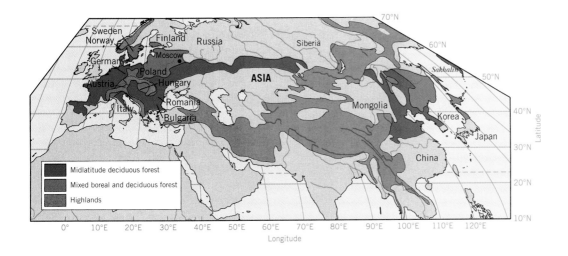

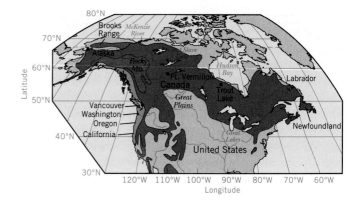

Figure 8.18 Northern hemisphere map of cold needleleaf forests, including coastal forest. (Data source same as Figure 8.6. Based on Goode Base Map.)

Boreal forest is the cold-climate needleleaf forest of high latitudes. It occurs in two great continental belts, one in North America and one in Eurasia (Figure 8.18). These belts span their landmasses from west to east in latitudes 45° to 75°, and they closely correspond to the region of boreal forest climate ⑪. The boreal forest of North America, Europe, and western Siberia is composed of such evergreen conifers as spruce and fir (Figure 8.19). The boreal forest of north central and eastern Siberia is dominated by larch. The larch tree sheds its needles in winter and is thus a deciduous needleleaf tree. Broadleaf deciduous trees, such as aspen, balsam poplar, willow, and birch, tend to take over rapidly in areas of needleleaf forest that have been burned over. These species can also be found bordering streams and in open places. Between the boreal forest and the midlatitude deciduous forest

lies a broad transition zone of mixed boreal and deciduous forest.

Needleleaf evergreen forest extends into lower latitudes wherever mountain ranges and high plateaus exist. For example, in western North America this formation class extends southward into the United States on the Sierra Nevada and Rocky Mountain ranges and over parts of the higher plateaus of the southwestern states (Figure 8.20, *left*). In Europe, needleleaf evergreen forests flourish on all the higher mountain ranges.

Needleleaf forests span the high latitudes of North America and Eurasia and extend southward on high mountains and plateaus.

Figure 8.19 A view of the Boreal forest, Denali National Park, Alaska, pictured here just after the first snowfall of the season. At this location near the northern limits of the Boreal forest, the tree cover is sparse. The golden leaves of aspen mark the presence of this deciduous species.

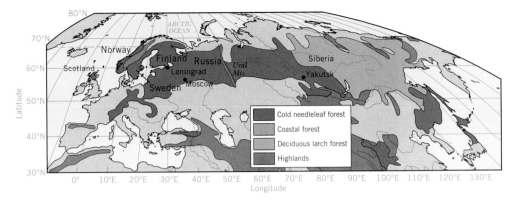

Figure 8.20 Two kinds of needleleaf forest of the western United States. (left) Open forest of western yellow pine (ponderosa pine), in the Kaibab National Forest, Arizona. (right) A grove of great redwood trees in Humboldt State Park, California.

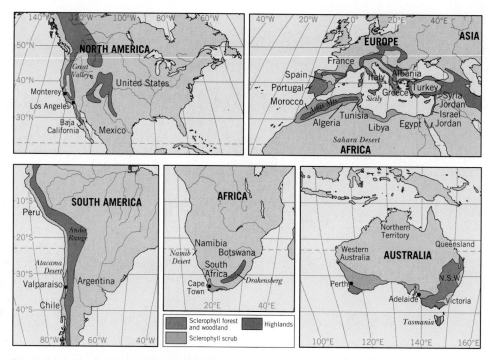

Figure 8.21 World map of sclerophyll vegetation. (Data source same as Figure 8.6. Based on Goode Base Map.)

In its northernmost range, the boreal needleleaf forest grades into cold woodland. This form of vegetation is limited to the northern portions of the boreal forest climate ⑪ and the southern portions of the tundra climate ⑫. Trees are low in height and spaced well apart. A shrub layer may be well developed. The ground cover of lichens and mosses is distinctive. Cold woodland is often referred to as *taiga*. It is transitional into the treeless tundra at its northern fringe.

Coastal forest is a distinctive needleleaf evergreen forest of the Pacific Northwest coastal belt, ranging in latitude from northern California to southern Alaska (Figure 8.18). Here, in a band of heavy orographic precipitation, mild temperatures, and high humidity, are perhaps the densest of all conifer forests, with the world's largest trees (Figure 8.20, *right*). Individual redwood trees attain heights of over 100 m (325 ft) and girths of over 20 m (65 ft).

Sclerophyll Forest

The native vegetation of the Mediterranean climate ⑦ is adapted to survival through the long summer drought. Shrubs and trees that can survive such drought are characteristically equipped with small, hard, or thick leaves that resist water loss through transpiration. As we noted earlier in the chapter, these plants are called sclerophylls.

Sclerophyll forest consists of trees with small, hard, leathery leaves. The trees are often low-branched and gnarled, with thick bark. The formation class includes *sclerophyll woodland,* an open forest in which only 25 to 60 percent of the ground is covered by trees. Also included are extensive areas of *scrub,* a plant formation type consisting of shrubs covering somewhat less than half of the ground area. The trees and shrubs are evergreen, retaining their thickened leaves despite a severe annual drought.

Our map of sclerophyll vegetation, Figure 8.21, includes forest, woodland, and scrub types. Sclerophyll forest is closely associated with the Mediterranean climate ⑦ and is narrowly limited to west coasts between 30° and 40° or 45° N and S latitude. In the Mediterranean lands, the sclerophyll forest forms a narrow, coastal belt ringing the Mediterranean Sea. Here, the Mediterranean forest consists of such trees as cork oak, live oak, Aleppo pine, stone pine, and olive. Over the centuries, human activity has reduced the sclerophyll forest to woodland or destroyed it entirely. Today, large areas formerly occupied by this forest consist of dense scrub.

The other northern hemisphere region of sclerophyll vegetation is the California coast ranges. Here, the sclerophyll forest or woodland is typically dominated by live oak and white oak. Grassland occupies the open ground between the scattered oaks (Figure

Figure 8.22   Evergreen oak woodland with grassland, Santa Ynez Valley, Santa Barbara County, California. (left) At the end of the long, dry summer the grasses are dormant, but the oak trees are green. (right) At the end of the cool, wet winter, grasses are a lush green, whereas a few deciduous oaks in the foreground are nearly leafless.

8.22). Much of the remaining vegetation is sclerophyll scrub or "dwarf forest," known as *chaparral* (Figure 8.23). It varies in composition with elevation and exposure. Chaparral may contain wild lilac, manzanita, mountain mahogany, poison "oak," and live oak.

In central Chile and in the Cape region of South Africa, sclerophyll vegetation has a similar appearance, but the dominant species are quite different. Important areas of sclerophyll forest, woodland, and scrub are also found in southeast, south central, and southwest Australia, including many species of eucalyptus and acacia.

Savanna Biome

The savanna biome is usually associated with the tropical wet-dry climate ③ of Africa and South America. It includes vegetation formation classes ranging from woodland to grassland. In *savanna woodland*, the trees are spaced rather widely apart, because soil moisture during the dry season is not sufficient to support a full tree cover. The open spacing permits development of a dense lower layer, which usually consists of grasses.

Figure 8.23   Chaparral, a vegetation type dominated by woody shrubs. Santa Lucia Mountains Wilderness, California.

The woodland has an open, parklike appearance. Savanna woodland usually lies in a broad belt adjacent to equatorial rainforest.

In the tropical savanna woodland of Africa, the trees are of medium height. Tree crowns are flattened or umbrella-shaped, and the trunks have thick, rough bark (see Figure 7.13). Some species of trees are xerophytic forms with small leaves and thorns. Others are broad-leaved deciduous species that shed their leaves in the dry season. In this respect, savanna woodland resembles monsoon forest.

Savanna vegetation is adapted to a strong wet-dry annual cycle. Grazing and periodic burning help maintain the openness of the savanna by suppressing tree seedlings.

Fire is a frequent occurrence in the savanna woodland during the dry season, but the tree species of the savanna are particularly resistant to fire. Many geographers hold the view that periodic burning of the savanna grasses maintains the grassland against the invasion of forest. Fire does not kill the underground parts of grass plants, but it limits tree growth to individuals of fire-resistant species. Many rainforest tree species that might otherwise grow in the wet-dry climate regime are prevented by fires from invading. The browsing of animals, which kills many young trees, is also a factor in maintaining grassland at the expense of forest.

The regions of savanna woodland are shown in Figure 8.13, along with monsoon forest (discussed earlier). In Africa, the savanna woodland grades into a belt of *thorntree-tall grass savanna*, a formation class transitional to the desert biome. The trees are largely of thorny species. Trees are more widely scattered, and the open grassland is more extensive than in the savanna woodland. One characteristic tree is the flat-topped acacia, seen in Figure 7.13. Elephant grass is a common species. It can grow to a height of 5 m (16 ft) to form an impenetrable thicket.

The thorntree-tall-grass savanna is closely identified with the semiarid subtype of the dry tropical and subtropical climates (④s, ⑤s). In the semiarid climate, soil water storage is adequate for the needs of plants only during the brief rainy season. The onset of the rains is quickly followed by the greening of the trees and grasses. For this reason, vegetation of the savanna

biome is described as rain-green, an adjective that also applies to the monsoon forest.

Grassland Biome

The grassland biome includes two major formation classes that we will discuss here—tall-grass prairie and steppe (Figure 8.24). *Tall-grass prairie* consists largely of tall grasses. *Forbs,* which are broad-leaved herbs, are also present. Trees and shrubs are absent from the prairie but may occur in the same region as narrow patches of forest in stream valleys. The grasses are deeply rooted and form a thick and continuous turf (Figure 8.25).

Prairie grasslands are best developed in regions of the midlatitude and subtropical zones with well-developed winter and summer seasons. The grasses flower in spring and early summer, and the forbs flower in late summer. Tall-grass prairies are closely associated with the drier areas of moist continental climate ⑩. Here, soil water is in short supply during the summer months.

When European settlers first arrived in North America, the tall-grass prairies were found in a belt extending from the Texas Gulf coast northward to southern Saskatchewan (Figure 8.24). A broad peninsula of tall-grass prairie extended eastward into Illinois, where conditions are somewhat more moist. Since the time of settlement, these prairies have been converted almost entirely to agricultural land. Another major area of tall-grass prairie is the Pampa region of South America, which occupies parts of Uruguay and eastern Argentina. The Pampa region falls into the moist subtropical climate ⑥, with mild winters and abundant precipitation.

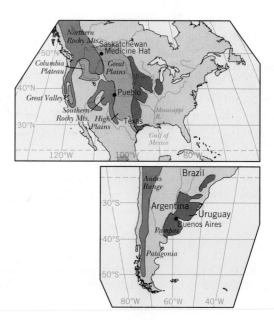

Figure 8.24 World map of the grassland biome in subtropical and midlatitude zones. (Data source same as Figure 8.6. Based on Goode Base Map.)

Figure 8.25 Tall-grass prairie, Iowa. In addition to grasses, tall-grass prairie vegetation includes many forbs, such as the flowering species shown in this photo.

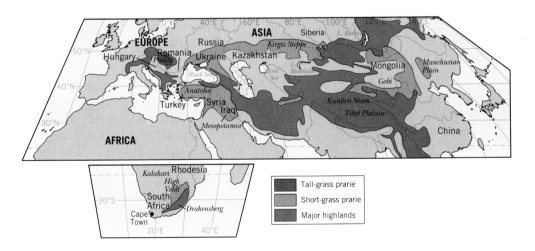

Eye on the Environment • Exploitation of the Low-Latitude Rainforest Ecosystem

Many of the world's equatorial and tropical regions are home to the rainforest ecosystem. This ecosystem is perhaps the most diverse on earth. That is, it possesses more species of plants and animals than any other. Very large tracts of rainforest still exist in South America, south Asia, and some parts of Africa. Ecologists regard this ecosystem as a genetic reservoir of many species of plants and animals. But as human populations expand and the quest for agricultural land continues, low-latitude rainforests are being threatened with clearing, logging, and cultivation for cash crops and animal grazing.

In the past, low-latitude rainforests were farmed by native peoples using the *slash-and-burn* method—cutting down all the vegetation in a small area, then burning it to ashes (see photo below). In a rainforest ecosystem, most of the nutrients are held within living plants rather than in the soil. Burning the vegetation on the site releases the trapped nutrients, returning a portion of them to the soil. Here, the nutrients are available to growing crops. The supply of nutrients derived from the original vegetation cover is small, however, and the harvesting of crops rapidly depletes the nutrients. After a few seasons of cultivation, a new field is cleared, and the old field is abandoned. Rainforest plants reestablish their hold on the abandoned area. Eventually, the rainforest returns to its original state. This cycle shows that primitive, slash-and-burn agriculture is fully compatible with the maintenance of the rainforest ecosystem.

On the other hand, modern intensive agriculture uses large areas of land and is not compatible with the rainforest ecosystem. When large areas are abandoned, seed sources are so far away that the original forest species cannot take hold. Instead, secondary species dominate, often accompanied by species from other vegetation types. These species are good invaders, and once they enter an area, they tend to stay. The dominance of these secondary species is permanent, at least on the human time scale. Thus, we can regard the rainforest ecosystem as a resource that, once cleared, will never return. The loss of low-latitude rainforest will result in the disappearance of thousands of species of organisms from the rainforest environment—a loss of millions of years of evolution, together with the destruction of the most complex ecosystem on earth.

In Amazonia, transformation of large areas of rainforest into agricultural land uses heavy machinery to carve out major highways, such as the Trans-Amazon Highway in Brazil, and innumerable secondary roads and trails (see photo to right). Large fields for cattle pasture or commercial crops are created by cutting, bulldozing, clearing, and

This rainforest in Maranhao, Brazil, has been felled and burned in preparation for cultivation.

burning the vegetation. In some regions, the great broadleaved rainforest trees are removed for commercial lumber.

What are the effects of large-scale clearing? A study using a computer model, completed in 1990 by scientists of the University of Maryland and the Brazilian Space Research Institute, indicated that when the Amazon rainforest is entirely removed and replaced with pasture, surface and soil temperatures will be increased 1 to 3°C (2 to 5°F). Precipitation in the region will decline by 26 percent and evaporation by 30 percent. The deforestation will change weather and wind patterns so that less water vapor enters the Amazon basin from outside sources, making the basin ever drier. In areas where a marked dry season occurs, that season will be lengthened. Although such models contain simplifications and are subject to error, the results confirm the pessimistic conclusion that once large-scale defor-

estation has occurred, artificial restoration of a rainforest comparable to the original one may be impossible to achieve.

According to a study published in 1990 by the World Resources Institute of Washington, D.C., the global total of low-latitude rainforest removed in 1989 was between 16 and 20 million hectares (60,000 to 80,000 sq mi). At the same time, figures released by the United Nations put the annual destruction rate at 17 million hectares (65,000 sq mi), an area about that of the state of Washington. The greatest loss of rainforest in 1989 was in Brazil—about 8 million hectares (30,000 sq mi). Deforestation rates in the decade 1980–1989 increased in India, Indonesia, and Costa Rica, the Philippines, Vietnam, Colombia, Ivory Coast, Laos, and Nigeria.

Although deforestation rates are very rapid in some regions, many nations are now working to reduce the rate of loss of rainforest environment. However, because

the rainforest can provide agricultural land, minerals, and timber, the pressure to allow deforestation continues

Questions

1. How do traditional agricultural practices in the low-latitude rainforest compare to present-day practices? What are the implications for the rainforest environment?

2. What are the effects of large-scale clearing on the rainforest environment?

The Trans-Amazon Highway in construction cut a great swath through the rainforest and required extensive earth moving to grade the roadbed and provide drainage.

Figure 8.26 Bison grazing on short-grass prairie in the northern Great Plains.

Tall-grass prairie provides rich agricultural land suited to cultivation and cropping. Short-grass prairie, or steppe, occupies vast regions of semidesert and is suited to grazing.

Recall from Chapter 7 that **steppe,** also called *short-grass prairie,* is a vegetation type consisting of short grasses occurring in sparse clumps or bunches (Figure 8.26). Scattered shrubs and low trees may also be found in the steppe. The plant cover is poor, and much bare soil is exposed. Many species of grasses and forbs occur. A typical grass of the American steppe is buffalo grass. Other typical plants are the sunflower and loco weed. Steppe grades into semidesert in dry environments and into prairie where rainfall is higher.

Our map of the grassland biome (Figure 8.24) shows that steppe grassland is concentrated largely in midlatitude areas of North America and Eurasia. The only southern hemisphere occurrence shown on the map is the "veldt" region of South Africa, a highland steppe surface in Orange Free State and Transvaal.

Steppe grasslands correspond well with the semiarid subtype of the dry continental climate ⑨. Spring rains nourish the grasses, which grow rapidly until early summer. By midsummer, the grasses are usually dormant. Occasional summer rainstorms cause periods of revived growth.

Desert Biome

The desert biome includes several formation classes that are transitional from grassland and savanna biomes into vegetation of the arid desert. Here we recognize two basic formation classes: semidesert and dry desert. Figure 8.29 is a world map of the desert biome.

Semidesert is a transitional formation class found in a wide latitude range—from the tropical zone to the midlatitude zone. It is identified primarily with the arid subtypes of all three dry climates. Semidesert consists of sparse xerophytic shrubs. One example is the

Figure 8.27   Sagebrush semidesert in Monument Valley, Utah.

Figure 8.28   A desert scene near Pheonix, Arizona. The tall, columnar plant is saguaro cactus; the delicate wandlike plant is ocotillo. Small clumps of prickly pear cactus are seen between groups of hard-leaved shrubs.

sagebrush vegetation of the middle and southern Rocky Mountain region and Colorado Plateau (Figure 8.27). Recently, as a result of overgrazing and trampling by livestock, semidesert shrub vegetation seems to have expanded widely into areas of the western United States that were formerly steppe grasslands.

Thorntree semidesert of the tropical zone consists of xerophytic trees and shrubs that are adapted to a climate with a very long, hot dry season and only a very brief, but intense, rainy season. These conditions are found in the semiarid and arid subtypes of the dry tropical ④ and dry subtropical ⑤ climates. The thorny trees and shrubs are known locally as thorn forest, thornbush, or thornwoods (see Figure 7.16). Many of these are deciduous plants that shed their leaves in the dry season. The shrubs may be closely intergrown to form dense thickets. Cactus plants are present in some localities.

Dry desert is a formation class of xerophytic plants that are widely dispersed over only a very small propor-

tion of the ground. The visible vegetation of dry desert consists of small, hard-leaved, or spiny shrubs, succulent plants (such as cactus), or hard grasses. Many species of small annual plants may be present but appear only after a rare, but heavy, desert downpour. Much of the world map area assigned to desert vegetation has no plant cover at all, because the surface consists of shifting-dune sands or sterile salt flats.

Tundra vegetation consists of low plants that are adapted to survival through a harshly cold winter. They grow, bloom, and set seed during a short summer thaw.

Desert plants differ greatly in appearance from one part of the world to another. In the Mojave and Sonoran deserts of the southwestern United States, plants are often large, giving the appearance of a

Figure 8.29 World map of the desert biome, including desert and semidesert formation classes. (Data source same as Figure 8.6. Based on Goode Base Map.)

woodland (Figure 8.28). Examples are the treelike saguaro cactus, the prickly pear cactus, the ocotillo, creosote bush, and smoke tree.

Tundra Biome

Arctic tundra is a formation class of the tundra climate ⑫. (See Figure 7.34 for a polar map of the arctic tundra.) In this climate, plants grow during the brief summer of long days and short (or absent) nights. At this time, air temperatures rise above freezing, and a shallow surface layer of ground ice thaws. The permafrost beneath, however, remains frozen, keeping the meltwater at the surface. These conditions create a marshy environment for at least a short time over wide areas. Because plant remains decay very slowly within the cold meltwater, layers of organic matter can build up in the marshy ground. Frost action in the soil fractures and breaks large roots, keeping tundra plants small. In winter, wind-driven snow and extreme cold also injure plant parts that project above the snow.

Plants of the arctic tundra are mostly low herbs, although dwarf willow, a small woody plant, occurs in places. Sedges, grasses, mosses, and lichens dominate the tundra in a low layer (Figure 7.36). Typical plant species are ridge sedge, arctic meadow grass, cotton grasses, and snow lichen. There are also many species of forbs that flower brightly in the summer. Tundra composition varies greatly as soils range from wet to well drained. One form of tundra consists of sturdy hummocks of plants with low, water-covered ground between. Some areas of arctic scrub vegetation composed of willows and birches are also found in tundra.

Tundra vegetation is also found at high elevations. This *alpine tundra* develops above the limit of tree growth and below the vegetation-free zone of bare rock and perpetual snow (Figure 8.30). Alpine tundra resembles arctic tundra in many physical respects.

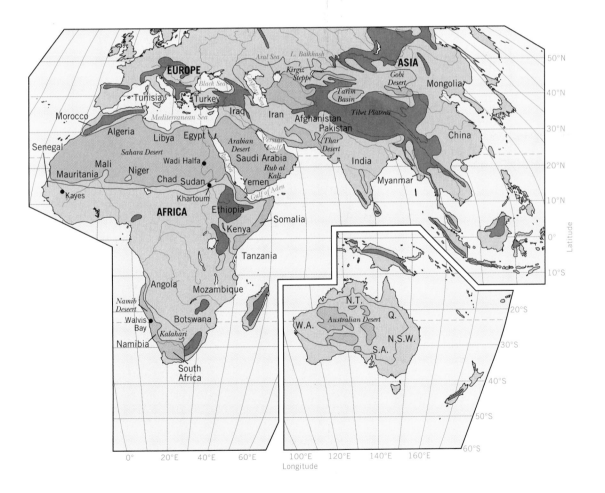

Altitude Zones of Vegetation

In earlier chapters, we described the effects of increasing elevation on climatic factors, particularly air temperature and precipitation. We noted that, with elevation, temperatures decrease and precipitation generally increases. These changes produce systematic changes in the vegetation cover as well, yielding a sequence of vegetation zones related to altitude.

The vegetation zones of the Colorado Plateau region in northern Arizona and adjacent states provide a strik-ing example of this altitude zonation. Figure 8.31 is a diagram showing a cross section of the land surface in this region. Elevations range from about 700 m (2000 ft) at the bottom of the Grand Canyon, to 3844 m (12,600 ft) at the top of San Francisco Peak. The vegetation cover and rainfall range are shown on the left. The vegetation zonation includes desert shrub, grassland, woodland, pine forest, Douglas fir forest, Engelmann spruce forest, and alpine meadow. Annual rainfall ranges from 12 to 25 cm (5 to 10 in.) in the desert shrub vegetation type to 80–90 cm (30–35 in.) in the Engelmann spruce forest.

Figure 8.30 Alpine tundra near the summit of the Snowy Range, Wyoming. Flag-shaped spruce trees (left), shaped by prevailing winds, mark the upper limit of tree growth.

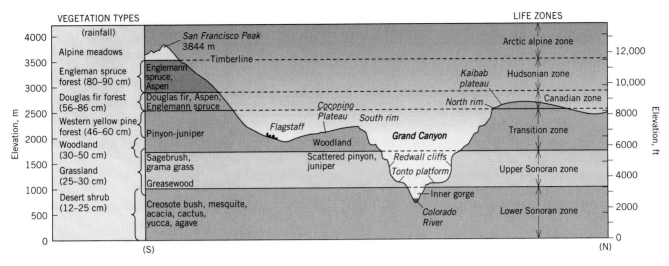

Figure 8.31   Altitude zones of vegetation in the arid southwestern United States. The profile shows the Grand Canyon–San Francisco Mountain district of northern Arizona. (Based on data of G. A. Pearson, C. H. Merriam, and A. N. Strahler.)

Ecologists have used the vegetation zonation to set up a series of *life zones,* which are also shown on the figure. These range from the lower Sonoran life zone, which includes the desert shrubs typical of the Sonoran Desert, to the arctic-alpine life zone, which includes the alpine meadows at the top of San Francisco Peak. Other life zones take their names from typical regions of vegetation cover. For example, the Hudsonian zone, 2900 to 3500 m (9500 to 11,500 ft), bears a needleleaf forest quite similar to needle-leaf boreal forest of the subarctic zone near Hudson's Bay, Canada.

Climatic Gradients and Vegetation Types

In discussing the major formation classes of vegetation, we have emphasized the importance of climate. As climate changes with latitude or longitude, vegetation will also change. Figure 8.32 shows three transects across portions of continents that illustrate this principle. (For these transects, we will ignore the effects of mountains or highland regions on climate and vegetation.)

The upper transect stretches from the equator to the tropic of cancer in Africa. Across this region, climate ranges through all four low-latitude climates: wet equatorial ①, monsoon and trade-wind coastal ②, wet-dry tropical ③, and dry tropical ④. Vegetation grades from equatorial rainforest, savanna woodland, and savanna grassland to tropical scrub and tropical desert.

The middle transect is a composite from the tropic of cancer to the arctic circle in Africa and Eurasia. Climates include many of the mid- and high-latitude

types: dry subtropical ⑤, Mediterranean ⑦, moist continental ⑩, boreal forest ⑪, and tundra ⑫. The vegetation cover grades from tropical desert through subtropical steppe, to sclerophyll forest in the Mediterranean. Further north is the midlatitude deciduous forest in the region of moist continental climate ⑩, which grades into boreal needleleaf forest, subarctic woodland, and finally tundra.

The lower transect ranges across the United States, from Nevada to Ohio. On this transect, the climate begins as dry midlatitude ⑨. Precipitation gradually increases eastward, reaching moist continental ⑩ near the Mississippi River. The vegetation changes from midlatitude desert and steppe to short-grass prairie, tall-grass prairie, and midlatitude deciduous forest.

The changes on these transects are largely gradational rather than abrupt. Yet, the global maps of both vegetation and climate show distinct boundaries from one region to the next. Which is correct? The true situation is gradational rather than abrupt. Maps must necessarily have boundaries to communicate information. But climate and vegetation know no specific boundaries. Instead, they are classified into specific types for convenience in studying their spatial patterns. When studying any map of natural features, keep in mind that boundaries are always approximate and gradational.

In this chapter, we have seen that the structure of the natural plant cover depends largely on climate. Each of the life-forms of plants has a capability for survival and a limit to survival as well. Forests exist where the environment is most favorable to plant growth

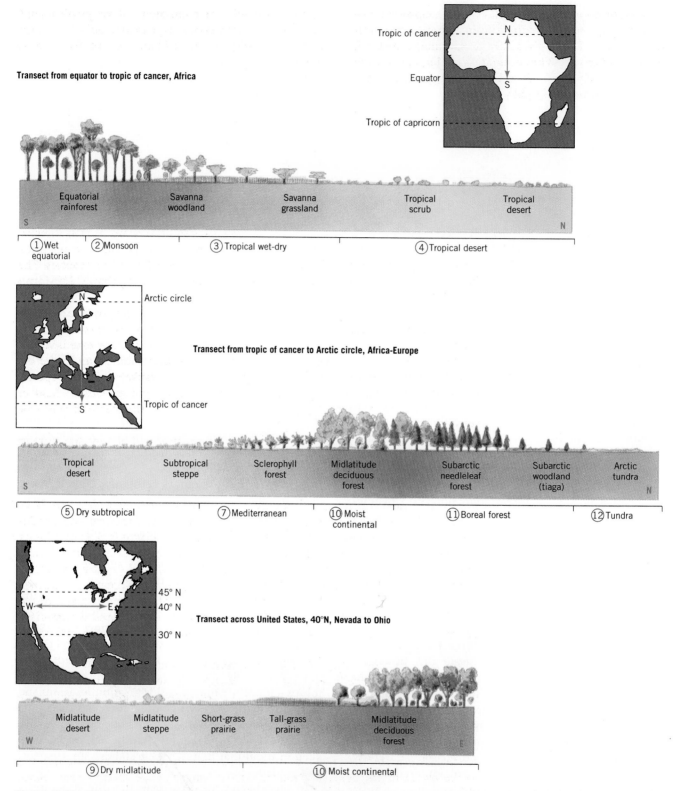

Transect from equator to tropic of cancer, Africa

Equatorial rainforest | Savanna woodland | Savanna grassland | Tropical scrub | Tropical desert

① Wet equatorial ② Monsoon ③ Tropical wet-dry ④ Tropical desert

Transect from tropic of cancer to Arctic circle, Africa-Europe

Tropical desert | Subtropical steppe | Sclerophyll forest | Midlatitude deciduous forest | Subarctic needleleaf forest | Subarctic woodland (tiaga) | Arctic tundra

⑤ Dry subtropical ⑦ Mediterranean ⑩ Moist continental ⑪ Boreal forest ⑫ Tundra

Transect across United States, 40°N, Nevada to Ohio

Midlatitude desert | Midlatitude steppe | Short-grass prairie | Tall-grass prairie | Midlatitude deciduous forest

⑨ Dry midlatitude ⑩ Moist continental

Figure 8.32 Three continental transects showing the succession of plant formation classes across climatic gradients.

because of favorable temperatures and abundant moisture during a long growing season. Where water is in short supply, forest gives way to savanna, grassland, steppe, and desert. Where climate is colder, low-latitude rainforest gives way to laurel forest, deciduous forest, needleleaf forest, and finally to tundra.

Plants are only one component of the physical landscape that is influenced by climate. Soils and landforms also respond to climate, as we will see in following chapters.

GLOBAL VEGETATION IN REVIEW

Natural vegetation is a plant cover that develops with little or no human interference. Although much vegetation appears to be in a natural state, humans influence the vegetation cover by fire suppression and introduction of new species. The life-form of a plant refers to its physical structure, size, and shape. Life-forms include trees, shrubs, lianas, herbs, and lichens.

Plants require water, which they take up and transpire. Xerophytes are plants adapted to dry environments, typically by small, wax-coated leaves or thickened, spongy stems. Other adaptations to drought include deep roots, or extensive and shallow roots, or a short life cycle. Deciduous plants drop their leaves in the dry or cold season. Temperature also limits plants. Only a relatively few plant species can endure freezing. In the process of ecological succession on cleared land or new ground, one vegetation type succeeds another until the stable climax type is reached.

Plants and animals, taken with all the nonliving matter and energy that support them, form an ecosystem. The largest unit of terrestrial ecosystems is the biome: forest, grassland, savanna, desert, and tundra.

The forest biome includes a number of important forest formation classes. The low-latitude rainforest exhibits a dense canopy and open floor with a very large number of species. Subtropical evergreen forest occurs in broadleaf and needleleaf forms in the moist subtropical climate ⑥. Monsoon forest is largely deciduous, with most species shedding their leaves after the wet season. Midlatitude deciduous forest is associated with the moist continental climate ⑩. Its species shed their leaves before the cold season. Needleleaf forest consists largely of evergreen conifers. It includes the coastal forest of the Pacific Northwest, the boreal forest of high latitudes, and needle-leaved mountain forests. Sclerophyll forest is comprised of trees with small, hard, leathery leaves, and is found in the Mediterranean climate ⑦ region.

The savanna biome consists of widely spaced trees with an understory, often of grasses. Dry-season fire is frequent in the savanna biome, limiting the number of trees and encouraging the growth of grasses. The grassland biome of midlatitude regions includes tall-grass prairie, in moister environments, and short-grass prairie, or steppe, in semiarid areas. Vegetation of the desert biome ranges from thorny shrubs and small trees to dry desert vegetation comprised of drought-adapted species. Tundra biome vegetation is limited largely to low herbs that are adapted to the severe drying cold experienced on the fringes of the Arctic Ocean. Since climate changes with altitude, vegetation typically occurs in altitudinal zones. Climate also changes gradually with latitude, and so biome changes are typically gradual, without abrupt boundaries.

KEY TERMS

natural vegetation
human-influenced vegetation
life-form
forest
woodland
transpiration

xerophyte
ecosystem
terrestrial ecosystem
biome
forest biome

grassland biome
savanna biome
desert biome
tundra biome
steppe

REVIEW QUESTIONS

1. What is natural vegetation? How do humans influence vegetation?

2. Plant geographers describe vegetation by its overall structure and by the life-forms of individual plants. Define and differentiate the following terms: forest, woodland, tree, shrub, herb, liana, perennial, deciduous, evergreen, broadleaf, needleleaf.

3. The climates of the globe provide great variations in annual moisture and temperature cycles. Discuss the various adaptations that plants have developed for life in stressful climates.

4. What is plant succession? Where and why does it occur? Describe the succession on old fields in the southeastern United States using a sketch of the vegetation sequence.

5. What are the five main biome types that ecologists and biogeographers recognize? Describe each briefly.

6. Low-latitude rainforests occupy a large region of the earth's land surface. What are the characteristics of these forests? Include forest structure, types of plants, diversity, and climate in your answer.

7. Monsoon forest and midlatitude deciduous forest are both deciduous, but for different reasons. Compare the characteristics of these two formation classes and their climates.

8. Subtropical broadleaf evergreen forest and tall-grass prairie are two vegetation formation classes that have been greatly altered by human activities. How was this done and why?

9. Distinguish among the types of needleleaf forest. What characteristics do they share? How are they different? How do their climates compare?

10. Which type of forest, with related woodland and scrub types, is associated with the Mediterranean climate? What are the features of these vegetation types? How are they adapted to the Mediterranean climate?

11. Describe the formation classes of the savanna biome. Where is this biome found and in what climate types? What role does fire play in the savanna biome?

12. Compare the two formation classes of the grassland biome. How do their climates differ?

13. Describe the vegetation types of the desert biome.

14. What are the features of arctic and alpine tundra? How does the cold tundra climate influence the vegetation cover?

15. How does elevation influence vegetation? Provide an example of how vegetation zonation is related to elevation.

IN-DEPTH QUESTIONS

1. Figure 8.33 presents a vegetation transect from Nevada to Ohio. Expand the transect so that it begins in Los Angeles, heads northeast from Ohio through Pennsylvania, New York, western Massachusetts, and New Hampshire, and ends in Maine. Sketch the vegetation types in your additions and label them, as in the diagram. Below your vegetation transect, draw a long bar subdivided to show the climate types.

2. Construct a similar transect of climate and vegetation from Miami to St. Louis, Minneapolis, and Winnipeg.

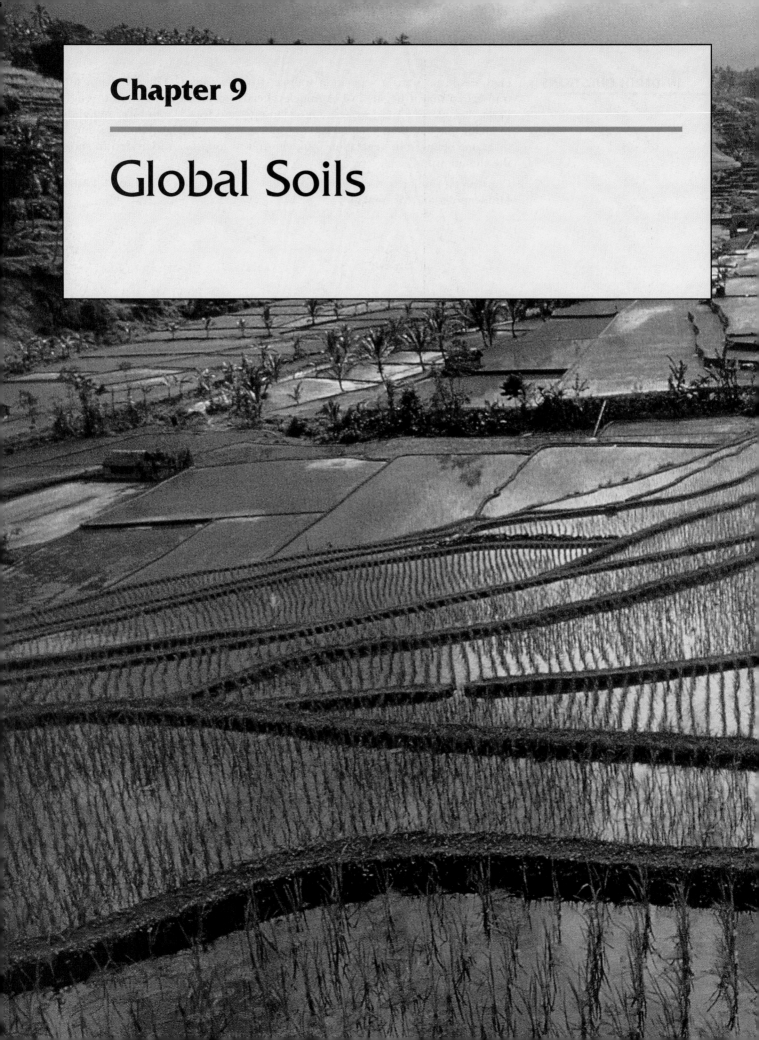

Chapter 9

Global Soils

When you think about soil, what comes to mind? Perhaps you are from New England, and recall digging behind the shed or garage for fishing worms. Your pitchfork turned over the soil's black surface layer, thick with decaying leaves of oaks and maples. If you are from the Southeast, you may picture a cut bank of red-orange, clayey soil, seen at the side of a country road. Or you may be from California and think of a field of grayish soil, with long green rows of young lettuce plants separated by deep furrows that are flooded by irrigation water. Whatever your thoughts about soil, you probably know that North American soils provide the base for our agriculture, which is the most efficient and productive in the world. But this has not always been the case.

At the close of the eighteenth century, the fertility of the agricultural lands of the Southeast was declining rapidly, a fact that was most evident to our nation's Founding Fathers. Romantically, we picture George Washington, happily retired on his Virginia estate Mount Vernon, reaping a richly deserved bounty from expansive farmlands under his personal supervision. This vision is far from the truth. In fact, agricultural yields at Mount Vernon, and elsewhere in the Southeast, were dropping.

Colonial farmers knew of the importance of natural fertilizers in maintaining high productivity. George Washington conserved animal manure to spread on his fields and had his field hands bring rich mud from creeks and marshes to spread on the soil to bring in a new supply of nutrients. Yet, still the decline continued. In 1834, 35 years after Washington's death, a visitor to Mount Vernon declared that "a more widespread and perfect agricultural ruin could not be imagined."

The mystery of failing agricultural fertility was solved by Edmund Ruffin (1794–1865), who owned lands on the coastal plain of Virginia. As with others, crop yields on his land were rapidly declining in the early 1800s. Quite by chance, Ruffin obtained a copy of Sir Humphry Davy's *Agricultural Chemistry*, published in 1813. Despite Ruffin's lack of formal education in science, he was quick to grasp the significance of one statement: "any acid matter . . . may be ameliorated by application of quicklime."

So it came about that on a February morning in 1818, Ruffin directed his field hands to haul marl from pits in low areas of his lands. Marl is a soft, lime-rich mud that occurs widely on the eastern coastal plain. The workers spread 200 bushels of marl over several acres of newly cleared ridge land of poor quality. In the spring, Ruffin planted this area in corn to test the effect of the marl. In the words of historian Avery Craven, this is what happened: "Eagerly he waited. As the season advanced, he found reason for joy. From the very start the plants on marled ground showed marked superiority, and at harvest time they yielded an advantage of fully forty per cent. The carts went back to the pits. Fields took on fresh life. A new era in agricultural history of the region had dawned."[1]

Why did the application of lime enhance soil fertility so dramatically? As we will see in this chapter, nutrients are held in soils on the surfaces of fine particles. In an acid environment, these surfaces do not hold nutrients well, and so downward-moving water from precipitation washes nutrients out of the reach of plant roots. Lime reduces acidity and restores the ability of soil particles to hold nutrients, so that they remain available to plants.

You may wonder why the American farmers did not understand the importance of liming their fields. The colonists who settled this region were largely from England, where their ancestors had farmed continuously and successfully for centuries. In England, soils are formed on freshly ground mineral matter left by the great ice sheets. The breakdown products of this fresh mineral matter act like lime to reduce soil acidity, so that liming is not needed. In contrast, upland soils of the eastern seaboard from Virginia to Georgia have been continually exposed to a mild, moist climate for tens of thousands of years. Fresh mineral matter in these soils has long been broken down, and the residual mineral matter that remains is highly acid. Thus, the colonists had no experience with these fundamentally different southeastern soils.

Soils are influenced by the rock material from which they develop, by the climate of the region (especially precipitation and temperature), by the vegetation cover that develops on them, and by the length of time that the soil has been in place. Your new knowledge of global climate and vegetation, acquired in the last two chapters, will be quite helpful in understanding soils and the processes that form them.

[1] Avery Craven, *Edmund Ruffin, Southerner* (New York: Appleton, 1932), p. 55.

Terraced fields, Bali, Indonesia.

Soil is the uppermost layer of the land surface that plants use and depend on for nutrients, water, and physical support. Soils can vary greatly from continent to continent, region to region, and even from field to field. This is because they are influenced by factors and processes that can vary widely from place to place. For example, a field near a large river that floods regularly may acquire a layer of nutrient-rich silt, making its soil very productive. A nearby field at a higher elevation, without the benefit of silt enrichment, may be sandy or stony and require the addition of fertilizers to grow crops productively.

Vegetation is an important factor in determining soil qualities. For example, America's richest soils developed in the Middle West under a cover of thick grass sod. The deep roots of the grass, in a cycle of growth and decay, deposited nutrients and organic matter throughout the thick soil layer. In the Northeast, conifer forests provided a surface layer of decaying needles that kept the soil quite acid. This acidity allowed nutrients to be washed below root depth, out of the reach of plants. When farmed today, these soils need applications of lime, which reduce the acidity and enhance soil fertility.

Climate, measured by precipitation and temperature, is also an important determiner of soil properties. Precipitation controls the downward movement of nutrients and other chemical compounds in soils. If precipitation is abundant, water tends to wash soluble compounds, including nutrients, deeper into the soil and out of reach of plant roots.

Temperature acts to control the rate of decay of organic matter that falls to the soil from the plant cover or that is provided to the soil by the death of roots. When conditions are warm and moist, decay organisms work efficiently, consuming organic matter readily. Thus, organic matter and nutrients in soils of the tropical and equatorial zones are generally low. Where conditions are cooler, decay proceeds more slowly, and organic matter is more abundant in the soil. Of course, if the climate is very dry or desertlike, then vegetation growth is slow or absent. No matter what desert temperatures are like, organic matter will be low.

Time is also an important factor. The characteristics and properties of soils require time for development. For example, a fresh deposit of mineral matter, like a sand dune, may require hundreds to thousands of years to acquire the structure and properties of a sandy soil.

Geographers are keenly interested in the differences in soils from place to place over the globe. The ability of the soils and climate of a region to produce food largely determines the size of the population it will support. In spite of the growth of cities, most of the world's inhabitants still live close to the soil that furnishes their food. And many of those same inhabitants die prematurely when the soil does not furnish enough food for all.

THE NATURE OF THE SOIL

Soil, as the term is used in soil science, is a natural surface layer that contains living matter and can support plants. The soil consists of matter in all three states—solid, liquid, and gas. It includes both *mineral matter* and *organic matter*. Mineral matter is largely derived from rock material, whereas organic matter is of biological origin and may be living or dead. Living matter in the soil consists not only of plant roots, but also of many kinds of organisms, including microorganisms.

Soil scientists use the term *humus* to describe finely divided, partially decomposed organic matter in soils. Some humus rests on the soil surface, and some is mixed through the soil. Humus particles of the finest size are gradually carried downward to lower soil layers by rainfall that sinks in and moves through the soil. When abundant, humus particles can give the soil a brown or black coloration.

Both air and water are found in soil. Water may tend to contain high levels of dissolved substances, such as nutrients. Air in soils may have high levels of such gases as carbon dioxide or methane, and low levels of oxygen.

The solid, liquid, and gaseous matter of the soil are constantly changing and interacting through chemical and physical processes. This makes the soil a very dynamic layer. Because of these processes, soil science, often called *pedology*, is a highly complex body of knowledge.

Although we may think of soil as occurring everywhere, large expanses of continents possess a surface layer that cannot be called soil. For example, dunes of moving sand, bare rock surfaces of deserts and high mountains, and surfaces of fresh lava near active volcanoes do not have a soil layer.

The characteristics of soils are developed over a long period of time through a combination of many processes acting together. Physical processes act to break down rock fragments of regolith into smaller and smaller pieces. An example is the growth of fine roots, or crystals of ice or salts, in rock fissures to wedge rock fragments apart. Chemical processes alter

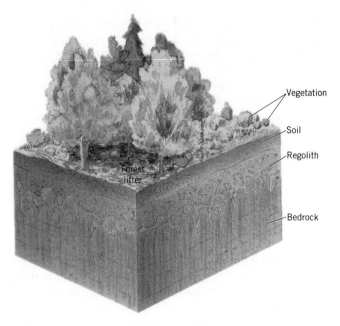

Figure 9.1 In this cross section through the land surface, vegetation and forest litter lie atop the soil. Below is regolith, produced by the breakup of the underlying bedrock.

the mineral composition of the original rock, producing new minerals. Taken together, these physical and chemical processes are referred to as *weathering.* Weathering occurs in soils and is part of the process by which soils develop their properties and characteristics. (We will return to the topic of weathering in Chapters 10 and 13.) In most soils, the inorganic material of the soil consists of fine particles of mineral matter. The term **parent material** describes all forms of mineral matter that are suitable for transformation into soil. Parent material may be derived from the underlying *bedrock*, which is solid rock below the soil layer (Figure 9.1). Over time, weathering processes soften, disintegrate, and break bedrock apart, forming a layer of *regolith*, or residual mineral matter. Regolith is one of the common forms of parent material. Other kinds of regolith consist of mineral particles transported to a place of rest by the action of streams, glaciers, waves and water currents, or winds. For example, dunes formed of sand transported by wind are a type of regolith on which soil may be formed.

Soil Color and Texture

The most obvious feature of a soil or soil layer is probably its color. Some color relationships are quite simple. For example, the soils of the Midwest prairies have a black or dark brown color because they contain abundant particles of partly decomposed plants. Red or yellow colors often mark the soils of the Southeat. These colors are created by the presence of iron-containing oxide compounds.

In some areas, soil color may be inherited from the mineral parent material, but, more generally, soil color is generated by soil-forming processes. For example, a white surface layer in soils of dry climates often indicates the presence of mineral salts, brought upward by evaporation. A pale, ash-gray layer near the top of soils of the boreal forest climate results when organic matter and various colored minerals are washed downward, leaving only pure, light-colored mineral matter behind. As we explain soil-forming processes and describe the various classes of soils, soil color will take on more meaning.

The major factors influencing soil and soil development are parent material, climate, vegetation, and time.

The mineral matter of the soil consists of individual mineral particles that vary widely in size. The term **soil texture** refers to the proportion of particles that fall into each of three size grades—*sand, silt,* and *clay.* The diameter range of each of these grades is shown in Figure 9.2. Millimeters are the standard units. Each unit on the scale represents a power of ten, so that clay particles of 0.000,001 millimeters diameter are one-millionth the size of sand grains 1 mm in diameter. The finest of all soil particles are termed *colloids.* In measuring soil texture, gravel and larger particles are eliminated, since these play no important role in soil processes.

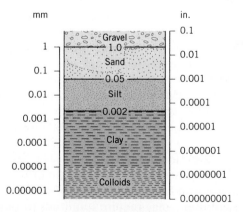

Figure 9.2 Size grades, which are names like sand, silt, and clay, refer to mineral particles within a specific size range. They are defined using the metric system. English equivalents are also shown.

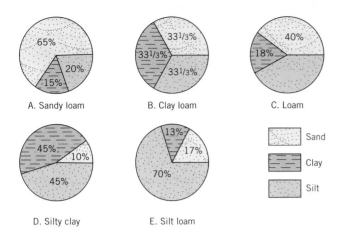

A. Sandy loam — 65%, 20%, 15%

B. Clay loam — 33⅓%, 33⅓%, 33⅓%

C. Loam — 40%, 18%

D. Silty clay — 45%, 45%, 10%

E. Silt loam — 13%, 17%, 70%

Sand
Clay
Silt

Figure 9.3 These diagrams show the proportions of sand, silt, and clay in five different soil texture classes.

Soil texture is described by a series of names that emphasize the dominant particle size, whether sand, silt, or clay (including colloids). Figure 9.3 gives examples of five soil textures with typical percentage compositions. A *loam* is a mixture containing a substantial proportion of each of the three grades. Loams are classified as sandy, silty, or clay-rich when one of these grades is dominant.

Why is soil texture important? Texture largely determines the ability of the soil to retain water. Coarse-textured (sandy) soils have many small passages between touching mineral grains that quickly conduct water through to deeper layers. If the soil consists of fine particles, passages and spaces are much smaller. Thus, water will penetrate more slowly and also tend to be retained. We will return to the important topic of the water-holding ability of soils in a later section.

Soil Colloids

Soil colloids consist of particles smaller than one hundred-thousandth of a millimeter (0.000,01 mm). Like other soil particles, some colloids are mineral, while others are organic. If you examine mineral colloids under a microscope, you will find that they consist of thin, platelike bodies (Figure 9.4). When well mixed in water, particles this small remain suspended indefinitely, giving the water a murky appearance. Mineral colloids are usually very fine particles of clay minerals. Organic colloids are tiny bits of organic matter that are resistant to decay.

Soil colloids are important because their surfaces attract soil nutrients, which are in the form of ions dissolved in soil water. An *ion* is an electrically charged atom or group of atoms. Among the many ions in soil water, one important group consists of **bases,** which

are ions of four elements: calcium (Ca^{++}), magnesium (Mg^{++}), potassium (K^+), and sodium (Na^+). Because plants require these elements, they are among the *plant nutrients*—ions or chemical compounds that are needed for plant growth.

Soil Acidity and Alkalinity

The soil solution also contains hydrogen (H^+) and aluminum (Al^{+++}) ions. But unlike the bases, they are not considered plant nutrients. The presence of these acid ions in the soil solution tends to make the solution acid in chemical balance.

An important principle of soil chemistry is that the acid ions have the power to replace the nutrient bases clinging to the surfaces of the soil colloids. As acid ions accumulate, the bases are released to the soil solution. They are gradually washed downward below rooting level, reducing soil fertility. When this happens, the soil acidity is increased.

Colloids are the finest particles in the soil. The surfaces of colloids attract base ions, which are used by plants as nutrients.

The degree of acidity or alkalinity of a solution is designated by the pH value. The lower the pH value,

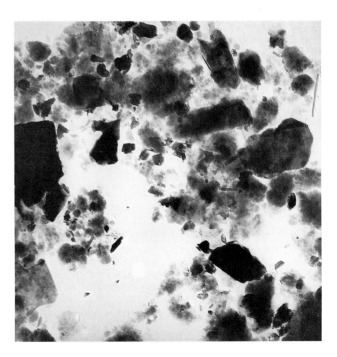

Figure 9.4 Seen here enlarged about 20,000 times are tiny flakes of clay minerals of colloidal dimensions. These particles have settled from suspension in San Francisco Bay.

the greater the degree of acidity. A pH value of 7 represents a neutral state—for example, pure water has a pH of 7. Lower values are in the acid range, while higher values are in the alkaline range.

Table 9.1 shows the natural range of acidity and alkalinity found in soils. High soil acidity is typical of cold, humid climates. In arid climates, soils are typically alkaline. Acidity can be corrected by the application of lime, a compound of calcium, carbon, and oxygen ($CaCO_3$), which removes acid ions and replaces them with the base calcium.

Soil Structure

Soil structure refers to the way in which soil grains are grouped together into larger masses, called *peds*. Peds range in size from small grains to large blocks. They are bound together by soil colloids. Small peds, roughly shaped like spheres, give the soil a granular structure or crumb structure (see Figure 9.5). Larger peds provide an angular, blocky structure. Peds form when colloid-rich clays shrink in volume as they dry out. Shrinkage results in formation of soil cracks, which define the surfaces of the peds.

Soils with a well-developed granular or blocky structure are easy to cultivate. This is an important agricultural factor in lands where primitive plows, drawn by animals, are still widely used. Soils with a high clay content can lack peds. These soils are sticky and heavy when wet and are difficult to cultivate. When dry, they become too hard to be worked.

Minerals of the Soil

Soil scientists recognize two classes of minerals abundant in soils: primary minerals and secondary minerals. The *primary minerals* are compounds present in unaltered rock. These are mostly silicate minerals—compounds of silicon and oxygen, with varying proportions of aluminum, calcium, sodium, iron, and magnesium. (The silicate minerals are described more fully in Chapter 10.) Primary minerals form a large fraction of the solid matter of many kinds of soils, but they play no important role in sustaining plant or animal life.

When primary minerals are exposed to air and water at or near the earth's surface, they are slowly altered in chemical composition. This process is part of *mineral alteration,* a chemical weathering process that is explained in more detail in Chapter 10. The primary minerals are altered into **secondary minerals,** which are essential to soil development and to soil fertility.

In terms of the properties of soils, the most important secondary minerals are the *clay minerals.* They form the majority of fine mineral particles in soils. From the viewpoint of soil fertility, the ability of a clay mineral to hold base ions is its most important property. This ability varies with the particular type of clay mineral; some hold bases tightly, and others loosely.

The nature of the clay minerals in a soil determines its *base status.* If the clay minerals can hold abundant base ions, the soil is of *high base status* and generally will be highly fertile. If the clay minerals hold a smaller supply of bases, the soil is of *low base status* and is generally less fertile. Humus colloids have a high capacity to hold bases so that the presence of humus is usually associated with potentially high soil fertility.

Mineral oxides are secondary minerals of importance in soils. They occur in many kinds of soils, particularly those that remain in place over very long periods of time (hundreds of thousands of years) in areas of warm, moist climates. Under these conditions, minerals are ultimately broken down chemically into simple oxides, compounds in which a single element is combined with oxygen.

Table 9.1   Soil Acidity and Alkalinity

pH	4.0	4.5	5.0	5.5	6.0	6.5	6.7	7.0	8.0	9.0	10.0	11.0
Acidity	Very strongly acid			Strongly acid	Moderately acid	Slightly acid	Neutral		Weakly alkaline	Alkaline	Strongly alkaline	Excessively alkaline
Lime requirements	Lime needed except for crops requiring acid soil			Lime needed for all but acid-tolerant crops		Lime generally not required			No lime needed			
Occurrence areas	Rare			Frequent	Very common in cultivated soils of humid climates				Common in sub-humid and arid climates		in deserts	Limited

Based on data of C. E. Millar, L. M. Turk, and H. D. Foth, *Fundamentals of Soil Science,* New York: John Wiley & Sons.

Oxides of aluminum and iron are the most important oxides in soils. Two atoms of aluminum are combined with three atoms of oxygen to form the *sesquioxide* of aluminum (Al_2O_3). (The prefix *sesqui-* means "one and a half" and refers to the chemical composition of one and one-half atoms of oxygen for every atom of aluminum.) In soils, aluminum oxide forms the mineral *bauxite,* which is a combination of aluminum sesquioxide and water molecules bound together. It occurs as hard, rocklike lumps and layers below the soil surface. Where bauxite layers are thick and uniform, they are sometimes strip-mined as aluminum ore.

Sesquioxide of iron (Fe_2O_3), again held in combination with water molecules, is *limonite,* a yellowish to reddish mineral that supplies the typical reddish to chocolate-brown colors of soils and rocks. Some shallow accumulations of limonite were formerly mined as a source of iron. Limonite and bauxite occur in close association in soils of warm, moist climates in low latitudes.

Soil Moisture

Besides providing nutrients for plant growth, the soil layer serves as a reservoir for the moisture that plants require. Soil moisture is a key factor in determining how the soils of a region support vegetation and crops.

The soil receives water from rain and from melting snow. Where does this water go? First, some of the water can run off the soil surface and not sink in. Instead, it flows into brooks, streams, and rivers, eventually reaching the sea. What of the water that sinks into the soil? Some of this water is returned to the atmosphere as water vapor. This happens when soil water evaporates and when transpiration by plants moves soil water from roots to leaves, where it evaporates. Taken together, we can term these last two losses *evapotranspiration.* Some water can also flow completely through the soil layer to recharge supplies of ground water at depths below the reach of plant roots.

When precipitation infiltrates the soil, the water wets the soil layer. This process is called *soil water recharge.* Eventually, the soil layer holds the maximum possible quantity of water, even though the larger pores may remain filled with air. Water movement then continues downward.

Suppose now that no further water enters the soil for a time. Excess soil water continues to drain downward, but some water clings to the soil particles. This water resists the pull of gravity because of the force of *capillary tension.* To understand this force, think about a droplet of condensation that has formed on the cold surface of a glass of ice water. The water droplet seems to be enclosed in a "skin" of surface molecules, drawing the droplet together into a rounded shape. The "skin" is produced by capillary tension. This force keeps the drop clinging to the side of the glass indefinitely, defying the force of gravity. Similarly, tiny films of water adhere to soil grains, particularly at the points of grain contacts. They remain until they evaporate or are absorbed by plant rootlets.

When a soil has first been saturated by water and then allowed to drain under gravity until no more water moves downward, the soil is said to be holding its *storage capacity* of water. For most soils drainage takes no more than two or three days. Most excess water is drained out within one day.

Figure 9.5 This soil shows a granular texture. The grains are refered to as peds.

Eye on the Environment • Death of a Civilization

What causes a civilization to collapse after flourishing for thousands of years? Wars and conquests come to mind, but sometimes the cause is less obvious. A case in point is the decline of the civilization of Sumer, and its successor, Babylon. Perhaps you have already learned something of these ancient Middle Eastern cultures, which were nourished by the Tigris and Euphrates rivers long before the birth of Christ. These two famous rivers drain the Zagros and Taurus mountains of Turkey and Iran, and flow out through a broad valley across the deserts of Iraq to the Persian Gulf. (See map at right.)

The Sumerian civilization evolved in the lower part of the Tigris-Euphrates Valley, beginning with a village culture dating to about five millennia B.C. By 3000 B.C., the Sumerian civilization was well established. The agriculture that supported Sumer depended on a system of irrigation canals utilizing the waters of the Tigris and Euphrates. Supported by this highly productive agricultural base, urban culture progressed to a high level, including the development of a written language. Skilled craftsmen produced pottery and—using gold, silver, and copper—jewelry and ornate weapons.

Trouble set in for the Sumerians in about 2400 B.C. It came in the form of a deterioration of their croplands caused by an accumulation of salt in the soil. In this process, today called salinization, a small increment of salt is added to the soil each year by the evaporation of irrigation water containing a low concentration of salts. Over time, the salt content of the soil increases until it reaches levels high enough to affect crop plants. Wheat, which was a staple crop of Sumer, is quite sensitive to salt, and its yield declines sharply as the salt concentration rises in the soil. Barley, another staple crop of that era, is somewhat less sensitive to salt.

Archaeologists and historians have inferred the onset of salinization in Sumerian agricultural lands by a shift in the proportions of wheat and barley in the national agricultural output. It seems that around 3500 B.C. these two grains were grown in about equal amounts. By 2500 B.C., however, wheat accounted for only about one-sixth of the total grain production. Evidently, the more salt-tolerant barley had replaced wheat over much of the area.

At this point in time, the Sumerian grain yield began to decline seriously. Ancient records show that in 2400 B.C., around the Sumerian city of Girsu, fields were yielding an average of about 300 kg of grain per hectare (270 lb/acre)—a high rate, even by modern standards. By 2100 B.C. the grain yield had declined to 180 kg per hectare (160 lb/acre), and by 1700 B.C., to a mere 100 kg/ha (90 lb/acre). Cities so declined in population that they finally dwindled into villages. As Sumerian civilization withered in the south, Babylonia, in the northern part of the valley, became more important. Soon, political control of the region passed from Sumer to Babylon.

Historians chronicle an event that also seems to have played a part in the decline of the Sumerian civilization. A group of Sumerian cities along the Euphrates River needed more irrigation water than they had. After a long and fruitless series of conflicts with upstream cities to gain additional water, the problem was solved by running a canal across the valley from the Tigris River. Now there was plenty of water. However, it was used so abundantly that a rise of the ground water table set in. As the water table came closer to the ground surface, capillary tension drew the ground water to the surface, where it evaporated rapidly in the hot sun. The result was an even faster accumulation of salt. When the salty water table came up still further, it saturated the roots of the barley plants, devastating the crops. Unknowingly, the Sumerians destroyed their own land in their quest for water.

The rising civilization of Babylon was doomed to fall in turn. Along the Tigris River, east of the modern city of Baghdad, an extremely elaborate system of irrigation canals was built. This system was begun in a pre-Babylonian period, 3000 to 2400 B.C. After a long history of abandonments and reconstructions, it was superseded between A.D. 200 and 500 by a final irrigation system based on a central canal, the Nahrwan Canal.

A problem soon developed, however. The irrigation system featured long, branching channels that proved to be traps for silt, which settled out of muddy river water. Frequent cleaning of silt from the channels was required, and silt was piled up in great embankments and mounds beside the channels. Gradually, silt from the mounds was washed by rains into nearby fields, adding layer upon layer to the land surface. According to one estimate, about 1 m (3 ft) of silt accumulated over the fields in a 500-year period. This rise in level of the farmland by silting made its irrigation all the more difficult. Until the strong central government authority collapsed, things went along well enough. Afterward, however, the entire system gradually broke down, and by the twelfth century it was abandoned completely. Not long after, Mongol hordes invaded the valley. By then, little remained of the once prosperous civilization that had dominated the region for over 4000 years.

Salinization, waterlogging, and silt accumulation have affected nearly every major irrigation system developed in a desert climate. Today, Pakistan struggles with salinization and waterlogging in a vast irrigation system in the Indus Valley. The lower Colorado River lands, including areas in Mexico, have suffered from salinization. Israel faces the threat of salinization of vital agricultural lands that have only recently been placed in irrigation. Some new methods may help. For example, soaking the ground through tubes, while keeping the soil covered under a plastic sheet, can reduce the buildup of salt.

Small wonder, then, that new proposals for desert irrigation systems arouse little enthusiasm. Perhaps a record of nearly total failure, spread over the entire span of civilization, is beginning to get its message across.

Questions

1. Describe the processes of salinization, waterlogging, and siltation and their impact on agriculture.
2. A technological "fix" for siltation is to build a dam upstream of an irrigation canal complex, so that silt will be trapped behind the dam. What other effects might building such a dam produce, both desirable and undesirable?

Sumer and Babylon are at the east end of the Fertile Crescent, the ancient cradle of Western civilization, which stretches from the Persian Gulf west to the coasts of Lebanon and Israel.

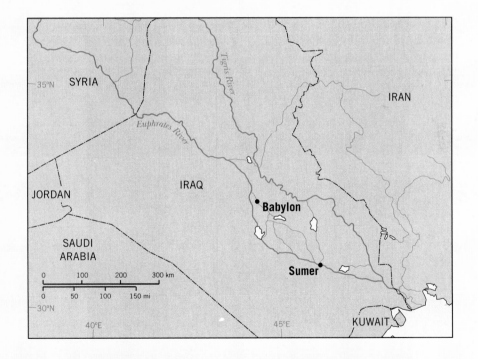

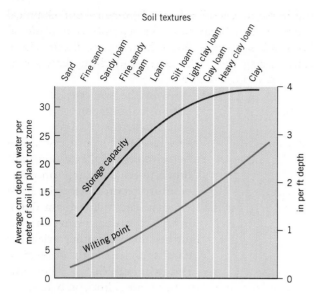

Figure 9.6 Storage capacity and wilting point vary according to soil texture. Finer textured soils hold more water. They also hold water more tightly, so that plants wilt more quickly.

Storage capacity is measured in units of depth, usually centimeters or inches, as with precipitation. It depends largely on the texture of the soil, as shown in Figure 9.6. Finer textures hold more water than coarser textures. This occurs because fine particles hold water more tightly than coarse particles. Thus, a sandy soil has a small storage capacity, while a clay soil has a large storage capacity.

The figure also shows the *wilting point,* which is the water storage level below which plants will wilt. The wilting point depends on soil texture. Because fine particles hold water more tightly, it is more difficult for plants to extract moisture from fine soils. Thus, plants can wilt in fine-textured soils even though more soil water is present than in coarse-textured soils.

SOIL DEVELOPMENT

How do soils develop their distinctive characteristics? Let's turn to the processes that act to form soils and soil layers. We begin with soil horizons.

Soil Horizons

Most soils possess **soil horizons**—distinctive horizontal layers that differ in physical composition, chemical composition, and/or organic content or structure (Figure 9.7). Soil horizons are developed by the interactions through time of climate, living organisms, and the configuration of the land surface. Horizons usually develop by either selective removal or accumulation of certain ions, colloids, and chemical compounds. The removal or accumulation is normally produced by water seeping down through the soil profile from the surface to deeper layers. Horizons are often distinguished by their color. The display of horizons on a cross section through the soil is termed a **soil profile.**

Let's review briefly the main types of horizons and their characteristics. For now, this discussion will apply to the types of horizons and processes that are found in moist forest climates. These are shown in Figure 9.8. Soil horizons are of two types: organic and mineral. Organic horizons, designated by the capital letter O, overlie the mineral horizons and are formed from accumulations of organic matter derived from plants and animals. Soil scientists recognize two possible layers. The upper O_i *horizon* contains decomposing organic matter that is recognizable as leaves or twigs. The lower O_a *horizon* contains material that is broken down beyond recognition by eye. This material is humus, which we mentioned earlier.

Figure 9.7 A column of soil will normally show a series of horizons, which are horizontal layers with different properties. The uppermost layers, within the reach of plant roots, are termed the soil solum.

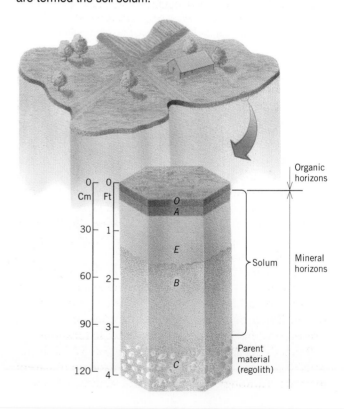

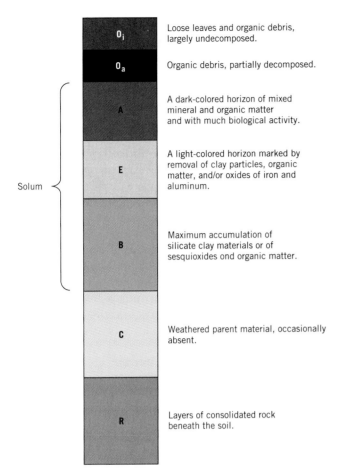

O$_i$ Loose leaves and organic debris, largely undecomposed.

O$_a$ Organic debris, partially decomposed.

A A dark-colored horizon of mixed mineral and organic matter and with much biological activity.

E A light-colored horizon marked by removal of clay particles, organic matter, and/or oxides of iron and aluminum.

B Maximum accumulation of silicate clay materials or of sesquioxides ond organic matter.

C Weathered parent material, occasionally absent.

R Layers of consolidated rock beneath the soil.

Solum { (bracket covering A, E, B)

Figure 9.8 A sequence of horizons that might appear in a forest soil developed under a cool, moist climate. (Soil Conservation Service, U.S. Department of Agriculture.)

this horizon, it is still affected by physical and chemical processes that break down its regolith. A transitional horizon is sometimes found between the *B* and *C* horizons. Below the *C* horizon, underlying bedrock is denoted as the *R* horizon.

Soil-Forming Processes

There are four classes of soil-forming processes. The first includes processes of *soil enrichment,* which add material to the soil body. For example, inorganic enrichment occurs when sediment is brought from higher to lower areas by overland flow. Stream flooding also deposits fine mineral particles on low-lying soil surfaces. Wind is another source of fine material that can accumulate on the soil surface. Organic enrichment occurs when humus accumulating in *O* horizons is carried downward to enrich the *A* horizon below.

The second class includes processes that remove material from the soil body. This *removal* occurs when surface erosion carries sediment away from the uppermost layer of soil. Another important process of loss is *leaching,* in which seeping water dissolves soil materials and moves them to deep levels or to ground water.

The third class of soil-forming processes involves *translocation,* in which materials are moved within the

Figure 9.9 A forest soil profile on outer Cape Cod. The pale grayish *E* horizon overlies a reddish *B* horizon. A thin layer of wind-deposited silt and dune sand (pale brown layer) has been deposited on top.

Mineral horizons lie below the organic horizons. Four main horizons are important—*A, E, B,* and *C.* Plant roots penetrate *A, E,* and *B* horizons and influence soil development within them. Soil scientists use the term *soil solum* to refer to the *A, E,* and *B* horizons. The *A horizon* is the uppermost mineral horizon. It is rich in organic matter, consisting of numerous plant roots and downwashed humus from the organic horizons above. Next is the *E horizon.* Clay particles and oxides of aluminum and iron are removed from the *E* horizon by downward-seeping water, leaving behind pure grains of sand or coarse silt.

The *B horizon* receives the clay particles, aluminum and iron oxides, as well as organic matter washed down from the *A* and *E* horizons. It is made dense and tough by the filling of natural spaces with clays and oxides. Transitional horizons are present at some locations.

Beneath the *B* horizon is the *C* horizon, which is not considered part of the true soil. It is made up of parent material. Although little biological activity occurs in

soil body, usually from one horizon to another. Two processes of translocation that operate simultaneously are eluviation and illuviation. **Eluviation** consists of the downward transport of fine particles, particularly the clays and colloids, from the uppermost part of the soil. Eluviation leaves behind grains of sand or coarse silt, forming the *E* horizon. **Illuviation** is the accumulation of materials that are brought down downward, normally from the *E* horizon to the *B* horizon. The materials that accumulate may be clay particles, humus, or sesquioxides of iron and aluminum.

Four classes of soil-forming processes include enrichment, removal, translocation, and transformation.

Figure 9.9 is a soil profile developed under a cool, humid forest climate. It shows the effects of both soil enrichment and translocation processes. The topmost layer of the soil is a thin deposit of wind-blown silt and dune sand, which has enriched the soil profile. In translocation processes, eluviation has removed colloids and sesquioxides from the whitened *E* horizon. Illuviation has added them to the *B* horizon, which displays the orange-red colors of iron sesquioxide.

The translocation of calcium carbonate is another important process. In pure form, this secondary mineral is calcite ($CaCO_3$). In many areas, the parent material of the soil contains a substantial proportion of calcium carbonate derived from the disintegration of limestone, a common variety of bedrock. Carbonic acid, which forms when carbon dioxide gas dissolves in rainwater or soil water, readily reacts with calcium carbonate. The products of this reaction remain dissolved in solution as ions.

In moist climates, a large amount of surplus soil water moves downward to the groundwater zone. This water movement leaches calcium carbonate from the entire soil in a process called *decalcification*. Soils that have lost most of their calcium are also usually acid in chemical balance and so are low in bases. Addition of lime or pulverized limestone not only corrects the acid condition, but also restores the calcium, used as a plant nutrient.

In dry climates, calcium carbonate is dissolved in upper layers of the soil during periods of rain or snowmelt when soil water recharge is taking place. The dissolved carbonate matter is carried down to the *B* horizon, where water penetration reaches its limits. Here, the carbonate matter is precipitated (deposited

in crystalline form) in the *B* horizon, a process called *calcification*. Calcium carbonate deposition takes the form of white or pale-colored grains, nodules, or plates in the *B* or *C* horizons.

A last process of translocation occurs in desert climates. In some low areas, a layer of ground water lies close to the surface, producing a flat, poorly drained area. Evaporation of water at or near the soil surface draws up a continual flow of ground water by capillary tension, much like a cotton wick draws kerosene upward in an oil lamp. Moreover, the ground water is often rich in dissolved salts. When evaporation occurs, the salts precipitate and accumulate as a distinctive, *salic horizon*. This process is called *salinization*. Most of the salts are compounds of sodium, of which ordinary table salt (sodium chloride, or halite, NaCl) is a familiar example. Sodium in large amounts is associated with highly alkaline conditions and is toxic to many kinds of plants. When salinization occurs in irrigated lands in a desert climate, the soil can be ruined for further agricultural use. (See *Eye on the Environment • Death of a Civilization*, in this chapter.)

The last class of soil-forming processes involves the *transformation* of material within the soil body. An example is the conversion of minerals from primary to secondary, which we have already described. Another example is decomposition of organic matter to produce humus, a process termed *humification*. In warm, moist climates, humification can decompose organic matter completely to yield carbon dioxide and water, leaving virtually no organic matter in the soil.

Soil Temperature

Soil temperature is another important factor in determining the chemical development of soils and the formation of horizons. Temperature acts as a control over biologic activity and also influences the intensity of chemical processes affecting soil minerals. Below 10°C (50°F), biological activity is slowed, and at or below the freezing point (0°C, 32°F), biological activity stops. Chemical processes affecting minerals are inactive. The root growth of most plants and germination of their seeds require soil temperatures above 5°C (41°F). For plants of the warm, wet low-latitude climates, germination of seeds may require a soil temperature of at least 24°C (75°F).

The temperature of the uppermost soil layer and the soil surface strongly affects the rate at which organic matter is decomposed by microorganisms. Thus, in cold climates, where decomposition is slow, organic matter in the form of fallen leaves and stems tends to accumulate to form a thick *O* horizon. As we have already described, this material becomes humus, which is carried downward to enrich the *A* horizon.

In warm, moist climates of low latitudes, the rate of decomposition of plant material is very high, so that nearly all the fallen leaves and stems are disposed of by bacterial activity. Under these conditions, the *O* horizon may be missing and the entire soil profile lacking in organic matter.

Soil and Surface Configuration

The configuration, or shape, of the ground surface is an important factor in soil formation. Configuration includes the steepness of the ground surface, or *slope,* as well as its compass orientation, or *aspect.* Generally speaking, soil horizons are thick on gentle slopes but thin on steep slopes. This is because the soil is more rapidly removed by erosion processes on the steeper slopes.

Aspect acts to influence soil temperatures and the soil water regime. Slopes facing away from the sun are sheltered from direct insolation and tend to have cooler, moister soils. Slopes facing toward the sun are exposed to direct solar rays, raising soil temperatures and increasing evapotranspiration.

Biological Processes in Soil Formation

The presence and activities of living plants and animals, as well as their nonliving organic products, have an important influence on soil. We have already noted the very important role that organic matter as humus plays in soil fertility. The colloidal structure of humus holds bases, which are needed for plant growth. In this way, humus helps keep nutrients cycling through plants and soils, and ensures that nutrients will not be lost by leaching to ground water. Humus also helps bind the soil into crumbs and clumps. This structure allows water and air to penetrate the soil freely, providing a healthy environment for plant roots.

Animals living in the soil include many species and come in many individual sizes—from bacteria to burrowing mammals. The total role of animals in soil formation is extremely important in soils that are warm and moist enough to support large animal populations. For example, earthworms continually rework the soil not only by burrowing, but also by passing soil through their intestinal tracts. They ingest large amounts of decaying leaf matter, carry it down from the surface and incorporate it into the mineral soil horizons. Many forms of insect larvae perform a similar function. Tubelike openings are made by larger

animals—moles, gophers, dogs, and many other specie

Human activity also infl chemical nature of the soil. L soils have been cultivated fo both the structure and comp tural soils have undergone altered soils are often recognize that are just as important as natu

THE GLOBAL SCOPE OF SOI

An important aspect of soil science phy is the classification of soils int subtypes that are recognized in term tion over the earth's land surfaces. particularly interested in the linkage o material, time, biologic process, and la distribution of types of soils. Geogra interested in the kinds of natural vegeta with each of the major soil classes. The soils is thus essential in determining the ronments of the globe. It is important be tility, along with availability of fresh wat measure of the ability of an environment produce food for human consumption.

Soils of the world are classified according developed by scientists of the U.S. Soil Co Service, in cooperation with soil scientists other nations. Here we are concerned only two highest levels of this classification system. level contains 11 **soil orders** summarized in T The second level consists of *suborders,* of which to mention only a few.

The presence of diagnostic horizons distin-guishes most soil orders and suborders.

Soil orders and suborders are often distinguished the presence of a diagnostic horizon. Each diagnos horizon has some unique combination of physic properties (color, structure, texture) or chemical pro erties (minerals present or absent). The two basi kinds of diagnostic horizons are (1) a horizon formed at the surface and called an *epipedon* and (2) a subsur-face horizon formed by the removal or accumulation of matter. In our descriptions of soil orders, we will refer to a number of diagnostic horizons.

We can recognize three groups of soil orders. The largest group includes seven orders with well-developed

rabbits,   badgers,   prairie

ences   the   physical   and
arge areas of agricultural
centuries. As a result,
osition of these agricul-
great   changes.   These
d as distinct soil classes
ral soils.

for physical geogra-
major types and
s of their distribu-
Geographers are
f climate, parent
ndform with the
phers are also
tion associated
geography of
uality of envi-
ause soil fer-
r, is a basic
al region to

to a system
nservation
of   many
with the
The top
ble 9.2
e need

they are found on recent deposits such as floodplains,
al landforms, sand dunes, marshlands, bogs, or
canic ash deposits.

Soil Orders

Table 9.3 explains the names of the soil orders. The formative element is a syllable used in the names of suborders and lower groups. Although each order has several suborders, we will refer to only a few.

Three soil orders dominate the vast land areas of low latitudes: Oxisols, Ultisols, and Vertisols. Soils of these orders have developed over long time spans in an envi-

SOILS OF THE WORLD

U.S. Comprehensive Soil Classification System.
Based on data of Soil Conservation Service,
U.S. Dept. of Agriculture.

S	Spodosols	**0**	Oxisols
A	Alfisols	**V**	Vertisols
A1	Boralfs	**M**	Mollisols
A2	Udalfs	**D**	Aridisols
A3	Ustalfs	**T**	Tundra soils
A4	Xeralfs	**H**	Highland (**I** Icesheet)
U	Ultisols		

Kilometers
0   1000   2000   3000

0   1000   2000
Miles

(True distances on mid-meridians and parallels 0 to 40)
Interrupted homolosine projection
Based on Goode Base Map

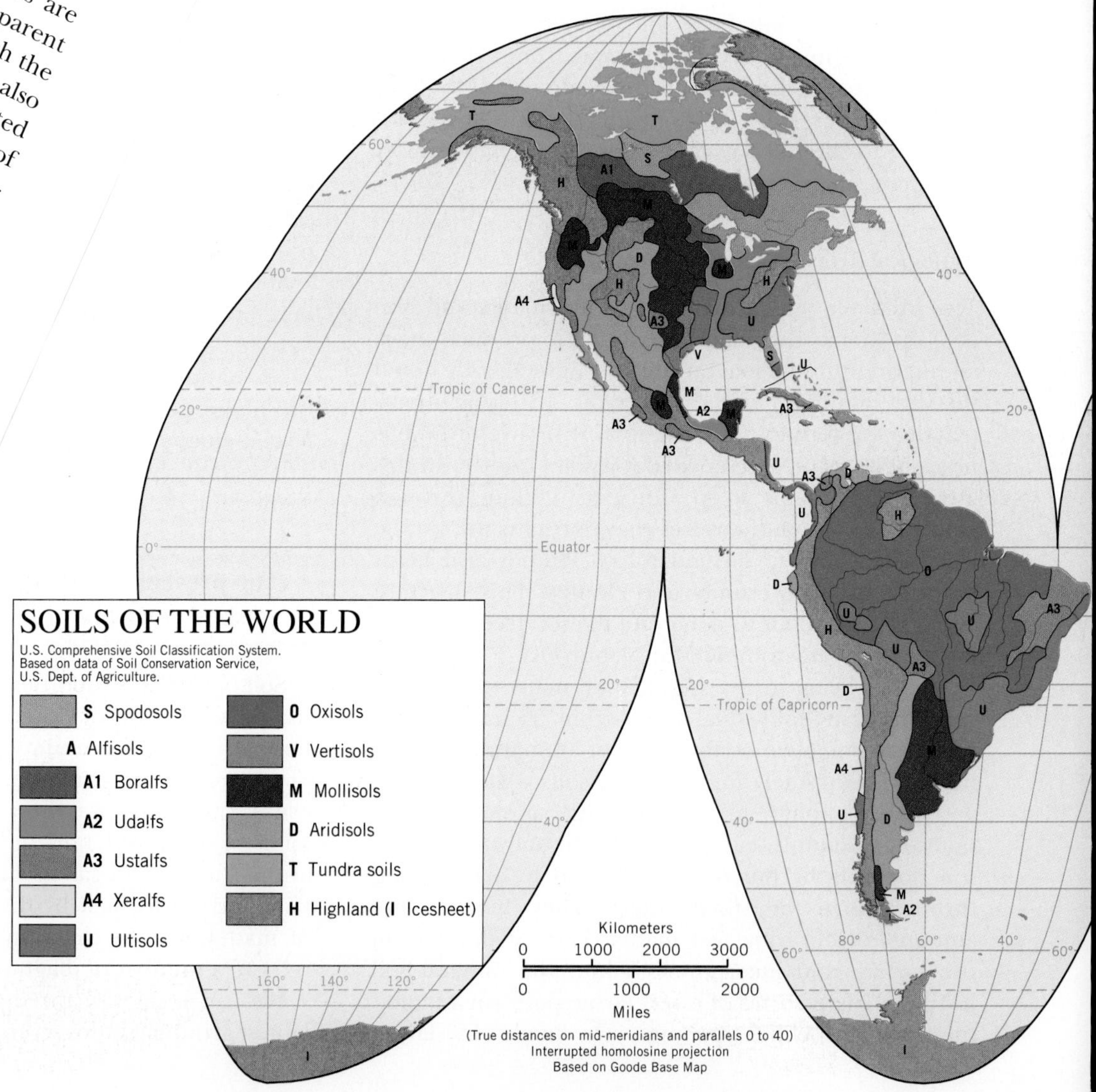

ronment of warm soil temperatures and soil water that is abundant in a wet season or throughout the year. We will discuss these orders first.

Oxisols

Oxisols have developed in equatorial, tropical, and subtropical zones on land surfaces that have been stable over long periods of time. During soil development, the climate has been moist, with a large water surplus. Oxisols have developed over vast areas of South America and Africa in the wet equatorial climate ①. Here the native vegetation is rainforest. The wet-dry tropical climate ③ with its large seasonal water surplus is also associated with Oxisols in South America and

Africa. Oxisols usually lack distinct horizons, except for darkened surface layers. Soil minerals are weathered to an extreme degree and are dominated by stable sesquioxides of aluminum and iron. Red, yellow, and yellowish-brown colors are normal (Figures 9.11*a* and 9.12). The base status of the Oxisols is very low, since nearly all the bases required by plants have been removed from the soil profile. A small store of nutrient bases occurs very close to the soil surface. The soil is quite easily broken apart and allows easy penetration by rainwater and plant roots.

In many areas, the Oxisol profile contains a subsurface horizon of sesquioxides. This horizon is capable of hardening to a rocklike material if it becomes exposed at the surface and is subjected to repeated

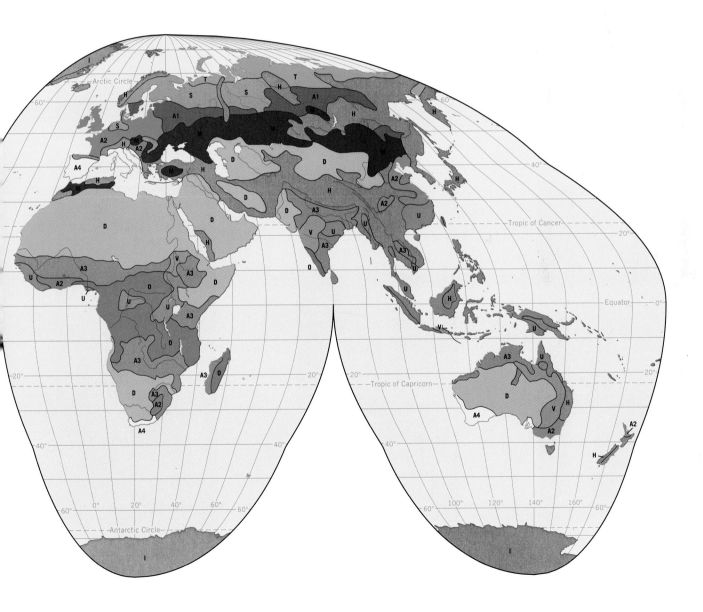

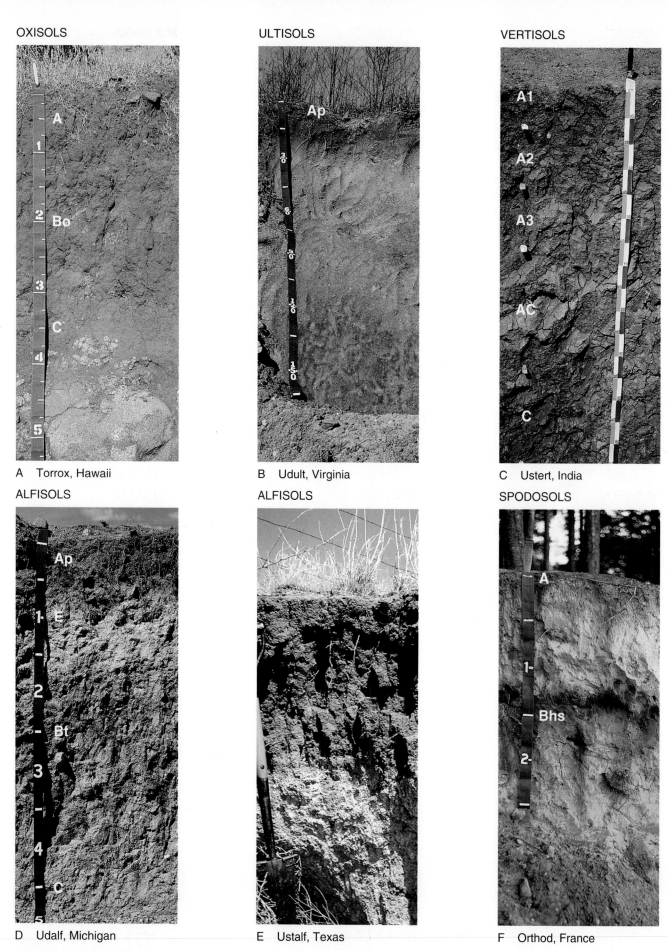

OXISOLS

A

1

2 Bo

3

C

4

5

A   Torrox, Hawaii

ULTISOLS

Ap

3,0

8,0

5,0

10

5,0

B   Udult, Virginia

VERTISOLS

A1

A2

A3

AC

C

C   Ustert, India

ALFISOLS

Ap

E

1

2

Bt

3

4

C

5

D   Udalf, Michigan

ALFISOLS

E   Ustalf, Texas

SPODOSOLS

A

1

Bhs

2

F   Orthod, France

Figure 9.11   Soil profiles of several soil orders.

248

MOLLISOLS

G   Boroll, USSR

MOLLISOLS

H   Udoll, Argentina

MOLLISOLS

I   Ustoll, Colorado

MOLLISOLS

J   Rendoll, Argentina

ARDISOLS

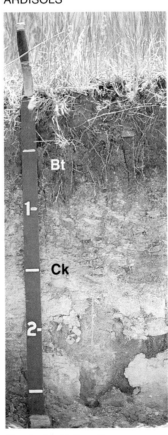

K   Argid, Colorado

HISTOSOLS

L   Fibrist, Minnesota

249

Table 9.2 Soil Orders

Group I

Soils with well-developed horizons or with fully weathered minerals, resulting from long-continued adjustment to prevailing soil temperature and soil water conditions.

Oxisols	Very old, highly weathered soils of low latitudes, with a subsurface horizon of accumulation of mineral oxides and very low base status.
Ultisols	Soils of equatorial, tropical, and subtropical latitude zones, with a subsurface horizon of clay accumulation and low base status.
Vertisols	Soils of subtropical and tropical zones with high clay content and high base status. Vertisols develop deep, wide cracks when dry, and the soil blocks formed by cracking move with respect to each other.
Alfisols	Soils of humid and subhumid climates with a subsurface horizon of clay accumulation and high base status. Alfisols range from equatorial to subarctic latitude zones.
Spodosols	Soils of cold, moist climates, with a well-developed *B* horizon of illuviation and low base status.
Mollisols	Soils of semiarid and subhumid midlatitude grasslands, with a dark, humus-rich epipedon and very high base status.
Aridisols	Soils of dry climates, low in organic matter, and often having subsurface horizons of accumulation of carbonate minerals or soluble salts.

Group II

Soils with a large proportion of organic matter.

Histosols	Soils with a thick upper layer very rich in organic matter.

Group III

Soils with poorly developed horizons or no horizons, and capable of further mineral alteration.

Entisols	Soils lacking horizons, usually because their parent material has accumulated only recently.
Inceptisols	Soils with weakly developed horizons, having minerals capable of further alteration by weathering processes.
Andisols	Soils with weakly developed horizons, having a high proportion of glassy volcanic parent material produced by erupting volcanoes.

Table 9.3 Formative Elements in Names of Soil Orders

Name of Order	Formative Element	Derivation of Formative Element	Pronunciation of Formative Element
Entisol	ent	Meaningless syllable	recent
Inceptisol	ept	L. *inceptum*, beginning	inept
Histosol	ist	Gr. *histos*, tissue	histology
Oxisol	ox	F. *oxide*, oxide	ox
Ultisol	ult	L. *ultimus*, last	ultimate
Vertisol	ert	L. *verto*, turn	invert
Alfisol	alf	Meaningless syllable	alfalfa
Spodosol	od	Gr. *spodos*, wood ash	odd
Mollisol	oll	L. *mollis*, soft	mollify
Aridisol	id	L. *aridus*, dry	arid
Andisol	and	Eng. *andesite*, a volcanic rock type	and

Figure 9.12 An Oxisol in Hawaii. Sugar cane is being cultivated here.

Figure 9.13 Laterite, formed by hardening of plinthite, is being quarried for building stone in this scene from India.

Figure 9.14 Ultisol profile in North Carolina. The thin, pale layer at the top is an *E* horizon, showing the effects of removal of materials by eluviation. Near the base of the thick, reddish *B* horizon is a blotchy-colored zone of plinthite.

wetting and drying. This material is referred to as *plinthite* (from the Greek word *plinthos*, meaning "brick"). In the hardened state, plinthite is referred to as *laterite* (from the Latin *later*, or brick). In Southeast Asia, plinthite is quarried and cut into building blocks (Figure 9.13). These blocks harden into laterite blocks when they are exposed to the air.

Ultisols

Ultisols are quite closely related to the Oxisols in outward appearance and environment of origin. Ultisols are reddish to yellowish in color (Figures 9.11*b* and 9.14). They have a subsurface horizon of clay accumulation, called an *argillic horizon*, which is not found in the Oxisols. It is a *B* horizon and has developed through accumulation of clay in the process of

illuviation. A plinthite horizon may also be present. Although forest is the characteristic native vegetation, the base status of the Ultisols is low. As in the Oxisols, most of the base cations are found in a shallow surface layer where they are released by the decay of plant matter. They are quickly taken up and recycled by the shallow roots of trees and shrubs.

Ultisols are widespread throughout Southeast Asia and the East Indies. Other important areas are in eastern Australia, Central America, South America, and the southeastern United States. Ultisols extend into the lower midlatitude zone in the United States, where they correspond quite closely in extent with the area of moist subtropical climate ⑥. In lower latitudes, Ultisols are identified with the wet-dry tropical climate ③ and the monsoon and trade-wind coastal climate ②. Note that all these climates have a dry season, even though it may be short.   Both Oxisols and Ultisols of low latitudes were used for centuries under shifting agriculture prior to the advent of modern agricultural technology. This primitive agricultural method, known as slash-and-burn, is still widely practiced (Chapter 8).

Figure 9.15 A Vertisol in Texas. The clay minerals that are abundant in Vertisols shrink when they dry out, producing deep cracks in the soil surface.

Without fertilizers, these soils can sustain crops on freshly cleared areas for only two or three years, at most, before the nutrient bases are exhausted and the garden plot must be abandoned. Substantial use of lime, fertilizers, and other industrial inputs is necessary for high, sustained crop yields. However, the exposed soil surface is vulnerable to devastating soil erosion, particularly on steep hill slopes.

Vertisols

Vertisols have a unique set of properties that stand in sharp contrast to the Oxisols and Ultisols. Vertisols are black in color and have a high clay content (Figures 9.11*c* and 9.15). Much of the clay consists of a particular mineral that shrinks and swells greatly with seasonal changes in soil water content. Wide, deep vertical cracks develop in the soil during the dry season. As the dry soil blocks are wetted and softened by rain, some fragments of surface soil drop into the cracks before they close, so that the soil "swallows itself" and is constantly being mixed.

Vertisols typically form under grass and savanna vegetation (Chapter 8) in subtropical and tropical climates with a pronounced dry season. These climates include the semiarid subtype of the dry tropical steppe climate ④ and the wet-dry tropical climate ③. Because Vertisols require a particular clay mineral as a parent material, the major areas of occurrence are scattered and show no distinctive pattern on the world map. An important region of Vertisols is the Deccan Plateau of western India, where basalt, a dark variety of igneous rock, supplies the silicate minerals that are altered into the necessary clay minerals.

Oxisols and Ultisols are soils of low-latitude regions. Ultisols have an argillic horizon, which Oxisols lack.

Vertisols are high in base status and are particularly rich in such nutrient bases as calcium and magnesium. The soil solution is nearly neutral in pH, and a moderate content of organic matter is distributed through the soil. The soil retains large amounts of water because of its fine texture, but much of this water is held tightly by the clay particles and is not available to plants. Where soil cultivation depends on human or animal power, as it does in most of the developing nations where the soil occurs, agricultural yields are low. This is because the moist soil becomes highly plastic and is difficult to till with primitive tools. For this reason, many areas of Vertisols have been left in grass or shrub cover, providing grazing for cattle. Soil scien-

tists think that the use of modern technology, including heavy farm machinery, could result in substantial production of food and fiber from Vertisols that are not now in production.

Inceptisols

Inceptisols are soils with horizons that are weakly developed, usually because the soil is quite young. These areas occur within some of the regions shown on the world map as Ultisols and Oxisols. Especially important are the Inceptisols of river floodplains and delta plains in Southeast Asia that support dense populations of rice farmers (see chapter opener photo).

In these regions, annual river floods cover low-lying plains and deposit layers of fine silt. This sediment is rich in primary minerals that yield bases as they weather chemically over time. The constant enrichment of the soil explains the high soil fertility in a region where uplands develop only Ultisols of low fertility. Inceptisols of these floodplain and delta lands are of a suborder called *Aquepts*—Inceptisols of wet places. Much closer to home is another prime example of Aquepts within the domain of the Ultisols—the lower Mississippi River floodplain and delta plain.

Andisols

Andisols are soils in which more than half of the parent mineral matter is volcanic ash, spewed high into the air from the craters of active volcanoes and coming to rest in layers over the surrounding landscape. The fine ash particles are glasslike shards. A high proportion of carbon, formed by the decay of plant matter, is also typical, so that the soil usually appears very dark in color. Andisols form over a wide range of latitudes and climates. They are for the most part fertile soils, and in moist climates they support a dense natural vegetation cover.

Andisols do not appear on our world map because they are found in small patches associated with individual volcanoes that are constructed of andesite (Chapter 10). They are mostly located in the "Ring of Fire"—a chain of volcanic mountains and islands that surrounds the great Pacific Ocean. Andisols are also found on the island of Hawaii, where volcanoes are presently active.

Alfisols

The **Alfisols** are soils characterized by an argillic horizon, produced by illuviation. Unlike the argillic horizon of the Ultisols, this *B* horizon is enriched by silicate clay minerals that have an adequate capacity to

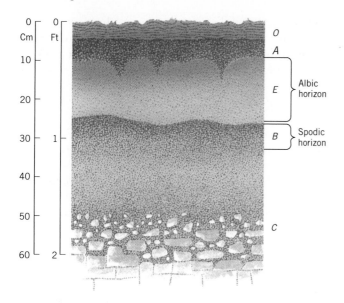

Figure 9.16 Diagram of a Spodosol profile.

hold bases such as calcium and magnesium. The base status of the Alfisols is therefore generally quite high.

Above the *B* horizon of clay accumulation is a horizon of pale color, the *E* horizon, that has lost some of the original bases, clay minerals, and sesquioxides by the process of eluviation. These materials have become concentrated by illuviation in the *B* horizon. Alfisols also have a gray, brownish, or reddish surface horizon.

The world distribution of Alfisols is extremely wide in latitude (see Figure 9.12). Alfisols range from latitudes as high as 60° N in North America and Eurasia to the equatorial zone in South America and Africa. Obviously, the Alfisols span an enormous range in climate types. For this reason, we need to recognize four of the important suborders of Alfisols, each with its own climate affiliation.

Alfisols exhibit an E horizon and an argillic horizon containing base-holding clay minerals.

Boralfs are Alfisols of cold (boreal) forest lands of North America and Eurasia. They have a gray surface horizon and a brownish subsoil. *Udalfs* are brownish Alfisols of the midlatitude zone. They are closely associated with the moist continental climate ⑩ in North America, Europe, and eastern Asia (Figure 9.11*d*).

Ustalfs are brownish to reddish Alfisols of the warmer climates (Figure 9.11*e*). They range from the subtropical zone to the equator and are associated with the wet-dry tropical climate ③ in Southeast Asia, Africa, Australia, and South America. *Xeralfs* are Alfisols of the Mediterranean climate ⑦, with its cool moist winter and dry summer. The Xeralfs are typically brownish or reddish in color.

Figure 9.17 Profile of a Histosol seen in the wall of a pit cut into a bog near Belle Glade, Florida. Water is being pumped from the floor of the pit.

Spodosols

Poleward of the Alfisols in North America and Eurasia lies a great belt of soils of the order **Spodosols,** formed in the cold boreal forest climate ⑪ beneath a needle-leaf forest. Spodosols have a unique property—a *B* horizon of accumulation of reddish mineral matter with a low capacity to hold base cations (Figure 9.11f). This horizon is called the spodic horizon (Figure 9.16). It is made up of a dense mixture of organic matter and compounds of aluminum and iron, all brought downward by eluviation from an overlying *E* horizon. Because of the intensive removal of matter from the *E* horizon, it has a bleached, pale gray to white appearance (Figure 9.9). This conspicuous feature led to the naming of the soil as *podzol* (ash-soil) by Russian peasants. In modern terminology, this pale layer is an *albic horizon*. The *O* horizon, a thin, very dark layer of organic matter, overlies the *A* horizon.

Spodosols are strongly acid, and are low in plant nutrients such as the base cations of calcium and magnesium. They are also low in humus. Although the base status of the Spodosols is low, forests of pine and spruce are supported through the process of recycling of the base cations.

Spodosols are closely associated with regions recently covered by the great ice sheets of the Pleistocene Epoch (Ice Age). These soils are therefore very young. Typically, the parent material is coarse sand consisting largely of the mineral quartz (silicon dioxide, SiO_2). This mineral cannot weather to form clay minerals. Spodosols are naturally poor soils in terms of agricultural productivity. Because they are acid, application of lime is essential. Heavy applications of fertilizers are also required. With proper management and the input of the required industrial products, Spodosols can be highly productive, if the soil texture is favorable. High yields of potatoes from Spodosols in Maine and New Brunswick are examples. Another factor unfavorable to agriculture is the shortness of the growing season in the more northerly parts of the Spodosol belt.

Histosols

Throughout the northern regions of Spodosols are countless patches of **Histosols.** This unique soil order has a very high content of organic matter in a thick, dark upper layer (Figures 9.11*l* and 9.17). Most Histosols go by such common names as peats or mucks. They have formed in shallow lakes and ponds by accumulation of partially decayed plant matter. In time, the water is replaced by a layer of organic matter, or *peat,* and becomes a bog (Figure 9.18). Peat bogs

Figure 9.18 This peat bog in Galway, Ireland, has been trenched to reveal a Histosol profile. Peat blocks, seen to the sides of the trench, are drying for use as fuel.

Figure 9.19   Garden crops cultivated on a Histosol of a former glacial lake bed in Southern Ontario, Canada.

Fig. 9.20   Schematic diagram of a Mollisol profile.

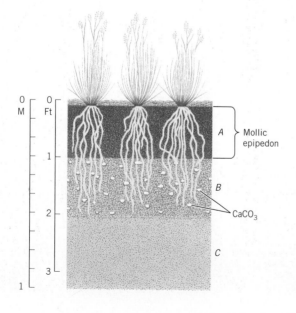

are used extensively for cultivation of cranberries (cranberry bogs). Sphagnum peat from bogs is dried and baled for sale as a mulch for use on suburban lawns and shrubbery beds. For centuries, Europe has used dried peat from bogs of glacial origin as a low-grade fuel.

Some Histosols are *mucks*—organic soils composed of fine black materials of sticky consistency. These are agriculturally valuable in midlatitudes, where they occur as beds of former lakes in glaciated regions. After appropriate drainage and application of lime and fertilizers, these mucks are remarkably productive for truck garden vegetables (Figure 9.19). Histosols are also found in low latitudes, where conditions of poor drainage have favored thick accumulations of plant matter.

Entisols

Entisols have in common the combination of a mineral soil and the absence of distinct horizons. Entisols are soils in the sense that they support plants, but they

Figure 9.21 A Mollisol developed on wind-blown silt. Dry prairie grasses are seen on the surface.

Mollisols

Mollisols are soils of grasslands that occupy vast areas of semiarid and subhumid climates in midlatitudes. Mollisols are unique in having a very thick, dark brown to black surface horizon called the *mollic epipedon* (Figures 9.11*g* to *j* and 9.20). This layer lies within the A horizon and is always more than 25 cm (10 in.) thick. The soil has a loose, granular structure (Figure 9.7, *left*) or a soft consistency when dry. Other important qualities of the Mollisols are the dominance of calcium among the base cations of the A and B horizons and the very high base status of the soil.

Most areas of Mollisols are closely associated with the semiarid subtype of the dry midlatitude climate ⑨ and the adjacent portion of the moist continental climate ⑩. In North America, Mollisols dominate the Great Plains region, the Columbia Plateau, and the northern Great basin. In South America, a large area of Mollisols covers the Pampa region of Argentina and Uruguay. In Eurasia, a great belt of Mollisols stretches from Rumania eastward across the steppes of Russia, Siberia, and Mongolia. The Russians refer to the

Figure 9.22 A salic horizon, appearing as a white layer, lies close to the surface in this Aridisol profile in the Nevada desert. The scale is marked in feet (1 ft = 30 cm).

may be found in any climate and under any vegetation. Entisols lack distinct horizons for two reasons. It may be the result of a parent material, such as quartz sand, in which horizons do not readily form. Or it may be the result of lack of time for horizons to form in recent deposits of alluvium or on actively reeroding slopes.

Entisols occur from equatorial to arctic latitude zones. From the standpoint of agricultural productivity, Entisols of the subarctic zone and tropical deserts (along with arctic areas of Inceptisols) are the poorest of all soils. In contrast, Entisols and Inceptisols of floodplains and delta plains in warm and moist climates are among the most highly productive agricultural soils in the world because of their favorable texture, ample nutrient content, and large soil water storage.

Mollisols as *chernozems,* a term that has gained widespread use throughout the Western world as well.

Because of their loose texture and very high base status, Mollisols are among the most naturally fertile soils in the world. They now produce most of the world's commercial grain crop. Most of these soils have been used for crop production only in the last century. Prior to that time, they were used mainly for grazing by nomadic herds. The Mollisols have favorable properties for growing cereals in large-scale mechanized farming and are relatively easy to manage. Production of grain varies considerably from one year to the next because seasonal rainfall is highly variable.

Aridisols are desert soils. They are low in organic matter and may contain horizons of calcium carbonate or salt accumulation.

A brief mention of four suborders of the Mollisols as they occur in the United States and Canada will help you to understand important regional soil differences related to climate. *Borolls,* the cold-climate suborder of the Mollisols, are found in a large area extending on both sides of the U.S.-Canadian border east of the Rocky Mountains (Figure 9.11*g*). *Udolls* are Mollisols of a relatively moist climate, as compared with the other suborders. Formerly, the Udolls supported tall-grass prairie, but today they are closely identified with the corn belt in the American Midwest (Figure 9.13H).

Ustolls are Mollisols of the semiarid subtype of the dry midlatitude climate ⑨, with a substantial soil water shortage in the summer months. The Ustolls underlie much of the Great Plains region east of the Rockies, a region of short-grass prairie (Figure 9.21). *Xerolls* are Mollisols of the Mediterranean climate ⑦, with its tendency to cool, moist winters and rainless summers.

Desert and Tundra Soils

Desert and tundra soils are soils of extreme environments. Aridisols characterize the desert climate. As might be expected, they are low in organic matter and high in salts. Tundra soils are poorly developed because they are formed on very recent parent material, left in place by glacial activity during the Ice Age. Cold temperatures have also restricted soil development in tundra regions.

Aridisols

Aridisols, soils of the desert climate, are dry for long periods of time. Because the climate supports only a very sparse vegetation, humus is lacking and the soil color ranges from pale gray to pale red (Figure 9.11*k*). Soil horizons are weakly developed, but there may be important subsurface horizons of accumulated calcium carbonate (petrocalcic horizon) or soluble salts (salic horizon) (Figure 9.22). The salts, of which sodium is a dominant constituent, give the soil a very high degree of alkalinity. The Aridisols are closely correlated with the arid subtype of the dry tropical ④, dry subtropical ⑤, and dry midlatitude ⑨ climates.

Most Aridisols are used for nomadic grazing, as they have been through the ages. This use is dictated by the limited rainfall, which is inadequate for crops without irrigation. Locally, where water supplies from moun-

Figure 9.23 This gray desert soil, an Aridisol, has proved highly productive when cultivated and irrigated. The locality is near Palm Springs, California, in the Coachella Valley.

Figure 9.24   Profile of a tundra soil (Cryaquept) in southern Yukon Teritory, Canada. Permafrost (perennially frozen soil water) appears at a depth of between 40 and 60 cm.

tain streams or ground water permit, Aridisols can be highly productive for a wide variety of crops under irrigation (Figure 9.23). Great irrigation systems, such as those of the Imperial Valley of the United States, the Nile Valley of Egypt, and the Indus Valley of Pakistan, have made Aridisols highly productive, but not without problems of salt buildup and waterlogging.

Tundra Soils

Soils of the arctic tundra fall largely into the order of Inceptisols, soils with weakly developed horizons that are usually associated with a moist climate. Inceptisols of the tundra climate belong to the suborder of Aquepts, which as we noted earlier are Inceptisols of wet places. More specifically, the tundra soils can be assigned to the *Cryaquepts*, a subdivision within the Aquepts. The prefix *cry* is derived from the Greek word *kryos*, meaning "icy cold." We may refer to these soils simply as *tundra soils.*

Tundra soils are formed largely of primary minerals ranging in size from silt to clay that are broken down by frost action and glacial grinding. Layers of peat are often present between mineral layers (Figure 9.24). Beneath the tundra soil lies perennially frozen ground (permafrost), described in Chapter 7. Because the annual summer thaw affects only a shallow surface layer, soil water cannot easily drain away. Thus, the soil is saturated with water over large areas. Repeated freezing and thawing of this shallow surface layer disrupts plant roots, so that only small, shallow-rooted plants can maintain a hold.

This chapter completes a sequence of chapters describing three interrelated topics—climate, vegetation, and soils. The climate of a region is determined by the characteristics of air masses and weather systems that it experiences through the year. The nature of the vegetation cover depends largely on the climate, varying from desert to tundra biomes. Soils depend on both climate and vegetation to develop their unique characteristics. But soils can also influence vegetation development. Similarly, the climate at a location can be influenced by its vegetation cover. Thus, there is a strong interaction between climate, soils, and vegetation.

The next three chapters also comprise a set that are strongly related. They deal with the lithosphere—the realm of the solid earth, as we described in the Prologue. Our survey of the lithosphere will begin with the nature of earth materials, then discuss how continents and ocean basins are formed and how they are continually changing, even today. Last, we turn to landforms occurring within continents that are produced by such lithospheric processes as volcanic activity and earthquake faulting.

GLOBAL SOILS IN REVIEW

The soil layer is a complex mixture of solid, liquid, and gaseous components. It is derived from parent material, or regolith, that is produced from rock by weathering. The major factors influencing soil and soil development are parent material, climate, vegetation, and time. Soil texture refers to the proportions of sand, silt, and clay that are present. Colloids are the finest particles in soils and are important because they help retain nutrients, or bases, that are used by plants. Soils show a wide range of pH values, from acid to alkaline. Soils with granular or blocky structures are most easily cultivated.

In soils, primary minerals are chemically altered to secondary minerals, which include oxides and clay minerals. The nature of the clay minerals determines the soil's base status. If base status is high, the soil retains nutrients. If low, the soil can lack fertility. When a soil is fully wetted by heavy rainfall or snowmelt and allowed to drain, it reaches its storage capacity. Evaporation from the surface and transpiration from plants draws down the soil water store until precipitation occurs again.

Soils possess distinctive horizontal layers called horizons. These layers are developed by processes of enrichment, removal, translocation, and transformation. In downward translocation, materials such as humus, clay particles, and mineral oxides are removed by eluviation from an upper horizon and accumulate by illuviation in a lower one. In salinization, salts are translocated upward by evaporating water to form a salic horizon. In humification, a transformation process, organic matter is broken down by bacterial decay. Where soil temperatures are warm, this process can be highly effective, leaving a soil low in organic content. Where they are abundant, animals, such as earthworms, can be very important in soil formation.

Global soils are classified into 11 soil orders, often by the presence of a diagnostic horizon. Oxisols are old, highly weathered soils of low latitudes. They have a horizon of mineral oxide accumulation and a low base status. Ultisols are also found in low latitudes. They have a horizon of clay accumulation and are also of low base status. Vertisols are rich in a type of clay mineral that expands and contracts with wetting and drying, and has a high base status. Like Ultisols, Alfisols have a horizon of clay accumulation, but they are of high base status. They are found in moist climates from equatorial to subarctic zones. Spodosols, found in cold, moist climates, exhibit a horizon of illuviation and low base status. Mollisols have a thick upper layer rich in humus. They are soils of midlatitude grasslands. Aridisols are soils of arid regions, marked by horizons of accumulation of carbonate minerals or salts. Histosols have a thick upper layer formed almost entirely of organic matter.

Three soil orders have poorly developed horizons or no horizons—Entisols, Inceptisols, and Andisols. Entisols are composed of fresh parent material and have no horizons. The horizons of Inceptisols are only weakly developed. Andisols are weakly developed soils occurring on fresh volcanic deposits.

KEY TERMS

soil	eluviation	Andisols
parent material	illuviation	Alfisols
soil texture	soil orders	Spodosols
soil colloids	Oxisols	Histosols
bases	Ultisols	Entisols
secondary minerals	Vertisols	Mollisols
soil horizons	Inceptisols	Aridisols
soil profile		

REVIEW QUESTIONS

1. Which important factors condition the nature and development of the soil?

2. Soil color, soil texture, and soil structure are used to describe soils and soil horizons. Identify each of these three terms, showing how they are applied.

3. Explain the concepts of acidity and alkalinity as they apply to soils.

4. Identify two important classes of secondary minerals in soils and provide examples of each class.

5. How does the ability of soils to hold water vary, and how does this ability relate to soil texture?

6. What is a soil horizon? How are soil horizons named? Provide two examples.

7. Identify four classes of soil-forming processes and describe each.

8. What are translocation processes? Identify and describe four translocation processes.

9. How many soil orders are there? Try to name them all.

10. Name three soil orders that are especially associated with low latitudes. For each order, provide at least one distinguishing characteristic and explain it.

11. Compare Alfisols and Spodosols. What features do they share? What features differentiate them? Where are they found?

12. Where are Mollisols found? How are the properties of Mollisols related to climate and vegetation cover? Name four suborders within the Mollisols.

13. Desert and tundra are extreme environments. Which soil order is characteristic of each environment? Briefly describe desert and tundra soils.

IN-DEPTH QUESTIONS

1. Document the important role of clay particles in soils. What is meant by the term *clay*? What are colloids? What are their properties? How does the type of clay mineral influence soil fertility? How does the amount of clay influence the water-holding capacity of the soil? What is the role of clay minerals in horizon development?

2. Using the world maps of global soils, global vegetation, and global climate, compare the pattern of soils on a transect along the 20° E longitude meridian with the patterns of vegetation and climate encountered along the same meridian. What conclusions can you draw about the relationship between soils, vegetation, and climate? Be specific.

Earth Materials

The "Visitor Parking" sign emerges quickly from the fog, and you nose the car to the left, pulling into a vacant space in the parking lot. You open the car door, step out, and stretch. It has been a long drive up the flanks of Kilauea from the Hilo airport. You expected better weather, but the view from the window of your airliner showed the whole eastern half of the "Big Island"—Hawaii—covered by trade-wind showers. Only the broad summit of Mauna Loa projected from the white blanket of clouds.

You are surprised how cool it is at this elevation—about 1200 m (4000 ft). You open the trunk to get the sweatshirt from your backpack and then change into your sturdy hiking boots. Your destination is the active Mauna Ulu crater, a nearly circular pit that contains a pool of molten lava.

At first, the trail winds through a scrubby forest of low trees and shrubs. You expected the vegetation to be more tropically lush, but then realize that little soil has developed on the lava-covered surface. Only a thin layer of organic matter cushions your feet. Holes and crevices are everywhere.

Soon you emerge from the woodland onto a surface of fresh lava. The tough rock is broken into angular blocks, separated by cracks and fissures. Although the trail is flat, the footing is difficult. Each step requires attention to keep your boot from becoming wedged into a crevice. The trail is clearly marked by small flags every 20 feet or so. The fog seems to thicken a bit, and you are glad that the flags are there. There is no vegetation anywhere on the fresh lava surface. Not a single blade of grass or speck of moss protrudes from a crack in the black jumble beneath your feet. As you walk from one flag to the next, you seem to be inside a dim,

gray bubble on a strange planet. There are no landmarks—nothing familiar within sight, as you pick your way across the rough terrain.

You gradually become aware of a noise, like a distant waterfall. As you continue along the trail, it gets louder. You stop for a moment to catch your breath and realize that the ground beneath you is vibrating. As you progress, the noise slowly builds, becoming a great booming rumble. Soon the ground is shaking beneath your feet.

The fog takes on a reddish glow. The acrid scent of sulfur catches in your nose and tightens your throat. Dimly, something emerges ahead. You are nearing the rim of the crater. You make out a wooden platform on the edge, with perhaps a half-dozen people clustered toward the back. As you reach the platform, the noise builds to a great roaring. You mount the trembling platform and approach the edge. The heat from the lava strikes your face like an unexpected wave. You blink and turn away, then turn once more toward the awesome display of primal force.

The sides of the crater drop away steeply. At the bottom of the great pit is the black surface crust of the molten lava. Long cracks gash the crust, glowing red from deep within. Suddenly, a huge rock mass falls from the side of the crater into the pit, fracturing the black surface into jagged pieces that are tossed outward by the impact. Molten rock splashes slowly upward, spattering the rock wall with red and white spray. Suddenly, the heat is unbearable, and you turn away, joining the others huddled behind you.

Over the next half hour, you take a dozen more looks. Each lasts no more than half a minute—the heat is simply too intense to stand longer at the edge. You notice from the pattern of cracks that the crust is moving slowly toward the far edge of the pit, where it is being drawn downward as the lava recedes to lower depths. Because the lava is receding, the crater is safe to visit, although the Park Ranger on duty tells you that this is the third platform built within the last few months. Its predecessors were undermined and finally fell into the pit.

The walk back to the car takes you slowly to the normal world again. As you walk, the roaring fades, and the fog loses its red glow. The lava landscape seems mysterious rather than ominous and threatening. The scrubby woods seems somehow welcoming, the twisted trees beckoning as outposts of earthly life on your return journey to familiar ground. You feel changed in some way, having confronted the brute strength of the earth's internal forces. Humbled by the immense power of nature to fashion rock and shape the surface of the earth, you feel small and insignificant.

The cooling of molten lava that is released at the earth's surface is one way in which new rock is formed. As we will see in this chapter, there are other ways that are perhaps not so dramatic, but equally important. Our objective will be to lay the groundwork necessary for understanding how the earth's crustal processes control landforms and landscapes, as well as provide the raw materials for the land-forming agents of water, wind, and glacial ice that we discuss in later chapters.

Lost City, King Canyon, Northwest Territories, Australia.

This chapter begins a sequence of three chapters that focus on the solid mineral realm of our planet. The first chapter (Chapter 10, "Earth Materials") deals with the materials that compose the earth's outer rock layer—minerals and rocks that are formed in various ways both deep within the earth and near or at its surface. In the second chapter (Chapter 11, "The Lithosphere and Plate Tectonics"), we will see that the outermost layer of the solid earth is a hardened crust of rock that is fractured, broken into large plates, and moved by forces deep within the earth. In the third chapter (Chapter 12, "Volcanic and Tectonic Landforms"), we will show how surface features reveal the history of the crust and the forces that have shaped it. The topics in these three chapters are drawn from *geology*, the science of the solid earth and its history.

Why are physical geographers concerned with the solid earth? The earth's outer layer serves as a platform, or base, for life on the lands. It provides the continental surfaces that are carved into landforms by moving water, wind, and glacial ice. Landforms, in turn, influence the distribution of ecosystems and exert strong controls over human occupation of the lands. Landforms made by water, wind, and ice are the subject of the closing chapters of this book.

We begin our study of the solid earth with an examination of earth materials—the minerals and rocks that are found at or near the earth's surface. We will be concerned only with a few of a great number of minerals and rocks, focusing on those that are most impor-

tant for a broad understanding of the continents, ocean basins, and their varied features.

THE CRUST AND ITS COMPOSITION

The thin, outermost layer of our planet is the *earth's crust*. This mineral skin ranges from about 8 to 40 km (5 to 25 mi) thick and contains the continents and ocean basins. It is the source of soil on the lands, of salts of the sea, of gases of the atmosphere, and of all the water of the oceans, atmosphere, and lands.

Figure 10.1 displays the eight most abundant elements of the earth's crust in terms of percentage by weight. Oxygen, the predominant element, accounts for a little less than half the total weight. Second is silicon, which accounts for a little more than a quarter. Together they account for 75 percent of the crust, by weight.

Aluminum accounts for approximately 8 percent and iron for about 5 percent of the earth's crust. These metals are very important to our industrial civilization, and, fortunately, they are relatively abundant. Four metallic elements that we recognized as bases (Chapter 9)—calcium, sodium, potassium, and magnesium—make up the remaining 12 percent. All four occur at about the same order of abundance (2 to 4 percent). These elements are essential for plant and animal life and are described in Chapter 9 as plant nutrients.

Rocks and Minerals

The elements of the earth's crust occur in chemical compounds that we recognize as minerals. A **mineral** is

Figure 10.1 The eight most abundant elements in the earth's crust, measured by percentage of weight. Oxygen and silicon dominate, with aluminum and iron following.

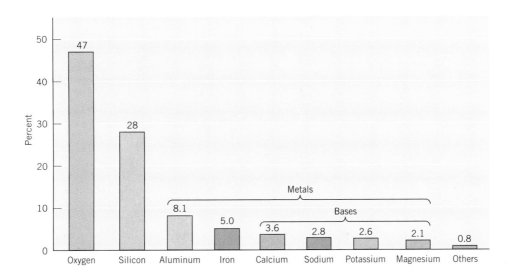

Figure 10.2 These large crystals of quartz form six-sided, translucent columns.

a naturally occurring, inorganic substance that usually possesses a definite chemical composition and characteristic atomic structure. Most minerals have a crystalline structure—gemstones such as diamonds or rubies are examples, although their crystalline shape is enhanced by the stonecutter. Perhaps you have seen some naturally occurring crystalline minerals. Quartz is a very common one and usually occurs as a clear, six-sided prism (Figure 10.2).

Minerals are combined into **rock,** which we can broadly define as an assemblage of minerals in the solid state. Rock comes in a very wide range of compositions, physical characteristics, and ages. A given variety of rock is usually composed of two or more minerals, and often many different minerals are present. However, a few rock varieties consist almost entirely of one mineral. Most rock of the earth's crust is extremely old by human standards, with the age of formation often ranging back many millions of years. But rock is also being formed at this very hour as active volcanoes emit lava that solidifies on contact with the atmosphere.

The three major categories of rocks are igneous, sedimentary, and metamorphic.

Rocks of the earth's crust fall into three major classes. (1) **Igneous rocks** are solidified from mineral matter in a high-temperature molten state. (2) **Sedimentary rocks** are layered accumulations of mineral particles derived mostly by weathering and erosion of preexisting rocks. (3) **Metamorphic rocks** are formed from igneous or sedimentary rocks that have been physically or chemically changed, usually by application of heat and pressure during mountain-making activities.

The three classes of rocks are constantly forming from one another in a continuous circuit—the **cycle of rock change**—through which the crustal minerals have been recycled during many millions of years of geologic time (Figure 10.3). In the process of melting, preexisting rock of any class is melted and then later cools to form igneous rock. In weathering and erosion, preexisting rock is broken down and accumulated in sedimentary layers that become rock. Heat and pressure convert igneous and sedimentary rocks to metamorphic rock. At the close of this chapter, after we have taken a more detailed look at rocks, minerals, and their formation processes, we will return to a more complete version of the cycle of rock change.

IGNEOUS ROCKS

Igneous rocks are formed when molten material moves from deep within the earth to a position within or atop the crust. There the molten material cools, forming rocks composed of mineral crystals.

Figure 10.3 The cycle of rock change. In this simplified version of the cycle of rock change, the three classes of rock are transformed into one another by weathering and erosion, melting, and exposure to heat and pressure.

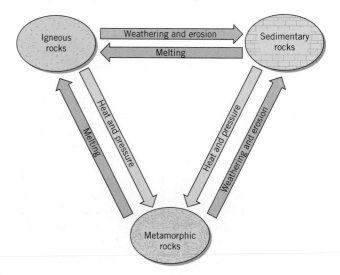

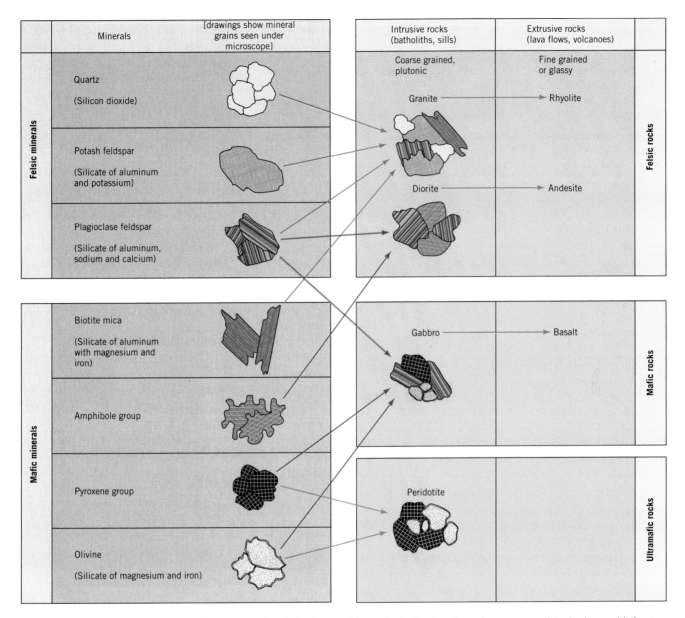

Figure 10.4 Silicate minerals and igneous rocks. Only the most important silicate mineral groups are listed, along with four common igneous rock types. The patterns shown for mineral grains indicate their general appearance through a microscope.

Most igneous rock consists of *silicate minerals,* chemical compounds that contain silicon and oxygen atoms. Most of the silicate minerals also have one, two, or more of the metallic elements listed in Figure 10.1—that is, aluminum, iron, calcium, sodium, potassium, and magnesium. Although there are many silicate minerals, we will only be concerned with seven of them, which are shown in Figure 10.4.

Among the most common minerals of all rock classes is **quartz,** which is silicon dioxide (SiO_2). It is quite hard and resists chemical breakdown. Two silicate-aluminum minerals called *feldspars* follow. One type, *potash feldspar,* contains potassium as the dominant metal besides aluminum. A second type, *plagio-clase feldspar,* is rich in sodium, calcium, or both. Quartz and feldspar form a silicate mineral group described as **felsic** ("fel" for feldspar; "si" for silicate).

Quartz and the feldspars are light in color (white, pink, or grayish) and lower in density than the other silicate minerals. By *density,* we mean the quantity of matter contained in a unit volume. Figure 10.5 shows the density of four earth materials: water, quartz, the mineral olivine (discussed below), and pure iron.

The next three silicate minerals are actually mineral groups, with a number of mineral varieties in each group. They are the *mica, amphibole,* and *pyroxene* groups. All three are silicates containing aluminum, magnesium, iron, and potassium or calcium. The

Table 10.1   Some Common Igneous Rock Types

Subclass	Rock Type	Description
Intrusive (cooling at depth, producing coarse crystal texture)	Granite	A felsic intrusive rock typically composed of quartz, feldspars, and mica
	Diorite	A felsic intrusive rock of plagioclase feldspar, pyroxene, and amphibole
	Gabbro	A mafic intrusive rock of plagioclase feldspar, pyroxene, and olivine
	Peridotite	An ultramafic rock of pyroxene and olivine
Extrusive (cooling at the surface, producing fine crystal texture)	Rhyolite	A felsic extrusive rock of granite composition
	Andesite	A felsic extrusive rock of diorite composition
	Basalt	A mafic extrusive rock of gabbro composition

seventh mineral, *olivine,* is a silicate of only magnesium and iron that lacks aluminum. Altogether, these minerals are described as **mafic** ("ma" for magnesium; "f" from the chemical symbol for iron, Fe). The mafic minerals are dark in color (usually black) and are denser than the felsic minerals.

Common Igneous Rocks

Igneous rocks solidify from rock in a hot, molten state, known as **magma.** From pockets a few kilometers below the earth's surface, magma makes its way upward through older solid rock and eventually solidifies as igneous rock. No single igneous rock is made up of all seven silicate minerals listed in Figure 10.4. Instead, a given rock variety contains three or four of those minerals as the major ingredients. Table 10.1 presents the more important igneous rock types that we will refer to in this chapter.

The column in the center of Figure 10.4 shows four common igneous rocks. Each rock is connected by arrows to the principal minerals it contains. These four rocks are carefully selected to be used in our later explanation of features of the earth's crust.

Granite and diorite are examples of felsic igneous rocks.

The first igneous rock is **granite.** The bulk of granite consists of quartz (27 percent), potash feldspar (40 percent), and plagioclase feldspar (15 percent). The remainder is mostly biotite and amphibole. Because most of the volume of granite is of felsic minerals, we classify granite as a *felsic igneous rock.* Granite is a mixture of white, grayish or pinkish, and black grains, but the overall appearance is a light gray or pink color (Figure 10.6).

Diorite, the second igneous rock on the list, lacks quartz. It consists largely of plagioclase feldspar (60 percent) and secondary amounts of amphibole and pyroxene. Diorite is a light-colored felsic rock that is only slightly denser than granite.

The third igneous rock is *gabbro,* in which the major mineral is pyroxene (60 percent). A substantial amount of plagioclase feldspar (20 to 40 percent) is present, and, in addition, there may be some olivine (0 to 20 percent). Since the mafic minerals pyroxene and olivine are dominant, gabbro is classed as a *mafic igneous rock.* It is dark in color and denser than the felsic rocks.

The fourth igneous rock, *peridotite,* is dominated by olivine (60 percent). The rest is mostly pyroxene (40 percent). Peridotite is classed as an *ultramafic igneous rock,* denser even than the mafic types. The mineral grains in igneous rocks are very tightly interlocked, and the rock is normally very strong.

Figure 10.5   The concept of density is illustrated by several cubes of the same size, but of different materials, hung from coil springs under the influence of gravity. The stretching of the coil spring is proportional to the density of each material, shown in grams per cubic centimeter. Quartz and olivine are common minerals. (Copyright © A. N. Strahler.)

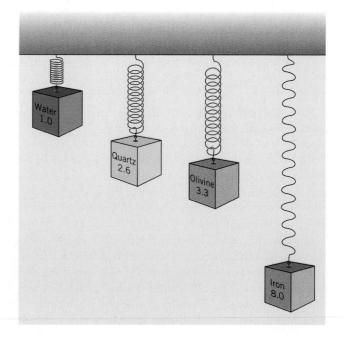

These four common varieties of igneous rock show an increasing range of density, from felsic, through mafic, to ultramafic types. This arrangement is duplicated on a grand scale in the principal rock layers that comprise the solid earth, with the least dense layer (mostly felsic rocks) near the surface and the densest layer (ultramafic rocks) deep in the earth's interior. We will stress this layered arrangement again in describing the earth's crust and the deeper interior zones.

Intrusive and Extrusive Igneous Rocks

Magma that solidifies below the earth's surface and remains surrounded by older, preexisting rock is called **intrusive igneous rock.** The process itself is *intrusion.* Where magma reaches the surface, it emerges as **lava,** which solidifies to form **extrusive igneous rock** (Figure 10.7). The process is one of *extrusion.*

Although both intrusive rock and extrusive rock can solidify from the same original body of magma, their outward appearances are quite different when you compare freshly broken samples of each. Intrusive igneous rocks cool very slowly and, as a result, develop large mineral crystals—that is, they are *coarse-textured.* A good example is the granite pictured in Figure 10.6.

In an intrusive igneous rock, you can see individual mineral crystals with the unaided eye or with the help of a simple magnifying lens. In an extrusive rock, which cools very rapidly, the individual crystals are very small. They can only be seen through a microscope. Rocks

Figure 10.6 This freshly broken piece of granite shows light grains of quartz, pinkish grains of feldspar, and dark grains of biotite crystals.

Figure 10.7 This fresh black lava shows a glassy surface and flow structures. Chain of Craters Road, Hawaii Volcanoes National Park, Hawaii.

with such small crystals are termed *fine-textured.* If the lava contains dissolved gases, it can cool to form a rock with a frothy, bubble-filled texture, called *scoria* (Figure 10.8*a*). Sometimes a lava cools to form a shiny natural volcanic glass (Figure 10.8*b*). Most lava solidifies simply as a dense, uniform rock with a dark, dull surface.

Sills and dikes are examples of plutons—bodies of intrusive igneous rocks that cool at depth.

Since the outward appearance of intrusive and extrusive rocks formed from the same magma are so different, they are named differently. The igneous rock types we have discussed so far—granite, diorite, gabbro, and peridotite—are intrusive rocks. Except for peridotite, all have counterparts as extrusive rocks. They are named in the right column of Figure 10.4. Each one has the same mineral composition as the plutonic rock named at the left. *Rhyolite* is the name for lava of the same composition as granite; **andesite** is lava with the mineral composition of diorite; and **basalt** is lava of the composition of gabbro. Andesite and basalt are the two most common types of lavas. Rhyolite and andesite are pale grayish or pink in color, whereas basalt is black. Lava flows, along with particles of solidified lava blown explosively from narrow vents, accumulate as isolated hills or mountains that we recognize as volcanoes. They are described in Chapter 12.

A body of intrusive igneous rock is called a **pluton.** Granite typically accumulates in enormous plutons, called *batholiths.* Figure 10.9 shows the relationship of a batholith to the overlying rock. As the hot fluid magma rises, it melts and incorporates the older rock lying above it. A single batholith extends down several kilometers and may occupy an area of several thousand square kilometers.

Figure 10.9 shows two other common forms of plutons. One is a *sill,* a platelike layer formed when magma forces its way between two preexisting rock layers. In the example shown, the sill has lifted the overlying rock layers to make room. A second kind of pluton is the *dike,* a wall-like body formed when a vertical rock fracture is forced open by magma. Commonly, these vertical fractures conduct magma to the land surface in the process of extrusion. Figure 10.10 shows a dike of mafic rock cutting across layers of older rock. The dike rock is fine-textured because of its rapid cooling. Magma entering small, irregular, branching fractures in the surrounding rock solidifies in a branching network of thin *veins.*

Chemical Alteration of Igneous Rocks

The minerals in igneous rocks are formed at high temperatures, and often at high pressures, as magma cools. When igneous rocks are exposed at or near the earth's surface, the conditions are quite different. Temperatures and pressures are low. Also, the rocks are exposed to soil and ground water solutions that contain dissolved oxygen and carbon dioxide. In this new environment, the minerals within an igneous rock may no longer be stable. Instead, most of these minerals undergo a slow chemical change that weakens their structure. Chemical change in response to this alien environment is called **mineral alteration.** It is a process of weathering, which we discussed in Chapter 9.

Weathering also includes the physical forces of disintegration that break up igneous rock into small fragments and separate the component minerals, grain from grain. This breakup, or fragmentation, is essential for the chemical reactions of mineral alteration. The reason is that fragmentation results in a great increase in mineral surface area exposed to chemically active solutions. (We will take up the processes of physical disintegration of rocks in Chapter 13.)

Oxygen dissolved in soil or groundwater can oxidize minerals. *Oxidation* occurs when oxygen is added in a chemical reaction. Oxidation is the normal fate of most silicate minerals exposed at the surface. With oxidation, the silicate minerals are converted to *oxides,* in which silicon and the metallic elements—such as calcium, magnesium, and iron—each bond completely with oxygen. Oxides are very stable. Quartz, with the composition silicon dioxide (SiO_2), is a mineral oxide that also occurs naturally in granite. Since it is stable, quartz is very long-lasting. As we will see shortly, it is a major constituent of sedimentary rocks.

Figure 10.8 (Left) Obsidian, or volcanic glass. The smooth, glassy appearance is acquired when a gas-free lava cools very rapidly. This sample shows red and black streaks, caused by minor variations in composition. (Right) A specimen of scoria, a form of lava containing many small holes and cavities produced by gas bubbles.

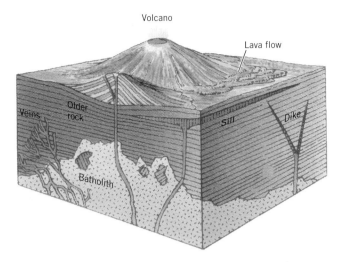

Figure 10.9 This block diagram illustrates various forms of intrusive igneous rock plutons as well as an extrusive lava flow.

Water combines with some silicate minerals in a reaction known as *hydrolysis*. This process is not merely a soaking or wetting of the mineral, but a true chemical change that produces a different mineral compound. The products of hydrolysis are stable and long-lasting, as are the products of oxidation.

Some of the alteration products of silicate minerals are clay minerals, described in Chapter 9. A **clay mineral** is one that has plastic properties when moist, because it consists of thin flakes of colloidal size that become lubricated by layers of water molecules. Clay minerals formed by mineral alteration are abundant in common types of sedimentary rocks.

When carbon dioxide dissolves in water, a weak acid—*carbonic acid*—is formed. Carbonic acid can dissolve certain minerals, especially calcium carbonate. In addition, where decaying vegetation is present, soil water contains many complex organic acids that are capable of reacting with minerals. Certain common minerals, such as rock salt (sodium chloride), dissolve directly in water, but simple solution is not particularly effective for the silicate minerals.

SEDIMENTS AND SEDIMENTARY ROCKS

We can now turn to the second great rock class, the sedimentary rocks. The mineral particles in sedimentary rocks can be derived from preexisting rock of any of the three rock classes as well as from newly formed organic matter. However, igneous rock is the most important original source of the inorganic mineral matter that makes up sedimentary rock. For example,

a granite can weather to yield grains of quartz and particles of clay minerals derived from feldspars, thus contributing sand and clay to a sedimentary rock. Sedimentary rocks include rock types with a wide range of physical and chemical properties. We will only touch on a few of the most important kinds of sedimentary rocks, which are shown in Table 10.2.

In the process of mineral alteration, solid rock is weakened, softened, and fragmented, yielding particles of many sizes and mineral compositions. When transported by a fluid medium—air, water, or ice—these particles are known collectively as **sediment.** Used in its broadest sense, sediment includes both inorganic and organic matter. Dissolved mineral matter in the form of ions in solution must also be included. (Ions were explained in Chapter 9.)

Streams and rivers carry sediment to lower land levels, where sediment can accumulate. The most favorable sites of sediment accumulation are shallow sea floors bordering continents. But sediments also accumulate in inland valleys, lakes, and marshes. Thick accumulations of sediment may become deeply buried under newer (younger) sediments. Wind and glacial ice also transport sediment, but not necessarily to lower elevations or to places suitable for accumulation. Over long spans of time, the sediments can undergo physical or chemical changes, becoming compacted and hardened to form sedimentary rock.

There are three major classes of sediment. First is **clastic sediment,** which consists of mineral particles derived by breakage from a parent rock source. Examples are the materials in a sand bar of a river bed,

Figure 10.10 A dike of mafic igneous rock with nearly vertical parallel sides, cutting across flat-lying sedimentary rock layers. Arrows mark the contact between igneous rock and sedimentary rock. Spanish Peaks region, Colorado.

Eye on the Environment • Battling Iceland's Heimaey Volcano

It was around 2:00 A.M., early on a January morning in 1973, that the eruption began on Heimaey, a small island very near the southern coast of Iceland. First, a fissure split the eastern side of the island from one coast to the other, sending up a curtain of fire in a pyrotechnical display nearly 2 km (1.2 mi) long. Soon, however, the spraying fountains of volcanic debris became restricted to a small area not far from Helgafell, an older volcanic cone. The lava and lava fragments of ash and cinders, called *tephra*, poured out at a rate of 100 cu m/sec (3500 cu ft/sec), building a cone that soon reached a height of 100 m (300 ft) above sea level. It was dubbed Kirkjufell after a farmstead, Kirkjubaer, which lay beneath the debris.

It wasn't long before strong easterly winds set in, and Iceland's main fishing port, Vestmannaeyjar, located within a kilometer (a half-mile) of the eruption, began to receive a "snowfall" of fine, black cinders. Houses on the east side of the town were buried under the tephra. Many collapsed under the weight of the hot ash. Still others were set afire. Lava flows soon reached the village, burying and burning still other buildings (see photo).

Over the passing weeks, the emissions of tephra ceased, but lava continued to flow from the cone, which soon reached 200 m (600 ft) in height. Unfortunately, the lava began to flow northward, narrowing the harbor and threatening the future livelihood of the town and its evacuated inhabitants. Fishing is a major industry in Iceland, accounting for nearly 80 percent of Iceland's foreign exchange. Vestmannaeyjar normally lands and processes 20 percent of the nation's fish catch.

This aerial photo shows a portion of Westmannaeyjar, on Heimaey, Iceland, after the eruption. New lava has invaded the town on the right. On the left, black tephra covers the ground. It has been swept from rooftops and streets.

Because of the importance of the harbor and fishing port to the nation's economy, Icelanders embarked on a bold plan to save the harbor by altering the course of the lava, diverting the flow eastward. This was to be accomplished by cooling the flow on its north edge with water streams, creating a natural wall to channel the flow alongside the harbor instead of across it. Within a few weeks, the first pumping of water onto the flowing lava was begun. By early March, a pump ship came into operation in the harbor, providing a steady flow of sea water to cool the slowly moving flows. It was joined in April by as many as 47 high-capacity pumps floating on barges, delivering in total as much as 1 cu m of sea water per second (35 cu ft/sec).

The most effective technique for slowing the lava at a particular location began with cooling the edge and nearby surface of the flow with water from hoses. This allowed bulldozers to build a crude road up and over the slowly moving flow, using nearby tephra as the road material. Large plastic pipes were then laid along the road and across the flow, with small holes spraying water on hot spots. As long as sea water was flowing in the plastic pipes, they remained cool enough not to melt. After a day or so, the flow typically began to slow. Pumping usually continued for about two weeks, until the lava stopped steaming at spray points. The result of the water applications was to build a broad wall of cool, rubbly lava with thickening lava behind it.

The huge undertaking successfully stabilized the northern front of the lava flow, keeping the harbor from becoming closed (see map). Although the lava had indeed reached the harbor, it had merely narrowed its entrance. In fact, the new flow improved the harbor's ability to shelter the boats within it. Within about five months, the eruption was over, and the digging out began in earnest. Within a year, life was back to normal on Heimaey.

The diversion of the lava and reconstruction of the village of Vestmannaeyjar were extremely costly. To pay for the cost, Iceland passed a special tax increase, requiring the average Icelandic family to pay about 10 percent of its annual income for one year. Generous foreign aid also helped cover the enormous expenses borne by the tiny nation.

This map of Heimaey Island, Iceland, shows the extent of lava flows from the eruptions of the Kirkjufell crater. (Modified from U.S. Geological Survey.)

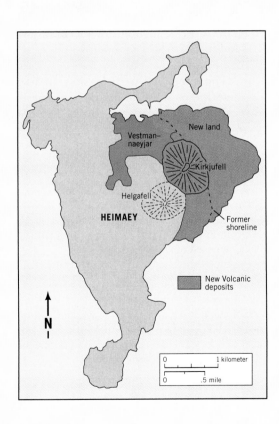

Questions

1. What were the initial effects of the eruption of Heimaey's volcano?
2. What important industry was threatened by the volcanic activity and why?
3. How was sea water used to keep lava from blocking the harbor?

or on a sandy ocean beach. Second is **chemically precipitated sediment,** which consists of inorganic mineral compounds precipitated from a saltwater solution or as hard parts of organisms. In the process of chemical precipitation, ions in solution combine to form solid mineral matter separate from the solution. Examples are a layer of rock salt and the white lime rock of a coral reef. A third class is **organic sediment** which consists of the tissues of plants and animals, accumulated and preserved after the death of the organism. An example is a layer of peat in a bog or marsh.

The three major classes of sediment are clastic, chemically-precipitated, and organic.

Sediment accumulates in more-or-less horizontal layers, called **strata,** or simply "beds" (see chapter opening photo). Individual strata are separated from those below and above by surfaces called stratification planes or bedding planes. These separation surfaces allow one layer to be easily removed from the next. Strata of widely different compositions can occur alternately, one above the next.

Clastic Sedimentary Rocks

Clastic sediments are derived from one or more of the rock groups—igneous, sedimentary, metamorphic—and thus include a very wide range of minerals for sedimentary rock formation. Silicate minerals are the most important, both in original form and as altered by oxidation and hydrolysis. Quartz and feldspar usually dominate. Because quartz is hard and is immune to

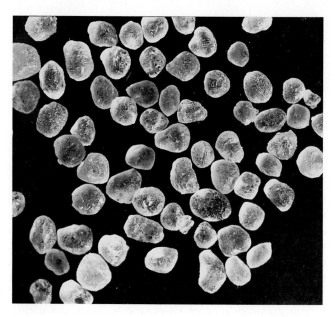

Figure 10.11 Rounded quartz grains from an ancient sandstone. The grains average about 1 mm (0.04 in.) in diameter.

alteration, it is the most important single component of the clastic sediments (Figure 10.11). Second in abundance are fragments of unaltered fine-grained parent rocks, such as tiny pieces of lava rock. Feldspar and mica are also commonly present. Clay minerals are major constituents of the finest clastic sediments.

The range of particle sizes in a clastic sediment determines how easily and how far the particles are transported by water currents. The finer the particles,

Table 10.2 Some Common Sedimentary Rock Types

Subclass	Rock Type	Description
Clastic (composed of rock and/or mineral fragments)	Sandstone	A rock composed of cemented sand grains
	Conglomerate	A sandstone containing pebbles of hard rock
	Mudstone	A rock composed of silt and clay, with some sand
	Claystone	A rock composed of clay
	Shale	A rock composed of clay that breaks easily into flat flakes and plates
Chemically precipitated (formed by chemical precipitation from sea water or salty inland lakes)	Limestone	A rock of calcium carbonate, formed by precipitation on sea or lake floors
	Dolomite	A rock of magnesium and calcium carbonates, similar to limestone
	Chert	A rock of silica, a noncrystalline form of quartz
	Evaporites	A class of rocks formed by evaporation of salty solutions in shallow inland lakes or coastal lagoons
Organic (formed from organic material)	Coal	A rock formed from peat or other organic deposits; may be burned as a mineral fuel
	Petroleum	A liquid hydrocarbon found in sedimentary deposits; not a true rock but a mineral fuel
	Natural gas	A gaseous hydrocarbon found in sedimentary deposits; not a true rock but a mineral fuel

the more easily they are held suspended in the fluid. On the other hand, the coarser particles tend to settle to the bottom of the fluid layer. In this way, a separation of size grades, called *sorting*, occurs. Sorting determines the texture of the sediment deposit and of the sedimentary rock derived from that sediment. Colloidal clay particles do not settle out unless they are made to clot together into larger clumps. This clotting process normally occurs when river water carrying clay mixes with the saltwater of the ocean.

When sediments accumulate in thick sequences, the lower strata are exposed to the pressure produced by the weight of the sediments above them. This pressure compacts the sediments, squeezing out excess water. Cementation occurs as dissolved minerals recrystallize where grains touch and in the spaces between mineral particles. Silicon dioxide (quartz, SiO_2) is very slightly soluble in water, and so the cement is often a form of quartz, called *silica*, which lacks a true crystalline form. Calcium carbonate ($CaCO_3$) is another common material that cements clastic sedimentary rocks. Compaction and cementation produce sedimentary rock.

The important varieties of clastic sedimentary rock are distinguished by the size of their particles. They include sandstone, conglomerate, mudstone, claystone, and shale. **Sandstone** is formed from fine to coarse sand (Figure 10.12). The cement may be silica or calcium carbonate. The sand grains are commonly of quartz, such as those shown in Figure 10.11. (Refer to Figure 9.2 for the names and diameters of the various grades of sediment particles.) Sandstone containing numerous rounded pebbles of hard rock is called *conglomerate* (Figure 10.13).

A mixture of water with particles of silt and clay, along with some sand grains, is called *mud*. The sedimentary rock hardened from such a mixture is called *mudstone*. Compacted and hardened clay layers become *claystone*. Sedimentary rocks of mud composition are commonly layered in such a way that they easily break apart into small flakes and plates. The rock is then described as being *fissile* and is given the name **shale**. Shale, the most abundant of all sedimentary rocks, is formed largely of clay minerals. The compaction of the mud to form mudstone and shale involves a considerable loss of volume as water is driven out of the clay.

Chemically Precipitated Sedimentary Rocks

Under favorable conditions, mineral compounds are deposited from the salt solutions of seawater and of salty inland lakes in desert climates. One of the most common sedimentary rocks formed by chemical precipitation is **limestone**, composed largely of the mineral *calcite*. Calcite is calcium carbonate ($CaCO_3$). Marine limestones—limestone strata formed on the sea floor—accumulated in thick layers in many ancient seaways in past geologic eras (Figure 10.14). A closely related rock is *dolomite*, composed of calcium-magnesium carbonate. Limestone and dolomite are grouped together as the *carbonate rocks*. They are dense rocks, some with a white color, while others are pale gray or even black.

Carbonate rocks—limestone and dolomite— are formed by chemical precipitation from seawater.

Sea water also yields sedimentary layers of silica in a hard, noncrystalline form called *chert*. Chert is a variety of sedimentary rock, but it also commonly occurs combined with limestone (cherty limestone).

Figure 10.12 An erosional remnant of massive sandstone in Utah. The thin neck of sandstone has been eroded from both sides to form a natural arch.

Figure 10.13 A piece of quartzitic conglomerate, cut through and polished, reveals rounded pebbles of quartz (clear and milky colors) and chert (grayish). It is about 12 cm (4 in.) in diameter.

Shallow water bodies acquire a very high level of salinity where evaporation is sustained and intense. One type of shallow water body is a bay or estuary in a coastal desert region. Another type is the salty lake of inland desert basins (see Figure 14.16). Sedimentary minerals and rocks deposited from such concentrated solutions are called **evaporites.** Ordinary rock salt, the mineral halite (sodium chloride), has accumulated in this way in thick sedimentary rock layers. These layers are mined as major commercial sources of halite.

Hydrocarbon Compounds in Sedimentary Rocks

Hydrocarbon compounds (compounds of carbon, hydrogen, and oxygen) form a most important type of organic sediment. These substances occur both as solids (peat and coal) and as liquids and gases (petroleum and natural gas). Only coal qualifies physically as a rock.

Peat is a soft, fibrous substance of brown to black color. It accumulates in a bog environment where the continual presence of water inhibits the decay of plant remains. We have already mentioned peat as an example of a Histosol in Chapter 9. (See Figure 9.18.)

At various times and places in the geologic past, plant remains accumulated on a large scale, accompanied by sinking of the area and burial of the compacted organic matter under thick layers of inorganic clastic sediments. *Coal* is the end result of this process (Figure 10.15). Individual coal seams are interbedded with shale, sandstone, and limestone strata. *Petroleum* (or *crude oil,* as the liquid form is often called) includes many hydrocarbon compounds. *Natural gas,* which is found in close association with accumulations of liquid petroleum, is a mixture of gases. The principal gas is methane (marsh gas, CH_4). Natural gas and petroleum commonly occupy open interconnected pores in a thick sedimentary rock layer—a porous sandstone, for example. They are not classed as minerals, but since they originated as organic compounds in sediments, they are classed as mineral fuels.

Hydrocarbon compounds in sedimentary rocks are important because they provide an energy resource on

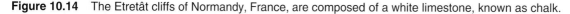

Figure 10.14 The Etretât cliffs of Normandy, France, are composed of a white limestone, known as chalk.

Figure 10.15 A coal seam near Sheridan, Wyoming, being strip mined by heavy equipment.

which modern human civilization depends. These **fossil fuels,** as they are called collectively, have required millions of years to accumulate. However, they are being consumed at a very rapid rate by our industrial society. These fuels are nonrenewable resources. Once they are gone, there will be no more, because the quantity produced by geologic processes even in a thousand years is scarcely measurable in comparison to the quantity stored through geologic time.

METAMORPHIC ROCKS

Any type of igneous or sedimentary rock may be altered by the tremendous pressures and high temperatures that accompany the mountain-building processes of the earth's crust. The result is a rock so changed in texture and structure as to be reclassified as **metamorphic rock.** Mineral components of the parent rock are, in many cases, reconstituted into different mineral varieties. Recrystallization of the original minerals can also occur. Our discussion of metamor-

phic rocks will mention only five common types—slate, schist, quartzite, marble, and gneiss (Table 10.3).

Slate is formed from shale that is heated and compressed by mountain-making forces. This fine-textured rock splits neatly into thin plates, which are familiar as roofing shingles and as patio flagstones. With application of increased heat and pressure, slate changes into **schist,** representing the most advanced stage of metamorphism. Schist has a structure called foliation, consisting of thin but rough, irregularly curved planes of parting in the rock (Figure 10.16). These are evidence of *shearing*—a compressional stress that pushes the layers sideways, like a deck of cards pushed into a fan with the sweep of a palm. Schist is set apart from slate by the coarse texture of the mineral grains, the abundance of mica, and occasionally the presence of scattered large crystals of newly formed minerals, such as garnet.

The metamorphic equivalent of conglomerate, sandstone, and siltstone is **quartzite,** which is formed by the addition of silica to fill completely the open spaces between grains. This process is carried out by the slow

Table 10.3   Some Common Metamorphic Rock Types

Rock Type	Description
Slate	A shale deposit exposed to heat and pressure that splits into hard, flat plates
Schist	A shale exposed to intense heat and pressure that shows evidence of shearing
Quartzite	A sandstone that is "welded" by a silica cement into a very hard rock of solid quartz
Marble	A limestone exposed to heat and pressure, resulting in larger, more uniform crystals
Gneiss	A rock resulting from the exposure of clastic sedimentary or intrusive igneous rocks to heat and pressure

Figure 10.16   This freshly exposed schist shows shiny surfaces where shearing has occurred. Known as a "greenschist," it represents the end product of deep deformation of black mud once deposited on the floor of an ancient oceanic trench.

movement of underground waters carrying silicate into the sandstone, where it is deposited.

Limestone, after undergoing metamorphism, becomes *marble*, a rock of sugary texture when freshly

Figure 10.17   This surf-washed rock surface exposes banded gneiss. Pemaquid Point, Maine.

broken. During the process of internal shearing, the calcite mineral of the limestone reforms into larger, more uniform crystals than before. Bedding planes are obscured, and masses of mineral impurities are drawn out into swirling bands.

Finally, the important metamorphic rock **gneiss** may be formed either from intrusive igneous rocks or from clastic sedimentary rocks that have been in close contact with intrusive magmas. A single description will not fit all gneisses because they vary considerably in appearance, mineral composition, and structure. One variety of gneiss is strongly banded into light and dark layers or lenses (Figure 10.17), which may be bent into wavy folds. These bands have differing mineral compositions. They are thought to be relics of sedimentary strata, such as shale and sandstone, to which new mineral matter has been added from nearby intrusive magmas.

THE CYCLE OF ROCK CHANGE

The processes that form rocks, when taken together, constitute a single system that over geologic time cycles and recycles earth materials from one form to another. The *cycle of rock change,* a concept that we introduced toward the beginning of this chapter, describes this system. Figure 10.18 shows a more complete version of this cycle. There are two environments—a surface environment of low pressures and temperatures and a deep environment of high pressures and temperatures. The surface environment is the site of rock alteration and sediment deposition. In this environment, igneous, sedimentary, and metamorphic rocks are uplifted and exposed to air and water. Their minerals are altered chemically and broken free from the parent rock, yielding sediment. The sediment accumulates in basins, where deeply buried sediment layers are compressed and cemented into sedimentary rock.

> **The cycle of rock change describes how earth materials are cycled and recycled by processes acting at the surface and at depth.**

In the deep environment, molten magma is heated by the slow radioactive decay of elements deep within the earth. Rocks lying close to the molten magma are exposed to heat and pressure, and are transformed into metamorphic rocks. Still other rocks are metamorphosed through compression and shearing in crustal mountain-building. Magma from the deep environment moves upward, melting and incorporating surrounding rock into its mass. Rising near the surface, the magma cools intrusively or extrusively, producing igneous rock.

The cycle of rock change has been active since the inception of our planet, forming and reforming rocks of all three major classes continuously. Igneous rocks of the common kinds we see around us today are by no means the "original" rocks of the earth's crust. Actually, there is no known record of the rocks that first formed the earth's crust. These rocks were consumed and recycled long ago.

This chapter has described the minerals and rocks of the earth's surface and the processes of their formation. As we look further into the earth's outermost layers in the next chapter, we will see that these processes do not occur everywhere. Instead, there is a grand plan that organizes the formation and destruction of rocks and distributes the processes of the cycle of rock change in a geographic pattern. This pattern is controlled by the pattern in which the solid earth's brittle outer layer is fractured into great plates that split and separate and also converge and collide. The grand plan is plate tectonics, the scheme for understanding the dynamics of the earth's crust over millions of years of geologic time.

Figure 10.18 The cycle of rock change.

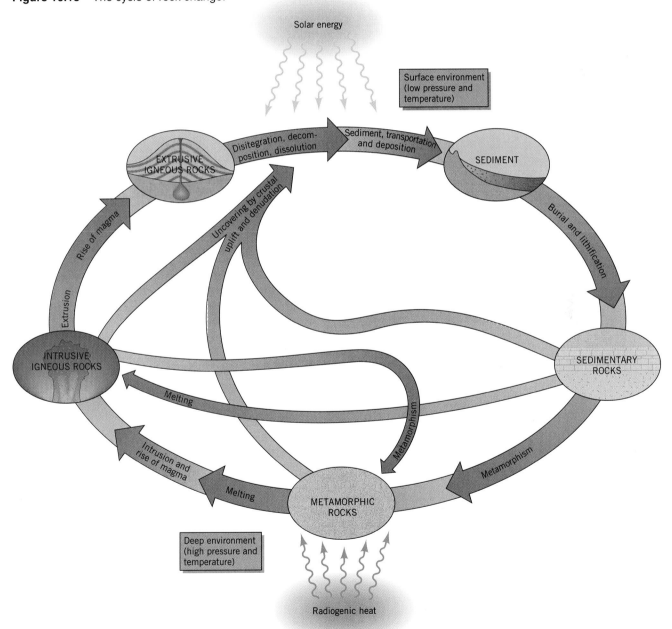

EARTH MATERIALS IN REVIEW

The elements oxygen and silicon dominate the earth's crust. Metallic elements, which include aluminum, iron, and the base elements, account for nearly all the remainder. These elements are found in minerals—naturally occurring, inorganic substances. Each has an individual chemical composition and atomic structure.

Silicates minerals make up the bulk of igneous rocks. They contain silicon and oxygen together with some of the metallic elements. There are three broad classes of igneous rocks, depending on their mineral content. Felsic rocks contain mostly felsic minerals and are least dense. Mafic rocks, containing mostly mafic minerals, are denser. Ultramafic rocks are most dense. Because felsic rocks are least dense, they are generally found in the upper layers of the earth's crust. Mafic and ultramafic rocks are more abundant in the layers below. If magma erupts on the surface to cool rapidly as lava, the rocks formed are extrusive and have a fine crystal texture. If the magma cools slowly below the surface as a pluton, the rocks are intrusive and the crystals are larger. Granite (felsic, intrusive), andesite (felsic, extrusive), and basalt (mafic, extrusive), are three very common igneous rock types.

Most silicate minerals found in igneous rocks undergo mineral alteration when exposed to air and moisture at the earth's surface. Mineral alteration occurs through oxidation, hydrolysis, or solution. Clay minerals are commonly produced by mineral alteration.

Sedimentary rocks are formed in layers, or strata. Clastic sedimentary rocks are composed of fragments of rocks and minerals that usually accumulate on ocean floors. As the layers are buried more and more deeply, water is pressed out and particles are cemented together. Sandstone and shale are common examples. Chemical precipitation also produces sedimentary rocks, such as limestone. Coal, petroleum, and natural gas are hydrocarbon compounds occurring in sedimentary rocks that are used as mineral fuels.

Metamorphic rocks are formed when igneous or sedimentary rocks are exposed to heat and pressure. Shale is altered to schist, sandstones become quartzite, and intrusive igneous rocks or clastic sediments are metamorphosed into gneiss.

In the cycle of rock change, rocks are exposed at the earth's surface, and their minerals are broken free and altered to form sediment. The sediment accumulates in basins, where the layers are compressed and cemented into sedimentary rock. Deep within the earth, the heat of radioactive decay melts preexisting rock into magma, which can move upward into the crust to form igneous rocks that cool at or below the surface. Rocks deep in the crust are exposed to the heat and pressure, forming metamorphic rock. Mountain-building forces move deep igneous, sedimentary, and metamorphic rocks upward to the surface, providing new material for surface alteration and breakup and completing the cycle.

KEY TERMS

mineral	intrusive igneous rock	organic sediment
rock	lava	strata
igneous rocks	extrusive igneous rock	sandstone
sedimentary rocks	andesite	shale
metamorphic rocks	basalt	limestone
cycle of rock change	pluton	evaporites
quartz	mineral alteration	fossil fuels
felsic	clay mineral	metamorphic rock
mafic	sediment	schist
magma	clastic sediment	quartzite
granite	chemically precipitated sediment	gneiss

REVIEW QUESTIONS

1. What is the earth's crust? What elements are most abundant in the crust?

2. Define the terms mineral and rock. Name the three major classes of rocks.

3. What are silicate minerals? Describe two classes of silicate minerals.

4. Name four types of igneous rocks and arrange them in order of density.

5. How do igneous rocks differ when magma cools (a) at depth, and (b) at the surface?

6. Sketch a cross section of the earth showing the following features: batholith, sill, dike, veins, lava, and volcano.

7. How are igneous rocks chemically altered? Identify and describe three processes of chemical alteration.

8. What is sediment? Define and describe three types of sediments.

9. Describe two processes that produce sedimentary rocks, and identify at least three important varieties of clastic sedimentary rocks.

10. How are sedimentary rocks formed by chemical precipitation?

11. What types of sedimentary deposits consist of hydrocarbon compounds? How are they formed?

12. What are metamorphic rocks? Describe at least three types of metamorphic rocks and how they are formed.

13. Sketch the cycle of rock change and describe the processes that act within it to form igneous, sedimentary, and metamorphic rocks.

IN-DEPTH QUESTION

A granite is exposed at the earth's surface, high in the Sierra Nevada mountain range. Describe how mineral grains from this granite might be released, altered, and eventually become incorporated in a sedimentary rock. Trace the route and processes that would incorporate the same grains in a metamorphic rock.

Chapter 11

The Lithosphere and Plate Tectonics

Perhaps it was in that eighth grade social studies class when you began to stare at the wall map of the world that hung in the front of the room, and you noticed that some parts of the map looked like a puzzle coming apart. Northeast Africa was particularly suspect. Twisting the continent toward the northeast would close the Red Sea and Persian Gulf, you thought, making the Arabian Peninsula join Africa and the Middle East in a continuous broad land bridge.

Then you noticed the Mediterranean. With a little squeezing, it could shut quite nicely, uniting Europe and Africa. Greece would have to tuck in under Turkey, which would move northward to fill in the Black Sea. Italy and its islands would have to move south, filling in the obvious gap in the African coast of Libya and Tunisia.

And then there was that funny bulge on the east coast of South America that seemed to tuck right under Africa's wing. The South American east coast would lie rather nicely along the west coast of southern Africa, with the Guianas butting against the West African republics from Liberia to Nigeria.

Perhaps, you wondered, all these lands actually did once fit together in a single world continent. Fun to think about, but utterly fantastic. Or was it?

The idea that the continents were once united and broken apart is an old one. In fact, in 1668, a Frenchman interpreted the matching coastlines of eastern South America and western Africa as proof that the two continents were separated during the biblical flood!

Within the last 30 years, geologists and physical geographers have accepted the idea that the continents have moved over long spans of geologic time. Evidence for the fracturing of the earth's outer rock layer and movement of the resulting rock plates has come from many sources—earthquake analysis, studies of ancient collections of plants and animals as they may have existed on united continents, and matching of geologic structures, such as mountain chains and rock sequences, from continent to continent. Most recently, the rates of movement of continents have been measured using laser beams bounced off orbiting satellites to determine with great accuracy the change in relative position of continental points.

As we will see in this chapter, the motions and histories of these vast rock plates, splitting and separating to form ocean basins, closing and colliding to form mountain ranges, can be taken as a grand plan to organize a great body of geologic and geographic knowledge. This is why this revolutionary plan—called plate tectonic theory—ranks with the theory of evolution as one of the great milestones in scientific study of the earth.

Arabian Peninsula, seen from space. Red Sea, with Gulfs of
Suez and Aqaba to left and right.

On the globes and maps we've seen since early childhood, the outline of each continent is so unique that we would never mistake one continent for another. But why are no two continents even closely alike? As we will see in this chapter, the continents have had a long history. In fact, in each of the continents some regions of metamorphic rocks date back more than 2 billion years. As part of that history, the continents have been fractured and split apart, as well as pushed together and joined.

Perhaps you've visited an old New England farmhouse that was constructed over hundreds of years—first the small two-room house, to which the kitchen shed was added at the back, and then the parlor wing at the side. Later the roof was raised for a second story, the tool house and barn were joined to the main house, the carriage house was built, and so it went. The earth's continents have that kind of history. They are composed of huge masses of continental crust that have been assembled at different times in each continent's history. The theory describing the motions and changes through time of the continents and ocean basins, and the processes that fracture and fuse them, is called *plate tectonics.*

The forces that move continents and cause them to collide are powered by energy sources deep within the earth. These forces are not influenced by the surface patterns of temperature, winds, precipitation, vegetation, or soils. Today we find volcanoes erupting in the cold desert of Antarctica as well as near the equator in African savannas. An alpine mountain range has been pushed up in the cold subarctic zone of Alaska, where it runs east-west. Yet another range lies astride the equator in South America and runs north-south. Both ranges lie in belts of crustal collision, where many strong earthquakes occur.

Although internal crustal processes operate independently, the processes that govern climate, vegetation, and soils are dependent on the major relief features and earth materials provided by the crustal processes. Thus, an understanding of plate tectonics is important to our understanding of the global patterns of the earth's landscapes—including its climate, soils, vegetation, and, ultimately, human activity.

In this chapter we will survey the major geologic features of our planet, starting with the layered structure of its deep interior. We will then examine the outermost layer, or crust, and compare the crust of the continents with the crust of the ocean basins. Lastly, we will turn to plate tectonics and describe how plate movements have created broad regions of igneous, sedimentary, and metamorphic rocks, as well as the uparching and downdropping of mountains and basins. We will find that plate tectonics furnishes us with a complete scenario of earth history on a grand scale in both time and spatial dimensions.

THE STRUCTURE OF THE EARTH

What lies deep within the earth? From studies of earthquake waves, reflected from deep earth layers, scientists have discovered that our earth is far from uniform from its outer crust to its center. Instead, it consists of a central core with several layers, or shells, surrounding it. The densest matter is at the center, and each layer above it is increasingly less dense. We will begin our examination of the earth's inner structure at the center and then work outward.

The Inner Structure and Crust

Figure 11.1 is a cutaway diagram of the earth showing its interior. The earth as a whole is an almost spherical body approximately 6400 km (4000 mi) in radius. The center is occupied by the **core,** which is about 3500 km (2200 mi) in radius. Because of the sudden change in behavior of earthquake waves upon reaching the core, scientists have concluded that the outer core has the properties of a liquid. However, the innermost part of the core is in the solid state. Based on earthquake waves (and other kinds of data), it has long been inferred that the core consists mostly of iron, with some nickel. The core is very hot—its temperature ranges from about 2800°C to 3100°C (5100°F to 5600°F).

The layers of the earth's interior include the crust, mantle, liquid core, and solid core.

Enclosing the metallic core is the **mantle,** a rock shell about 2900 km (1800 mi) thick, composed of ultramafic mineral matter. Judging from the behavior of earthquake waves, we can conclude that the mantle is composed of mafic minerals similar to olivine (a silicate of iron and magnesium). Thus, the mantle rock may resemble the ultramafic igneous rock peridotite, which is found exposed here and there on the continental surface. Temperatures in the mantle range from about 2800°C (5100°F) near the core to about 1800°C (3300°F) near the crust.

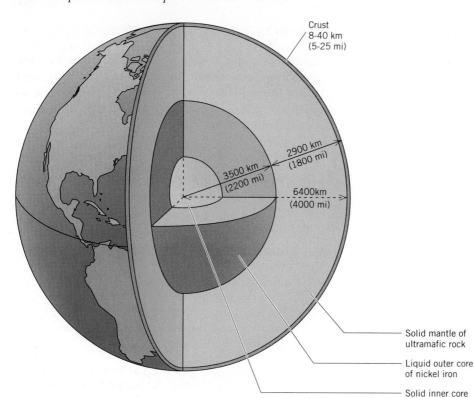

Crust
8-40 km
(5-25 mi)

3500 km
(2200 mi)

2900 km
(1800 mi)

6400km
(4000 mi)

Solid mantle of
ultramafic rock

Liquid outer core
of nickel iron

Solid inner core

Figure 11.1 This cutaway diagram of the earth shows the inner core of iron, which is solid, surrounded by a liquid outer core. The mantle, which surrounds the core, is a thick layer of ultramafic rock. The crust is too thin to show to correct scale.

The outermost and thinnest of the earth shells is the **crust,** a layer normally about 8 to 40 km (5 to 25 mi) thick (Figure 11.2). It is formed largely of igneous rock, but it also contains substantial proportions of both sedimentary and metamorphic rock. The base of the crust, where it contacts the mantle, is sharply defined. This contact is detected by the way in which earthquake waves abruptly change velocity at that level. The boundary surface between crust and mantle is called the *Moho,* a simplification of the name of the seismologist, Andrija Mohorovicic, who discovered it in 1909.

The continental crust is quite different from the crust beneath the oceans. From their study of earthquake waves, geologists have concluded that the **continental crust** consists of two continuous zones—a lower, continuous rock zone of mafic composition, which is more dense, and an upper zone of felsic rock, which is less dense (Figure 11.2). Because the felsic portion has a chemical composition similar to that of granite, it is commonly described as being *granitic rock.* Much of the granitic rock is metamorphic rock. There is no sharply defined separation between the felsic and mafic zones.

The crust of the ocean basins is sharply different from continental crust. **Oceanic crust** consists almost entirely of the mafic rocks basalt and gabbro. Basalt, as lava, forms an upper zone, whereas gabbro, an intrusive rock of the same composition, lies beneath the basalt.

Now, add to the above description the fact that the crust is much thicker beneath the continents than beneath the ocean floors, as Figure 11.2 shows. Whereas 35 km (20 mi) is a good average figure for crustal thickness beneath the continents, 7 km (4 mi) is an average figure for thickness of the basalt/gabbro crust beneath the deep ocean floors. The differences in both thickness and rock composition between continental and oceanic crust are attributable to the processes that have created the crust, which we discuss later in this chapter.

THE LITHOSPHERE

Geologists use the term **lithosphere** to mean an outer earth layer, or shell, of rigid, brittle rock. This usage of the term is different from that found in our Prologue, which identified the lithosphere as one of the four great realms of earth—atmosphere, hydrosphere, lithosphere, and biosphere. The word does not carry any meaning in terms of chemical makeup of the rock.

The lithosphere is a much thicker layer than the crust, as shown in Figure 11.2. It includes not only the crust, but also the cooler, upper part of the mantle that is composed of brittle rock. Since mantle rock beneath the brittle lithosphere is highly heated, it is in a plastic physical state. You can think of it as being much like white-hot iron on the verge of melting and capable of being shaped with little difficulty. In this state, iron is easily pressed into a form or drawn out into wire. Using the same analogy, we can say that the

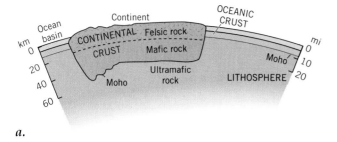

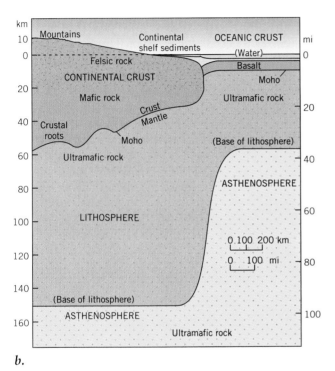

Figure 11.2 (*a*) Idealized cross section of the earth's crust and upper mantle. (*b*) Details of the crust and mantle at the edge of a continent, including the types of rocks found there. Also shown are the lithosphere and asthenosphere. (Copyright © A. N. Strahler.)

lithosphere resembles cold cast iron that responds to strong twisting or sharp bending by breaking abruptly ("snapping") along sharp fractures.

The lithosphere is the solid, brittle outer portion of the earth. It includes the crust and the uppermost part of the mantle.

Some tens of kilometers deep in the earth, the brittle condition of the lithospheric rock gives way gradually to a plastic, or "soft," layer named the **asthenosphere** (Figure 11.2). (This word is derived from the Greek root *asthenes*, meaning "weak.") However, at still further depth in the mantle, the strength of the rock

material again increases. Thus, the asthenosphere is a soft layer sandwiched between the "hard" lithosphere above and a "hard" mantle rock layer below. In terms of states of matter, the asthenosphere is not a liquid, even though its temperature reaches 1400°C (2600°F). Its melting point is raised by the immense pressure from the weight of overlying rocks.

As shown in Figure 11.2, the lithosphere ranges in thickness from 60 to 150 km (40 to 95 mi). It is thickest under the continents and thinnest under the ocean basins. Figure 11.3 idealizes the lithosphere and asthenosphere as two simple global layers of uniform thickness. We use the thickness of the oceanic lithosphere (60 km, or 150 mi) in this diagram. We present the layers this way only to show their true-scale dimensions in relation to the entire earth. The asthenosphere extends down to a depth of at least 300 km (185 mi), but both upper and lower boundaries are actually gradational. Under the ocean floors, the

Figure 11.3 The lithosphere and asthenosphere drawn to true scale. The diagram illustrates the extreme thinness of the mobile lithospheric plates that move over the asthenosphere. The brown line at the top of the upper diagram is scaled to represent a thickness of 10 km (6 mi). It will accommodate about 98 percent of the earth's surface features, from ocean floors to high mountains and plateaus. Only a few lofty mountains would project above the brown line, and only a few deep ocean trenches would project below the line.

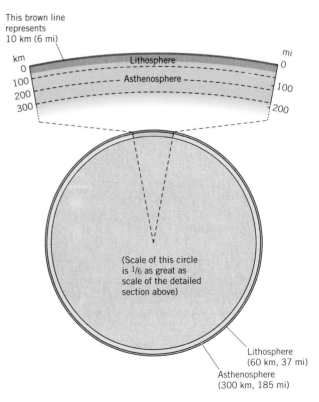

weakest portion of the asthenosphere lies at a depth of roughly 200 km (125 mi). Here the ultramafic mantle rock is close to its melting point.

The rigid, brittle lithosphere (or "hard shell") can move bodily over the soft, plastic asthenosphere. The yielding of the asthenosphere to allow this "gliding" motion is distributed through a thickness of many tens of kilometers. Also, the lithospheric shell is broken into large pieces called **lithospheric plates.** A single plate can be as large as a continent and can move independently of the plates that surround it. Like great slabs of floating ice on the polar sea, lithospheric plates can separate from one another at one location, while elsewhere they may collide in crushing impacts that raise great ridges. Along these collision ridges, one plate can be found diving down beneath the edge of its neighbor. These varied sorts of plate movements are the primary focus of this chapter.

THE GEOLOGIC TIMETABLE

To place the great movements of lithospheric plates in their correct positions in historical sequence, we will need to refer to some major units in the scale of geologic time. Table 11.1 lists the major geologic time divisions. All time older than 570 million years (m.y.) before the present is *Precambrian time.* Three *eras* of time follow: *Paleozoic, Mesozoic,* and *Cenozoic.* These eras saw the evolution of life-forms in the oceans and on the lands. The geologic eras are subdivided into *periods.* Their names, ages, and durations are also given in Table 11.1.

The Cenozoic Era is particularly important in terms of the continental surfaces, because nearly all landscape features seen today have been produced in the 65 million years since that era began. The Cenozoic Era is comparatively short in duration, scarcely more than the average duration of a single period in older eras. For that reason, it is subdivided directly into seven lesser time units called *epochs.* Details of the Pleistocene and Holocene epochs are given in Chapter 18.

The human genus *Homo* evolved during the late Pliocene Epoch and throughout the Pleistocene Epoch. As you can see, the period of human occupation of the earth's surface—a few million years at best—is but a fleeting moment in the vast duration of our planet's history.

Table 11.1 Table of Geologic Time

Era	Period	Epoch	Duration millions of yrs	Age millions of years	Orogenies
Cenozoic		Holocene	(10,000 yr)		
		Pleistocene	2		
				2	
		Pliocene	3		
				5	
		Miocene	19		
				24	
		Oligocene	13		
				37	
		Eocene	21		
				58	
		Paleocene	8		
				66	Cordilleran
Mesozoic	Cretaceous		78		
				144	
	Jurassic		64		
				208	
	Triassic		37		
				245	Alleghany, or Hercynian
Paleozoic	Permian		41		
				286	
	Carboniferous		74		
				360	
	Devonian		48		
				408	Caledonian
	Silurian		30		
				438	
	Ordovician		67		
				505	
	Cambrian		65		
				570	

Precambrian Time (Extends to oldest known rocks, about 3.6 billion years)
Age of earth as a planet: 4.6 to 4.7 billion years.
Age of universe: 17 to 18 billion years.

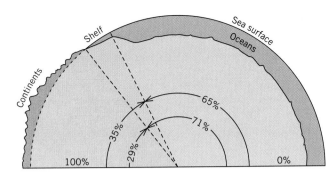

Figure 11.4 Actual global percentages of land and ocean areas compared with percentages if sea level were to drop 180 m (600 ft), exposing the continental shelves.

CONTINENTS AND OCEAN BASINS

The major relief features of the earth are the continents and ocean basins (Figure 11.4). Detailed global maps show that about 29 percent of the earth's surface is land and 71 percent oceans. If the seas were to drain away, however, we would see that broad areas lying close to the continental shores are actually covered by shallow water, less than 150 m (500 ft) deep. From these relatively shallow continental shelves, the ocean floor drops rapidly to depths of thousands of meters. In that respect, the ocean basins seem to be more than "brim full" of water. That is, the oceans have spread over the margins of ground that would otherwise be assigned to the continents. If the ocean level were to drop by 150 m (500 ft), the shelves would be exposed, adding about 6 percent to the area of the continents. Then, the surface area of continents would be increased to 35 percent, and the ocean basin area decreased to 65 percent. These revised figures represent the true relative proportions of continents and oceans.

Relief Features of the Continents

Broadly viewed, the continental masses consist of two basic subdivisions: (1) active belts of mountain-making and (2) inactive regions of old, stable rock. The mountain ranges in the active belts grow through one of two very different geologic processes. First is *volcanism,* the formation of massive accumulations of volcanic rock by extrusion of magma. Many lofty mountain ranges consist of chains of volcanoes built of extrusive igneous rocks. The second mountain-building process is *tectonic activity,* the breaking and bending of the earth's crust under internal earth forces. This tectonic activity usually occurs when great lithospheric plates come together in titanic collisions. (We will explain

this topic later in this chapter.) Crustal masses that are raised by tectonic activity form mountains and plateaus. Masses that are lowered form crustal depressions. In some instances, volcanism and tectonic activity have combined to produce a mountain range. Landforms produced by volcanic and tectonic activity are the subject of Chapter 12.

The two basic subdivisions of continental masses are active belts of mountain-making and inactive regions of old, stable rock.

Alpine Chains

Active mountain-making belts are narrow zones that are usually found along continental margins. These belts are sometimes referred to as *alpine chains,* because they are characterized by high, rugged mountains, such as the Alps of Central Europe. These mountain belts were formed in the Cenozoic Era by volcanism, or tectonic activity, or a combination of both. Alpine mountain-building continues even today in many places.

The alpine chains are characterized by broadly curved patterns on the world map (Figure 11.5). Each curved section of an alpine chain is referred to as a **mountain arc.** The arcs are linked in sequence to form the two principal mountain belts. One is the *circum-Pacific belt,* which rings the Pacific Ocean basin. In North and South America, this belt is largely on the continents and includes the Andes and Cordilleran ranges. In the western part of the Pacific basin, the mountain arcs lie well offshore from the continents and take the form of **island arcs.** Partly submerged, they join the Aleutians, Kurils, Japan, the Philippines, and other smaller islands. These island arcs are the result of volcanic activity. Between the larger islands, the arcs are represented by volcanoes rising above the sea as small, isolated islands.

The second chain of major mountain arcs forms the *Eurasian-Indonesian belt,* shown in Figure 11.5. It starts in the west at the Atlas Mountains of North Africa and continues through the European Alps and the ranges of the Near East and Iran to join the Himalayas. The belt then continues through Southeast Asia into Indonesia, where it abruptly meets the circum-Pacific belt. Later we will return to these active belts of mountain-making and explain them in terms of lithospheric plate motions.

Continental Shields

Belts of recent and active mountain-making account for only a small portion of the continental crust. The

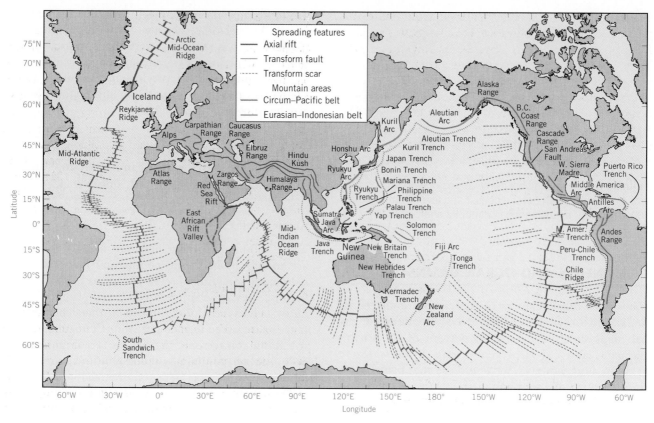

Figure 11.5 Principal mountain arcs, island arcs, and trenches of the world and the midoceanic ridge. (Midoceanic ridge map copyright © A. N. Strahler.)

remainder consists of comparatively inactive regions under which lie much older rock. Within these stable regions we recognize two structural types of crust— shields and mountain roots. **Continental shields** are low-lying continental surfaces beneath which lie igneous and metamorphic rocks in a complex arrangement. Figure 11.6 is a very generalized map showing the shield areas of the continents. Two classes of shield are shown: exposed shields and covered shields. *Exposed shields* include very old rocks, mostly from Precambrian time, and have had a very complex geologic history. An example of an exposed shield is the Canadian Shield of North America. Exposed shields are also extensive in Scandinavia, South America, Africa, peninsular India, and Australia.

The exposed shields are largely regions of low hills and low plateaus, although there are some exceptions where large crustal blocks have been recently uplifted. Many thousands of meters of rock have been eroded from these shields during their continuous exposure throughout the past half-billion or more years.

Large areas of the continental shields are covered by younger sedimentary layers, ranging in age from Paleozoic through Cenozoic eras. These strata accumulated at times when the shields subsided and were inundated by shallow seas. Marine sediments were laid down on the ancient shield rocks in thicknesses ranging from hundreds to thousands of meters. These shield areas were then broadly arched and again became land surfaces. Erosion has since removed large sections of the sedimentary cover, but it still remains intact over vast areas. We refer to such areas as *covered shields* in order to distinguish them from the exposed shields in which the Precambrian rocks lie bare. The covered shields are shown in Figure 11.6.

Some core areas of the shields are composed of rock as old as early Precambrian time, dating back to a time period called the Archean Eon, 2.5 to 3.5 billion years ago. On our map, these ancient areas are shown encircled by bold lines. The ancient cores are exposed in some areas but covered in others.

Ancient Mountain Roots

Remains of older mountain belts lie within the shields in many places. These *mountain roots* are mostly formed of Paleozoic and early Mesozoic sedimentary rocks that have been intensely bent and folded, and in some locations changed into metamorphic rocks—slate, schist, and quartzite, for example. Thousands of meters of overlying rocks have been removed from these old tectonic belts, so that only lowermost structures remain. Roots appear as chains of long, narrow ridges, rarely rising over a thousand meters above sea level. (Landforms of mountain roots are described in Chapter 16.)

One important system of mountain roots was formed in the Paleozoic Era, during a great collision between two enormous lithospheric plates that took place about 400 million years ago. What were then high alpine mountain chains have since been worn down to belts of subdued mountains and hills. Today these roots, called Caledonides, form a highland belt across the northern British Isles and Scandinavia. They are also present in the Maritime Provinces of eastern Canada and in New England. A second, but younger, root system was formed during another great collision of plates near the close of the Paleozoic Era, about 250 million years ago. In North America, this highland system is represented by the Appalachian Mountains. The Caledonides and Appalachians are shown as mountain roots in Figure 11.6.

Mountain roots, long eroded to low highlands and ridges, mark ancient plate collisions.

Relief Features of the Ocean Basins

Crustal rock of the ocean floors consists almost entirely of basalt, which is covered over large areas by a comparatively thin accumulation of sediments. Age determinations of the basalt and its sediment cover show that the oceanic crust is quite young, geologically speaking. Much of that crust was formed during the Cenozoic Era and is less than 60 million years old. Over some large areas, however, the rock is of Mesozoic age, mostly from 65 to about 135 million years old. When we consider that the great bulk of the continental crust is of Precambrian age—mostly over 1 billion years old—the young age of the oceanic crust is quite remarkable. However, we will soon see how plate tectonic theory explains this young age.

The Midoceanic Ridge

Figure 11.7 shows the important relief features of ocean basins. The ocean basins are characterized by a central ridge structure that divides the basin in about half. The *midoceanic ridge* consists of submarine hills that rise gradually to a rugged central zone. Precisely in the center of the ridge, at its highest point, is the *axial rift*, which is a narrow, trenchlike feature. The

Figure 11.6 Generalized world map of continental shields, exposed and covered. Continental centers of early Precambrian age (Archean) lie within the areas encircled by a bold line. Lines of bold dots show mountain roots of Caledonian and Hercynian orogenies. (Based in part on data of R. E. Murphy, P. M. Hurley, and others. Copyright © A. N. Strahler.)

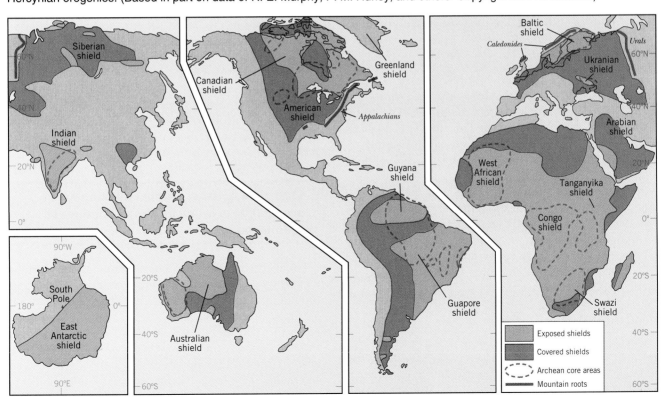

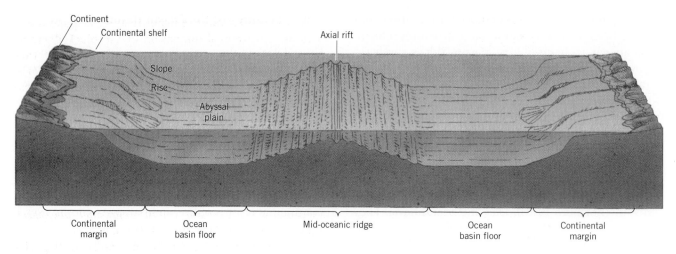

Figure 11.7 This schematic block diagram shows the main features of ocean basins. It applies particularly well to the North and South Atlantic oceans.

location and form of this rift suggest that the crust is being pulled apart along the line of the rift.

The midoceanic ridge and its principal branches can be traced through the ocean basins for a total distance of about 60,000 km (37,000 mi). Figure 11.5 shows the extent of the ridge. Beginning in the Arctic Ocean, it then divides the Atlantic Ocean basin from Iceland to the South Atlantic. Turning east, it enters the Indian Ocean, where one branch penetrates Africa. The other branch continues east between Australia and Antarctica, and then swings across the South Pacific. Nearing South America, it turns north and reaches North America at the head of the Gulf of California.

The Ocean Basin Floor

On either side of the midoceanic ridge are broad, deep plains and hill belts that belong to the *ocean basin floor* (Figure 11.7). Their average depth below sea level is about 5000 m (16,000 ft). The flat surfaces are called *abyssal plains*. They are extremely smooth because they have been built up of fine sediment that has settled slowly and evenly from ocean water above.

Many details of the ocean basins and their submarine landforms are shown in Figure 11.8. This map of the ocean floor was constructed from thousands of bottom profiles made by automatic depth-recording equipment carried on oceanographic research vessels. The apparatus uses reflected sound waves to produce a picture of the ocean floor.

Continental Margins

The *continental margin*, labeled in Figure 11.7 as the third form element of the typical ocean basin, can be defined as the narrow zone in which oceanic lithosphere is in contact with continental lithosphere (see

Figure 11.2*b*). Thus, the continental margin is a feature shared by the continent and its adjacent ocean basin.

As the continental margin is approached, the ocean floor begins to slope gradually upward, forming the *continental rise* (Figure 11.7). The floor then steepens greatly on the *continental slope*. At the top of this slope we arrive at the brink of the *continental shelf*, a gently sloping platform some 120 to 160 km (75 to 100 mi) wide along the eastern margin of North America. Water depth is about 150 m (500 ft) at the outer edge of the shelf.

The symmetrical model illustrated in Figure 11.7 is nicely shown in the North Atlantic and South Atlantic Ocean basins. It also applies rather well to the Indian Ocean and Arctic Ocean basins. The margins of these symmetrical basins are described as *passive continental margins*, which means they have not been subjected to Cenozoic tectonic and volcanic activity. Both continental and oceanic lithosphere at a passive continental margin are part of the same lithospheric plate and move together, away from the axial rift.

Over time, passive continental margins accumulate great quantities of continental sediments.

Great thicknesses of sedimentary strata, derived from the continents, have accumulated at passive continental margins. The strata range in age from the late Mesozoic (Jurassic, Cretaceous) through the Cenozoic. Near the continent, the shelf strata form a wedge-shaped deposit, thickening oceanward. A block diagram, Figure 11.9, shows details of this deposit. The sediments have been brought from the land by rivers and spread over the shallow sea floor by currents.

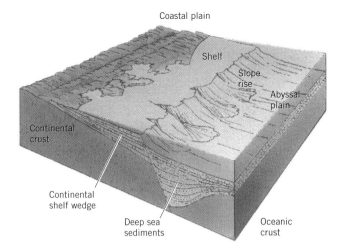

Coastal plain

Shelf

Slope rise

Abyssal plain

Continental crust

Continental shelf wedge

Deep sea sediments

Oceanic crust

Figure 11.9 This block diagram shows an inner wedge of sediments beneath the continental shelf and an outer wedge of deep-sea sediments beneath the continental rise and abyssal plain.

Below the continental rise and its adjacent abyssal plain is another thick sediment deposit. It is formed of deep-sea sediments carried down the continental slope by swift muddy currents. The weight of these sediments causes them to sink, depressing the oceanic crust at its margin with the continental crust.

The Pacific Ocean basin differs from our ideal model in the layout of its margins. Although it has a midoceanic ridge with ocean basin floors on either side, it has quite different continental margins. They are characterized by mountain arcs or island arcs with deep offshore *oceanic trenches*. Geologists refer to these trenched ocean basin limits as *active continental margins*.

The locations of the major trenches are shown in Figure 11.5. Trench floors can reach depths of nearly 11,000 m (36,000 ft), although most range from 7,000 to 10,000 m (23,000 to 33,000 ft) (see Figure 11.8). Many lines of scientific evidence show that the oceanic crust is bent down sharply to form these trenches and that they mark the boundary between two lithospheric plates that are being brought together.

PLATE TECTONICS

Both crustal spreading along the axial rift of the midoceanic ridge and crustal downbending beneath oceanic trenches involve the entire thickness of lithospheric plates. The general theory of lithospheric plates with their relative motions and boundary interactions is called **plate tectonics. Tectonics** is a noun meaning "the study of tectonic activity." As defined earlier, tectonic activity refers to all forms of breaking and bending of the entire lithosphere, including the crust.

Figure 11.10 shows the major features of plate interactions. The vertical dimension of the block diagram (*a*) is greatly exaggerated, as are the landforms. A true-scale cross section (*b*) shows the correct relationships between crust and lithosphere, but surface relief features can scarcely be shown.

There are two very different kinds of lithospheric plates (see Figure 11.10*b*). Plates that lie beneath the ocean basins consist of *oceanic lithosphere*. The diagrams show two plates, plates X and Y, both made up of oceanic lithosphere, which is comparatively thin (about 50 km, 30 mi thick). Plate Z, bearing thick continental crust, is made up of *continental lithosphere*, which is much thicker (about 150 km, 95 mi).

An important scientific principle applies to the relative surface heights of the two kinds of lithosphere. The lithosphere can be thought of as "floating" on the soft asthenosphere. Consider two blocks of wood, one thicker than the other, floating in a pan of water. The surface of the thick block will ride higher above the water surface than that of the thin block. This principle explains why the continental surfaces rise so high above the ocean floors.

Plate Motions and Interactions

As shown in Figure 11.10, plates X and Y are pulling apart along their common boundary, which lies along the axis of a midoceanic ridge. This pulling apart tends to create a gaping crack in the crust, but magma continually rises from the mantle beneath to fill it. The magma appears as basaltic lava in the floor of the rift and quickly congeals. At greater depth under the rift, magma solidifies into gabbro, an intrusive rock of the same composition as basalt. Together, the basalt and gabbro continually form new oceanic crust. This type of boundary between plates is termed a *spreading boundary.*

At the right, the oceanic lithosphere of plate Y is moving toward the thick mass of continental lithosphere that comprises plate Z. Where these two plates collide, they form a *converging boundary.* Because the oceanic plate is comparatively thin and dense, in contrast to the thick, buoyant continental plate, the oceanic lithosphere bends down and plunges into the soft layer, or asthenosphere. The process of downplunging of one plate beneath another is called **subduction.**

The leading edge of the descending plate is cooler than the surrounding asthenosphere—sufficiently cooler, in fact, that this descending slab of brittle rock is denser than the surrounding soft hot rock. Consequently, once subduction has begun, the slab "sinks under its own weight," so to speak. However, the slab is gradually heated by the surrounding hot rock and thus eventually softens. The underportion, which is mantle rock in composition, simply reverts to mantle

Figure 11.8 A portion of *Map of the World Ocean Floor* by Bruce C. Heezen and Marie Tharp. Based on Mercator map projection. (Copyright © 1977 by Marie Tharp. Used by permission.)

ICELAND

Reykjavik

AEGIR RIDGE

PLATEAU

ICELAND-FAEROE
RIDGE

FAEROE
ISLANDS

Oslo
Stockholm

Hel.

REYKJANES RIDGE

ORKNEY
IS.

Copenhagen

ADOR SEA

ROCKALL
PLATEAU

NORTH SEA

EUR

Dublin

Northwest Atlantic
Mid-Ocean Canyon

FRACTURE ZONE

PORCUPINE
PLATEAU

London

Rotterdam

Rhine

CARPATHIAN

ORPHAN KNOLL

CELTIC
SHELF

Seine

FARACA FRACTURE ZONE

BISCAY
ABYSSAL
PLAIN

ALPS
MOUNTAINS

Venice

FLEMISH
CAP

Marseille

Rome

GRAND BANKS

Barcelona

NEWFOUNDLAND
SEAMOUNTS

IBER
ABYSSAL PLAIN

Lisbon

BALEARIC
A.P.

LAURENTIAN
CONE

NEWFOUNDLAND
RIDGE

OCEAN

AZORES F.Z.

Tunis

Oran

MALTA
PLATEAU

OHM ABYSSAL PLAIN

ZONE

OR MER RIDGE

Casablanca

Tripoli

CORNER
RIDGE

SEINE A.P.

MADIERA
A.P.

CANARY
A.P.

CANARY
ISLANDS

AFRICA

ZONE

CAPE VERDE
ISLANDS

Sénégal

Dakar

GAMBIA
A.P.

Freetown

DEMERARA
PLATEAU

Amazon
Canyon

PARA A.P.

SIERRA LEONE
RIDGE

Accra

Lagos

Niger

AMAZON
CONE

SIERRA LEONE
A.P.

CEARA

FRACTURE ZONE

ZONE

Congo

Amazon

Iquatorial
Mid-Ocean
Canyon

ROMA

LUANDA
PLATEAU

Luanda

Belém

Fortaleza

Recife

São Francisco

PERNAMBUCO
A.P.

ZONE

ANGOLA
ABYSSAL
PLAIN

AMERICA

Eye on the Environment • Alfred Wegener and Continental Drift

Although modern plate tectonic theory is only a few decades old, the concept of breakup of an early supercontinent into fragments that drifted apart dates back to the nineteenth century and beyond. Almost as soon as good navigational charts became available to show the continental outlines, natural scientists became intrigued with the close correspondence in outline between the eastern coastline of South America and the western coastline of Africa. In 1858 Antonio Snider-Pelligrini produced a map to show that the American continents nested closely against Africa and Europe. He went beyond the purely geometrical fitting to suggest that the reconstructed single continent explains the close similarity of fossil plant types in coal-bearing rocks in both Europe and North America.

Moving ahead to the early twentieth century, we come to the ideas of two Americans, Frank B. Taylor and Howard B. Baker, whose published articles presented evidence favoring the hypothesis that the New World and Old World continents had drifted apart. Nevertheless, credit for a full-scale hypothesis of the breakup of a single supercontinent and the drifting apart of individual continents belongs to Alfred Wegener, the German meteorologist and geophysicist who became interested in the various lines of geologic evidence that the continents had once been united. He first presented his ideas in 1912, and his major work on the subject appeared in 1922. One of his maps is shown at right. Without any inkling of the existence of lithospheric plates, Wegener had reconstructed a supercontinent named *Pangaea,* which existed intact as early as about 300 million years ago in the Carboniferous Period. His map is strikingly similar to the one we have shown as Figure 11.20*a*.

A storm of controversy followed, and many American geologists denounced the "continental drift" hypothesis. However, Wegener had loyal supporters in Europe, South Africa, and Australia. Several lines of hard geologic evidence presented by Wegener and his followers strongly favored the existence of Pangaea. Those arguments remain strong today. But the actual physical process of separation of the continents was strongly criticized on valid physical grounds.

Wegener had proposed that a continental layer of less dense rock had moved like a great floating "raft" through a "sea" of denser oceanic crustal rock. Note that both the "raft" and the "sea" are composed of strong, brittle rock. Geologists could show by use of established principles of physics that this proposed mechanism was impossible, because two strong, rigid crustal rock layers, one

---

rock as it softens. The thin upper crust, formed of less dense mineral matter, can melt and become magma. This magma tends to rise because it is less dense than the surrounding material. Figure 11.10 shows some magma pockets formed from the upper edge of the slab. They are pictured as rising like hot-air balloons through the overlying continental lithosphere. When they reach the earth's surface, they form a volcano chain lying about parallel with the deep oceanic trench that marks the line of descent of the oceanic plate.

Sea-floor spreading builds lithospheric plates by accretion. Plate material is lost to consumption when it sinks into plastic mantle rock and melts.

Viewing plate Y as a unit in Figure 11.10, we see that this single lithospheric plate is simultaneously undergoing *accretion* (growth by addition) and *consumption* (by softening and melting). If rates of accretion and consumption are equal, the plate will maintain its over-

all size. If consumption is slower, the plate will expand. If accretion is slower, the plate will shrink.

We have yet to consider a third type of lithospheric plate boundary. Two lithospheric plates may be in contact along a common boundary on which one plate merely slides past the other with no motion that would cause the plates either to separate or to converge (Figure 11.11). This is a *transform boundary*. The plane along which motion occurs is a nearly vertical fracture extending down through the entire lithosphere, and it is called a *transform fault*. A *fault* is a rock plane along which there is motion of the rock mass on one side with respect to that on the other. (More about faults will appear in Chapter 12.) Transform boundaries are often associated with midoceanic ridges and are shown in Figures 11.8 and 11.10.

Plate Boundaries Summarized

In summary, there are three major kinds of active plate boundaries:

embedded in the other, could not behave in such a fashion.

Wegener's scenario of continental drift took on a new context in the 1960s and 1970s, when plate tectonics emerged as a leading theory. The modern interpretation is that continental drift involves entire lithospheric plates, much thicker than merely the outer crust of either the continents or the ocean basins. Plate motions over a soft, plastic asthenosphere have allowed the continents and the ocean basins to be carried along merely as passengers. Plate tectonics is a whole new scientific theory in itself.

Questions

1. Provide a brief history of the idea of "drifting continents."
2. What was Wegener's theory about "continental drift?" Why was it opposed at the time?

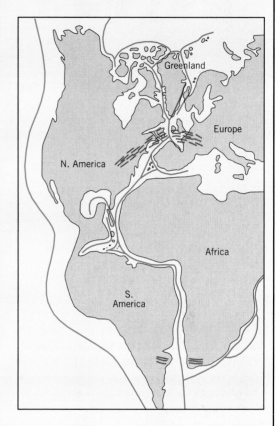

Alfred Wegener's 1915 map fitting together the continents that today border the Atlantic Ocean basin. The sets of short lines show the fit of Paleozoic tectonic structures between Europe and North America and between southernmost Africa and South America. (From A. Wegener, 1915, *De Entstehung der Kontinente und Ozeane*, F. Vieweg, Braunschweig.)

Spreading boundaries. New lithosphere is being formed by accretion. Example: Sea-floor spreading along the axial rift.

Converging boundaries. Subduction is in progress, and lithosphere is being consumed. Example: Active continental margin.

Transform boundaries. Plates are gliding past one another on a transform fault. Example: Transform boundary associated with midoceanic ridge.

Let us put these three boundaries into a pattern to include an entire lithospheric plate. Visualize the sunroof of an automobile, in which a portion of the roof slides to open. Just as the sunroof can move by sliding past the fixed portion of the roof at its sides, so a lithospheric plate can move by sliding past other plates on transform faults. Where the sunroof opens, the situation is similar to a spreading boundary. Where the sunroof slides under the rear part of the roof, the situation is similar to a converging boundary. Boundaries of lithospheric plates can be curved as well as straight, and individual plates can pivot as they move. There are

many geometric variations in the shapes and motions of individual plates.

The Global System of Lithospheric Plates

The global system of lithospheric plates consists of six great plates. These are listed in Table 11.2 and are shown on a world map in Figure 11.12. Several lesser plates and subplates are also recognized. They range in size from intermediate to comparatively small. Plate boundaries are shown by symbols, explained in the key accompanying the map.

The great Pacific plate occupies much of the Pacific Ocean basin and consists almost entirely of oceanic lithosphere. Its relative motion is northwesterly, so that it has a subduction boundary along most of the western and northern edge. The eastern and southern edge is mostly a spreading boundary. A sliver of continental lithosphere is included and makes up the coastal portion of California and all of Baja California. The California portion is bounded by an active transform fault (the San Andreas Fault).

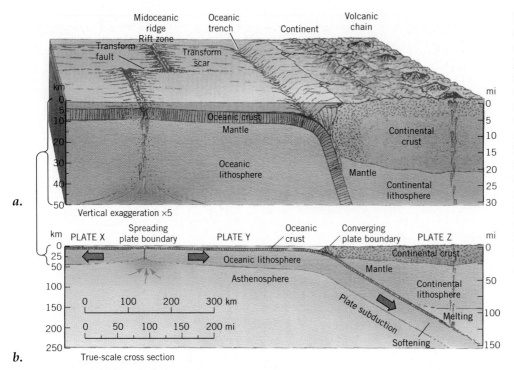

The Nazca and Cocos plates of the eastern Pacific are plunging beneath the western edges of Central and South America, generating mountain ranges.

The American plate includes most of the continental lithosphere of North and South America as well as the entire oceanic lithosphere lying west of the midoceanic ridge that divides the Atlantic Ocean basin down the middle. For the most part, the western edge of the American plate is a subduction boundary, with oceanic lithosphere diving beneath the continental lithosphere. The eastern edge is a spreading boundary. (Some classifications recognize separate North American and South American plates.) The Eurasian plate is mostly continental lithosphere, but it is fringed on the west and north by a belt of oceanic lithosphere.

The African plate has a central core of continental lithosphere nearly surrounded by oceanic lithosphere. The Austral-Indian plate takes the form of a long rectangle. It is mostly oceanic lithosphere but contains two cores of continental lithosphere—Australia and peninsular India. The Antarctic plate has an elliptical shape and is almost completely enclosed by a spreading plate boundary. This means that the other plates are moving away from the pole. The continent of Antarctica forms a central core of continental lithosphere completely surrounded by oceanic lithosphere.

Of the nine lesser plates, the Nazca and Cocos plates of the eastern Pacific are rather simple fragments of oceanic lithosphere bounded by the Pacific midoceanic spreading boundary on the west and by a subduction boundary on the east. The Philippine plate is noteworthy as having subduction boundaries on both east and west edges. Two small but distinct lesser plates—Caroline and Bismark—lie to the southeast of the Philippine plate, but these can be included within the Pacific plate. The Arabian plate has two transform fault boundaries, and its relative motion is northeasterly. The Caribbean plate also has important transform fault boundaries. The tiny Juan de Fuca plate is steadily diminishing in size and will eventually

Figure 11.11 A transform fault involves the horizontal motion of two adjacent lithospheric plates, one sliding past the other. (Copyright © by A. N. Strahler.)

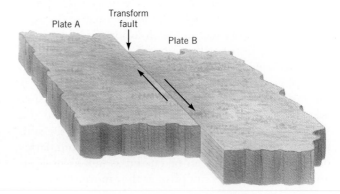

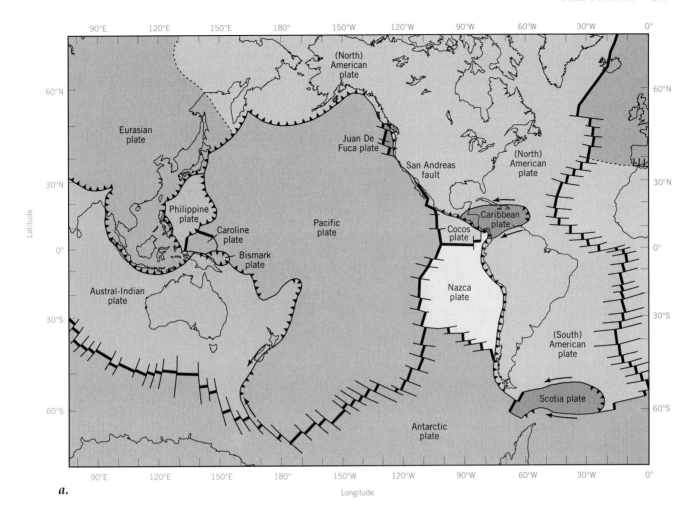

a.

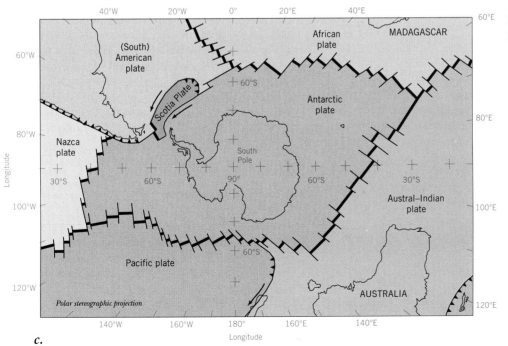

c.

Figure 11.12 World map of lithospheric plates.

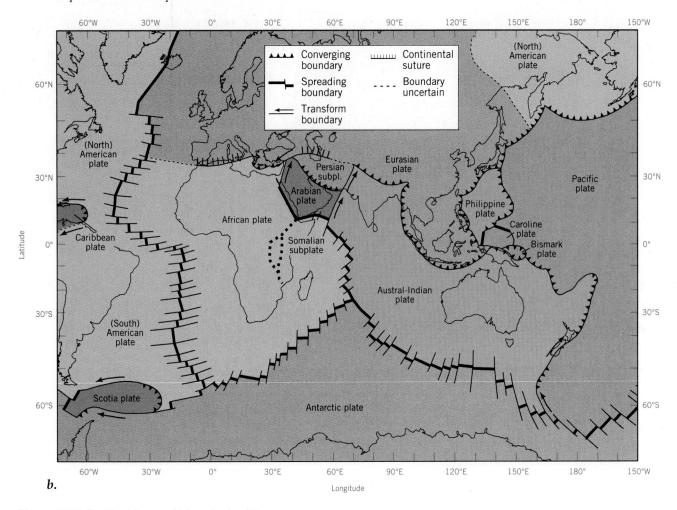

Figure 11.12B World map of lithospheric plates.

Table 11.2 The Lithospheric Plates

Great plates	Lesser plates
Pacific	Nazca
American (North, South)	Cocos
Eurasian	Philippine
Persian subplate	Caribbean
African	Arabian
Somalian subplate	Juan de Fuca
Austral-Indian	Caroline
Antarctic	Bismark
	Scotia

the equator. The section crosses the African plate, heads northeast across the Eurasian plate through the Himalayas to Japan and Korea, then dips southeast across the Pacific plate, cutting across the South American plate, and finally returns to Africa. Three spreading boundaries at midoceanic ridges are encountered, with two subduction zones (the Japan and Peru trenches) and a continent-to-continent collision, where the Austral-Indian plate dives under the Eurasian plate.

Subduction Tectonics

Converging plate boundaries, with subduction in progress, are zones of intense tectonic and volcanic activity. An active continental margin is a subduction boundary where oceanic lithosphere plunges below continental lithosphere. (See plates Y and Z in Figure 11.10.) Let's look at this boundary in more detail (Figure 11.14).

disappear by subduction beneath the American plate. Similarly, the Scotia plate is being consumed by the American and Antarctic plates.

Figure 11.13 is a schematic circular cross section of the lithosphere along a great circle in low latitudes. It shows several of the great plates and their boundaries. The great circle is tilted by about 30° with respect to

Figure 11.13 Schematic circular cross section of the major plates on a great circle tilted about 30 degrees with respect to the equator. (A. N. Strahler.)

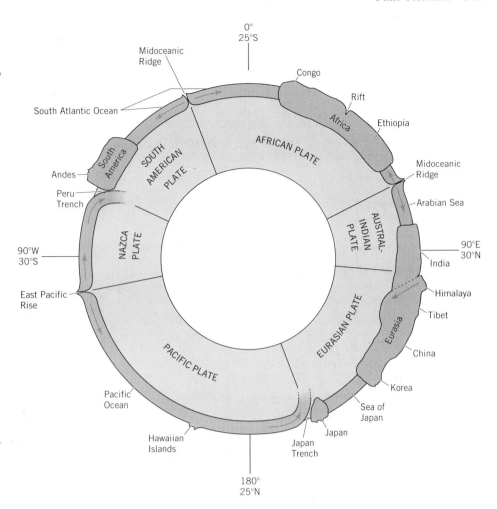

The oceanic trench receives sediment coming from two sources. Carried along on the moving oceanic plate is deep-ocean sediment—fine clay and ooze—that has settled to the ocean floor. From the continent comes terrestrial sediment in the form of sand and mud brought by streams to the shore and then swept into deep water by currents. In the bottom of the trench, both types of sediment are intensely deformed and are dragged down with the moving plate. The deformed sediment is then scraped off the plate and

Figure 11.14 Some typical features of an active subduction zone. The diagram uses a great vertical exaggeration to show surface and crustal details. Sediments scraped off the moving plate form tilted wedges that accumulate in a rising tectonic mass. Near the mainland is a shallow trough in which sediment brought from the land is accumulating. Metamorphic rock is forming above the descending plate. Magma rising from the top of the descending plate reaches the surface to build a chain of volcanoes. (Copyright © A. N. Strahler.)

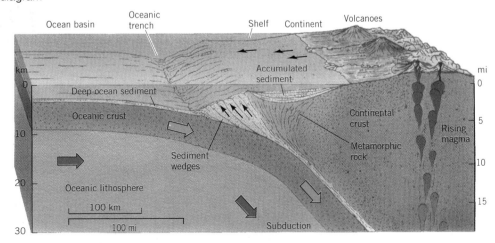

shaped into wedges that ride up, one over the other, on steep fault planes. The sediment wedges accumulate into a large mass in which metamorphism takes place. In this way, the continental margin is built outward as new continental crust of metamorphic rock is formed.

Intense heating of the upper surface of the descending plate melts the oceanic crust, forming magma. Some of this magma rises and reaches the surface to form volcanoes.

Tectonic Processes

Earlier in the chapter, we introduced the concept of tectonic activity as the bending and breaking of the lithosphere, including the crust that lies at the top of the lithosphere. We can apply this concept to plate tectonics. Once set in motion, huge lithospheric plates can cause enormous "damage" to one another when they collide. Sedimentary strata that lay flat and undisturbed for tens of millions of years on ocean floors and continental margins are crumpled and sheared into fragments by such collisions. The collision process itself may take millions of years to complete, but in the perspective of geologic time, it's only a brief episode.

Generally speaking, prominent mountain masses and mountain chains (other than volcanic mountains) are formed by one of two basic tectonic processes: *compression* and *extension*. Compressional tectonic activity—"squeezing together" or "crushing"—acts at converging plate boundaries. Extensional tectonic activity—"pulling apart"—occurs where oceanic plates are separating or where a continental plate is undergoing breakup into fragments. First, we look at the compressional tectonic processes.

Alpine mountain chains typically consist of intensely deformed strata of marine origin. The strata are tightly compressed into wavelike structures called *folds*. A simple experiment you can do will demonstrate how folding occurs in mountain-building. First, take an ordinary towel of thick, limp cloth and lay it out on a smooth hard table top. Using both hands, palms down, bring the ends of the cloth slowly together. First, a simple up-fold will develop and grow, until it overturns to one side or the other. Then more folds will form, grow, and overturn, giving you a simple model of a new mountain range.

Typically, alpine folds can be traced through a history. First, they are overturned, and then they become *recumbent*, as the folds are overturned upon themselves (Figure 11.15). Accompanying the folding is a variety of fault motion in which slices of rock move over the underlying rock on glide surfaces of low inclination.

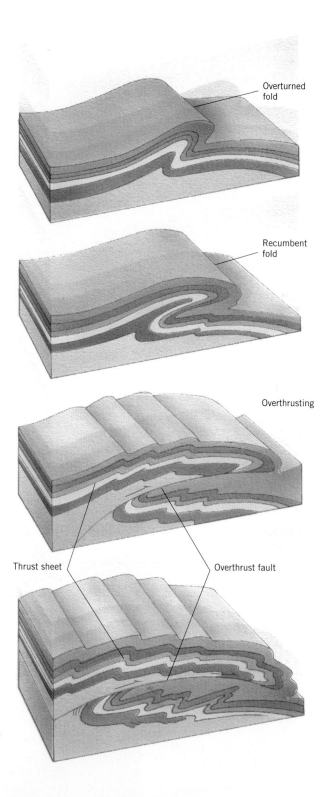

Figure 11.15 Schematic diagrams to show the development of a recumbent fold, broken by a low-angle overthrust fault to produce a thrust sheet, or nappe, in alpine structure. (Based on diagrams by A. Heim, *Geologie der Schweiz*, vol. II–1, Tauschnitz, Leipzig.)

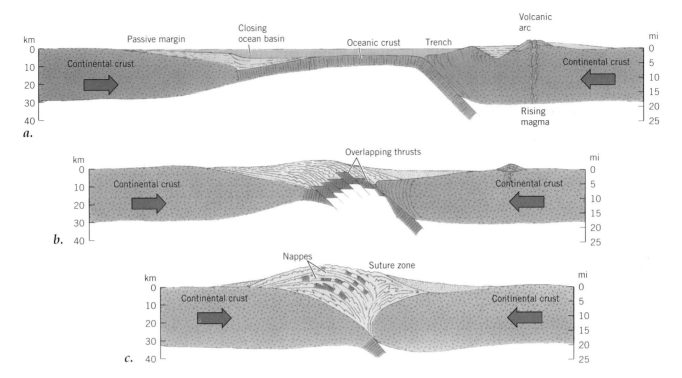

Figure 11.16 Schematic cross sections showing continent–continent collision and the formation of a suture zone with nappes. (Copyright © A. N. Strahler.)

These are *overthrust faults.* Individual rock slices, called *thrust sheets,* are carried many tens of kilometers over the underlying rock. In the European Alps, thrust sheets of this kind were named *nappes* (from the French word meaning "cover sheet" or "tablecloth"). Nappes may be thrust one over the other to form a great pile. The entire deformed rock mass produced by such compressional mountain-making is called an *orogen.* The historical event during which it was produced is an *orogeny.*

Orogens and Collisions

Visualize, now, a situation in which two continental lithospheric plates converge along a subduction boundary. Ultimately, the two masses must collide, because the impacting masses are too thick and too buoyant to allow either plate to slip under the other. The result is an orogeny in which various kinds of crustal rocks are crumpled into folds and sliced into nappes. Collision permanently unites the two plates, terminating further tectonic activity along that collision zone. Appropriately, the collision zone has been named a **continental suture.** Our world map of plates, Figure 11.12, uses a special symbol for these sutures, shown in northern Africa and near the Aral Sea.

Continent-to-continent collisions occurred in the Cenozoic Era along a great tectonic line that marks the southern boundary of the Eurasian plate. (See Figure 11.13, map *b.*). The line begins with the Atlas Mountains of North Africa, and it runs across the Aegean Sea region into western Turkey. Beyond a major gap in Turkey, the line takes up again in the Zagros Mountains of Iran. Jumping another gap in southeastern Iran and Pakistan, the collision line sets in again in the great Himalayan Range.

Orogens are huge deformed rock masses formed by continent-to-continent collisions.

Each segment of this collision zone represents the collision of a different north-moving plate against the single and relatively immobile Eurasian plate. A European segment containing the Alps was formed by collision of the African plate with the Eurasian plate in the Mediterranean region. A Persian segment resulted from the collision of the Arabian plate with the Eurasian plate. A Himalayan segment represents the collision of the Indian continental portion of the Austral-Indian plate with the Eurasian plate.

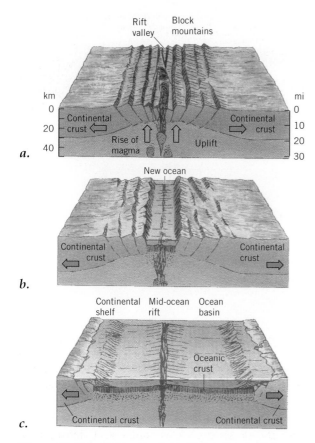

Figure 11.17 Stages in continental rupture and the opening up of a new ocean basin. The vertical scale is greatly exaggerated. (*a*) The crust is uplifted and stretched apart, causing it to break into blocks. (*b*) A narrow ocean is formed, floored by new oceanic crust. (*c*) The ocean basin widens, while the passive continental margins subside. (Copyright © A. N. Strahler.)

Figure 11.16 is a series of cross sections in which the tectonic events of a typical continent-to-continent collision are reconstructed. Diagram *a* shows a passive margin at the left and an active subduction margin at the right. As the ocean between the converging continents is eliminated, a succession of overlapping thrust faults cuts through the oceanic crust in diagram *b*. The thrust slices ride up, one over the other, telescoping the oceanic crust and the sediments above it. As the slices become more and more tightly squeezed, they are forced upward. The upper part of each thrust sheet assumes a horizontal attitude to form a nappe in diagram *c*, which then glides forward under gravity on a low downgrade. A mass of metamorphic rock is formed between the joined continental plates, welding them together. This new rock mass is the continental suture. It is a distinctive type of orogen.

Continent-to-continent collisions have occurred many times since the late Precambrian time. Several ancient sutures have been identified in the continental shields. The Ural Mountains, which divide Europe from Asia, are one such suture, formed near the end of the Paleozoic Era.

Continental Rupture and New Ocean Basins

We have already noted that the continental margins bordering the Atlantic Ocean basin on both its eastern and western sides are very different from the active margin of a subduction zone. At present the Atlantic margins have no important tectonic activity and are passive continental margins. Even so, they represent the contact between continental lithosphere and oceanic lithosphere, with continental crust meeting oceanic crust, as shown in Figure 11.9. Passive margins are formed when a single plate of continental lithosphere is rifted apart. This process is called *continental rupture*.

How Continental Rupture Occurs

Figure 11.17 uses three schematic block diagrams to show how continental rupture takes place and leads to the development of passive continental margins. At first, the crust is both lifted and stretched apart as the lithospheric plate is arched upward. The crust fractures and moves along faults—upthrown blocks form mountains, while downdropped blocks form basins.

Eventually a long narrow valley, called a *rift valley*, appears (block *a*). The widening crack in its center is continually filled in with magma rising from the mantle below. The magma solidifies to form new crust in the floor of the rift valley. Crustal blocks on either side slip down along a succession of steep faults, creating a mountainous landscape. As separation continues, a narrow ocean appears, with a spreading plate boundary running down its center (block *b*). Plate accretion takes place in the central rift to produce new oceanic crust and lithosphere.

The Red Sea is a narrow ocean formed by a continental rupture. Its straight coasts mark the edges of the rupture. The widening of such an ocean basin can continue until a large ocean has formed and the continents are widely separated (Figure 11.17, block *c*).

The Red Sea is a good example of a recent continental rupture.

Deep-Sea Cones

Notice in block *c* of Figure 11.17 that the overall appearance of the ocean basin and its continental margins resembles the schematic diagram in Figure 11.7. The passive margins accumulate terrestrial sediment in the form of a continental shelf that rests on the continental crust, while the oceanic crust accumulates a wedge of deep-sea sediments. The continental margin gradually sinks as these sediment wedges thicken, until the total sediment thickness reaches several kilometers (Figure 11.9). A wide, shallow continental shelf is typical of passive continental margins. Large deltas built by rivers contribute a great deal of the shelf sediment. Tonguelike turbid currents thickened with fine silt and clay carry sediment down the steep continental slope and spread it out on the continental rise, building *deep-sea cones* (Figure 11.18). In Figure 11.8, you can find deep-sea cones of the Mississippi and Congo rivers.

Transform Scars

During the process of opening of an ocean basin, the spreading boundary develops a series of offsets, one of which is shown in the upper left-hand part of diagram *a* of Figure 11.10. The offset ends of the axial rift are connected by an active transform fault. As spreading continues, a scarlike feature is formed on the ocean floor as an extension of the transform fault. These *transform scars* take the form of narrow ridges or clifflike features and may extend for hundreds of kilometers across the ocean floors. Before the true nature of these scars was understood, they were called "fracture zones." That name still persists and can be seen on maps of the ocean floor. The scars are not, for the most part, associated with active faults.

The Power Source for Plate Movements

The system of lithospheric plates in motion represents a huge flow system of dense mineral matter. Its operation requires enormous power within an internal energy system. It is generally agreed that the source of this energy system lies in the phenomenon of radioactivity.

Radioactive elements in the crust and upper mantle constantly give off heat in the spontaneous process of radioactive decay. As the temperature of mantle rock rises, the rock expands. Recall that in the case of the atmosphere, upward motion of warmer, less dense air takes place by convection. It is thought that, in a somewhat similar type of convection, mantle material rises steadily beneath spreading plate boundaries. Geologists do not yet have a full understanding of how

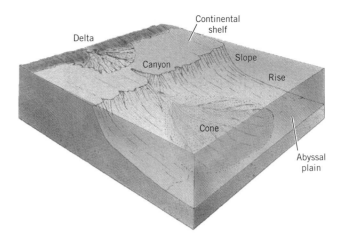

Figure 11.18 Block diagram of a deep-sea cone and its relationship to the continental shelf. (Copyright © A. N. Strahler.)

this rise of heated rock causes plates to move. One hypothesis states that, as the rising mantle lifts the lithospheric plate to a higher elevation, the lithospheric plate tends to move horizontally away from the spreading axis under the influence of gravity.

At the opposite edge of the plate, subduction occurs because the oceanic plate is colder and denser than the asthenosphere through which it is sinking. The motion of the oceanic plate exerts a drag on the underlying asthenosphere, setting in motion flow currents in the upper mantle. Thus, slow convection currents probably exist in the asthenosphere beneath the moving plates, but their pathways and depths of operation are not well understood.

Continents of the Past

The continents are moving today. Rates of separation, or convergence, between two plates are on the order of 5 to 10 centimeters (2 to 4 in.) per year, or 50 to 100 km per million years. At that rate, global geography was very different in past geologic eras than it is today. Many continental riftings and many plate collisions have taken place over the past billion or so years. Single continents have fragmented into smaller ones, while at the same time small continents have merged to form large ones.

A brief look at the changing continental arrangements and locations over the past 250 million years (m.y.) is well worthwhile in our study of physical geography because those changes brought with them changes in climate, soils, and vegetation on each continent as the continents moved across the parallels of latitude. Geography of the geologic past bears the name "paleogeography." If you decide to become a "paleogeographer," you will need to carry with you into the past all you have learned about the physical geography of the present.

Modern reconstructions of the global arrangements of past continents have been available since the mid-1960s, when the modern theory of plate tectonics came into wide acceptance. Global maps drawn for each geologic period have been repeatedly revised in the light of new evidence, but differences of interpretation affect only the minor details.

Figure 11.19 is a recent version of the history of our continents, starting in the Mesozoic Era. Geologists now agree that in the Permian Period, ending about

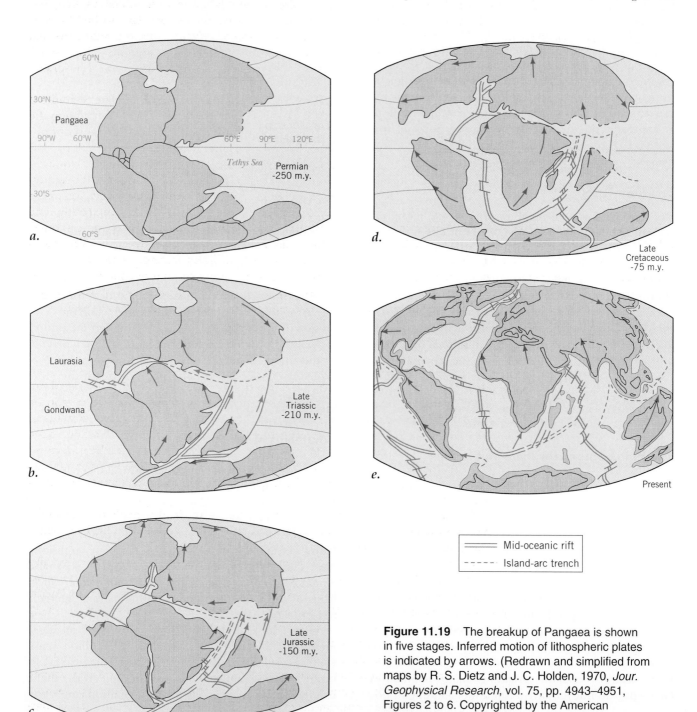

Figure 11.19 The breakup of Pangaea is shown in five stages. Inferred motion of lithospheric plates is indicated by arrows. (Redrawn and simplified from maps by R. S. Dietz and J. C. Holden, 1970, *Jour. Geophysical Research*, vol. 75, pp. 4943–4951, Figures 2 to 6. Copyrighted by the American Geophysical Union. Used by permission.)

250 million years ago, there existed a single great supercontinent named *Pangaea*. Thus, there was only one worldwide superocean, named *Panthalassa*. In map *a* Pangaea lies astride the equator and extends nearly to the two poles. Regions that are now North America and western Eurasia lie in the northern hemisphere. Jointly they are called *Laurasia*. Regions that are now South America, Africa, Antarctica, Australia, New Zealand, Madagascar, and peninsular India lie south of the equator. Jointly, they go by the name *Gondwana*. Subsequent maps (*b* through *d*) show the breaking apart and dispersal of the Laurasia and Gondwana plates to yield their modern components and locations, seen in map *e*.

A single great ocean, Panthalassa, once surrounded a single great continent, Pangaea, that broke apart during millions of years to form the present-day continents.

Note in particular that North America traveled from a low-latitude location into high latitudes, finally closing off the Arctic Ocean as a largely landlocked sea. This change may have been a major factor in bringing on the ice age in late Cenozoic time. (See Chapter 18.) Another traveler was the Indian Peninsula, which started out from a near-subarctic southerly location in Permian time and streaked northeast across the Thethys Sea to collide with Asia in the northern tropical savanna zone. India's bedrock still bears the grooves and scratches of a great Permian glaciation.

Our planet's surface is shaped and reshaped endlessly through geologic time. Oceans open and close, mountains rise and fall. Rocks are folded, faulted, and fractured. Arcs of volcanoes spew lava to build chains of lofty peaks. Deep trenches consume sediments that are carried beneath adjacent continents. These activities are byproducts of the motions of lithospheric plates and are described by plate tectonics.

For geographers, the great value of plate tectonics is that it provides a grand scheme for understanding the nature and distribution of the largest and most obvious features of our planet's surface—its continents and ocean basins and their major relief features. However, large features, such as mountain ranges, are made up of many smaller features—individual peaks, for example. In the next chapter, we will look at the continental surface in more detail, examining the volcanic and tectonic landforms that result when volcanoes erupt and rock layers are folded and faulted. As we will see, the formation of these landforms is often marked by powerful earthquakes—phenomena that can produce major impacts, indeed, on modern technological society.

THE LITHOSPHERE AND PLATE TECTONICS IN REVIEW

At the center of the earth lies the core—a dense mass of liquid iron and nickel that is solid at the very center. Enclosing the metallic core is the mantle, composed of ultramafic rock. The outermost layer is the crust. Continental crust consists of two zones—a lighter zone of felsic rocks atop a denser zone of mafic rocks. Oceanic crust consists only of denser, mafic rocks. The lithosphere, the outermost shell of rigid, brittle rock, includes the crust and an upper layer of the mantle. Below the lithosphere is the asthenosphere, a region of the mantle in which mantle rock is soft, or plastic.

Geologists trace the history of the earth through the geologic timetable. Precambrian time includes the earth's earliest history. It is followed by three major divisions—the Paleozoic, Mesozoic, and Cenozoic eras.

Continental masses consist of active belts of mountain-making and inactive regions of old, stable rock. Mountain-building occurs by volcanism and tectonic activity. Alpine chains occur in two principal mountain belts—the circum-Pacific and Eurasian-Indonesian belts. Continental shields are regions of low-lying igneous and metamorphic rocks. They may be exposed or covered by layers of sedimentary rocks. Ancient mountain roots lie within some shield regions.

The ocean basins are marked by a midoceanic ridge with its central axial rift. This ridge occurs at the site of crustal spreading. Most of the ocean floor is abyssal plain, covered by fine sediment. As passive continental margins are approached, the continental rise, slope, and shelf are encountered. At active continental margins, deep oceanic trenches lie offshore.

Continental lithosphere includes the thicker, lighter continental crust and a rigid layer of mantle rock beneath. Oceanic lithosphere is comprised of the thinner, denser oceanic crust and rigid mantle below. The lithosphere is fractured and broken into a set of plates, large and small, that move with respect to each other. Where plates move apart, a spreading boundary occurs. At converging boundaries, plates collide. At transform boundaries, plates move past one another on a transform fault. There are six major lithospheric plates.

When oceanic lithosphere and continental lithosphere collide, the denser oceanic lithosphere plunges beneath the continental lithospheric plate, a process called subduction. A trench marks the site of downplunging. Some subducted oceanic crust melts and rises to the surface, producing volcanoes.

Compression resulting from the collision of lithospheric plates can shape rock layers into folds. Under the severe compression that occurs with continent-to-continent collision, the layers can break and move atop one another along thrust faults. The two continental plates are welded together in a zone of metamorphic rock named a continental suture. In continental rupture, extensional tectonic forces move a continental plate in opposite directions, creating a rift valley. Eventually the rift valley widens and opens to the ocean, and new oceanic crust forms as spreading continues.

Plate movements are thought to be powered by convection currents in the plastic mantle rock of the asthenosphere. During the Permian Period, the continents were joined in a single, large supercontinent—Pangaea—that broke apart, leading eventually to the present arrangement of continents and ocean basins.

KEY TERMS

core	lithosphere	continental shields
mantle	asthenosphere	plate tectonics
crust	lithospheric plates	tectonics
continental crust	mountain arc	subduction
oceanic crust	island arcs	continental suture

REVIEW QUESTIONS

1. Describe the earth's inner structure, from the center outward. What types of crust are present? How are they different?

2. How do geologists use the term lithosphere? What layer underlies the lithosphere, and what are its properties? Define the term lithospheric plate.

3. More recent geologic time is divided into three eras. Name them in order from oldest to youngest. How do geologists use the terms period and epoch? What age is applied to time before the earliest era?

4. What proportion of the earth's surface is in ocean? In land? Do these proportions reflect the true proportions of continents and oceans? If not, why not?

5. What are the two basic subdivisions of continental masses?

6. What term is attached to belts of active mountain-making? What are the two basic processes by which mountain belts are constructed? Provide examples of mountain arcs and island arcs.

7. What is a continental shield? How old are continental shields? What two types of shields are recognized?

8. Sketch a cross section of an ocean basin with passive continental margins. Label the following features: midoceanic ridge, axial rift, abyssal plain, continental rise, continental slope, and continental shelf.

9. Identify and describe two types of lithospheric plates. Sketch a cross section showing a collision between the two types. Label the following features: oceanic crust, continental crust, mantle, oceanic trench, and rising magma. Indicate where subduction is occurring.

10. Name the six great lithospheric plates. Identify an example of a spreading boundary by general geographic location and the plates involved. Do the same for a converging boundary.

11. Describe the process of subduction as it occurs at a converging boundary of continental and oceanic lithospheric plates. How is the continental margin extended? How is subduction related to volcanic activity?

12. Describe how compressional mountain-building produces folds, faults, overthrust faults, and thrust sheets (nappes).

13. Describe the formation of a continental suture, keying your description to a sketch of a continent-to-continent collision. Provide a present-day example where a continental suture is being formed, and give an example of an ancient continental suture.

14. How does continental rupture produce passive continental margins? Describe the process of rupturing and its various stages.

15. What are transform faults? Where do they occur?

16. What forces are thought to power plate tectonics? Explain.

17. Briefly summarize the history of our continents that geologists have reconstructed beginning with the Permian period.

IN-DEPTH QUESTIONS

1. Figure 11.12 is a Mercator map of lithospheric plates, whereas Figure 11.13 presents a cross section of plates on a great circle. Construct a similar cross section of plates on the 30°S parallel of latitude. As in Figure 11.13, label the plates and major geographic features. Refer to Figure 11.5 for arcs, trenches, and mid-ocean ridges.

2. Suppose astronomers discover a new planet that, like earth, has continents and oceans. They dispatch a reconnaissance satellite to photograph the new planet. What features would you look for, and why, to detect past and present plate tectonic activity on the new planet?

Chapter 12

Volcanic and Tectonic Landforms

Figure 12.6 Basaltic shield volcanoes of Hawaii. At lower left is the inactive Halemaumau fire pit, formed in the floor of the central depression of Kilauea volcano. On the distant skyline is the snow-capped summit of Mauna Kea volcano, its elevation over 4000 m (13,000 ft).

the asthenosphere. Directly above a mantle plume, crustal basalt is heated to the point of melting and produces a magma pocket. The site of rising magma is called a *hot spot.* Magma of basaltic composition makes its way through the overlying lithosphere to emerge at the surface as lava.

> **When a huge volcanic eruption blows the top off a volcano, the crater that remains is a caldera.**

Where mantle plumes create hot spots in the oceanic lithosphere, the emerging basalt builds a class of initial landforms known as **shield volcanoes.** These broad basaltic domes are constructed on the deep ocean floor, far from plate boundaries, and may be built high enough to rise above sea level as volcanic islands. As a lithospheric plate drifts slowly over a mantle plume beneath, a succession of shield volcanoes is formed. Thus, a chain of basaltic volcanic islands comes into existence. Several such chains exist in the Pacific Ocean basin, including the Midway Islands and the Hawaiian group.

A few basaltic volcanoes also occur along the mid-oceanic ridge, where sea-floor spreading is in progress. Perhaps the outstanding example is Iceland, in the North Atlantic Ocean. Iceland is constructed entirely of basalt. Basaltic flows are superimposed on older basaltic rocks as dikes and sills formed by magma emerging from deep within the spreading rift. Mount Hekla, an active volcano on Iceland, is a shield volcano somewhat similar to those of Hawaii. In Chapter 10, *Eye on the Environment • Battling Iceland's Heimaey Volcano* documents how valiant Icelanders coped with a recent eruption. Other islands consisting of basaltic volcanoes located along or close to the axis of the Mid-Atlantic Ridge are the Azores, Ascension, and Tristan da Cunha.

Hawaiian Volcanoes

The shield volcanoes of the Hawaiian Islands are characterized by gently rising, smooth slopes that flatten near the top, producing a broad-topped volcano (Figure 12.6). Domes on the island of Hawaii rise to summit elevations of 4000 m (13,000 ft) above sea level. Including the basal portion lying below sea level, they are more than twice that high. In width they range from 16 to 80 km (10 to 50 mi) at sea level and up to 160 km (100 mi) wide at the submerged base. The basalt lava of the Hawaiian volcanoes is highly fluid and travels far down the gentle slopes (Figure 12.7). Most of the lava flows issue from fissures (long, gaping cracks) on the flanks of the volcano (see chapter opening photo).

Hawaiian lava domes have a wide, steep-sided central depression that may be 3 km (2 mi) or more wide and

Figure 12.7   Basalt lava flows exposed in cliffs bordering the Columbia River in Washington. Each set of cliffs is a major lava flow. In cooling, the lava acquires a vertical structure that makes it crack into tall columns.

several hundred meters deep. These large depressions are a type of collapse caldera. Molten basalt is sometimes seen in the floors of deep pit craters that occur on the floor of the central depression or elsewhere over the surface of the lava dome.

Flood Basalts

Where a mantle plume lies beneath a continental lithospheric plate, the hot spot may generate enormous volumes of basaltic lava that emerge from numerous vents and fissures and accumulate layer upon layer. The basalt may ultimately attain a thickness of thousands of meters and cover thousands of square kilometers. These accumulations are called *flood basalts.*

An important American example is found in the Columbia Plateau region of southeastern Washington, northeastern Oregon, and westernmost Idaho. Here, basalts of Cenozoic age cover an area of about 130,000 sq km (50,000 sq mi)—about the same area as the state

Figure 12.8   A young cinder cone surrounded by rough-surfaced basalt lava flows. Lava Beds National Monument, northern California.

Figure 12.9 Mammoth Hot Springs, an example of geothermal activity in Yellowstone National Park, Wyoming. Terraces ringed by mineral deposits hold steaming pools of hot water.

of New York. Individual basalt flows, exposed in the walls of river gorges, are expressed as cliffs, in which vertical joint columns are conspicuous (Figure 12.7).

Cinder Cones

Associated with flood basalts, shield volcanoes, and scattered occurrences of basaltic lava flows is a small volcano known as a *cinder cone* (Figure 12.8). Cinder cones form when frothy basalt magma is ejected under high pressure from a narrow vent, producing tephra. The rain of tephra accumulates around the vent to form a roughly circular hill with a central crater. Cinder cones rarely grow to heights of more than a few hundred meters. An exceptionally fine example of a cinder cone is Wizard Island, built on the floor of Crater Lake long after the caldera was formed (Figure 12.5).

Hot Springs and Geysers

Where hot rock material is near the earth's surface, it can heat nearby ground water to high temperatures.

When the ground water reaches the surface, it provides *hot springs* at temperatures not far below the boiling point of water (Figure 12.9). At some places, jetlike emissions of steam and hot water occur at intervals from small vents—producing *geysers* (Figure 12.10). Since the water that emerges from hot springs and geysers is largely ground water that has been heated in contact with hot rock, this water is recycled surface water. Little, if any, is water that was originally held in rising bodies of magma.

Volcanic Eruptions as Environmental Hazards

Volcanic eruptions are environmental hazards that can take a heavy toll of plant and animal life and devastate human habitations. Loss occurs principally from sweeping clouds of glowing gases that descend the volcano slopes like great avalanches, and from relentlessly advancing lava flows. Showers of ash, cinders, and bombs, as well as violent local earthquakes associated with volcanic activity, can also be as destructive. The inhabitants of low-lying coasts face the additional peril of great seismic sea waves, some of which are generated by explosive eruptions of distant coastal or submarine volcanoes.

In 1985 a violent eruption of Nevado del Ruiz, a volcano in the Colombian Andes, caused the rapid melting of ice and snow in the summit area. Mixing with volcanic ash, the water formed a great surging tongue of mud and boulders, known as a *lahar*. Rushing downslope at speeds up to 145 km (90 mi) per hour,

Figure 12.10 Old Faithful Geyser in Yellowstone National Park, Wyoming.

the lahar became channeled into a valley on the lower slopes where it engulfed a town and killed more than 20,000 persons.

LANDFORMS OF TECTONIC ACTIVITY

Recall from Chapter 11, our introduction to global plate tectonics, that there are basically two different forms of tectonic activity: compressional and extensional. Along converging lithospheric plate boundaries, tectonic activity is basically that of compression, illustrated schematically in Figure 12.11. In subduction zones, sedimentary layers of the ocean floor are subject to compression within a trench as the descending plate forces them against the overlying plate. In continental collision, compression is of the severest kind.

In zones of rifting of continental plates, explained in Chapter 11, the brittle continental crust is pulled apart and yields by faulting. This motion is described as extensional. In the simple model shown in Figure 12.11, rifting is expressed in a pair of opposite-facing faults. The crustal block between them moves down to form a depressed area.

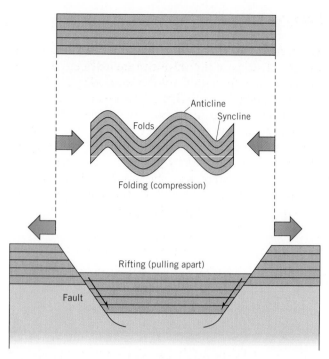

Figure 12.11 Two basic forms of tectonic activity that produce initial landforms. Flat-lying rock layers may be compressed to form folds or pulled apart to form faults by rifting.

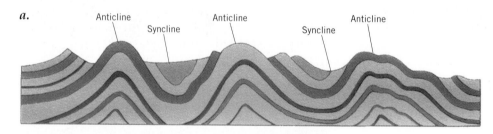

a.

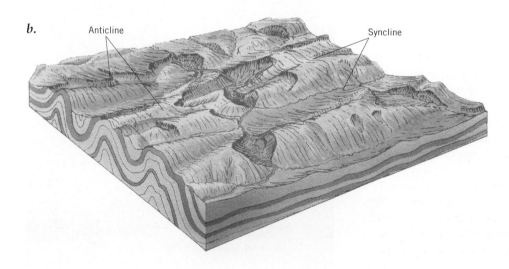

b.

Figure 12.12 Anticlines and synclines of the Jura Mountains, France and Switzerland. A cross section shows the folds and ground surface (*a*). The landscape developed on the folds is shown in the block diagram (*b*). (After E. Raisz.)

Imagine yourself on the balcony of an exclusive beach hotel at Waikiki. It is the last day of your vacation, and lying on your lounge chair, the Hawaiian sun still feels as good on your back as it did on the first day. You stretch and roll over, sitting up and taking a sip of the cold, sparkling mineral water that the room service steward has just delivered. Could there be anything more glorious than this?

The view from your balcony is postcard-perfect. The arc of hotels stretches to your right and left, fringed with palms and pools. Beyond is the beach, lined with sun worshippers. Heads and arms of swimmers bob and dunk in the surf zone just beyond. Sailboats glide lazily back and forth in the lagoon, while outrigger canoes move more deliberately under paddle power.

As you gaze out beyond the surfriders and sailors, you notice a white foaming of the surf out at the edge of the reef. Slowly the foaming intensifies, and within a few moments the reef is a white crescent of churning, seething water. Something very wrong is happening. You rise to your feet and stand at the edge of the balcony. The frothing mass of water is surging across the lagoon toward the beach. The confused surf topples the sailboats easily. The outrigger canoes are quickly swamped. You are suddenly aware of a roaring noise, created by the wave-tossed, on-rushing surf. On the beach below, people are running toward the hotels as the wall of white water approaches.

The advancing surf quickly floods the beach and surges toward the hotel terraces. Palm trees bend and fall under the onslaught. Tables and chairs follow the fleeing bathers and diners as they are swept landward. As the first waves hit your hotel, you feel a shaking motion, like an earthquake. You grab the railing hard and hold on for dear life. The beach has completely disappeared, and the waves, no longer slowed by the barrier reef or beach, are crashing full-force against the hotel, shattering glass windows and doors and breaking interior partitions. Terrified, you can feel the thumping and tearing of steel and concrete as the building shudders with the blows. The water is flooding into the streets behind the hotels now, rushing in furious torrents between the buildings.

Just as you are beginning to think about what to do, the water begins to retreat, first slowly, then as a great outgoing flood. Splintered woodwork, uprooted plants, and even floating cars flash past as the ebbing waters carry them seaward. There is a different rhythm to the shaking now, and the crash of the surf has been replaced by a rushing sound. The beach is visible once again, dissected by broad torrents of ocean-bound water carving deep gullies. The shaking subsides, and the surf withdraws far out in the lagoon, leaving a vast sandy plain littered with debris and shallow water pools, glistening in the sun. Slowly you sink down to the floor of the terrace, awestruck by the devastation spread in front of you. About a minute later, you regain control and rise, heading for the sliding glass door to your room. But then you hear the roaring noise again, slowly building. You turn and stare seaward. The reef is awash with foam once more. Another great sea wave is headed for the beach!

Seem far-fetched? Not a bit. On April 1, 1946, a major earthquake close to the Aleutian Islands generated a seismic sea wave, or tsunami, that struck the Hawaiian Islands five hours later. An observer on the north shore of Oahu documented the wave as a rise and fall of sea level, repeated several times at 12-minute intervals. The first wave inundated the coast to a height of 3.9 m (13 ft) above tide level. The water then retreated, but it was followed minutes later by an even higher landward surge, this time reaching 5.1 m (17 ft) above tide level. A third, even higher surge, followed the second.

On the island of Hawaii, the city of Hilo was ravaged by the tsunami. Breaking waves reached a height of 9 m (30 ft) above sea level. A large section of the city was destroyed as houses were floated from their foundations and swept inland. Some buildings were crushed into tangled masses of debris, and others floated almost intact to new locations. The sudden rise of water and its pounding surf caught many of Hilo's inhabitants by surprise, with 83 persons known dead and 13 missing. The death toll, including other island cities in the Hawaiian chain, was 170.

A tsunami is a very dramatic example of earth forces at work—in this case, a secondary effect of a huge, submarine earthquake. Earth forces act to shape the landscape in other ways beyond the occasional effects of seismic sea waves flooding coasts. Upwelling magma spews forth explosively from vents and fissures, creating steep-sided volcanoes or flooding vast areas with molten rock. Compressional stresses fold flat-lying rocks into wavelike forms, producing long ridges from upfolds and long valleys from downfolds. Earthquake faults mark the lines on which rock layers break and move past one other, producing huge uplifted mountain blocks next to deep, downdropped valleys. These landscape features, created by internal earth forces, are the subject of this chapter.

An eruption of Kilauea Volcano, Hawaii.

Have you ever seen a volcano? If you live on the west coast, you may have driven by California's Mount Lassen or Mount Shasta. Or perhaps you've visited Mount St. Helens, to see the devastation of its recent explosion, or even skied on the slopes of Mount Rainier. If you've been to the island of Hawaii, you may have seen flowing lava at first-hand. Even if you've never visited a real volcano, you've probably seen pictures of volcanoes, either on the television news as they erupt, or perhaps on postcards or calendars of volcanoes as scenic attractions. These lofty mountain forms are unmistakable and are one example of the landforms we will study in this chapter.

Even if you've never seen a real volcano, you've probably seen mountains. Some mountains are simply masses of rock that are resistant to erosion, and so they stand out against a surrounding area of weaker rocks. Others, however, are the direct result of tectonic forces, which have recently uplifted them. The mountains of the western United States contain many ranges that are examples of uplifted mountain blocks bounded by faults—planes of slippage along which rock masses move in opposite directions. Tectonic activity can also produce mountains by bending rock layers into long upfolded ridges, with downfolded valleys in between.

By looking in detail at landforms created by volcanic and tectonic activity, we are moving down from the global perspective of enormous lithospheric plates that collide to form vast mountain arcs or separate to form yawning ocean basins. We are now "zooming in" to examine landscape features that are actually small enough to view from a single vantage point. As we will see, the global perspective helps explain how these landscape features are created and where they occur.

LANDFORMS

Landforms are the surface features of the land—for example, mountain peaks, cliffs, canyons, plains, beaches, and sand dunes. Landforms are created by many processes, and much of the remainder of this book describes landforms and how they are produced. **Geomorphology** is the scientific study of landforms, including their history and the processes that continually shape them. In this chapter, we will examine landforms produced directly by volcanic and tectonic processes.

The shapes of continental surfaces reflect the "balance of power," so to speak, between two sets of forces. Internal earth forces, acting through volcanic and tectonic processes, move crustal materials upward and

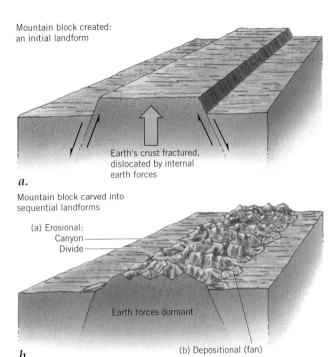

Figure 12.1 Initial and sequential landforms. (Drawn by A. N. Strahler.)

create new rock at the earth's surface. In opposition are processes and forces that act to lower continental surfaces by removing and transporting mineral matter through the action of running water, waves and currents, glacial ice, and wind. We can refer to these processes of land sculpture collectively as **denudation.**

Seen in this perspective, landforms in general fall into two basic groups—initial landforms and sequential landforms. *Initial landforms* are produced directly by volcanic and tectonic activity. They include volcanoes and lava flows, as well as downdropped rift valleys and elevated mountain blocks in zones of recent crustal deformation. Figure 12.1a shows a mountain block uplifted by crustal activity. The energy for lifting molten rock and rigid crustal masses to produce the initial landforms has an internal heat source. As we explained in Chapter 11, this heat energy is generated largely by natural radioactivity in rock of the earth's crust and mantle. It is the fundamental energy source for the motions of lithospheric plates.

Landforms are produced by the endless struggle between internal earth forces that lift crustal masses upward and the external forces of erosion and denudation that wear them down.

Landforms shaped by processes and agents of denudation belong to the group of *sequential landforms.* The word "sequential" means that they follow in sequence after the initial landforms have been created, and a crustal mass has been raised to an elevated position. Figure 12.1*b* shows an uplifted crustal block (an initial landform) that has been attacked by agents of denudation and carved up into a large number of sequential landforms.

You can think of any landscape as representing the existing stage in a great contest of opposing forces. As lithospheric plates collide or pull apart, internal earth forces periodically elevate parts of the crust to create initial landforms. The external agents of denudation—running water, waves, wind, and glacial ice—persistently wear these masses down and carve them into vast numbers of smaller sequential landforms.

Today we can observe the many stages of this endless struggle between internal and external forces by traveling to different parts of the globe. Where we find high alpine mountains and volcanic chains, internal earth forces have recently dominated the contest. In the rolling low plains of the continental interiors, the agents of denudation have won a temporary victory. At other locations, we can find many intermediate stages. Because the internal earth forces act repeatedly and violently, new initial landforms keep coming into existence as old ones are subdued.

VOLCANIC ACTIVITY

We have already identified *volcanism,* or volcanic activity, as one of the forms of mountain-building. The extrusion of magma builds landforms, and these can accumulate in a single area both as volcanoes and as thick lava flows. Through these volcanic activities, imposing mountain ranges are constructed.

A **volcano** is a conical or dome-shaped initial landform built by the emission of lava and its contained gases from a constricted vent in the earth's surface (Figure 12.2). The magma rises in a narrow, pipelike conduit from a magma reservoir lying beneath. Upon reaching the surface, igneous material may pour out in tonguelike lava flows. Magma may also be ejected in the form of solid fragments driven skyward under pressure of confined gases. Ejected fragments, ranging in size from boulders to fine dust, are collectively called *tephra.* Forms and dimensions of a volcano are quite varied, depending on the type of lava and the presence or absence of tephra.

Stratovolcanoes

The nature of volcanic eruption, whether explosive or quiet, depends on the type of magma. Recall from Chapter 10 that there are two main types of igneous rocks: felsic and mafic. The felsic lavas (rhyolite and andesite) have a high degree of viscosity; that is, they are thick and gummy, and resist flow. So, volcanoes of felsic composition typically have steep slopes, and lava usually does not flow long distances from the volcano's vent. Also, felsic lavas usually hold large amounts of gas under high pressure. As a result, these lavas can produce explosive eruptions. The eruption of Mount St. Helens in 1980 is an example of an explosive eruption of felsic lavas (Figure 12.3).

Tall, steep-sided volcanic cones are produced by felsic lavas. These cones usually steepen toward the summit, where a bowl-shaped depression—the *crater*—is located. When the volcano erupts, tephra falls on the area surrounding the crater and contributes to the structure of the cone. *Volcanic bombs* are also included in the tephra. These solidified masses of lava range up to the size of large boulders and fall close to the crater. Very fine volcanic dust can rise high into the troposphere and stratosphere, traveling hundreds or thousands of kilometers before settling to the earth's surface (Figure 12.3).

The interlayering of ash strata and sluggish streams of felsic lava produces a **stratovolcano.** Lofty, conical stratovolcanoes are well known for their scenic beauty. Fine examples are Mount Hood and Mount St. Helens in the Cascade Range (Figure 12.3), Fujiyama in Japan, Mount Mayon in the Philippines (Figure 12.4), and Mount Shishaldin in the Aleutian Islands.

Another important form of emission from explosive stratovolcanoes is a cloud of white-hot gases and fine ash. This intensely hot cloud, or "glowing avalanche,"

Figure 12.2 Idealized cross section of a stratovolcano with feeders from magma chamber beneath. The steep-sided cone is built up from layers of lava and tephra. (Copyright © A. N. Strahler.)

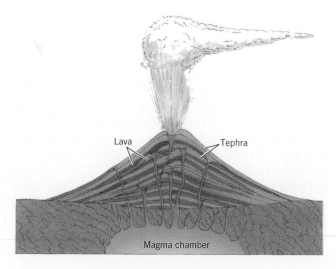

Lava   Tephra

Magma chamber

Figure 12.3   Mount St. Helens, a stratovolcano of the Cascade Range in southwestern Washington, erupted without warning on the morning of May 18, 1980, emitting a great cloud of condensed steam, heated gases, and ash from the summit crater. Within a few minutes, the plume had risen to a height of 20 km (12 mi), and its contents were being carried eastward by stratospheric winds. The eruption was initiated by explosive demolition of the northern portion of the cone, concealed from this viewpoint.

Figure 12.4   Mount Mayon, in southeastern Luzon, the Philippines, is often considered the world's most nearly perfect stratovolcanic cone. An active volcano, its summit rises to an altitude of nearly 2400 m (8000 ft). A series of eruptions began here in February, 1993.

travels rapidly down the flank of the volcanic cone, searing everything in its path. On the Caribbean island of Martinique, in 1902, a glowing cloud issued without warning from Mount Pelée. It swept down on the city of St. Pierre, destroying the city and killing all but two of its 30,000 inhabitants.

Calderas

One of the most catastrophic of natural phenomena is a volcanic explosion so violent that it destroys the entire central portion of the volcano. Vast quantities of ash and dust are emitted and fill the atmosphere for many hundreds of square kilometers around the volcano. There remains only a great central depression named a *caldera*. Although some of the upper part of the volcano is blown outward in fragments, most of it settles back into the cavity formed beneath the volcano by the explosion.

Krakatoa, a volcanic island in Indonesia, exploded in 1883, leaving a huge caldera. It is estimated that 75 cu km (18 cu mi) of rock was blown out of the crater during the explosion. Great seismic sea waves generated by the explosion killed many thousands of persons living in low coastal areas of Sumatra and Java.

A classic example of a caldera produced in prehistoric times is Crater Lake, Oregon (Figure 12.5). The former volcano, named Mount Mazama, is estimated to have risen 1200 m (4000 ft) higher than the present caldera rim. The great explosion and collapse occurred about 6600 years ago.

Stratovolcanoes and Subduction Arcs

Most of the world's active stratovolcanoes lie within the circum-Pacific belt. Here, subduction of the Pacific, Nazca, Cocos, and Juan de Fuca plates is active. In Chapter 11, we explained how andesitic magmas rise beneath volcanic arcs of active continental margins and island arcs (see Figure 11.5). One good example is the volcanic arc of Sumatra and Java, lying over the subduction zone between the Australian plate and the Eurasian plate. Another is the Aleutian volcanic arc, located where the Pacific plate dives beneath the North American plate. The Cascade Mountains of northern California, Oregon, and Washington form a similar chain. Important segments of the Andes Mountains in South America consist of stratovolcanoes.

Shield Volcanoes

In contrast to thick, gassy felsic lava, mafic lava (basalt) is often highly fluid. It typically has a low viscosity and holds little gas. As a result, eruptions of basaltic lava are usually quiet, and the lava can travel long distances to spread out in thin layers. Typically, then, large basaltic volcanoes are broadly rounded domes with gentle slopes. Hawaiian volcanoes are of this type.

Volcanic activity involving basaltic lavas can occur in the midst of oceanic or continental lithospheric plates. Here, in isolated spots far from active plate boundaries, basaltic lavas are created by *mantle plumes*—isolated columns of molten material rising slowly within

Figure 12.5 Crater Lake, Oregon, is surrounded by the high, steep wall of a great caldera. Wizard Island (center foreground) is an almost perfect basaltic cinder cone with basalt lava flows. It was built on the floor of the caldera after the major explosive activity had ceased.

Fold Belts

When continental collision begins to take place, broad wedges of strata of a passive continental margin come under strong forces of compression (Figure 11.15). The strata, which were originally more or less flat-lying, experience **folding,** as shown in Figure 12.11. The wavelike shapes imposed on the strata consist of alternating archlike upfolds, called *anticlines,* and troughlike downfolds, called *synclines.* Thus, the initial landform associated with an anticline is a broadly rounded mountain ridge, and the landform corresponding to a syncline is an elongate, open valley.

Compressional tectonic activity produces folding and overthrust faulting. Extensional tectonic activity produces normal faults, uplifted blocks, and rift valleys.

An example of open folds of comparatively young geologic age that has long attracted the interest of geographers is the Jura Mountains of France and Switzerland. Figure 12.12 is a block diagram of a small portion of that fold belt. The rock strata are mostly limestone layers and were capable of being deformed by bending with little brittle fracturing. Folding occurred in late Cenozoic (Miocene) time. Notice that each mountain crest is associated with the axis of an anticline, while each valley lies over the axis of a syncline. Some of the anticlinal arches have been partially removed by erosion processes. The rock structure can be seen clearly in the walls of the winding gorge of a major river that crosses the area. The Jura folds lie just to the north of the main collision orogen of the Alps. Because of their location near a mountain mass, they are called *foreland folds.*

Faults and Fault Landforms

A **fault** in the brittle rocks of the earth's crust occurs when rocks suddenly yield to unequal stresses by fracturing. Faulting is accompanied by a displacement—a slipping motion—along the plane of breakage, or *fault plane.* Faults are often of great horizontal extent, so that the surface trace, or fault line, can sometimes be followed along the ground for many kilometers. Most major faults extend down into the crust for at least several kilometers.

Faulting occurs in sudden slippage movements that generate earthquakes. A single fault movement may result in a slippage of as little as a centimeter or as much as 15 m (50 ft). Successive movements may occur many years or decades apart, even several centuries apart.

Over long time spans, the accumulated displacements can amount to tens or hundreds of kilometers. In some places, clearly recognizable sedimentary rock layers are offset on opposite sides of a fault, allowing the total amount of displacement to be measured accurately.

Normal Faults

One common type of fault associated with crustal rifting is the **normal fault** (Figure 12.13*a*). The plane of slippage, or fault plane, is steeply inclined. The crust

Figure 12.13 Four types of faults.

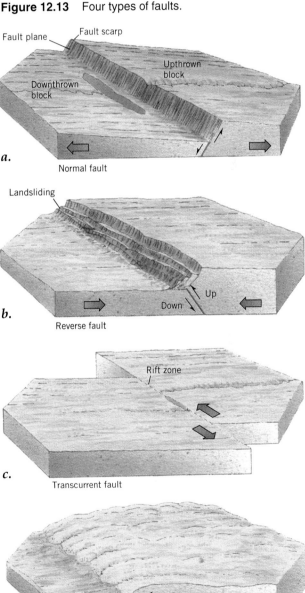

a. Normal fault

b. Reverse fault

c. Transcurrent fault

d. Overthrust fault

Figure 12.14 This fault scarp was formed during the Hebgen Lake, Montana, earthquake of 1959. In a single instant, a displacement of 6 m (19 ft) took place on a normal fault.

on one side is raised, or upthrown, relative to the other, which is downthrown. A normal fault results in a steep, straight, clifflike feature called a *fault scarp* (Figure 12.13). Fault scarps range in height from a few meters to a few hundred meters (Figure 12.14). Their length is usually measurable in kilometers. In some cases they attain lengths as great as 300 km (200 mi).

Normal faults are not usually isolated features. Commonly, they occur in multiple arrangements, often as a set of parallel faults. This gives rise to a grain or pattern of rock structure and topography. A narrow block dropped down between two normal faults is a *graben* (Figure 12.15). A narrow block elevated between two normal faults is a *horst*. Grabens make

conspicuous topographic trenches, with straight, parallel walls. Horsts make blocklike plateaus or mountains, often with a flat top but steep, straight sides.

In rifted zones of the continents, regions where normal faulting occurs on a grand scale, mountain masses called *block mountains* are produced. The up-faulted mountain blocks can be described as either tilted or lifted (Figure 12.16). A tilted block has one steep face—the fault scarp—and one gently sloping side. A lifted block, which is a type of horst, is bounded by steep fault scarps on both sides.

Transcurrent Faults

Recall that lithospheric plates slide past one another along major transform faults and that these features comprise one type of lithospheric plate boundary. Long before the principles of plate tectonics became known, geologists referred to such faults as **transcurrent faults** (Figure 12.13*c*), or sometimes as *strike-slip faults*. In a transcurrent fault, the movement is predominantly horizontal. Where the land surface is nearly flat, no scarp, or a very low one at most, results. Only a thin fault line is traceable across the surface. In some places a narrow trench, or rift, marks the fault.

On a transcurrent fault, rock masses move in a horizontal direction past one another.

The best known of the active transcurrent faults is the great San Andreas Fault, which can be followed for a distance of about 1000 km (600 mi) from the Gulf of California to Cape Mendocino, a location well north of

Figure 12.15 Initial landforms of normal faulting. A graben (*a*) is a downdropped block, often forming a long, narrow valley. A horst (*b*) is an upthrown block, forming a plateau, mesa, or mountain.

a.

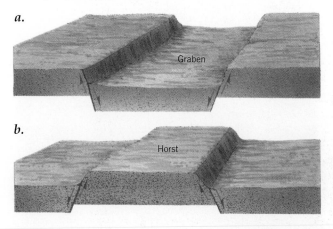

Graben

b.

Horst

Figure 12.16 Fault block mountains may be of tilted type (*left*) or lifted type (*right*). (After W. M. Davis.)

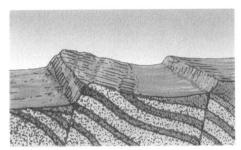

the San Francisco area, where it heads out to sea. It is a transform fault that marks the active boundary between the Pacific plate and the North American plate (see Figure 11.13). The Pacific plate is moving toward the northwest, which means that a great portion of the state of California and all of Lower (Baja) California is moving bodily northwest with respect to the North American mainland.

Throughout many kilometers of its length, the San Andreas Fault appears as a straight, narrow scar. In some places this scar is a trenchlike feature, and elsewhere it is a low scarp (Figure 12.17). Frequently, a stream valley takes an abrupt jog—for example, first right, then left—when crossing the fault line. This offset in the stream's course shows that many meters of movement have occurred in fairly recent time.

Figure 12.17 The San Andreas Fault in Southern California. The fault is marked by a narrow trough. The fault is slightly offset in the middle distance.

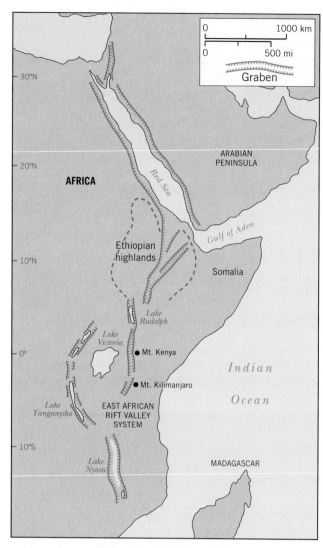

Figure 12.18 A sketch map of the East African rift valley system and the Red Sea to the north.

Reverse and Overthrust Faults

In a *reverse fault,* the inclination of the fault plane is such that one side rides up over the other and a crustal shortening occurs (Figure 12.13*b*). Reverse faults produce fault scarps similar to those of normal faults, but the possibility of landsliding is greater because an overhanging scarp tends to be formed. The San Fernando, California, earthquake of 1971 was generated by slippage on a reverse fault.

The *low-angle overthrust fault* (Figure 12.13*d*) involves predominantly horizontal movement. One slice of rock rides over the adjacent ground surface. A thrust slice may be up to 50 km (30 mi) wide. The evolution of low-angle thrust faults was explained in Chapter 11 and illustrated in Figure 11.15.

The Rift Valley System of East Africa

Rifting of continental lithosphere is the very first stage in the splitting apart of a continent to form a new ocean basin. The process is beautifully illustrated by the East African rift valley system. This region has attracted the attention of geologists since the early 1900s. They gave the name *rift valley* to what is basically a graben, but with a more complex history that includes the building of volcanoes on the graben floor.

Figure 12.18 is a sketch map of the East African rift valley system. It is about 3000 km (1900 mi) long and extends from the Red Sea southward to the Zambezi River. Along this axis, the earth's crust is being lifted upward and spread apart in a long, ridgelike swell. The rift valley system consists of a number of grabenlike troughs. Each is a separate rift valley ranging in width from about 30 to 60 km (20 to 40 mi). As geologists noted in early field surveys of this system, the rift valleys are like keystone blocks of a masonry arch

Figure 12.19 The rift valley wall in Ethiopia. Multiple fault scarps give the landscape a stepped appearance.

that have slipped down between neighboring blocks because the arch has spread apart somewhat. Thus, the floors of the rift valleys are above the elevation of most of the African continental surface. Major rivers and several long, deep lakes—Lake Nyasa and Lake Rudolph, for example—occupy some of the valley floors. The sides of the rift valleys typically consist of multiple fault steps (Figure 12.19). Sediments, derived from the high plateaus that form the flanks of troughs, make thick fills in the floors of the valleys.

EARTHQUAKES

You have probably seen television news accounts of disastrous earthquakes and their destructive effects (Figure 12.20). Californians know about severe earthquakes from first-hand experience, but several other areas in North America have also experienced strong earthquakes, and a few of these have been very severe. An **earthquake** is a motion of the ground surface, ranging from a faint tremor to a wild motion capable of shaking buildings apart.

The earthquake is a form of energy of wave motion transmitted through the surface layer of the earth. Waves move outward in widening circles from a point

of sudden energy release, called the *focus*. Like ripples produced when a pebble is thrown into a quiet pond, these *seismic waves* gradually lose energy as they travel outward in all directions. (The term *seismic* means "pertaining to earthquakes.")

Earthquakes are produced by sudden slip movements along faults. They occur when rock on both sides of the fault is slowly bent over many years by tectonic forces. Energy accumulates in the bent rock, just as it does in a bent archer's bow. When a critical point is reached, the strain is relieved by slippage on the fault, and the rocks on opposite sides of the fault move in different directions. A large quantity of energy is instantaneously released in the form of seismic waves, which shake the ground. In the case of a transcurrent fault, on which movement is in a horizontal direction, slow bending of the rock that precedes the shock takes place over many decades. Sometimes a slow, steady displacement known as *fault creep* occurs, which tends to reduce the accumulation of stored energy.

The devastating San Francisco earthquake of 1906 resulted from slippage along the San Andreas Fault, which is dominantly a transcurrent fault. This fault also passes about 60 km (40 mi) inland of Los Angeles, placing the densely populated metropolitan Los Angeles region in great jeopardy. Associated with the

Figure 12.20 Earthquake devastation in Mexico City following the earthquake of September, 1985.

Eye on the Environment • The Loma Prieta Earthquake of 1989

On October 17, 1989, the San Francisco Bay area was severely jolted by an earthquake with a Richter magnitude of 7.1. The quake's epicenter was located near Loma Prieta peak, about 80 km (50 mi) southeast of San Francisco, at a point only 12 km (7 mi) from the city of Santa Cruz, on Monterey Bay. The displacement that caused the quake occurred deep beneath the surface not far from the San Andreas Fault, which has not slipped since the great San Francisco earthquake of 1906 (see Figure 12.22). The 1989 slippage on the Loma Prieta fault amounted to about 1.8 m (6 ft) horizontally and 1.2 m (4 ft) vertically, but did not break the ground surface above it. The city of Santa Cruz suffered severe structural damage to older buildings. Many ground fissures (cracks) appeared on nearby mountain slopes, along with landslides that blocked roads.

Destructive ground shaking in the distant San Francisco Bay area proved surprisingly severe. Many areas near the bay are underlain by deposits of water-saturated, unconsolidated sand and mud that have been used to fill in the edges of the bay, extending land outward. With shaking, these sediments can undergo "spontaneous liquifaction"—that is, change to a near-liquid condition. In this condition, earthquake ground motion is greatly amplified, and heavy structures built on landfill can sink, tilt, or crack. This phenomenon had caused major structural damage in the great 1906 earthquake, and sensitive areas of landfill had been carefully mapped.

In the Loma Prieta earthquake, buildings, bridges, and viaducts on such landfills were particularly hard hit. The Marina area of San Francisco again experienced severe damage, similar to that of 1906. Quite unexpected was the collapse of the upper level of the Cypress section of Interstate 880 across the Bay, in Oakland (see photo). Equally unexpected was the collapse of a section of the nearby Bay Bridge. Altogether, 62 lives were lost in this earthquake, and the damage was estimated to be about $6 billion. In comparison, the 1906 earthquake took a toll of 700 lives and property damage equivalent to 20 billion 1987 dollars.

The San Andreas Fault has not slipped near San Francisco since 1906, when the lateral displacement was 6 m (20 ft). Geologic history indicates that the average rate of lateral motion between the two opposing lithospheric plates is 3.4 cm (1.34 in.) per year. At that rate, 175 years is required to produce the same amount of stored energy that was released in the 1906 earthquake. Geologists state that the Loma Prieta slippage, although near the San Andreas fault, probably has not relieved more than a small portion of the strain on the San Andreas.

A report on the Loma Prieta earthquake, issued by the U.S. Geological Survey (Circular 1045, 1989), contained this commentary:

The recurring theme of the Loma Prieta earthquake is that geologic conditions strongly influence damage. In other words, the geology determines where fault ruptures are likely to be, how hard the ground will shake, where landslides will occur, and where the ground will sink and crack. A parallel theme is that the pattern of damage from shaking and geologic effects ob-

San Andreas Fault are several important parallel and branching transcurrent faults, all of which are capable of generating severe earthquakes.

The Richter scale measures the energy released by an earthquake.

A scale of earthquake magnitudes was devised in 1935 by the distinguished seismologist Charles F. Richter. Now called the *Richter scale,* it describes the quantity of energy released by a single earthquake. Scale numbers range from 0 to 9, but there is really no upper limit other than nature's own energy release limit. For each whole unit of increase (say, from 5.0 to 6.0), the quantity of energy release increases by a factor of 33. A value of 9.5 is the largest observed to date—the

Chilean earthquake of 1960. The great San Francisco earthquake of 1906 is now rated as magnitude 7.9.

Earthquakes and Plate Tectonics

Seismic activity—the repeated occurrence of earthquakes—occurs primarily near lithospheric plate boundaries. Figure 12.21, which shows the location of all earthquake centers during a typical seven-year period, clearly reveals this pattern. The greatest intensity of seismic activity is found along converging plate boundaries where oceanic plates are undergoing subduction. Strong pressures build up at the downslanting contact of the two plates, and these are relieved by sudden fault slippages that generate earthquakes of large magnitude. This mechanism explains the great earthquakes experienced in Japan, Alaska, Chile, and other

served in 1989 is very similar to that witnessed in 1906. Thus, many of the lessons taught by the 1906 shock have been forgotten or ignored. As philosopher-poet George Santayana aptly noted "Those who cannot remember the past are prone to repeat it."

Questions

1. Where and near what fault did the Loma Prieta earthquake occur? How much movement occurred? How does that compare with earlier earthquakes?

2. What regions of the San Francisco Bay are most likely to experience earthquake damage? Why? What unexpected damage occurred in the Loma Prieta earthquake?

A collapsed section of the double-decked Nimitz Freeway (Interstate 880) in Oakland, California. Shaking by the Loma Prieta earthquake caused the upper level of the freeway to collapse upon the lower deck, crushing at least 39 persons in their cars.

narrow zones close to trenches and volcanic arcs of the Pacific Ocean basin.

Transform boundaries that cut through the continental lithosphere are also sites of intense seismic activity, with moderate to strong earthquakes. The most familiar example is the San Andreas Fault.

Spreading boundaries are the location of a third class of narrow zones of seismic activity related to lithospheric plates. Most of these boundaries are identified with the midoceanic ridge and its branches. For the most part, earthquakes in this class are limited to moderate intensities.

Earthquakes also occur at scattered locations over the continental plates, far from active plate boundaries. In many cases, no active fault is visible, and the geologic cause of the earthquake is uncertain. For example, the great New Madrid earthquake of 1811

was centered in the Mississippi River floodplain in Missouri. It produced three great shocks in close succession, rated from 8.1 to 8.3 on the Richter scale.

Seismic Sea Waves

An important environmental hazard often associated with a major earthquake centered on a subduction plate boundary is the *seismic sea wave,* or **tsunami,** as it is known to the Japanese. A train of these water waves is often generated in the ocean by a sudden movement of the sea floor at a point near the earthquake source. The waves travel over the ocean in ever-widening circles, but they are not perceptible at sea in deep water. (Seismic sea waves are sometimes referred to as "tidal waves," but since they have nothing to do with tides, the name is quite misleading.)

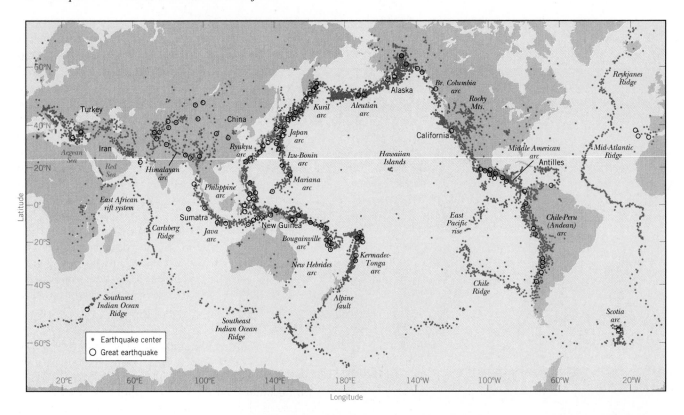

Figure 12.21 World map of earthquake center locations, 1961–1967, and centers of great earthquakes. Center locations of all earthquakes originating at depths of 0 to 100 km (62 mi) are shown by color dots. Each dot represents a single location or a cluster of centers. Open circles identify centers of earthquakes of Richter magnitude 8.0 or greater, 1897–1976. (Compiled by A. N. Strahler from data of U.S. government. Copyright © A. N. Strahler.)

When a tsunami arrives at a distant coastline, the effect is to cause a rise of water level. Normal wind-driven waves, superimposed on the heightened water level, attack places inland that are normally above their reach. For example, in this century several destructive seismic sea waves in the Pacific Ocean attacked ground as high as 10 m (30 ft) above normal high-tide level, causing widespread destruction and many deaths by drowning in low-lying coastal areas. It is thought that the coastal flooding that occurred in Japan in 1703, with an estimated loss of life of 100,000 persons, may have been caused by seismic sea waves.

Earthquakes along the San Andreas Fault

More than eight decades have passed since the great San Francisco earthquake of 1906 was generated by movement on the San Andreas Fault. The maximum horizontal displacement of the ground was about 6 m (20 ft). Since then, this sector of the fault has been locked—that is, the rocks on the two sides of the fault have been held together without sudden slippage (Figure 12.22). In the meantime, however, the two lithospheric plates that meet along the fault have been moving steadily with respect to one another. This means that a huge amount of unrelieved strain energy

has already accumulated in the crustal rock on either side of the fault. The Loma Prieta earthquake of 1989, although producing much damage in the San Francisco region, did little to relieve this strain. While the occurrence of another major earthquake cannot be predicted with precision, it is inevitable. As each decade passes, the probability of that event becomes greater.

A recent estimate placed at about 50 percent the likelihood that a very large earthquake will occur within the next 30 years somewhere along the southern California portion of the San Andreas Fault. In 1992 three severe earthquakes occurred in close succession along local faults a short distance north of the San Andreas Fault within the area labeled the southern active area on the map. These events have led to speculation that the likelihood of a major slip on the San Andreas Fault in the near future has substantially increased.

For residents of the Los Angeles area, an additional serious threat lies in the large number of active faults close at hand. Movements on these local faults have produced more than 40 damaging earthquakes since 1800, including the Long Beach earthquakes of the 1930s and the San Fernando earthquake of 1971. The latter measured 6.6 on the Richter scale and produced

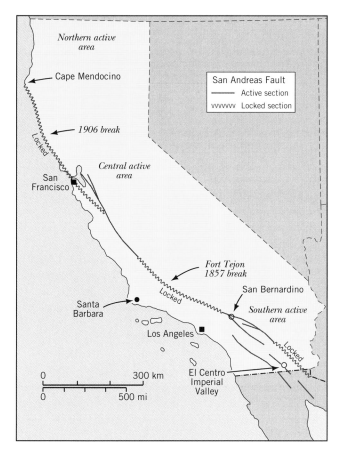

Figure 12.22 A sketch map of the San Andreas fault showing the locked sections alternating with active sections of frequent small earthquakes and slow, creeping motion. (Based on data of the U.S. Geological Survey.)

In the last three chapters, we have surveyed the composition, structure, and geologic activity of the earth's crust. We began with a study of the rocks and minerals that make up the earth's crust and core. We saw how the rock cycle describes a continuous cycling and recycling of rocks and minerals that has occurred over some 3 billion years or more of geologic time. In the second chapter, we developed plate tectonics as the mechanism powering the rock cycle. The global pattern of plate tectonics also explains the geographical distribution of mountain ranges, ocean basins, and continental shields occurring on the earth's surface. In this chapter, we described the landforms that result directly from volcanic and tectonic activity, which occur primarily at the boundaries of spreading or colliding lithospheric plates.

With this survey of the earth's crust and the geologic processes that shape it now completed, we can turn to other landform-creating processes in the following chapters. First will be the processes of weathering, which breaks rock into small particles, and mass wasting, which moves them downhill as large and small masses under the influence of gravity. Next, we turn to running water in three chapters that first describe the behavior of rivers and streams, then their work in shaping landforms, and lastly how running water dissects rock layers to reveal their inner structures. In our two concluding chapters, we will examine landforms created by waves, wind, and glacial ice.

Figure 12.23 Severe structural damage and the collapse of buildings of the Veterans Administration Hospital in Sylmar, Los Angeles County, caused by the San Fernando earthquake of 1971.

severe structural damage near the earthquake center (Figure 12.23). In 1987 an earthquake of magnitude 6.1 struck the vicinity of Pasadena and Whittier, located within about 20 km (12 mi) of downtown Los Angeles. Known as the Whittier Narrows earthquake, it was generated along a local fault system that had not previously shown significant seismic activity. The brief but intense primary shock and aftershocks that followed damaged beyond repair many older structures built of unreinforced brick masonry.

A slip along the San Andreas Fault, some 50 km (30 mi) to the north of the densely populated region of Los Angeles, will release an enormously larger quantity of energy than these local earthquakes. On the other hand, the destructive effects of a San Andreas earthquake in downtown Los Angeles will be somewhat moderated by the greater travel distance. Although the intensity of ground shaking might not be much different from that of the San Fernando earthquake, for example, it will last much longer and cover a much wider area of the Los Angeles region. The potential for damage and loss of life is enormous.

VOLCANIC AND TECTONIC LANDFORMS IN REVIEW

Landforms are the surface features of the land, and geomorphology is the scientific study of landforms. Initial landforms are shaped by volcanic and tectonic activity, while sequential landforms are sculpted by agents of denudation, including running water, waves, wind, and glacial ice.

Volcanoes are landforms marking the eruption of lava at the earth's surface. Stratovolcanoes, formed by the emission of thick, gassy, felsic lavas, have steep slopes and tend to explosive eruptions that can form calderas. Most active stratovolcanoes lie along the Pacific rim, where subduction of oceanic lithospheric plates is occurring. At hot spots, rising mantle material provides mafic magma that erupts as basaltic lavas. Because these lavas are more fluid and contain little gas, they form broadly rounded shield volcanoes. Hot spots occurring beneath continental crust can also provide vast areas of flood basalts. Some basaltic volcanoes occur along the midoceanic ridge.

The two forms of tectonic activity are compressional and extensional. Compressional activity occurs at lithospheric plate collisions. At first, the compression produces folding—anticlines (upfolds) and synclines (downfolds). If compression continues, folds may be overturned and eventually overthrust faulting can occur. Extensional activity occurs where lithospheric plates are spreading apart, generating normal faults. These can produce upthrown and downdropped blocks that are sometimes as large as mountain ranges or rift valleys. Transcurrent faults occur where two rock masses move horizontally past each other.

Earthquakes occur when rock layers, bent by tectonic activity, suddenly fracture and move. The sudden motion at the fault produces earthquake waves that shake and move the ground surface in the adjacent region. The energy released by an earthquake is measured by the Richter scale. Large earthquakes occurring near developed areas can cause great damage. Most severe earthquakes occur near plate collision boundaries. The San Andreas fault is a major transcurrent fault located near two great urban areas—Los Angeles and San Francisco. The potential for a severe earthquake on this fault is high, and the probability of a major earth movement increases every year.

KEY TERMS

landform	stratovolcano	normal fault
geomorphology	shield volcano	transcurrent fault
denudation	folding	earthquake
volcano	fault	tsunami

REVIEW QUESTIONS

1. Distinguish between initial and sequential landforms. How do they represent the balance of power between internal earth forces and external forces of denudation agents?

2. What is a stratovolcano? What is its characteristic shape, and why does that shape occur? Where do stratovolcanoes generally occur and why?

3. What is a shield volcano? How is it distinguished from a stratovolcano? Where are shield volcanoes found, and why? Give an example of a shield volcano. How are flood basalts related to shield volcanoes?

4. How can volcanic eruptions become natural disasters? Be specific about the types of volcanic events that can devastate habitations and extinguish nearby populations.

5. What types of landforms characterize foreland fold belts? Where do these belts occur, and how are they formed?

6. Sketch a cross section through a normal fault, labeling the fault plane, upthrown side, downthrown side, and fault scarp.

7. Briefly describe the rift valley system of East Africa as an example of normal faulting.

8. How does a transcurrent fault differ from a normal fault? What landforms are expected along a transcurrent fault? How are transcurrent faults related to plate tectonic movements?

9. What is an earthquake, and how does it arise? How are the locations of earthquakes related to plate tectonics?

10. Describe the tsunami, including its origin and effects.

11. Briefly summarize the geography and recent history of the San Andreas fault system in California. What are the prospects for future earthquakes along the San Andreas fault?

IN-DEPTH QUESTIONS

1. Write a fictional news account of a volcanic eruption. Select a type of volcano—stratovolcano or shield volcano—and a plausible location. Describe the eruption and its effects as it was witnessed by observers. Make up any details you need, but be sure they are scientifically correct.

2. How are mountains formed? Provide the plate tectonic setting for mountain formation, and then describe how specific types of mountain landforms arise.

Weathering and Mass Wasting

The pull of gravity is relentless and unforgiving. It acts to move all things closer to the earth's center, and for objects on the earth's surface, that generally means downhill. If you are a skier, you may be grateful for that downhill pull (providing the ski lift is working to handle the uphill part of skiing). As a hiker, the downhill force makes the trek home from the summit a lot easier (if your knees and leg muscles are in good shape). But if you are an inhabitant of a peaceful valley below a steep mountain slope that suddenly gives way, you may not be so grateful for the force that has buried your home under tons of rocky debris.

Landslide, earthflow, mudflow, soil creep—these words describe the different ways that earth materials move downhill under the force of gravity. Sometimes the materials move very quickly, other times very slowly. Perhaps you've walked the sidewalks of a town or city built on a steep slope and seen retaining walls leaning over or even toppled by the slow downhill creep of soil. While annoying and expensive for a homeowner, a failing retaining wall is not likely to result in death or destruction. But a landslide, earthflow, or mudflow can easily take lives and destroy property.

In 1903 an enormous mass of limestone rock slid from the face of Turtle Mountain, near Frank, Alberta, into the valley below (see Figure 13.19). The volume of the landslide was estimated at about 27 million cu m (35 million cu yd). It moved down through 900 m (3000 ft) of elevation at great speed and slid uphill on the opposite side of the valley to regain over 100 m (300 ft) of elevation. The rocky debris buried part of the town of Frank, Alberta, taking 70 lives.

Earthflows are slower but no less destructive. For residents of the Palos Verdes Hills, a coastal peninsula in Los Angeles County, the Portuguese Bend "landslide" was a nightmare that wouldn't go away. About 160 hectares (400 acres) of residential land settled and moved downslope some 20 m (70 ft) over a three-year period. Houses were slowly tilted, cracked, and broken apart as the land underneath them crept along a shale layer softened by water from the septic systems and abundant lawn-watering of the residential area.

The Nicolet earthflow of 1955 occurred when a clay layer beneath the city of Nicolet, Quebec, unexpectedly became liquid. A large chunk of the town quickly flowed into the Nicolet River. Only three lives were lost, but damage was in the millions.

Some large mudflows can follow stream and river courses for long distances. Immediately following the eruption of Mount St. Helens in 1980, a mudflow of fresh ash and melted snow hurtled down the Toutle River, destroying fields and forests. It traveled a total distance of more than 120 km (75 mi), eventually reaching the Columbia River.

Mass movements of earth under the influence of gravity are one of the two main subjects of this chapter. These movements create distinctive landforms of an obvious nature. However, the subject we will present first is weathering—a set of physical and chemical processes that break rocks apart to create large and small particles and to alter their mineral composition. Weathering prepares the way for mass movements, and it can also create its own distinctive landforms, as we will see.

Napali Coast, Island of Hawaii, Hawaii.

ow that we have completed our study of the earth's crust, its mineral composition, its moving lithospheric plates, and its tectonic and volcanic landforms, we can focus on the shallow life layer itself. At this sensitive interface, the external processes of denudation carve sequential landforms from the earth materials uplifted by the earth's internal processes. Our investigation of what happens to rock upon exposure at the surface began in Chapter 10 with a close look at the mineral alteration of rock and the production of sediment. This is an essential part of the cycle of rock transformation. In this chapter, we look further at the softening and breakup of rock, termed weathering, and how the resulting particles move downhill under the force of gravity, a process termed mass wasting.

Weathering is the general term applied to the combined action of all processes that cause rock to disintegrate physically and decompose chemically because of exposure near the earth's surface. In Chapter 10, we described the processes of mineral alteration (oxidation, hydrolysis, and carbonic acid solution) that transform minerals from types that were stable when the rocks were formed to types that are now stable at surface temperatures and pressures. These processes are part of chemical weathering, by which rock particles are altered chemically at the surface. A second type of weathering is physical weathering, in which rock particles are fractured and broken apart. Weathering also leads to a number of distinctive landforms, which we will describe in this chapter.

This chapter also includes a related process that creates landforms. It is the spontaneous downhill movement of soil, regolith, and rock under the influence of gravity, but without the action of moving water, air, or ice. This downhill movement is referred to as *mass wasting*. Movement of a mass of soil or rock to lower levels takes place when the internal strength declines to a critical point below which the force of gravity cannot be resisted. This failure of strength under the ever-present force of gravity takes many forms and scales, and we will see that human activity causes or aggravates several forms of mass wasting.

SLOPES AND REGOLITH

Mass wasting occurs on slopes. As used in physical geography, the term *slope* designates a small strip or patch of the land surface that is inclined from the horizontal. Thus, we speak of "mountain slopes," "hill slopes," or "valley-side slopes" to describe some of the inclined ground surfaces that we might encounter in traveling

across a landscape. Slopes guide the downhill flow of surface water and fit together to form drainage systems within which surface-water flow converges into stream channels. Nearly all natural surfaces slope to some degree. Very few are perfectly horizontal or vertical.

Wasting of slopes refers to the downhill movement of soil and regolith under the force of gravity.

As described in Chapter 9, most slopes are mantled with *regolith* (Figure 13.1), a surface layer of weathered rock particles. The regolith grades downward into solid, unaltered rock, known simply as **bedrock.** Regolith, in turn, provides the source for **sediment,** consisting of detached mineral particles that are transported and deposited in a fluid medium. This fluid may be water, air, or even glacial ice. As we saw in Chapter 9, both regolith and sediment comprise parent materials for the formation of the true soil, which is a surface layer capable of supporting the growth of plants.

Figure 13.1 shows a typical hill slope that forms one wall of the valley of a small stream. Soil and regolith blanket the bedrock, except in a few places where the bedrock is particularly hard and projects in the form of *outcrops. Residual regolith* is derived directly from the rock beneath and moves very slowly down the slope toward the stream. Beneath the valley bottom are layers of transported regolith, called **alluvium,** which is sediment transported and deposited by the stream. This sediment had its source in regolith prepared on hill slopes many kilometers or miles upstream. All accumulations of sediment on the land surface, whether deposited by streams, waves and currents, wind, or glacial ice, can be designated *transported regolith,* in contrast to residual regolith. Formation of regolith is greatly aided by the presence of innumerable bedrock cracks, called *joints.* Water can move easily through joints to promote rock weathering.

The thickness of soil and regolith is quite variable. Although the soil is rarely more than 1 or 2 m thick (about 3 to 6 ft), residual regolith on decayed and fragmented rock may extend down to depths of 5 to 100 m (about 15 to 300 ft), or more at some locations. On the other hand, soil or regolith, or both, may be missing. In some places, everything is stripped off down to the bedrock, which then appears at the surface as an outcrop. In other places, following cultivation or forest fires, the fertile soil is partly or entirely eroded away, and severe erosion exposes the regolith.

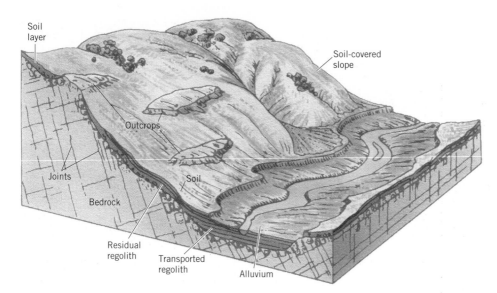

Figure 13.1 Soil, regolith, and outcrops on a hill slope. Alluvium, a form of transported regolith, lies in the floor of an adjacent stream valley. (After A. N. Strahler.)

PHYSICAL WEATHERING

Physical weathering produces fine particles of regolith from massive rock by the action of forces strong enough to fracture the rock. The physical weathering processes discussed in this chapter include frost action, salt-crystal growth, unloading, expansion-contraction, and wedging by plant roots.

Frost Action

One of the most important physical weathering processes in cold climates is *frost action*, the repeated growth and melting of ice crystals in the pore spaces of soil and in rock fractures. This activity can rupture even extremely hard rocks. Frost action produces a number of conspicuous effects and forms in all climates that have cold winters. Features caused by frost action and the buildup of ice below the surface are particularly conspicuous in the tundra climate of arctic coastal fringes and islands, and above timberline in high mountains.

Frost shattering of hard rocks exposed above timberline leads to surface accumulations of large angular fragments, including huge boulders (Figure 13.2). Frost action on cliffs of bare rock in high mountains detaches rock fragments that fall to the cliff base. Where production of fragments is rapid, they accumulate to form *talus slopes*. Most cliffs are notched by narrow ravines that funnel the rock fragments into separate tracks. Each track, or chute, feeds a growing, conelike talus body. Talus cones are arranged side by side along the base of the cliff (Figure 13.3). Fresh talus slopes are unstable, so that the disturbance created by walking across the slope or dropping a large rock

fragment from the cliff above will easily set off a sliding or rolling motion within the surface layer of fragments.

In fine-textured soils and sediments, composed largely of silt and clay, freezing of soil water takes place in horizontal layers or lens-shaped bodies. As these

Figure 13.2 Frost-shattered blocks, Sierra Nevada, California. Ice crystal growth within the joint planes of rock can cause the rock to split apart. This boulder is an example.

Figure 13.3   Talus cones near the shore of Lake Louise in the Canadian Rockies.

ice layers thicken, the overlying soil layer is heaved upward. Prolonged frost heaving can produce minor irregularities and small mounds on the soil surface. Where a rock fragment lies at the surface, perpendicular ice needles grow beneath the fragment and raise it above the surface (Figure 13.4). The same process acting on a rock fragment below the soil surface will eventually bring the fragment to the surface.

A related process of ice-crystal growth acts in coarse-textured regolith of the barren tundra to cause the coarsest fragments—pebbles and cobbles—to move horizontally and to become sorted out from the finer particles. This type of sorting produces ringlike arrangements of coarse fragments. Linked with adjacent rings, the gross pattern becomes netlike to form a system of *stone rings* (Figure 13.5).

Figure 13.4   Needle ice growth. At night, water in the soil freezes at the surface, creating ice needles that can lift particles of soil or move larger stones.

In silty alluvium, such as that formed on river flood-plains and delta plains in the arctic environment, ice accumulates in vertical wedge-forms in deep cracks in the sediment. These ice wedges are interconnected into a system of polygons, called *ice-wedge polygons* (Figure 13.6). Ice wedges are thought to originate as shrinkage cracks formed during extreme winter cold. During the spring melt, surface water enters the cracks and becomes frozen, adding to the thickness of the wedges.

Salt-Crystal Growth

Closely related to the growth of ice crystals is the weathering process of rock disintegration by growth of salt crystals in rock pores. This process, *salt-crystal growth*, operates extensively in dry climates and is responsible for many of the niches, shallow caves, rock arches, and pits seen in sandstone formations. During long drought periods, ground water is drawn to the surface of the rock by capillary force. As evaporation of the water takes place in the porous outer zone of the sandstone, tiny crystals of salts such as halite (sodium chloride), calcite (calcium carbonate), or gypsum (calcium sulfate) are left behind. The growth force of these crystals produces grain-by-grain breakup of the

sandstone, which crumbles into a sand and is swept away by wind and rain.

Especially susceptible are zones of rock lying close to the base of a cliff, because there the ground water seeps outward to reach the rock surface (Figure 13.7). In the southwestern United States, many of the deep niches or cavelike recesses formed in this way were occupied by Native Americans. Their cliff dwellings gave them protection from the elements and safety from armed attack (Figure 13.8).

Salt-crystal growth can break rocks apart grain by grain, producing niches, shallow caves, and rock arches.

Salt crystallization also damages masonry buildings, as well as concrete sidewalks and streets. Brick and concrete in contact with moist soil are highly susceptible to grain-by-grain disintegration from salt crystallization. The salt crystals can be seen as a soft, white, fibrous layer on damp basement floors and walls. The deicing salts spread on streets and highways can be quite destructive. Sodium chloride (rock salt), widely used for this purpose, is particularly damaging to concrete pavements and walks, curbstones, and other exposed masonry structures.

Figure 13.5   Sorted circles of gravel form a system of netlike stone rings on this nearly flat land surface where water drainage is poor. The circles in the foreground are 3 to 4 m (10 to 13 ft) across; the gravel ridges are 20 to 30 cm (8 to 12 in.) high. Broggerhalvoya, western Spitsbergen, latitude 78°N.

Figure 13.6 These polygons are formed by the growth of ice wedges. On the Alaskan north slope, near the border of Alaska and Yukon Territory, Canada.

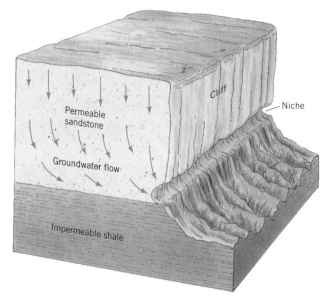

Figure 13.7 In dry climates there is a slow seepage of water from the cliff base. Salt crystal growth separates the grains of permeable sandstone, breaking them loose and creating a niche. (After A. N. Strahler.)

Figure 13.8 The White House Ruin, a former habitation of Native Americans, occupies a large niche in sandstone in the lower wall of Canyon de Chelly, Arizona.

Unloading

A widespread process of rock disruption related to physical weathering results from *unloading*, the relief of confining pressure of overlying rock. Unloading occurs as rock is brought near the surface by erosion of overlying layers. Rock formed at great depth beneath the earth's surface (particularly igneous and metamorphic rock) is in a slightly compressed state because of the confining pressure of overlying rock. As the rock above is slowly worn away, the pressure is reduced, and the rock expands slightly in volume. Expansion causes thick shells of rock to break free from the parent mass below. The new surfaces of fracture are a form of jointing called *sheeting structure*. The rock sheets show best in massive rocks such as granite and marble.

Where sheeting structure has formed over the top of a single large body of massive rock, an *exfoliation dome* is produced (Figure 13.9). Domes are among the largest of the landforms shaped primarily by weathering. In Yosemite Valley, California, where domes are spectacularly displayed, the individual rock sheets may be as thick as 15 m (50 ft). (See *Eye on the Environment • Marvelous, Majestic, Monolithic Domes,* in Chapter 16.)

Figure 13.9 North Dome and Royal Arches, Yosemite National Park, California. Thick exfoliation rock layers are visible in the lower part of the photo.

Other Physical Weathering Processes

Most rock-forming minerals expand when heated and contract when cooled. Where rock surfaces are exposed daily to the intense heating of the sun alternating with nightly cooling, the resulting expansion and contraction exerts powerful disruptive forces on the rock. Although first-hand evidence is lacking, it seems likely that daily temperature changes can cause the breakup of a surface layer of rock already weakened by other agents of weathering.

Another mechanism of rock breakup is the growth of plant roots, which can wedge joint blocks apart. You may have seen a tree whose lower trunk and roots are firmly wedged between two great joint blocks of massive rock. Whether the tree has actually been able to spread the blocks farther apart or has merely occupied the available space is sometimes open to question. However, it is certain that pressure exerted by growth of tiny rootlets in joint fractures causes the loosening of countless small rock scales and grains.

CHEMICAL WEATHERING AND ITS LANDFORMS

We investigated **chemical weathering** processes in Chapter 10 under the heading of mineral alteration. Recall that the dominant processes of chemical change affecting silicate minerals are oxidation, hydrolysis, and carbonic acid action. While feldspars and the mafic minerals are very susceptible to chemical change, quartz is a highly stable mineral, almost immune to decay.

Hydrolysis and Oxidation

Decomposition by hydrolysis and oxidation changes strong rock into very weak regolith. This change allows erosion to operate with great effectiveness wherever the regolith is exposed. Weakness of the regolith also makes it susceptible to natural forms of mass wasting. In warm, humid climates of the equatorial, tropical, and subtropical zones, hydrolysis and oxidation often result in the decay of igneous and metamorphic rocks to depths as great as 100 m (more than 300 ft). To the construction engineer, deeply weathered rock is of major concern in the building of highways, dams, or other heavy structures. Although the regolith is soft and can often be removed by power shovels with little blasting, foundations on regolith can fail under heavy loads. This regolith also has undesirable plastic properties because of its high content of clay minerals.

> In chemical weathering, the minerals that make up rocks are chemically altered or dissolved.

The hydrolysis of exposed granite surfaces is accompanied by the grain-by-grain breakup of the rock. This process creates many interesting boulder and pinnacle forms by the rounding of angular joint blocks (Figure 13.10). These forms are particularly conspicuous in arid regions. In most deserts there is ample moisture for hydrolysis to act, given sufficient time. The products of grain-by-grain breakup form a fine desert gravel, which consists largely of quartz and partially decomposed feldspar crystals.

Figure 13.10 Large joint blocks of granite are gradually rounded into smooth forms by grain-by-grain disintegration in a desert environment. Alabama Hills, Owens Valley, California.

Carbonic Acid Action

Atmospheric carbon dioxide is dissolved in all surface waters of the lands, including rainwater, soil water, and stream water. The solution of carbon dioxide in water produces a weak acid, called *carbonic acid,* which can dissolve some types of minerals. Carbonate sedimentary rocks, such as limestone and marble, are particularly susceptible to the acid action. In this process, the mineral calcium carbonate is dissolved and carried away in solution in stream water.

Carbonic acid reaction with limestone produces many interesting surface forms, mostly of small dimensions. Outcrops of limestone typically show cupping, rilling, grooving, and fluting in intricate designs (Figure 13.11). In a few places, the scale of deep grooves and high wall-like rock fins reaches proportions that keep people and animals from passing through. Dissolution of limestone by carbonic acid in groundwater can produce underground caverns as well as distinctive landscapes that are formed when underground caverns collapse. These landforms and landscapes will be described in Chapter 16.

In urban areas, air pollution by sulfur and nitrogen oxides commonly occurs. When these gases dissolve in rainwater, the result is acid precipitation. (See *Eye on the Environment • Acid Deposition and Its Effects* in Chapter 4.) The acids rapidly dissolve limestone and chemically weather other types of building stones. The result can be very damaging to stone sculptures, building decorations, and tombstones (Figure 13.12).

In the wet low-latitude climates, mafic rock, particularly basaltic lava, undergoes rapid removal under attack by soil acids and produces landforms quite similar to

Figure 13.11 This outcrop of pure limestone shows rills and cups formed by solution weathering. County Clare, Ireland.

Eye on the Environment • The Great Hebgen Lake Disaster

For some 200 vacationers camping near the Madison River in a deep canyon not far west of Yellowstone Park, the night of August 17, 1959, began quietly, with almost everyone safely bedded down in their tents or camping trailers. Up to a certain point, it was everything a great vacation should be—that point in time was 11:37 P.M., Mountain Standard Time. At precisely that instant, not one but four terrifying forms of disaster were set loose on the sleeping vacationers—earthquake, landslide, hurricane-force wind, and raging flood. The earthquake, which measured 7.1 on the Richter scale, was the cause of it all. The first shock, lasting several minutes, rocked the campers violently in their trailers and tents. Those who struggled to go outside could scarcely stand up, let alone run for safety.

Then came the landslide. A dentist and his wife watched through the window of their trailer as a mountain seemed to move across the canyon in front of them, trees flying from its surface like toothpicks in a gale. Then, as rocks began to bang against the sides and top of their trailer, they got out and raced for safer ground. Later, they found that the slide had stopped only about 25 m (80 ft) from the trailer. Pushed by the moving mountain came a vicious blast of wind. It swept upriver, tumbling trailers end over end.

Then came the flood. Two women schoolteachers, sleeping in their car only about 5 m (15 ft) from the river bank, awoke to the violent shaking of the earthquake. Like other campers, they first thought they had a marauding bear

on their hands. Puzzled and frightened, they started the engine and headed the car for higher ground. As they did so, they were greeted by a great roar coming from the mountainside above and behind them. An instant later, the car was completely engulfed by a wall of water that surged up the river bank, then quickly drained back. With the screams of drowning campers in their ears, the two women managed to drive the car to safe ground, high above the river.

After the first surge of water, generated as the landslide mass hit the river, the river began a rapid rise. This rise was aided by great surges of water topping the Hebgen Dam, located upstream, as earthquake aftershocks rocked the water of Hebgen Lake back and forth along its length of about 50 km (30 mi). But, of course, in the darkness of night, the terrified victims of the flood had no idea of what was happening. In the panic that ensued, a 71-year-old man performed an almost unbelievable act of heroism to save his wife and himself from drowning. As the water rose inside their house trailer, he forced open the door, pulled his wife with him to the trailer roof, then carried her up on the branches of a nearby pine tree. Here they were finally able to reach safety. The water had risen 10 m (about 30 ft) above ground level in just minutes.

The Madison Slide, as the huge earth movement was later named, had a bulk of 28 million cu m (37

A reporter examines a camper's automobile that was crushed by the landslide, then swept along the riverbed by the flood wave. The campground is buried beneath the slide in the background.

A violent earthquake caused this huge landslide. Rocky debris slid from the steep mountain slope into the river valley below.

million cu yd) of rock (see photo). It consisted of a chunk of the south wall of the canyon, measuring over 600 m (about 2000 ft) in length and 300 m (about 1000 ft) in thickness. The mass descended over a half of a kilometer (about a third of a mile) to the Madison River, its speed estimated at 160 km (100 mi) per hour. Pulverized into bouldery debris, the slide crossed the canyon floor, its momentum carrying it over 120 m (about 400 ft) in vertical distance up the opposite canyon wall. At least 26 persons died beneath the slide, and their bodies have never been recovered. Acting as a huge dam, the slide caused the Madison River to back up, forming a new lake. In three weeks' time, the lake was nearly 100 m (330 ft) deep. Today it is a permanent feature, named Earthquake Lake.

Question

Describe the formation of Earthquake Lake. What four forms of natural environmental disaster were involved?

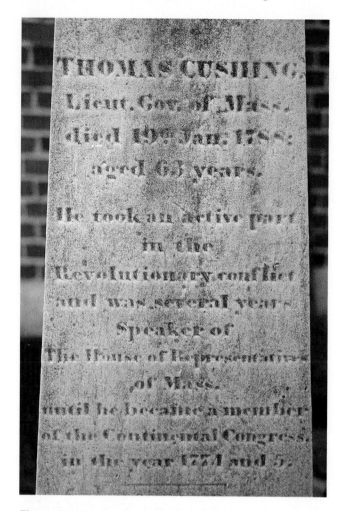

Figure 13.12 Weathered tombstones from a burying ground in Boston, Massachusetts. The marker on the left, carved in limestone, has been strongly weathered, weakening the lettering. The marker on the right, made of slate, is much more resistant to erosion.

those formed by carbonic acid on massive limestones in the moist climates of higher latitudes. The effects of solution removal of basaltic lava are displayed in spectacular grooves, fins, and spires on the walls of deep alcoves in part of the Hawaiian Islands (see chapter opening photo).

MASS WASTING

Everywhere on the earth's surface, gravity pulls continually downward on all materials. Bedrock is usually so strong and well supported that it remains fixed in place. However, when a mountain slope becomes too steep, bedrock masses break free and fall or slide to new positions of rest. In cases where huge masses of bedrock are involved, the result can be catastrophic in loss to life and property in towns and villages in the path of the slide. Such slides are a major form of environmental hazard in mountainous regions.

Because soil, regolith, and many forms of sediment are held together poorly, they are much more susceptible to movement under the force of gravity than hard, massive bedrock. Abundant evidence shows that on most slopes at least a small amount of downhill movement is going on constantly. Although much of this motion is imperceptible, the regolith sometimes slides or flows rapidly.

Taken together, the various kinds of downhill movements of rock and regolith that occur under the pull of gravity are collectively called **mass wasting**—an important process in the lowering of continental surfaces. Humans have added to the natural forms of mass wasting by moving enormous volumes of rock and soil on construction sites of dams, canals, highways, and buildings.

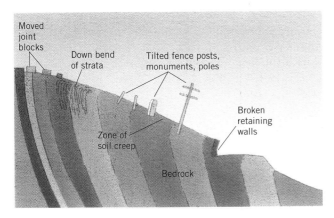

Figure 13.13 The slow, downhill creep of soil and regolith shows up in many ways on a hillside. (After C.F.S. Sharpe.)

Labels in figure: Moved joint blocks; Down bend of strata; Tilted fence posts, monuments, poles; Zone of soil creep; Broken retaining walls; Bedrock

Soil Creep

On almost any soil-covered slope, you can find evidence of extremely slow downhill movement of soil and regolith, a process called **soil creep.** Figure 13.13 shows some of the evidence that the process is going on. Joint blocks of distinctive rock types are found moved far downslope from the outcrop. In some layered rocks such as shales or slates, edges of the strata seem to "bend" in the downhill direction (Figure 13.14). This is not true plastic bending, but is the result of the slight movement of many rock pieces on small joint cracks. Creep causes fence posts and utility poles to lean downslope and even shift measurably out of line. Roadside retaining walls can buckle and break under the pressure of soil creep

Figure 13.14 Slow, downhill creep of regolith on this mountainside near Downieville, California, has caused vertical rock layers to seem to "bend over."

Soil creep refers to the slow downhill movement of soil and regolith.

What causes soil creep? Alternate drying and wetting of the soil, growth of frost needles, heating and cooling of the soil, trampling and burrowing by animals, and shaking by earthquakes all produce some disturbance of the soil and regolith. Because gravity exerts a downhill pull on every such rearrangement, the particles very gradually work their way downslope.

Earthflow

In regions of humid climate, a mass of water-saturated soil, regolith, or weak shale may move down a steep slope during a period of a few hours in the form of an **earthflow.** Figure 13.15 is a sketch of an earthflow showing how the material slumps away from the top, leaving a steplike terrace bounded by a curved, wall-like scarp. The saturated material flows sluggishly to form a bulging toe.

Shallow earthflows, affecting only the soil and regolith, are common on sod-covered and forested slopes that have been saturated by heavy rains. An earthflow may affect a few square meters, or it may cover an area of several hectares. If the bedrock of a mountainous region is rich in clay (derived from shale or deeply weathered volcanic rocks), earthflows sometimes involve millions of metric tons of bedrock moving by plastic flowage like a great mass of thick mud.

Earthflows are a common cause of blockage of highways and railroad lines, usually during periods of heavy

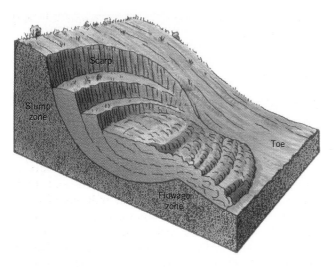

Figure 13.15 An earthflow with slump features well developed in the upper part. Flowage has produced a bulging toe. (Drawn by A. N. Strahler.)

rains. Generally, the rate of flowage is slow, so that the flows are not a threat to life. Property damage to buildings, pavements, and utility lines is often severe where construction has taken place on unstable soil slopes.

A special variety of earthflow found in the arctic tundra is *solifluction* (from Latin words meaning "soil" and "flow"). In early summer, when thawing has penetrated the upper few decimeters, the soil is fully saturated. Soil water cannot escape downward because the per-

mafrost layer will not allow water to drain through it. Flowing almost imperceptibly, this saturated soil forms solifluction terraces and lobes that give the tundra slope a stepped appearance (Figure 13.16).

Mudflow

Mudflow is one of the most spectacular forms of mass wasting and is a potentially serious environmental hazard. This mud stream of fluid consistency pours swiftly down canyons in mountainous regions (Figure 13.17). In deserts, where vegetation does not protect the mountain soils, local thunderstorms produce rain much faster than it can be absorbed by the soil. As the water runs down the slopes, it forms a thin mud, which flows down to the canyon floors. Following stream courses, the mud continues to flow until it becomes so thick it must stop. Great boulders are carried along, buoyed up in the mud. Roads, bridges, and houses in the canyon floor are engulfed and destroyed. Where the mudflow emerges from the canyon and spreads across an alluvial fan, severe property damage and even loss of life may be the result (Figure 13.18).

As explained in Chapter 12, mudflows that occur on the slopes of erupting volcanoes are called lahars. Freshly fallen volcanic ash and dust are turned into mud by heavy rains or melting snows and flow down the slopes of the volcano. Herculaneum, a city at the base of Mount Vesuvius, was destroyed by a mudflow during

Figure 13.16 Solifluction lobes in the Richardson Mountains, Northwest Territories, Canada. Bulging masses of water-saturated regolith have slowly moved downslope, overriding the ground surface below while bearing intact their covers of plants and soil.

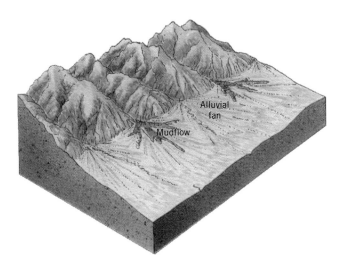

Figure 13.17 Thin, streamlike mudflows issue occasionally from canyon mouths in arid regions. The mud spreads out on the slopes below in long, narrow tongues. (Drawn by A. N. Strahler.)

the eruption of A.D. 79. At the same time, the neighboring city of Pompeii was buried under volcanic ash.

Mudflows vary in consistency, from a mixture like concrete emerging from a mixing truck to consistencies similar to that of turbid river floodwaters. The watery type of mudflow is called a *debris flood* or *debris flow* in the western United States, and particularly in

Southern California, where it occurs commonly and with disastrous effects.

Landslide

A **landslide** is the rapid sliding of large masses of bedrock. Wherever mountain slopes are steep, there is a possibility of large, disastrous landslides (Figure 13.19). In Switzerland, Norway, or the Canadian Rockies, for example, villages built on the floors of steep-sided valleys have been destroyed and their inhabitants killed by the sliding of millions of cubic meters of rock set loose without warning. Severe earthquakes in mountainous regions are a major cause of landslides and earthflows. (See *Eye on the Environment • The Great Hebgen Lake Disaster* in this chapter.)

In a landslide, a large mass of rock suddenly moves from a steep mountain slope to the valley below.

Aside from occasional local catastrophes, landslides have rather limited environmental influence because they occur only sporadically and usually in thinly populated mountainous regions. Small slides can, however, repeatedly block or break important mountain highways or railway lines.

Figure 13.18 This mudflow, carrying numerous large boulders, issued from a steep mountain canyon in the Wasatch Mountains, Utah.

Figure 13.19   A great landslide descended from the high mountain summit at the upper left. The rock became pulverized into loose debris, which rode across the valley floor at great speed on a cushion of trapped air. Turtle Mountain slide, Frank, Alberta.

INDUCED MASS WASTING

Human activities can induce mass wasting in forms ranging from mudflow and earthflow to landslide. These activities include (1) piling up of waste soil and rock into unstable accumulations that can move spontaneously, and (2) removal of support by undermining natural masses of soil, regolith, and bedrock. We can refer to mass movements produced by human activities as *induced mass wasting.*

In Los Angeles County, California, real estate development has been carried out on very steep hillsides and mountainsides by the process of bulldozing roads and homesites out of the deep regolith. The excavated regolith is pushed out into adjacent embankments where its instability poses a threat to slopes and stream channels below. When saturated by heavy winter rains, these embankments can give way, producing earthflows, mudflows, and debris floods that travel far

down the canyon floors and spread out on the alluvial fan surfaces below, burying streets and yards in bouldery mud.

Many debris floods of this area are also produced by heavy rains falling on mountain slopes denuded of vegetation by fire in the preceding dry season. Some of these fires are set carelessly or deliberately by humans.

Scarification of the Land

Industrial societies now possess enormous machine power and explosives capable of moving great masses of regolith and bedrock from one place to another. One use of this technology is the extraction of mineral resources. Another is the movement of earth in the construction of highway grades, airfields, building foundations, dams, canals, and various other large structures. Both activities involve removal of earth materials, a process that destroys the preexisting ecosystems and habitats of plants and animals. When the materials are used to build up new land on adjacent surfaces, ecosystems and habitats are also destroyed—by burial. What distinguishes artificial forms of mass wasting from the natural forms is that machinery is used to raise earth materials against the force of gravity. Explosives used in blasting can produce disruptive forces many times more powerful than the natural forces of physical weathering.

Scarification is a general term for excavations and other land disturbances produced to extract mineral resources. It includes the accumulation of rock waste as spoil or tailings. Among the forms of scarification are open-pit mines, strip mines, quarries for structural materials, borrow pits along highway grades, sand and gravel pits, clay pits, phosphate pits, scars from hydraulic mining, and stream gravel deposits reworked by dredging.

Figure 13.20 Contour strip mining near Lynch, Kentucky. A highway makes use of the winding bench at the base of the high rock wall.

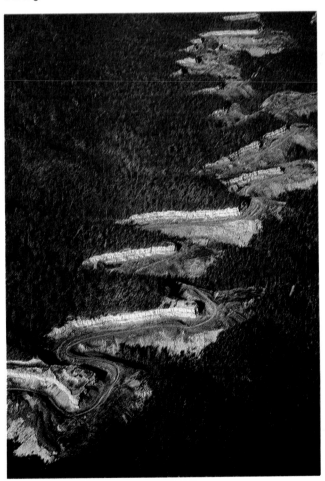

Strip mining is a particularly destructive scarification activity. Strip mining of coal is practiced where coal seams lie close to the surface or actually appear as out-croppings along hillsides (Figure 13.20). Here, earth-moving equipment removes the covering strata, or overburden, to bare the coal, which is lifted out by power shovels. The piled-up overburden is referred to as spoil. When saturated by heavy rains and melting snows, the spoil generates earthflows and mudflows that descend on houses, roads, and forest. The spoil also supplies sediment that clogs stream channels far down the valleys. As a result of these problems, strip mining is under strict control in most locations. Mine operators are not permitted to create hazardous spoil slopes, and they must restore spoil banks and ridges to a natural condition.

Scarification is on the increase. Driven by an ever increasing human population, demands for coal and industrial minerals used in manufacturing and construction are on the rise. At the same time, as the richer and more readily available mineral deposits are consumed, industry turns to poorer and less easily accessible grades. As a result, the rate of scarification is further increased.

In this chapter, we have examined two related topics—weathering and mass wasting. In the weathering process, rock near the surface is broken up into smaller fragments and often altered in chemical composition. In the mass wasting process, weathered rock and soil move downhill in slow to sudden mass movements. The landforms of mass wasting are produced by gravity acting directly on soil and regolith. Gravity also powers another landform-producing agent—running water—which we take up in the next three chapters. The first deals with water in the hydrologic cycle, in soil, and in streams. The second deals specifically with how streams and rivers erode regolith and deposit sediment to create landforms. The third describes how stream erosion strips away rock layers of different resistance, providing large landforms that reveal underlying rock structures.

WEATHERING AND MASS WASTING IN REVIEW

Weathering is the action of processes that cause rock to disintegrate and decompose because of surface exposure. Mass wasting is the spontaneous downhill motion of soil, rock, or regolith under gravity. Regolith, the surface layer of weathered rock particles, may be residual or transported.

Physical weathering produces regolith from solid rock by breaking bedrock into pieces. Frost action breaks rock apart by the repeated growth and melting of ice crystals in rock fractures and joints. Needle ice and ice lenses push rock fragments upward. Frost action also produces stone rings and ice-wedge polygons in tundra climates. In dry climates, salt-crystal growth breaks individual grains of rock free, and can damage brick and concrete. Unloading of the weight of overlying rock layers can cause some types of rock to expand and break loose into thick shells, producing exfoliation domes. Daily temperature cycles in arid environments are thought to cause rock breakup. Wedging by plant roots also forces rock masses apart.

Chemical weathering results from mineral alteration. Igneous and metamorphic rocks can decay to great depths through hydrolysis and oxidation, producing a regolith that is often rich in clay minerals. Carbonic acid reacts with and dissolves limestone. In warm, humid environments, basaltic lavas can also show features of solution weathering.

Soil creep is a process of mass wasting in which soil moves down slopes almost imperceptibly under the influence of gravity. In an earthflow, water-saturated soil or regolith slowly flows downhill. A mudflow is much swifter and follows stream courses, becoming thicker as it descends. A landslide is a rapid sliding of large masses of bedrock, sometimes triggered by an earthquake. Scarification of land by human activities, such as mining of coal or ores, can heap up soil and regolith into unstable masses that produce earthflows or mudflows. Mass wasting can also be caused by removal of support, undermining the natural support of soil or regolith. These actions are termed induced mass wasting.

KEY TERMS

weathering	physical weathering	earthflow
bedrock	chemical weathering	mudflow
sediment	mass wasting	landslide
alluvium	soil creep	scarification

REVIEW QUESTIONS

1. What is meant by the term weathering? What types of weathering are recognized?

2. Define the terms regolith, bedrock, sediment, and alluvium. Sketch a cross section through a part of the landscape showing these features and label them on the sketch.

3. How does frost action break up rock? Describe some landforms created by frost action and how they are formed.

4. How does salt-crystal growth break up rock? Give an example of a landform that arises from salt-crystal growth.

5. What is an exfoliation dome, and how does it arise? Provide an example.

6. Name three types of chemical weathering. Describe how limestone is often altered by a chemical weathering process.

7. Define mass wasting and identify the processes it includes.

8. What is soil creep, and how does it arise?

9. What is an earthflow? What features distinguish it as a landform? What special type of earthflow is encountered on the tundra?

10. Contrast earthflows and mudflows, providing an example of each.

11. Define the term landslide. How does a landslide differ from an earthflow?

12. Define and describe induced mass wasting. Provide some examples.

13. Explain the term scarification. Provide an example of an activity that produces scarification.

IN-DEPTH QUESTIONS

1. A landscape includes a range of lofty mountains elevated above a dry desert plain. Describe the processes of weathering and mass wasting that might be found on this landscape and identify their location.

2. Imagine yourself as the newly appointed director of public safety and disaster planning for your state. One of your first jobs is to identify locations where human populations are threatened by potential disasters, including those of mass wasting. Where would you look for mass wasting hazards and why? In preparing your answer, you may want to consult maps of your state.

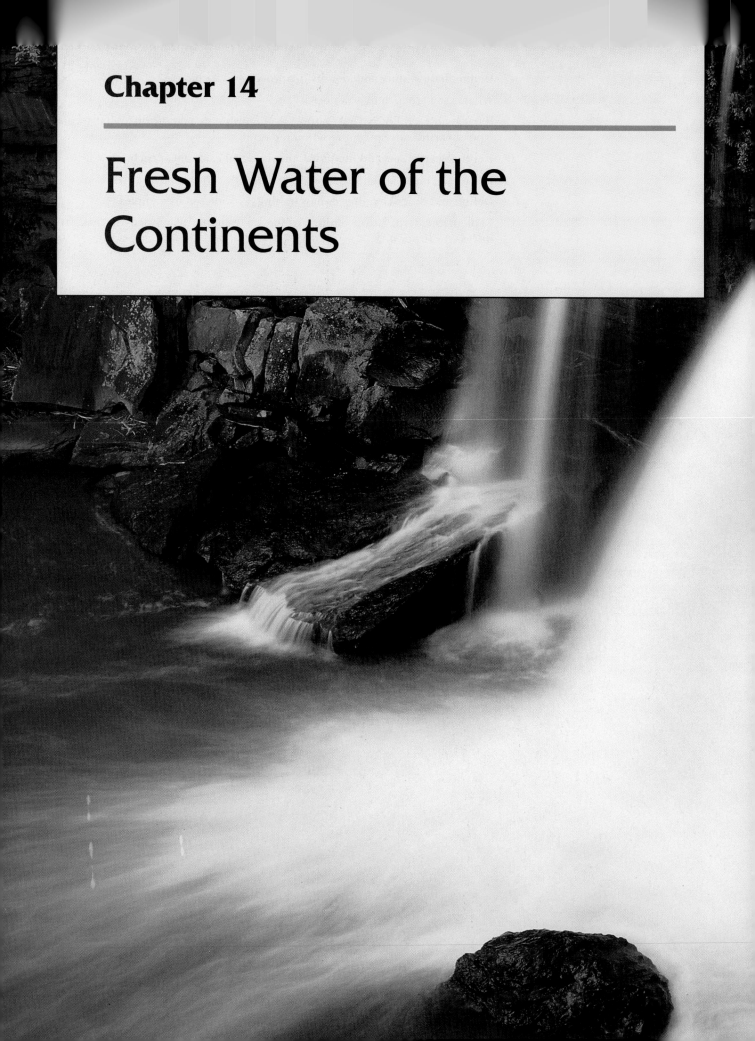

Chapter 14

Fresh Water of the Continents

East of the Caspian Sea, astride the former Soviet republics of Kazakhstan and Uzbekistan, lies an immense saline lake—the Aral Sea. Fed by meltwaters of high glaciers and snowfields in the lofty Hindu Kush, Pamir, and Tien Shan ranges, the lake endured through thousands of years as an oasis for terrestrial and aquatic wildlife deep in the heart of the central Asian desert.

But in the last 30 years, the Aral Sea, once larger than Lake Huron, has shrunk to a shadow of its former extent. The volume of its waters has decreased by 66 percent, and its salinity has increased from 1 percent to nearly 3 percent, making it almost as salty as sea water. Twenty of the 24 fish species native to the lake have disappeared. Its catch of commercial fish, which once supplied 10 percent of the total for the Soviet Union, has dwindled to zero. The deltas of the Amu Darya and Syr Darya rivers, which enter the south and east sides of the lake, were islands of great ecological diversity, teeming with fingerling fishes, birds, and their predators. Now only about half of the species of nesting birds remain. Many species of aquatic plants, shrubs, and grasses have vanished. Commercial hunting and trapping have almost ceased.

What caused this ecological catastrophe? The answer is simple—the lake's water supply was cut off. As an inland lake with no outlet, the Aral Sea receives water from the Amu Darya and Syr Darya, as well as a small amount from direct precipitation, but loses water by evaporation. Its gains balanced its losses, and, although these varied from year to year, the area, depth, and volume of the lake remained nearly constant until about 1960.

In the late 1950s the Soviet government embarked on the first phases of a vast irrigation program, using water from the Amu Darya and Syr Darya for cotton cropping on the region's desert plain. The diversion of water soon became significant as more and more land came under irrigation. As a result, the influx fell to nearly zero by the early 1980s. The surface level of the Aral was sharply lowered and its area reduced. The sea became divided into two separate parts.

As the lake's shoreline receded, the exposed lakebed became encrusted with salts. The flourishing fishing port of Muynak became a ghost town, 50 km (30 mi) from the new lake shoreline. Strong winds now blow salt particles and mineral dusts in great clouds southwestward over the irrigated cotton fields and westward over grazing pastures. These salts—particularly the sodium chloride and sodium sulfate components—are toxic to plants. The salt dust permanently poisons the soil and can only be flushed away with more irrigation water.

Although not inevitable, the fate of the Aral Sea appears grim. By the year 2000 the lake's salinity is predicted to reach 4 percent. Public outcry against the prevailing irrigation policy has increased sharply, but with the breakup of the Soviet Union, it has become an international problem. Help, if there is any, may come too late to save the Aral Sea.

The Aral Sea is a small, but significant, part of the world's supply of water on the continents. Although the Aral Sea is now saline, this supply is largely fresh water found in lakes and streams, as well as in soils and deep underground reservoirs. Much of this water is evaporated from the oceans by solar energy, condensing or depositing over the continents to fall as precipitation and provide the fresh water on which terrestrial life depends. In this chapter, we trace the fate of precipitation as it sinks into the ground to recharge the stock of underground water, or as it runs off the land to feed the network of streams and rivers that blankets the landscape.

Little River Falls, De Soto State Park, Alabama.

Before we continue with our survey of landforms and the processes that create them, we need to look more closely at the universal environmental agent that plays the dominant role in shaping the landscape—water. In this chapter we will focus first on two parts of the hydrologic cycle—the parts that involve water at the land surface and water that lies within the ground.

Recall from Chapter 4 that fresh water on the continents in surface and subsurface water constitutes only about 3 percent of the hydrosphere's total water. Most of this fresh water is locked into ice sheets and mountain glaciers. Ground water accounts for a little more than half of 1 percent, and fresh water in lakes, streams, and rivers constitutes only three-hundredths of 1 percent of the total water. Although very small proportionally, the fresh, liquid water at the land surface is a vital part of the environment of plants and animals dwelling on continents.

In Chapter 4, we discussed atmospheric moisture and precipitation, describing the part of the hydrologic cycle in which water evaporates from ocean and land surfaces, condenses, or deposits, and then precipitates as rain or snow. What happens then to this precipitation? Figure 14.1 provides the answer. As the diagram shows, a portion of the precipitation returns directly to the atmosphere through evaporation from the soil. Another portion travels downward, moving through the soil under the force of gravity to become part of the underlying ground water body. Following underground flow paths, this subsurface water eventually emerges to become surface water, or it may emerge directly in the shore zone of the ocean. A third portion flows over the ground surface as runoff to lower levels. As it travels, the water flow becomes collected into streams, which eventually conduct the running water to the ocean.

In this chapter, we will trace the parts of the hydrologic cycle that include both the subsurface and surface

Figure 14.1   The hydrologic cycle traces the various paths of water from the oceans, through the atmosphere, to land, and its return to the oceans.

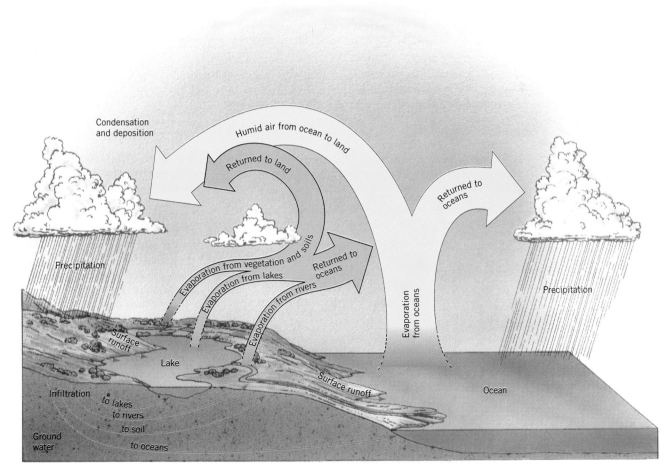

pathways of water flow. The study of these flows is part of the science of *hydrology*, which is the study of water as a complex but unified system on the earth.

Figure 14.2 shows what happens to water from precipitation as it first reaches the land surface. Most soil surfaces in their undisturbed, natural states are capable of absorbing the water from light or moderate rains by **infiltration.** In this process, water enters the small natural passageways between poorly fitting soil particles, as well as the larger openings in the soil surface. These openings are formed by the borings of worms and animals, earth cracks produced by soil drying, cavities left from decay of plant roots, or spaces made by the growth and melting of frost crystals. A mat of decaying leaves and stems breaks the force of falling water drops and helps to keep these openings clear.

The precipitation that infiltrates the soil is temporarily held in the soil layer as soil water, occupying the *soil water belt* (Figure 14.2). Water within this belt can be returned to the surface and then to the atmosphere through a process that combines two components—evaporation and transpiration. As we explained in Chapter 9, these two forms of water vapor transport are combined in the term *evapotranspiration.*

When rain falls too rapidly to be passed downward through soil openings, **runoff** occurs and a surface water layer runs over the surface and down the direction of ground slope. This surface runoff is called **overland flow.** In periods of heavy, prolonged rain or rapid snowmelt, streams are fed directly by overland flow.

Overland flow also occurs when rainfall or snowmelt provides water to a soil that is already saturated. Since soil openings and pores are already filled, water can infiltrate the soil only as quickly as it can drain down to deeper layers. Under these conditions, nearly all of the precipitation will run off.

Overland flow is a form of runoff in which water flows across the soil surface to reach stream channels.

Runoff as overland flow moves surface particles from hills to valleys, and so it is an agent that shapes landforms. Because runoff supplies water to streams and rivers, it also allows rivers to cut canyons and gorges, and carry sediment to the ocean—but we are getting ahead of our story. We'll return to landforms carved by running water in Chapter 15.

GROUND WATER

As shown in Figure 14.2, water derived from precipitation can continue to flow downward beyond the soil water belt. This slow downward flow under the influence of gravity is termed *percolation.* Eventually, the percolating water reaches ground water. **Ground water** is the part of the subsurface water that fully saturates the pore spaces in bedrock, regolith, or soil, and so occupies the *saturated zone* (Figure 14.3). The **water table** marks the top of this zone. Above it is the *unsaturated zone,* in which water does not fully saturate the pores. Here, water is held by capillary tension—as thin films of water adhering to mineral surfaces. This zone also includes the soil water belt.

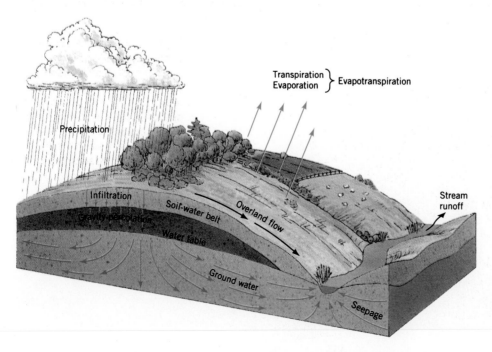

Figure 14.2 Paths of precipitation falling on the land. Some is returned to the atmosphere through evapotranspiration. Some runs off the soil surface, while the remainder sinks into the soil water belt, where it is accessible to plants. A portion of the infiltrating water passes through the soil water belt and percolates down to the ground water zone.

Figure 14.3 Zones of sub-surface water.

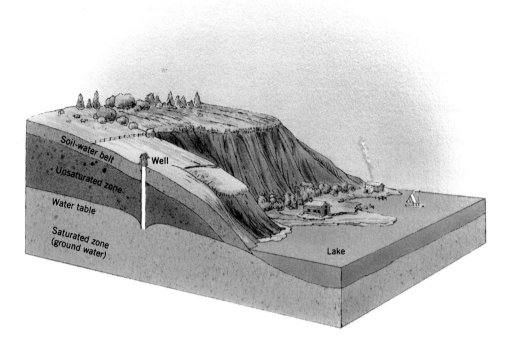

Ground water moves slowly in deep paths, eventually emerging by seepage into streams, ponds, lakes, and marshes. In these places the land surface dips below the water table. Streams that flow throughout the year—perennial streams—derive much of their water from ground water seepage.

The Water Table

Where there are many wells in an area, the position of the water table can be mapped in detail (Figure 14.4). This is done by plotting the water heights and noting the trend of change in elevation from one well to the other. The water table is highest under the highest areas of land surface—hilltops and divides. The water table declines in elevation toward the valleys, where it appears at the surface close to streams, lakes, or marshes.

The reason for this water table configuration is that water percolating down through the unsaturated zone tends to raise the water table, while seepage into lakes, streams, and marshes tends to draw off ground water

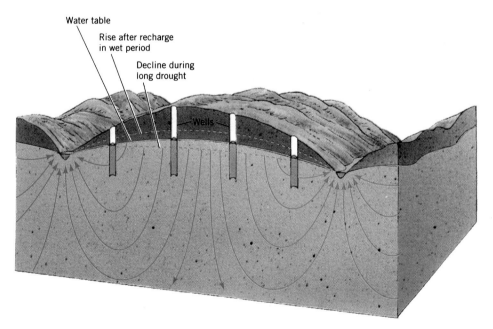

Figure 14.4 The configuration of the water table surface conforms broadly with the land surface above it. It varies in response to prolonged wet and dry periods. Ground water flow paths circulate water to deep levels in a very slow motion and eventually feed streams by seepage.

and to lower its level. Because ground water moves extremely slowly, a difference in water table level is built up and maintained between high and low points on the water table. In periods of high precipitation, the water table rises under divide areas. In periods of water deficit, or during a drought, the water table falls (Figure 14.4).

The water table is highest under hilltops and divides. It slopes to intersect the surface at streams, lakes, or marshes.

Figure 14.4 also shows paths of ground water flow. Water that enters the hillside midway between the hilltop and the stream flows rather directly toward the stream. Water reaching the water table midway between streams, however, flows almost straight down to great depths before recurving and rising upward again. Progress along these deep paths is incredibly slow, while flow near the surface is much faster. The most rapid flow is close to the stream, where the arrows converge. Over time, the level of the water table tends to remain stable, and the flow of water released to streams and lakes must balance the flow of water percolating down into the water table.

PROBLEMS OF GROUND WATER MANAGEMENT

Rapid withdrawal of ground water has seriously impacted the environment in many places. Increased urban populations and industrial developments require larger water supplies—needs that cannot always be met by constructing new surface-water reservoirs. To fill these needs, vast numbers of wells using powerful pumps draw great volumes of ground water to the surface, greatly altering nature's balance of ground water recharge and discharge.

In dry climates, agriculture is often heavily dependent on irrigation water from pumped wells—especially since major river systems are likely to be already fully utilized for irrigation. Wells are also convenient water sources. They can be drilled within the limits of a given agricultural or industrial property and can provide immediate supplies of water without any need to construct expensive canals or aqueducts.

In earlier times, the small well that supplied the domestic and livestock needs of a home or farmstead was actually dug by hand and sometimes lined with masonry. By contrast, a modern well supplying irrigation and industrial water is drilled by powerful machinery that can bore a hole 40 cm (16 in.) or more in diameter to depths of 300 m (1000 ft) or more. Drilled

wells are sealed off by metal casings that exclude impure near-surface water and prevent clogging of the tube by caving of the walls. Near the lower end of the hole, in the ground water zone, the casing is perforated to admit the water. The yields of single wells range from as low as a few hundred liters or gallons per day in a domestic well to many millions of liters or gallons per day for large industrial or irrigation wells.

Water Table Depletion

As water is pumped from a well, the level of water in the well drops. At the same time, the surrounding water table is lowered in the shape of a downward-pointing cone, termed the *cone of depression* (Figure 14.5). The difference in height between the cone tip and the original water table is the *drawdown*. The cone of depression may extend out as far as 16 km (10 mi) or more from a well where heavy pumping is continued. Where many wells are in operation, their intersecting cones will produce a general lowering of the water table.

Water table depletion often greatly exceeds recharge—the rate at which infiltrating water moves downward to the saturated zone. In arid regions, much of the ground water for irrigation is drawn from wells driven into thick sands and gravels. These deposits are often recharged by the seasonal flow of streams that

Figure 14.5 Drawdown and cone of depression in a pumped well. As the well draws water, the water table is depressed in a cone shape centered on the well.

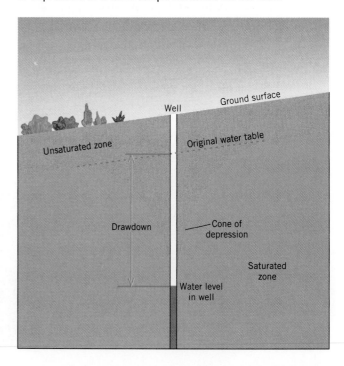

head high in adjacent mountains. Fanning out across the dry lowlands, the streams lose water, which sinks into the sands and gravels and eventually percolates to the water table below. The extraction of ground water by pumping can greatly exceed this recharge by stream flow, lowering the water table. Deeper wells and more powerful pumps are then required. The result is exhaustion of a natural resource that is not renewable except over long periods of time

Contamination of Ground Water

Another major environmental problem related to ground water withdrawal is contamination of wells by pollutants that infiltrate the ground and reach the water table. Both solid and liquid wastes are responsible. Disposal of solid wastes poses a major environmental problem in the United States because our advanced industrial economy provides an endless source of garbage and trash. Traditionally, these waste products were trucked to the town dump and burned there in continually smoldering fires that emitted foul smoke and gases. The partially consumed residual waste was then buried under earth.

In recent decades, a major effort has been made to improve solid-waste disposal methods. One method is high-temperature incineration, but it often leads to air pollution. Another is the sanitary landfill method in which waste is not allowed to burn. Instead, layers of waste are continually buried, usually by sand or clay available on the landfill site. The waste is thus situated in the unsaturated zone. Here it can react with rainwater that infiltrates the ground surface. This water picks up a wide variety of chemical compounds from the waste body and carries them down to the water table (Figure 14.6).

A common source of ground water contamination is the sanitary landfill.

Once in the water table, the pollutants follow the flow paths of the ground water. As the arrows in the

figure indicate, the polluted water may flow toward a supply well, which is drawing in ground water from a large radius. Once the polluted water has reached the well, the water becomes unfit for human consumption. Polluted water may also move toward a nearby valley, causing pollution of the stream flowing there (left side of Figure 14.6).

SURFACE WATER

So far, we have examined how water moves below the land surface. Now, we turn to tracing the flow paths of surplus water that runs off the land surface and ultimately reaches the sea. Here, we will be concerned primarily with rivers and streams. (In general usage, we speak of "rivers" as large watercourses and "streams" as smaller ones. However, the word "stream" is also used as a scientific term designating channeled flow of surface water of any amount.)

Overland Flow and Stream Flow

As we saw earlier in this chapter, runoff that flows down the slopes of the land in broadly distributed sheets is overland flow. We can distinguish overland flow from *stream flow*, in which the water occupies a narrow channel confined by lateral banks. Overland flow can take several forms. It may be a continuous thin film, called *sheet flow*, where the soil or rock surface is smooth (Figure 14.7). Where the ground is rough or pitted, flow may take the form of a series of tiny rivulets connecting one water-filled hollow with another. On a grass-covered slope, overland flow is subdivided into countless tiny threads of water, passing around the stems. Even in a heavy and prolonged rain, you might not notice overland flow in progress on a sloping lawn. On heavily forested slopes, overland flow may pass entirely concealed beneath a thick mat of decaying leaves.

Overland flow eventually contributes to a stream, which is a much deeper, more concentrated form of runoff. We can define a **stream** as a long, narrow body

Figure 14.6 Polluted water, leached from a waste disposal site, moves toward a supply well (*right*) and a stream (*left*). (Copyright © A. N. Strahler.)

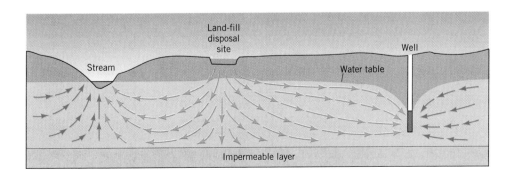

Figure 14.7 Overland flow taking the form of a thin sheet of water covers the nearly flat plain in the middle distance. This water is converging into stream flow in a narrow, steep-sided gully (left). The photograph was taken shortly after a summer thunderstorm had deluged the area. The locality, near Raton, New Mexico, shows steppe grassland vegetation.

of flowing water occupying a trenchlike depression, or channel, and moving to lower levels under the force of gravity. The **channel** of a stream is a narrow trough, shaped by the forces of flowing water to be most effective in moving the quantities of water and sediment supplied to the stream (Figure 14.8). Channels may be so narrow that a person can jump across them, or, in the case of the Mississippi River, as wide as 1.5 km (1 mi).

As a stream flows under the influence of gravity, the water encounters resistance—a form of friction—with the channel walls. As a result, water close to the bed and banks moves slowly, and that in the deepest and most centrally located zone flows fastest. If the channel is straight and symmetrical, the single line of maximum velocity is located in midstream. If the stream

Figure 14.8 Stream flow within a channel is most rapid near the center.

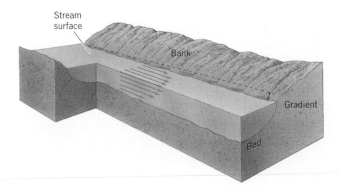

curves, the maximum velocity is found toward the bank on the outside of the curve.

Actually, the arrows in Figure 14.8 only show average velocity. In all but the most sluggish streams, the water is affected by *turbulence*, a system of countless eddies that are continually forming and dissolving. If we follow a particular water molecule it will travel a highly irregular, corkscrew path as it is swept downstream. Motions include upward, downward, and sideward directions. Only if we measure the water velocity at a certain fixed point for a long period of time, say, several minutes, will the average motion at that point be downstream and in a line parallel with the surface and bed.

Stream Discharge

Stream flow at a given location is measured by its **discharge,** which is defined as the volume of water per unit time passing through a cross section of the stream at that location. It is measured in cubic meters (cu ft) per second. The cross-sectional area and average velocity of a stream can change within a short distance, even though the stream discharge does not change (Figure 14.9). These changes occur because of changes in the *gradient* of the stream channel. The gradient is the rate of fall in elevation of the stream surface in the downstream direction (Figure 14.8).

When the gradient is steep, the force of gravity will act more strongly and flow velocity will be greater. When the stream channel has a gentle gradient, the velocity will be slower. However, in a short stretch of

Figure 14.9 In order to preserve a uniform discharge, a stream becomes wider and deeper in pools, where flow velocity is slower. In rapids, flow is swifter, and the stream becomes narrower and shallower.

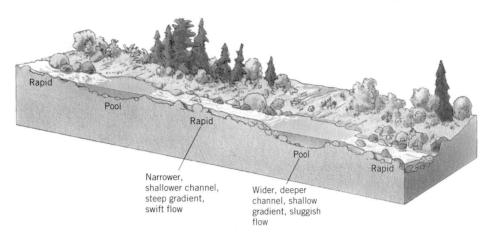

Narrower, shallower channel, steep gradient, swift flow

Wider, deeper channel, shallow gradient, sluggish flow

stream, the discharge will remain constant. As shown in Figure 14.9, this means that in stretches of rapids, where the stream flows swiftly, the stream channel will be shallow and narrow. In pools, where the stream flows more slowly, the stream channel will be wider and deeper to maintain the same discharge. Sequences of pools and rapids can be found along streams of all sizes.

The discharges of streams and rivers change from day to day, and records of daily and flood discharges of major streams and rivers are important information. They are used in planning the development and distri-

bution of surface waters, as well as in designing flood-protection structures and predicting floods as they progress down a river system. An important activity of the U.S. Geological Survey is the measurement, or gauging, of stream discharge in the United States. In cooperation with states and municipalities, this organization maintains over 6,000 gauging stations on principal streams and their tributaries.

Figure 14.10 is a map showing the relative discharge of major rivers of the United States. The mighty Mississippi with its tributaries dwarfs all other North

Figure 14.10 This schematic map shows the relative magnitude of U.S. rivers. Width of the color band is proportional to mean annual discharge. (After U.S. Geological Survey.)

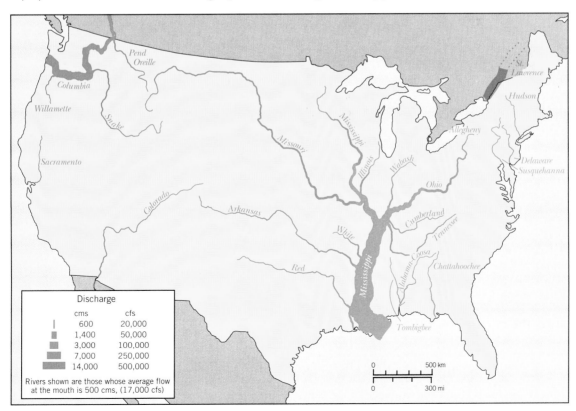

Eye on the Environment • Sinking Cities

Ground water is a resource on which human civilization now depends to meet much of its demand for fresh water. However, it may take thousands of years to accumulate large reservoirs of underground water. When this water is removed rapidly by pumping, the rate of removal far exceeds the rate of recharge. The volume of water in the reservoir is reduced, and the water table falls. What are the effects of this change?

One important environmental effect of excessive ground water withdrawal is subsidence of the ground surface. Venice, Italy, provides a dramatic example of this side effect. Venice was built in the eleventh century A.D. on low-lying islands in a coastal lagoon, sheltered from the ocean by a barrier beach. Underlying the area are some 1000 m (3000 ft) of layers of sand, gravel, clay, and silt, with some layers of peat. Compaction of these soft layers has been going on gradually for centuries under the

Land subsidence has subjected Venice to episodes of flooding by waters of the Adriatic Sea. Here, high water has flooded the Piazza San Marco and an outdoor café.

America rivers. The Great Lakes discharge through the St. Lawrence River is also of major proportions. The Colorado River, a much smaller stream, crosses a vast semiarid and arid region in which little tributary flow is added to its snowmelt source high in the Rocky Mountains.

The larger the cross section of a river, the lower will be its gradient.

As the figure shows, discharge of major rivers increases downstream as a natural consequence of the way that streams and rivers combine to deliver runoff and sediment to the oceans. The gradient also changes in a downstream direction. The general rule is the larger the cross-sectional area of the stream, the lower the gradient. Great rivers, such as the Mississippi and Amazon, have gradients so low that they can be described as "flat." For example, the water surface of the lower Mississippi River falls in elevation about 3 cm for each kilometer of downstream distance (1.9 in. per mi).

Rivers with headwaters in high mountains have characteristics that are especially desirable for utilizing river flow as irrigation and for preventing floods. The higher ranges serve as snow storage areas, storing the

heavy load of city buildings. However, ground water withdrawal, which has been greatly accelerated in recent decades, has aggravated the condition.

Many ancient buildings in Venice now rest at lower levels and have suffered severe damage as a result of flooding during winter storms on the adjacent Adriatic Sea. Sea level is normally raised by the effects of coastal storms, and when high tides occur at the same time, water rises even higher. The problem of flooding during storms is aggravated by the fact that the canals of Venice receive raw sewage, so that the floodwater is contaminated.

Most of the subsidence in recent decades has been attributed to withdrawals of large amounts of ground water from industrial wells at Porto Marghere, the modern port of Venice, located a few kilometers distant on the mainland shore. This pumping has now been greatly curtailed, reducing the rate of subsidence to a very small natural rate (about 1 mm per year). However, the threat of flooding and damage to churches and other buildings of great historical value remains. Flood control now depends on the construction of seawalls and floodgates on the bar-

rier beach that lies between Venice and the open ocean.

By the early 1980s the port city of Bangkok, located on the Chao Phraya River a short distance from the Gulf of Thailand, had become the world's most rapidly sinking city as a result of massive ground water withdrawals from soft marine sediments beneath the city. During the 1980s the rate of subsidence reached 14 cm (6 in.) per year, and major floods were frequent. Some reduction in the rate of ground water withdrawal has produced a modest decrease in the rate of subsidence, but as in the case of Venice, the flood danger remains.

Ground water withdrawal has affected several regions in California, where ground water for irrigation has been pumped from basins filled with alluvial (stream and lake-deposited) sediments. Water table levels in these basins dropped over 30 m (100ft), with a maximum drop of 120 to 150 m (400 to 500 ft) being recorded in one locality of California's San Joaquin Valley. In the Los Baños-Kettleman City area, ground subsidence of as much as 4.2 to 4.8 m (14 to 16 ft) was measured at some locations in a 35-year period.

Another important area of ground subsidence accompanying

water withdrawal is beneath Houston, Texas, where the ground surface has subsided from 0.3 to 1 m (1 to 3 ft) in a metropolitan area 50 km (30 mi) across. Damage has resulted to buildings, pavements, airport runways, and other structures.

Perhaps the most celebrated case of ground subsidence is that affecting Mexico City. Carefully measured ground subsidence has ranged from about 4 to 7 m (13 to 23 ft). The subsidence resulted from withdrawal of ground water from an aquifer system beneath the city and has caused many serious engineering problems. Clay beds overlying the aquifer have contracted greatly in volume as water has been drained out. To combat the ground subsidence, recharge wells were drilled to inject water into the aquifer. In addition, new water supplies from sources outside the city area were developed to replace local groundwater use.

Questions

1. Why has land subsidence occurred in Venice? What are the effects?
2. Identify four other locations that have suffered land subsidence caused by ground water withdrawal.

winter and spring precipitation until early or midsummer, when it is released slowly through melting. As melting proceeds to successively higher levels, the meltwater is supplied to the river. In this way, a continuous river flow is maintained. Among the snow-fed rivers of the western United States are the Columbia, Snake, Missouri, Arkansas, and Colorado.

Drainage Systems

As runoff moves to lower and lower levels and eventually to the sea, it becomes organized into a **drainage system.** The system consists of a branched network of

stream channels, as well as the sloping ground surfaces that contribute overland flow to those channels. Between the channels on the ridges are *drainage divides*, which mark the boundary between slopes that contribute water to different streams or drainage systems. The entire system is bounded by a drainage divide that outlines a more-or-less pear-shaped **drainage basin** (Figure 14.11).

A drainage basin contains a branched network of stream channels and adjacent slopes.

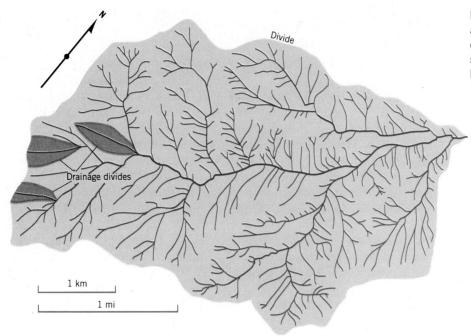

Figure 14.11 Channel network of a small drainage basin. Drainage divides mark the boundaries of stream basins. (Data of U.S. Geological Survey and Mark A. Melton.)

A typical stream network within a drainage basin is shown in Figure 14.11. Each fingertip tributary receives runoff from a small area of land surface surrounding the channel. The entire surface within the outer divide of the drainage basin constitutes the watershed for overland flow. The drainage system provides a converging mechanism that funnels overland flow and smaller streams into larger ones.

River Floods

You've probably seen enough media coverage of river floods to have a good idea of the appearance of floodwaters and the damage caused by flood erosion and deposition of silt and clay. Even so, it is not easy to define the term **flood.** Perhaps it is enough to say that a flood condition exists when the discharge of a river cannot be accommodated within its normal channel. As a result, the water spreads over the adjoining ground, which is normally cropland or forest. Sometimes, however, the ground is occupied by houses, factories, or transportation corridors.

Most rivers of humid climates have a *floodplain,* a broad belt of low flat ground bordering the channel on one or both sides that is flooded by stream waters about once a year. This flood usually occurs in the season when abundant surface runoff combines with the effects of a high water table to supply more runoff than can be carried in the channel. This annual inundation is considered a flood, even though its occurrence is expected and does not prevent the cultivation of crops after the flood has subsided. The seasonal inundation does not interfere with the growth of dense forests, which are widely distributed over low, marshy floodplains in all humid regions of the world. Still higher discharges of water, the rare and disastrous floods that may occur as rarely as once in 30 or 50 years, inundate ground lying well above the floodplain (Figure 14.12).

For practical purposes, the National Weather Service, which provides a flood-warning service, designates a particular river surface height, or *stage,* at a given place as the *flood stage.* Above this critical level, inundation of the floodplain will occur.

Flood Prediction

How often can floodwaters be expected, and how high will they be? The answer, of course, is different for each river and each location on it. Hydrologists have developed a graphical method to present the flood history of a river. Figure 14.13 shows a graph of the flood history of the Mississippi River at Vicksburg and explains the meaning of the bar symbols. The Mississippi responds largely to spring floods, producing a simple annual cycle.

Flood forecasts rely on precipitation patterns, past history, and present river levels to predict when and how high a flood will crest.

The National Weather Service operates a River and Flood Forecasting Service through 85 offices located at strategic points along major river systems of the United

Figure 14.12 The Mississippi River flood of 1993 inundated this riverfront area in eastern Iowa.

Figure 14.13 Maximum monthly stages of the Mississippi River at Vicksburg, Mississippi. The river shows an annual cycle of spring flooding, with lowest water occurring in the fall. (After National Weather Service.)

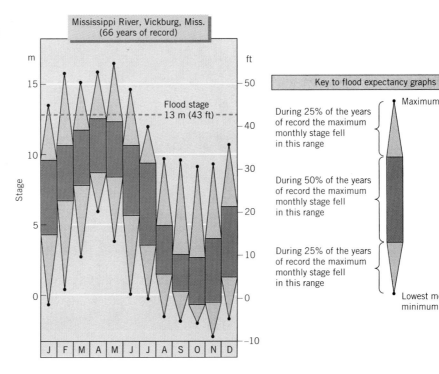

States. When a flood threatens, forecasters analyze precipitation patterns and the progress of high waters moving downstream. Examining the flood history of the rivers and streams concerned, they develop specific flood forecasts. These are delivered to communities within the associated district, which usually covers one or more large watersheds. Flood warnings are publicized by every possible means. Close cooperation is maintained with various agencies to plan evacuation of threatened areas and the removal or protection of damageable property.

Lakes

A **lake** is a water body that has an upper surface exposed to the atmosphere and no appreciable gradient. The term *lake* includes a wide range of water bodies. Ponds (which are small, usually shallow water bodies), marshes, and swamps with standing water can all be included under the definition of a lake. Lakes receive water input from streams, overland flow, and ground water, and so are included as parts of drainage systems. Many lakes lose water at an outlet, where water drains over a dam (natural or constructed) to become an outflowing stream. Lakes also lose water by evaporation. Lakes, like streams, are landscape features but are not usually considered to be landforms.

Lakes are quite important from the human viewpoint. They are frequently used as sources of fresh water, and they also support ecosystems that provide food for humans. Where dammed to a high level above the outlet stream, they can provide hydroelectric power as well. Lakes and ponds are also important as recreation sites and sources of natural beauty.

Where lakes are not naturally present in the valley bottoms of drainage systems, we create lakes as needed by placing dams across the stream channels. Many regions that formerly had almost no natural lakes are now abundantly supplied. Some are small ponds made to serve ranches and farms, while others cover hundreds of square kilometers. In some areas, the number of artificial lakes is large enough to have significant effects on the region's hydrologic cycle.

Basins occupied by lakes show a wide range of origins as well as a vast range in dimensions. Lake basins, like stream channels, are true landforms. Basins are created by a number of geologic processes. For example, the tectonic process of crustal faulting creates many large, deep lakes. Lava flows often form a dam in a river valley, causing water to back up as a lake. Landslides suddenly create lakes, as we saw in the case of the Madison Slide. (See *Eye on the Environment • The Great Hebgen Lake Disaster* in Chapter 13.)

If climate changes and becomes more arid, lakes can shrink in size or dry up completely.

An important point about lakes in general is that they are short-lived features on the geologic time scale. Lakes disappear by one of two processes, or a combination of both. First, lakes that have stream outlets will be gradually drained as the outlets are eroded to lower levels. Where a strong bedrock threshold underlies the outlet, erosion will be slow but nevertheless certain. Second, lakes accumulate inorganic sediment carried by streams entering the lake and organic matter produced by plants within the lake. Eventually, they fill up, forming a boggy wetland with little or no free water surface.

Lakes can also disappear when climate changes. If precipitation is reduced within a region, or temperatures and net radiation increase, evaporation can exceed input and the lake will dry up. Many former lakes of the southwestern United States flourished in moister periods of glacial advance during the Pleistocene Epoch. Today, they are greatly shrunken or have disappeared entirely under the present arid regime.

In moist climates, the water level of lakes and ponds coincides closely with the water table in the surrounding area. Seepage of ground water into the lake, as well as direct runoff of precipitation, maintains these free water surfaces permanently throughout the year. Examples of such freshwater ponds are found widely distributed in glaciated regions of North America and

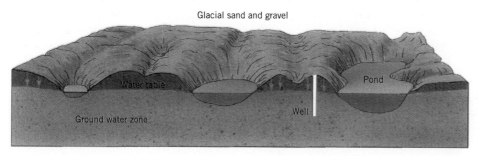

Figure 14.14 Freshwater ponds in sandy glacial deposits on Cape Cod, Massachusetts. (Redrawn from *A Geologist's View of Cape Cod*, Doubleday & Co., New York. Copyright © A. N. Strahler. Used by permission of Doubleday & Co.)

Figure 14.15   A freshwater pond in Wisconsin. Vegetation is slowly growing inward at the edges. Eventually, the pond will become a bog supporting wet forest.

Europe. Here, plains of glacial sand and gravel contain natural pits and hollows left by the melting of stagnant ice masses that were buried in the sand and gravel deposits (see Chapter 18). Figure 14.14 is a block diagram showing small freshwater ponds on Cape Cod. The surface elevation of these ponds coincides closely with the level of the surrounding water table.

Many former freshwater water table ponds have become partially or entirely filled by organic matter from the growth and decay of water-loving plants (Figure 14.15). The ultimate result is a bog with its surface close to the water table.

Freshwater marshes and swamps, in which water stands at or close to the ground surface over a broad area, also represent the appearance of the water table at the surface. Such areas of poor surface drainage are included under the name of *wetlands*.

Saline Lakes and Salt Flats

Lakes with no surface outlet are characteristic of arid regions. Here, the average rate of water loss by evaporation balances the average rate of stream inflow. If the rate of inflow increases, the lake level will rise. At the same time, the lake surface will increase in area, allowing a greater rate of evaporation. A new balance can then be achieved. Similarly, if the region becomes more arid, reducing input and increasing evaporation, the water level will fall to a lower level. The Aral Sea, which has shrunk in response to the diversion of large amounts of water from the rivers that feed it, is a prime example.

Lakes without outlets often show salt buildup. Dissolved solids are brought into the lake by streams—usually streams that head in distant highlands where a

Figure 14.16 These salt encrustations at the edge of Great Salt Lake, Utah, were formed when the lake level dropped during a dry period.

water surplus exists. Since evaporation removes only pure water, the salts remain behind and the salinity of the water slowly increases (Figure 14.16). Salinity, or degree of "saltiness," refers to the abundance of certain common ions in the water. Eventually, salinity levels reach a point where salts are precipitated as solids. (See the section "Evaporites" in Chapter 10.)

Sometimes the surfaces of such lakes lie below sea level. An example is the Dead Sea, with surface elevation of –396 m (–300 ft). The largest of all lakes, the Caspian Sea, has a surface elevation of –25 m (80 ft). Both of these large lakes are saline.

In regions where climatic conditions consistently favor evaporation over input, the lake may be absent. Instead, a shallow basin covered with salt deposits (see Figure 15.29), otherwise known as a *salt flat* or dry lake, will occur. In Chapter 15 we will describe dry lakebeds as landforms. On rare occasions, these flats are covered by a shallow layer of water, brought by flooding streams heading in adjacent highlands.

A number of desert salts are of economic value and have been profitably extracted from salt flats. In shallow coastal estuaries in the desert climate, sea salt for human consumption is commercially harvested by allowing it to evaporate in shallow basins. One well-known source of this salt is the Rann of Kutch, a coastal lowland in the tropical desert of westernmost India, close to Pakistan. Here the evaporation of shallow water of the Arabian Sea has long provided a major source of salt for inhabitants of the interior.

Desert Irrigation

Human interaction with the tropical desert environment is as old as civilization itself. Two of the earliest sites of civilization—Egypt and Mesopotamia—lie in the tropical deserts. Successful occupation of these deserts requires irrigation with large supplies of water from nondesert sources. For Egypt and Mesopotamia, the water sources of ancient times were the rivers that cross the desert but derive their flow from regions that have a water surplus. These are referred to as *exotic rivers*, since their flows are derived from an outside region.

Salinization and waterlogging are undesirable side effects of continued irrigation.

Irrigation systems in arid lands divert the discharge of an exotic river such as the Nile, Indus, Jordan, or Colorado into a distribution system that allows the water to infiltrate the soil of areas under crop cultivation. Ultimately, such irrigation projects can suffer from two undesirable side effects: salinization and waterlogging of the soil.

Salinization occurs when salts build up in the soil to levels that inhibit plant growth. This happens because an irrigated area within a desert loses large amounts of soil water through evapotranspiration. Salts contained in the irrigation water remain in the soil and increase to high concentrations. Prevention or cure of salinization may be possible by flushing the soil salts downward to lower levels by the use of more water. This remedy requires greater water use than for crop growth alone. In addition, new drainage systems must be installed to dispose of the excess saltwater.

Waterlogging occurs when irrigation with large volumes of water causes a rise in the water table, bringing the zone of saturation close to the surface. Most food crops cannot grow in perpetually saturated soils. When the water table rises to the point at which upward movement under capillary action can bring water to the surface, evaporation is increased and salinization is intensified.

Agricultural areas of major salinization include the Indus River Valley in Pakistan, the Euphrates Valley in Syria, the Nile Delta of Egypt, and the wheat belt of western Australia. In the United States, extensive regions of heavily salinized agriculture are found in the San Joaquin and Imperial valleys of California. Other areas of salinization occur throughout the entire semiarid and arid regions of the western United States. In Chapter 9, *Eye on the Environment • Death of a Civilization* documents the effects of salinization and waterlogging on the civilizations of ancient Mesopotamia.

Pollution of Surface Water

Streams, lakes, bogs, and marshes are specialized habitats of plants and animals. Their ecosystems are particularly sensitive to changes induced by human activity in the water balance and in water chemistry. Not only does our industrial society make radical physical changes in water flow by construction of engineering works (dams, irrigation systems, canals, dredged channels), but we also pollute and contaminate our surface waters with a large variety of wastes.

The sources of water pollutants are many and varied. Some industrial plants dispose of toxic metals and organic compounds by discharging them directly into streams and lakes. Many communities still discharge untreated or partly treated sewage wastes into surface waters. In urban and suburban areas, pollutant matter entering streams and lakes includes deicing salt and lawn conditioners (lime and fertilizers). These ions

can also contaminate ground water. In agricultural regions, important sources of pollutants are fertilizers and the body wastes of livestock. Mining and processing of mineral deposits are major sources of water pollution. In addition to chemical pollution, there is thermal pollution from discharge of heated water from electric power generating plants. The possibility of contamination by radioactive substances released from nuclear power and processing plants also exists.

Among the common chemical pollutants of both surface water and groundwater are sulfate, chloride, sodium, nitrate, phosphate, and calcium ions. (Ions were explained in Chapter 9.) Sulfate ions enter runoff both by fallout from polluted urban air and as sewage effluent. Chloride and sodium ions are contributed both by fallout from polluted air and by deicing salts used on highways. In some locations, community water supplies located close to highways have become polluted from deicing salts. Important sources of nitrate ions are fertilizers and sewage effluent. Excessive concentrations of nitrate in freshwater supplies are highly toxic, and, at the same time, their removal is difficult and expensive. Phosphate ions are contributed in part by fertilizers and by detergents in sewage effluent.

Water pollutants include various types of common ions and salts, as well as heavy metals, organic compounds, and acids.

Phosphate and nitrate are plant nutrients and can lead to excessive growth of algae and other aquatic plants in streams and lakes. Applied to lakes, this process is known as *eutrophication*, which is often described as the "aging" of a lake. In eutrophication, the accumulation of nutrients stimulates plant growth, producing a large supply of dead organic matter in the lake. Microorganisms break down this organic matter but require oxygen in the process. However, oxygen dissolves only slightly in water, and so it is normally present only in low concentrations. The added burden of oxygen use by the decomposers reduces the oxygen level to the point where other organisms, such as desirable types of fish, cannot survive. After a few years of nutrient pollution, the lake can take on the characteristics of a shallow pond that results when a lake is slowly filled with sediment and organic matter over thousands of years by natural "aging" processes.

A particular form of chemical pollution of surface water goes under the name of *acid mine drainage*. It is an important form of environmental degradation in parts of Appalachia where abandoned coal mines and strip-mine workings are concentrated (Figure 14.17). Ground water emerges from abandoned mines and as soil water percolating through strip-mine waste banks. This water contains sulfuric acid and various salts of metals, particularly of iron. Acid of this origin in stream waters can have adverse effects on animal life. In sufficient concentrations, it is lethal to certain species of fish and has at times caused massive fish kills.

Toxic metals, among them mercury, along with pesticides and a host of other industrial chemicals, are introduced into streams and lakes in quantities that are locally damaging or lethal to plant and animal communities. In addition, sewage introduces live bacteria and viruses that are classed as biological pollutants. These pose a threat to the health of humans and animals.

Thermal pollution is a term applied generally to the discharge of heat into the environment from combustion of fuels and from the conversion of nuclear energy into electric power. We have described thermal pollution of the atmosphere and its effects in Chapter 4. Thermal pollution of water is different in its environmental effects because it takes the form of heavy discharges of heated water locally into streams, estuaries, and lakes. The thermal environmental impact may thus be quite drastic in a small area.

SURFACE WATER AS A NATURAL RESOURCE

Fresh surface water is a basic natural resource essential to human agricultural and industrial activities. Runoff held in reservoirs behind dams provides water supplies for great urban centers, such as New York City and Los Angeles. When diverted from large rivers, it provides irrigation water for highly productive lowlands in arid lands, such as the Imperial Valley of California and the Nile Valley of Egypt. To these uses of runoff are added hydroelectric power, where the gradient of a river is steep, or routes of inland navigation, where the gradient is gentle.

Our heavily industrialized society requires enormous supplies of fresh water for its sustained operation. Urban dwellers consume water in their homes at rates of 150 to 400 liters (50 to 100 gallons) per person per day (Figure 14.18). Large quantities of water are used for cooling purposes in air conditioning units and power plants. Much of this water is obtained from surface-water supplies, and the demand increases daily.

Unlike ground water, which represents a large water storage body, fresh surface water in the liquid state is stored only in small quantities. (An exception is the Great Lakes system.) Recall from Chapter 4 that the quantity of available ground water is about 20 times as

Figure 14.17 This small stream is badly polluted by acid waters that have percolated through strip mine waste.

large as that stored in freshwater lakes and that the water held in streams is only about one one-hundredth of that in lakes. Because of small natural storage capacities, surface water can be drawn only at a rate comparable with its annual renewal through precipitation. Dams are built to develop useful storage capacity for runoff that would otherwise escape to the sea, but, once the reservoir has been filled, water use must be scaled to match the natural supply rate averaged over the year. Development of surface-water supplies brings on many environmental changes, both physical and biological, and these must be taken into account in planning for future water developments.

This chapter has focused on water, including the precipitation that runs off the land and flows into the sea or accumulates in inland basins. Think for a moment about the drainage system of a stream that conducts this flow of runoff. The smallest streams catch runoff from slopes, carrying the runoff into larger streams with which they join. The larger streams, in turn, receive runoff from their side slopes and also pass their flow on to still larger streams. However, this cannot happen unless the gradients of slopes and streams are adjusted so that water keeps flowing downhill. This means that the landscape is shaped and organized into landforms that are an essential part of the drainage system. The shaping of landforms within the drainage system occurs as running water erodes the landscape, which is the subject of the next chapter.

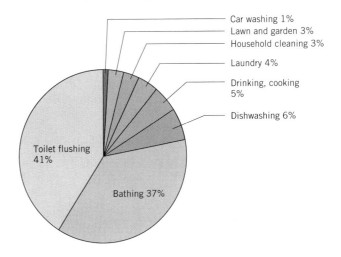

Figure 14.18 How water is used in an average home in Akron, Ohio. (U.S. Geological Survey.)

Car washing 1%
Lawn and garden 3%
Household cleaning 3%
Laundry 4%
Drinking, cooking 5%
Dishwashing 6%
Toilet flushing 41%
Bathing 37%

FRESH WATER OF THE CONTINENTS IN REVIEW

The fresh water of the lands accounts for only a small fraction of the earth's water. Since it is produced by precipitation over land, it depends on the continued operation of the hydrologic cycle for its existence. The soil layer plays a key role in determining the fate of precipitation by diverting it in three ways: to the atmosphere as evapotranspiration, to ground water through percolation, and to streams and rivers as runoff.

Ground water occupies the pore spaces in rock and regolith. The water table marks the upper surface of the saturated zone of ground water, where pores are completely full of water. Ground water moves in slow paths deep underground, recharging rivers, streams, ponds, and lakes by upward seepage and thus contributing to runoff. Wells draw down the water table and, in some regions, lower the water table more quickly than it can be recharged. Ground water contamination can occur when precipitation percolates through contaminated soils or waste materials. Landfills and dumps are common sources of ground water contaminants.

Runoff includes overland flow, moving as a sheet across the land surface, and flow in streams and rivers, which is confined to a channel. Rivers and streams are organized into a drainage network that moves runoff from slopes into channels, and from smaller channels into larger ones. The discharge of a stream measures the flow rate of water moving past a given location. Discharge increases downstream as tributary streams add additional runoff.

Floods occur when river discharge increases, and the flow can no longer be contained within the river's usual channel. Water spreads over the floodplain, inundating low fields and forests adjacent to the channel. When discharge is high, floodwaters can rise beyond normal levels to inundate nearby areas of development, causing damage and sometimes taking lives.

Lakes are especially important parts of the drainage system because they are sources of fresh water. They are also used for recreation and, in many cases, can supply hydroelectric power. Where lakes occur in inland basins, they are often saline. When climate changes, such lakes can dry up, creating salt flats.

Irrigation is the diversion of fresh water from streams and rivers to supply the water needs of crops. In desert regions, where irrigation is most needed, problems of salinization and waterlogging can occur, reducing productivity and eventually creating unusable land.

Water pollution arises from many sources, including industrial sites, sewage treatment plants, agricultural activities, mining, and processing of mineral deposits. Sulfate, nitrate, phosphate, chloride, sodium, and calcium ions are frequent contaminants. Toxic metals, pesticides, and industrial chemicals are also hazards.

Ground water and surface water resources are essential for human activities. Human civilization is dependent on abundant supplies of fresh water for many uses. But because the fresh water of the continents is such a small part of the global water pool, utilization of water resources takes careful planning and management.

KEY TERMS

infiltration	water table	drainage system
runoff	stream	drainage basin
overland flow	channel	flood
ground water	discharge	lake

REVIEW QUESTIONS

1. What happens to precipitation falling on soil? What processes are involved?

2. How and under what conditions does precipitation reach ground water?

3. How and where does ground water flow? Sketch a diagram showing the flow paths of ground water.

4. How do wells affect the water table? What happens when pumping exceeds recharge?

5. How is ground water contaminated? Describe how a well might become contaminated by a nearby landfill dump.

6. Define discharge (of a stream) and the two quantities that determine it. How does discharge vary in a downstream direction? How does gradient vary in a downstream direction?

7. What is a drainage system? How are slopes and streams arranged in a drainage basin?

8. Define the term flood. What is the floodplain? What factors are used in forecasting floods?

9. How are lakes defined? What are some of their characteristics? What factors influence the size of lakes?

10. Describe some of the problems that can arise in long-continued irrigation of desert areas.

11. Identify common surface-water pollutants and their sources.

12. How is surface water utilized as a natural resource?

IN-DEPTH QUESTIONS

1. A thundershower causes heavy rain to fall in a small region near the headwaters of a major river system. Describe the flow paths of that water as it returns to the atmosphere and ocean. What human activities influence the flows? In what ways?

2. Imagine yourself a recently elected mayor of a small city located on the banks of a large river. What issues might you be concerned with that involve the river? In developing your answer, choose and specify some characteristics for this city—such as its population, its industries, its sewage systems, and the present uses of the river for water supply or recreation.

Chapter 15

Landforms Made by Running Water

The rickety school bus turns off the pavement and bumps down the dirt road toward the riverbank. It pitches violently from side to side as it rolls through the potholes, and you hold on to the top of the seat in front with a tight grip. Finally, the bus stops with a lurch on a wide, gravel-covered spot next to the river, and you descend, glad to leave the cramped quarters. Behind you is the ancient truck carrying the rafts, paddles, and gear that you and your group will need for your 12-day float trip through the Grand Canyon down the Colorado River. The sun is high in the sky, and unloading the rafts and gear is hot work. Soon, however, you are afloat, seven to a raft, slowly drifting with the current down the wide river.

"Left side, forward!" "Right side, back!" The river guide barks the orders as you practice paddling together as a team. Sitting on the outer tube of the raft, you twist sideways, dipping your canoe paddle deeply into the blue-green water and pushing hard. This is real work! The river guide explains that safe passage through the hundred or more rapids you'll encounter on the trip will depend on how well you and your raftmates can move your craft across the swift currents of the mighty river.

Then, almost too quickly, you are at Paria Riffle, where the Paria Canyon empties a muddy stream into the Colorado. The raft glides easily through the rough water, and you think that perhaps this won't be so difficult after all. Then your river guide tells you that before the day is done, you'll be shooting the Badger Creek and Soap Creek Rapids, which have claimed several lives over the years. Back to paddle exercises, this time in earnest!

Through the past week, the river has become your life—not just your highway into the scenic beauty of the Grand Canyon, but the source of all peace and harmony, as well as

excitement and adventure. The thrill of wild white-water rides has alternated with the calm of languid drifting in the bright summer sun. Although you've taken foot trips to explore the side canyons and their waterfalls, as well as historic sites commemorating the Colorado's rich history of river riders, you've always returned to the river to continue your journey, and you feel a close bond with it. But this comfortable feeling about the river is due to change abruptly.

For two or three days, there's been talk among the river guides about Lava Falls—usually in hushed tones of reverence. For the last 40 km (25 mi), the river has been relatively calm, as if gathering its energy for the 11 m (37 ft) drop, the highest in the canyon and the last major rapids on the trip. Perched on the edge of your raft, you listen intently as the river guide goes over the route through the rapids for the last time. A booming noise begins to strengthen, telling you that the rapids are approaching. The smooth, but swift, tongue of water that leads into the rapids accelerates the raft. You and your crew paddle furiously across the current, aiming for the side of a huge drop that lies ahead. Suddenly, you're falling over the edge and down into the froth below. Gliding and twisting, the raft splashes into the white water and is thrown violently to the side. Flying spray hits you, hard. The shock of the cold water on your face and arms makes you gasp for breath. The bow rises as the raft hits the huge standing wave on the downstream side of the hole.

"Pull! Pull!" The guide's urgent shout rises above the roar. Your paddle splits the water, backed by every ounce of strength you can muster. Perched on top of the standing wave, the raft seems for a sickening moment about to slip back into the pitching waters of the hole behind. Then, suddenly, the raft's bow drops and you're yanked forward as the raft falls into the trough beyond. Then it's a wild roller coaster ride through the haystacks of waves below.

"Left side forward, right side back!" The raft rebounds to the right off a huge pillow of foam that is lifted by a boulder just below the water surface. The raft bounces into a side eddy, a patch of calmer water well below the roaring fall. You catch your breath, your heart pounding with the excitement, surrounded by white water on every side. This is a trip you'll never forget! A few moments later, you are pulling on your paddle once again as the raft plunges back into the current on the final leg of its torturous trip through the twisting torrent of Lava Falls.

That night around the campfire, you relive the experience with your raftmates. You've all felt the mighty power of the Colorado, and your lives are forever changed.

The power of rivers is awesome, especially when you experience it first-hand. Rivers are agents that shape the landscape by carving deep canyons and by winding across broad valleys. They carry sediment to the ocean, both as fine particles in suspension and as larger cobbles and boulders rolled and bumped along the river's bed. Rivers act together with small streams and runoff to erode the landscape as a fluvial system, creating distinctive landforms. The processes of erosion, transportation, and deposition by fluvial systems, and the landforms they create, are the subjects of this chapter.

An entrenched meander on the Colorado River. Horseshoe Overlook, Page, Arizona.

R unning water is a powerful land-forming agent. Flowing as a sheet across a land surface, running water picks up particles and moves them downslope into a stream channel. When rainfall is heavy, streams and rivers swell, lifting large volumes of sediment and carrying them downstream. In this way, running water erodes mountains and hills, carves valleys, and deposits sediment. This chapter describes the work of running water and the landforms that are shaped by running water.

The four agents of denudation are running water, waves, glacial ice, and wind.

Running water is one of four flowing substances that erode, transport, and deposit mineral and organic matter. The other three are waves, glacial ice, and wind. These four fluid agents carry out the processes of *denudation,* which we discussed in Chapter 12. Recall that denudation is the total action of all processes by which the exposed rocks of the continents are worn away and the resulting sediments are transported to the sea or closed inland basins.

Denudation is an overall lowering of the land surface. Denudation processes, if left unchecked to operate over geologic time, will reduce a continent to a nearly featureless, sea-level surface. However, that fate has always been averted, since plate tectonic activity has kept continental crust elevated well above the ocean basins. The result is that running water, waves, glacial ice, and wind have always had plenty of raw material available to create the many landforms that we see around us.

Thanks to denudation processes and plate tectonics, the land environments of life have always been in constant change, even as plants and animals have undergone their evolutionary development. Wind, water, waves, and ice have produced, maintained, and changed a wide variety of landforms, and these have been the habitats for evolving life-forms. In turn, the life-forms have become adapted to those habitats and have diversified to a degree that matches the diversity of the landforms themselves.

FLUVIAL PROCESSES AND LANDFORMS

Landforms shaped by running water are conveniently described as **fluvial landforms** in order to distinguish them from landforms made by the other flowing agents—glacial ice, wind, and waves. Fluvial landforms are shaped by the **fluvial processes** of overland flow and stream flow, which we described in Chapter 14. Weathering and the slower forms of mass wasting (Chapter 13), such as soil creep, operate hand in hand with overland flow, providing the rock and mineral fragments that are carried into stream systems.

Fluvial landforms and fluvial processes dominate the continental land surfaces the world over. Throughout geologic history, glacial ice has been present only in continental areas located in mid- and high latitudes and in high mountains. Landforms made by wind action are found only in very small parts of the continental surfaces. And landforms made by waves and currents are restricted to the narrow contact zone between oceans and continents. That is why, in terms of area, the fluvial landforms dominate the environment of terrestrial life.

Areas of fluvial landforms are also the major source areas of human food resources through the practice of agriculture. Except for areas in the northern hemisphere that were formerly occupied by glacial ice, most land areas used in crop cultivation or for grazing have been shaped by fluvial processes.

Erosional and Depositional Landforms

All agents of denudation perform the geological activities of erosion, transportation, and deposition. Consequently, there are two major groups of landforms—erosional landforms and depositional landforms. When an initial landform, such as an uplifted crustal block, is created, it is attacked by the processes of denudation and especially by fluvial action. Valleys are formed where rock is eroded away by fluvial agents. Between the valleys are ridges, hills, or mountain summits, representing the remaining parts of the crustal block that is as yet uncarved by running water. These sequential landforms, shaped by progressive removal of the bedrock mass, are *erosional landforms.*

Erosional landforms are shaped by progressive removal of rock material. Depositional landforms are built of material transported and deposited by fluid agents.

Fragments of soil, regolith, and bedrock that are removed from the parent rock mass are transported by the fluid agent and deposited elsewhere to make an entirely different set of surface features—the *depositional landforms.* Figure 15.1 illustrates the two groups of landforms as produced by fluvial processes. The ravine, canyon, peak, spur, and col are erosional landforms. The fan, built of rock fragments below the

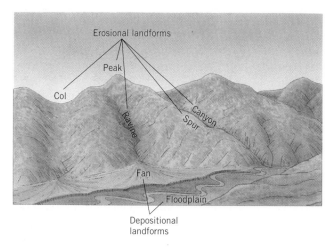

Figure 15.1 Erosional and depositional landforms. (A. N. Strahler.)

mouth of the ravine, is a depositional landform. The floodplain, built of material transported by a stream, is also a depositional landform.

SLOPE EROSION

Fluvial action starts on the uplands as *soil erosion*. By exerting a dragging force over the soil surface, overland flow picks up particles of mineral matter ranging in size from fine colloidal clay to coarse sand or even gravel. The size grade selected depends on the speed of the current and the degree to which the particles are bound by plant rootlets or held down by a mat of leaves. Added to this solid matter is dissolved mineral matter in the form of ions produced by acid reactions or direct solution.

This ongoing removal of soil is part of the natural geological process of denudation. It occurs everywhere that precipitation falls on land. Under stable natural conditions in a humid climate, the erosion rate is slow enough that a soil with distinct horizons is formed and maintained. Each year a small amount of soil is washed away, while a small amount of solid rock material becomes altered to new regolith and soil. These conditions also enable plant communities to maintain themselves in a stable equilibrium. Soil scientists refer to this state of activity as the *geologic norm*.

Accelerated Erosion

In contrast, the rate of soil erosion may be enormously speeded up by human activities or by rare natural erosional events to produce a state of *accelerated erosion*. What happens then is that the soil is removed much faster than it can be formed, and the uppermost soil horizons are progressively exposed. Accelerated ero-

sion arises most commonly when the plant cover and the physical state of the ground surface change. Destruction of vegetation by the clearing of land for cultivation or by forest fires sets the stage for a series of drastic changes. No foliage remains to intercept rain, and protection of a ground cover of fallen leaves and stems is removed. Consequently, the raindrops fall directly on the mineral soil.

The direct force of falling drops on bare soil causes a geyserlike splashing in which soil particles are lifted and then dropped into new positions. This process is termed *splash erosion* (Figure 15.2). Soil scientists estimate that a torrential rainstorm has the ability to disturb as much as 225 metric tons of soil per hectare (100 U.S. tons per acre). On a sloping ground surface, splash erosion shifts the soil slowly downhill. An even more important effect is to cause the soil surface to become much less able to absorb water. This occurs because the natural soil openings become sealed by

Figure 15.2 A large raindrop (above) lands on a wet soil surface, producing a miniature crater (below). Grains of clay and silt are thrown into the air and the soil surface is disturbed.

particles shifted by raindrop splash. Reduced infiltration, in turn, permits a much greater depth of overland flow to occur from a given amount of rain. So, the rate of soil erosion is intensified.

Another effect of the destruction of vegetation is to reduce greatly the resistance of the ground surface to the force of erosion under overland flow. On a slope covered by grass sod, even a deep layer of overland flow causes little soil erosion. This protective action is present because the energy of the moving water is dissipated in friction with the grass stems, which are tough and elastic. On a heavily forested slope, the surface layer of leaves, twigs, roots, and even fallen tree trunks take up the force of overland flow. Without such a cover, the eroding force is applied directly to the bare soil surface, easily dislodging the grains and sweeping them downslope.

Sheet Erosion and Rilling

Accelerated soil erosion is a constant problem in cultivated regions with a substantial water surplus. When the natural cover of forest or prairie grasslands is first removed and the soil is plowed for cultivation, little erosion will occur until the action of rain splash has broken down the soil aggregates and sealed the

larger openings. Then, however, overland flow begins to remove the soil in rather uniform thin layers, a process termed *sheet erosion*. Because of seasonal cultivation, the effects of sheet erosion are often little noticed until the upper horizons of the soil are removed or greatly thinned.

Where land slopes are steep, runoff from torrential rains produces a more destructive activity, *rill erosion*, in which many closely spaced channels are scored into the soil and regolith. If the rills are not destroyed by soil tillage, they may soon begin to join together into still larger channels. These deepen rapidly and soon become *gullies*—steep-walled, canyonlike trenches whose upper ends grow progressively upslope (Figure 15.3). Ultimately, a rugged, barren topography results from accelerated soil erosion that is allowed to proceed unchecked.

Colluvium and Alluvium

With an increased depth of overland flow and no vegetation cover to absorb the eroding force, soil particles are easily picked up and moved downslope. Eventually, they reach the base of the slope, where the surface slope becomes more gentle and meets the valley bottom. There the particles come to rest and accumulate

Figure 15.3 Deep branching gullies have carved up an overgrazed pasture near Shawnee, Oklahoma. Contour terracing and check dams have halted the headward growth of the gullies.

in a thickening layer termed *colluvium*. Because this deposit is built by overland flow, it has a sheetlike distribution and may be little noticed, except where it eventually buries fence posts or tree trunks.

Deposits of sediment that are laid down by streams are referred to as alluvium.

If not deposited as colluvium, sediment carried by overland flow eventually reaches a stream in the adjacent valley floor. Once in the stream, it is carried farther downvalley and may accumulate as alluvium in layers on the valley floor. The term **alluvium** is used to describe any stream-laid sediment deposit. Deposition of alluvium can bury fertile floodplain soil under infertile, sandy layers. Coarse alluvium chokes the channels of small streams and can cause the water to flood broadly over the valley bottoms.

Slope Erosion in Steppelands and Savannas

Thus far, we've discussed slope erosion in moist climates with a natural vegetation of forest or a dense prairie grassland. Conditions are quite different in a midlatitude semiarid climate with summer drought. Here, the natural plant cover consists of short-grass prairie (steppe). Although it is sparse and provides a rather poor ground cover of plant litter, the grass cover is normally strong enough that a slow pace of erosion can be sustained.

Much the same conditions are also found in the tropical savanna grasslands. In these semiarid environments, however, the natural equilibrium is highly sensi-

tive and easily upset. Depletion of the plant cover by fires or the grazing of herds of domesticated animals can easily set off rapid erosion. These sensitive, marginal environments require cautious use, because they lack the potential to recover rapidly from accelerated erosion once it has begun.

Badlands

Erosion at a very high rate by overland flow is actually a natural process in certain favorable locations in semiarid and arid lands. Here, the erosion produces *badlands*. Badlands are underlain by clay formations, which are easily eroded by overland flow. Erosion rates are too fast to permit plants to take hold, and no soil can develop. A maze of small stream channels is developed, and ground slopes are very steep (Figure 15.4).

One well-known area of badlands in the semiarid short-grass prairie is the Big Badlands of South Dakota, along the White River. Badlands such as these are self-sustaining and have been in existence on continents throughout much of geologic time. Badlands can also result from poor agricultural processes, especially when the vegetation cover of clay formations is disturbed by plowing or overgrazing.

THE WORK OF STREAMS

The work of streams consists of three closely related activities—erosion, transportation, and deposition. **Stream erosion** is the progressive removal of mineral material from the floor and sides of the channel, whether bedrock or regolith. **Stream transportation**

Figure 15.4 Badlands at Zabriskie Point, Death Valley National Monument, California.

Figure 15.5 These potholes in lava bedrock attest to the abrasion that takes place on the bed of a swift mountain stream. McCloud River, California.

consists of movement of the eroded particles dragged over the stream bed, suspended in the body of the stream, or held in solution as ions. **Stream deposition** is the accumulation of transported particles on the stream bed and floodplain, or on the floor of a standing body of water into which the stream empties. Erosion cannot occur without some transportation taking place, and the transported particles must eventually come to rest. Thus, erosion, transportation, and deposition are simply three phases of a single activity.

Stream Erosion

Streams erode in various ways, depending on the nature of the channel materials and the tools with which the current is armed. The force of the flowing water not only sets up a dragging action on the bed and banks, but also causes particles to impact the bed and banks. Dragging and impact can easily erode alluvial materials, such as gravel, sand, silt, and clay. This form of erosion, called *hydraulic action,* can excavate enormous quantities in a short time. The undermining of the banks causes large masses of alluvium to slump into the river, where the particles are quickly separated and become part of the stream's load. This process of bank caving is an important source of sediment during high river stages and floods.

Where rock particles carried by the swift current strike against bedrock channel walls, chips of rock are detached. The large, strong fragments become rounded as they travel. The rolling of cobbles and boulders over the stream bed further crushes and grinds the smaller grains to produce a wide assortment of grain sizes. This process of mechanical wear is called *abrasion.* It is the

principal means of erosion in bedrock too strong to be affected by simple hydraulic action. A striking example of abrasion is the erosion of a *pothole.* This occurs when a shallow depression in the bedrock of a stream bed acquires one or several grinding stones, which are spun around and around by the flowing water and carve a deep depression (Figure 15.5).

Finally, the chemical processes of rock weathering—acid reactions and solution—are effective in removing rock from the stream channel. The process is called *corrosion.* Effects of corrosion are conspicuous in limestone, which develops cupped and fluted surfaces.

Stream Transportation

The solid matter carried by a stream is the **stream load.** It is carried in three forms (Figure 15.6). Dissolved matter is transported invisibly in the form of chemical ions. All streams carry some dissolved ions resulting from mineral alteration. Sand, gravel, and larger particles move as *bed load* close to the channel floor by rolling or sliding. Clay and silt are carried in *suspension*—that is, they are held up in the water by the upward elements of flow in turbulent eddies in the stream. This fraction of the transported matter is the *suspended load* (Figure 15.7). Of the three forms, suspended load is generally the largest.

A large river such as the Mississippi carries as much as 90 percent of its load in suspension. Most of the suspended load comes from its great western tributary, the Missouri River, which is fed from semiarid lands, including the Dakota Badlands. (Figure 14.10 shows the Mississippi River system and its major branches.)

> **Rivers carry most of their sediment load in suspension. Land cultivation can greatly increase that load by increasing soil erosion.**

The Yellow (Huang) River of China heads the world list in annual suspended sediment load, and its watershed sediment yield is one of the highest known for a large river basin. This is because much of its basin consists of cultivated upland surfaces of wind-deposited silt that is very easily eroded. (See Chapter 17 and Figure 17.35.) In addition, much of the river's upper watershed is in a semiarid climate with dry winters. Vegetation is sparse, and the runoff from heavy summer rains sweeps up a large amount of sediment.

Capacity of a Stream to Transport Load

The maximum solid load of debris that can be carried by a stream at a given discharge is a measure of the

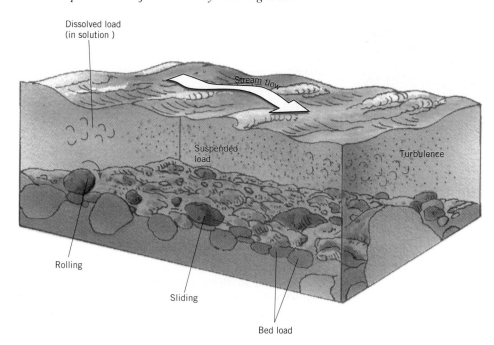

Dissolved load
(in solution)

Stream flow

Suspended load

Turbulence

Rolling

Sliding

Bed load

Figure 15.6 Streams carry their load as dissolved, suspended, and bed load. Suspended load is kept in suspension by turbulence. Bed load moves by sliding or rolling.

stream capacity. This load is usually measured in units of metric tons per day passing downstream at a given location. Total solid load includes both the bed load and the suspended load.

Figure 15.7 This turbulent stream in Madagascar carries a heavy load of suspended sediment. Gullying erodes the denuded hills in the foreground.

A stream's capacity to carry suspended load increases sharply with an increase in the stream's velocity, because the swifter the current, the more intense is the turbulence. The capacity to move bed load also increases with velocity—the faster water motion produces a stronger dragging force against the bed. In fact, the capacity to move bed load increases according to the third to fourth power of the velocity. In other words, when a stream's velocity is doubled in flood, its ability to transport bed load is increased from eight to sixteen times. Thus, most of the conspicuous changes in the channel of a stream occur in a flood stage.

When water flow increases, a stream flowing in a channel that is cut into thick layers of silt, sand, and gravel will easily widen and deepen its channel. When the flow slackens, the stream will deposit material in the bed, filling the channel again. Where a stream flows in a channel of hard bedrock, the channel cannot be quickly deepened in response to rising waters and may not change much during a single flood. Such conditions exist in streams that occupy deep canyons and have steep gradients.

STREAM GRADATION

A main stream, fully developed within its drainage basin, has normally undergone thousands of years of adjustment of its channel. It can discharge not only the surplus runoff produced by the basin, but also the solid load that the tributary channels supply. This transport capability is related to the gradient of the channel (discussed in Chapter 14). The steeper the gradient,

Figure 15.8 Schematic diagram of gradation of a stream. Originally, the channel consists of a succession of lakes, falls, and rapids. (Copyright © A. N. Strahler.)

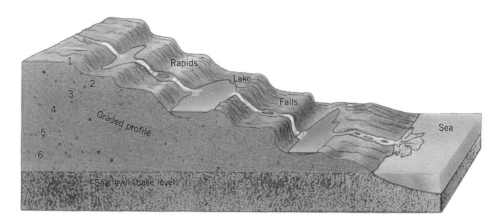

the higher the stream velocity and the greater the ability of the stream to carry sediment. The gradient of a stream adjusts over time to achieve an average balanced state of operation, year in and year out and from decade to decade. In this equilibrium condition, the stream is referred to as a **graded stream.**

To develop the concept of a graded stream system, we will investigate the changes that take place along a stretch of stream that is initially very poorly adjusted to the transport of its load. Such an ungraded channel is illustrated by an example in Figure 15.8. The side of the block shows a series of *stream profiles*—plots of elevation of the stream with distance from the sea. The starting profile (1) is produced by crustal uplift in a series of fault steps, bringing to view a land surface that was formerly beneath the ocean and exposing it to fluvial processes for the first time. Overland flow collects in shallow depressions, which fill and overflow from higher to lower levels. In this way, a through-flowing channel originates and begins to conduct runoff to the sea. As time passes, the landscape is slowly eroded by fluvial action, and the stream profile is smoothed out into a uniform curve, shown as profile line 3. The profile has now been graded. From that point on into time, this *graded profile* is steadily lowered in elevation as the landscape is further eroded (curves 4 through 6).

Landscape Evolution of a Graded Stream

Early Stages

Figure 15.9 illustrates the stream gradation process in a series of block diagrams. In (*a*), we see *waterfalls* and *rapids,* which are simply portions of the channel with steep gradients. Flow velocity at these points is greatly increased, and abrasion of bedrock is therefore most intense. As a result, the falls are cut back and the rapids are trenched. At the same time, the ponded

stretches of the stream are first filled by sediment and are later lowered in level as the lake outlets are cut down. In time, the lakes disappear and the falls are transformed into rapids.

In the early stages of gradation and tributary extension, the capacity of the stream exceeds the load supplied to it, so that little or no alluvium accumulates in the channels. Abrasion continues to deepen the major channels, with the result that they come to occupy steep-walled *gorges* or *canyons* (Figure 15.10). Weathering and mass wasting of these rock walls contribute an increasing supply of rock debris to the channels. Also on the increase is debris shed from land surfaces that contribute overland flow to the newly developed branches.

Over time, a stream develops a graded profile in which the gradient is just sufficient to carry the average annual load of water and sediment produced by its drainage basin.

The erosion of rapids reduces the gradient to a slope angle that more closely approximates the average gradient of that section of the stream (Figure 15.9*b*). At the same time, branches of the main stream are being extended into higher parts of the original land mass. These carve out many new small drainage basins. Thus, the original tectonic landscape is transformed into a complete fluvial landform system.

Attaining a Graded Condition

As landscape change continues, a gradual decrease in a stream's capacity to move bed load results from the gradual reduction in the channel gradient. Simultaneously, the load supplied to the stream from the entire upstream area is on the increase. So, a time will come when the supply of load exactly matches

a.

b.

c.

d.

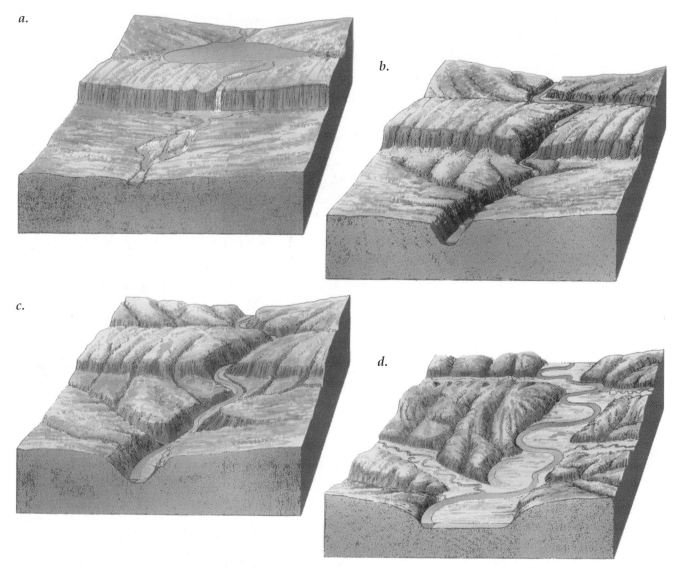

Figure 15.9   Evolution of a stream and its valley. (Drawn by E. Raisz. Copyright © A. N. Strahler.) (*a*) Stream established on a land surface dominated by landforms of recent tectonic activity. (*b*) Gradation in progress. The lakes and marshes drained. The gorge is deepening, and tributary valleys are extending. (*c*) Graded profile attained. Floodplain development is beginning, and widening of the valley is in progress. (*d*) Floodplain widened to accommodate meanders. Floodplains now extend up tributary valleys.

the stream's capacity to transport it. Figure 15.9*c* shows the landscape at this point in time. Now, all the major streams have achieved the graded condition and possess graded profiles that descend smoothly and uniformly.

The first indication that a stream has attained a graded condition is the beginning of floodplain development. On the outside of each bend, the channel shifts laterally (sidewise) to make a curve of larger radius (Figure 15.11). This forced change requires the stream to undercut and steepen the valley wall. On the inside of the bend, alluvium accumulates as a long, curving deposit of sediment—termed a *point bar.* Widening of the bar deposit produces a crescent-

shaped area of low ground, which is the first stage in floodplain development. This stage is illustrated in Figure 15.9*c*. As lateral cutting continues, the floodplain strips are widened, and the channel develops sweeping bends. These winding river bends are called **alluvial meanders** (*d*). In this way, the floodplain is widened into a continuous belt of flat land between steep valley walls. Figure 15.12 illustrates these changes.

Floodplain development reduces the frequency with which the river attacks and undermines the adjacent valley wall. Weathering, mass wasting, and overland flow can then act to reduce the steepness of the valley-side slopes (Figure 15.12). As a result, in a humid climate, the gorgelike aspect of the valley gradually disappears

Figure 15.10 The Grand Canyon of the Yellowstone River has been carved in volcanic rock (rhyolite) of the Yellowstone Plateau. Yellowstone National Park, Wyoming.

Figure 15.11 Idealized map and cross profile of a meander bend of a large alluvial river, such as the lower Mississippi. Arrows show the position of the swiftest current. (Copyright © A. N. Strahler.)

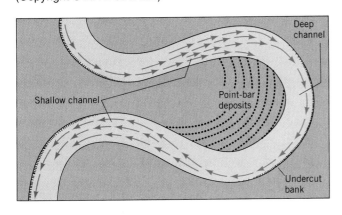

and eventually gives way to an open valley with soil-covered slopes protected by a dense plant cover. The time required for the stream to reach a graded condition and erode a broad valley will be a few tens of millions of years.

Great Waterfalls

Although small, high waterfalls are common features of alpine mountains carved by glacial erosion (Chapter 18), large waterfalls on major rivers are comparatively rare the world over. Faulting and dislocation of large crustal blocks have caused spectacular waterfalls on several east African rivers (Figure 15.13). As we explained in Chapter 12, this is the Rift Valley region, where block faulting has been taking place.

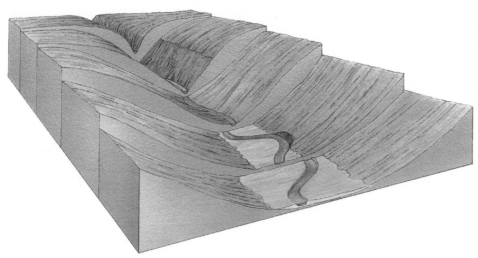

Figure 15.12 Following stream gradation, the valley walls become gentler in slope, and the bedrock is covered by soil and weathered rock. (After W. M. Davis. Copyright © A. N. Strahler.)

Another class of large waterfalls involves new river channels resulting from glacial activity in the Pleistocene Epoch. Erosion and deposition by large moving ice sheets greatly disrupted drainage patterns in northern continental regions, creating lakes and causing river courses to be shifted to new locations. Niagara Falls is a good example. The outlet from Lake Erie into Lake Ontario, the Niagara River, is situated over a gently inclined layer of limestone, beneath which lies easily eroded shale (Figure 15.14). The fall is maintained by continual undermining of the limestone by erosion of the less resistant shale at the base of the fall (Figure 15.15). The height of the falls is now 52 m (170 ft), and its discharge is about 17,000 cubic m per second (200,000 cu ft per second). The drop of Niagara Falls is utilized for the production of hydroelectric power by the Niagara Power Project. Water is withdrawn upstream from the falls and carried in tunnels to generating plants located about 6 km (4 mi) downstream from the falls.

Figure 15.13 Victoria Falls is located on the Zambesi River, which forms the border between Zambia and Zimbabwe in southern Africa. The water falls 128 m (420 ft) to the bottom of a deep, narrow gorge excavated along a fault line.

Figure 15.14 A bird's eye view of the Niagara River with its falls and gorge carved in strata of the Niagara Escarpment. View is toward the southwest from a point over Lake Ontario. (After a sketch by G. K. Gilbert. Redrawn from A. N. Strahler. Copyright © A. N. Strahler.)

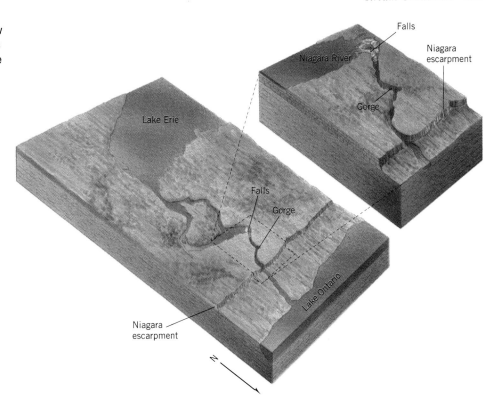

Dams and Resources

Because most large rivers of steep gradient do not have falls, dams are necessary to create the vertical drop required to spin the turbines of electric power generators. An example is the Hoover Dam, behind which lies Lake Mead, occupying the lower canyon of the Colorado River.

Hydroelectric power is cheap, nonpolluting, and renewable, and large dams provide fresh water for urban use and irrigation. However, lakes behind dams drown river valleys and can inundate the gorges, rapids, and waterfalls of major rivers. Scenic and recreational resources such as these can be lost. White-water boating and rafting, for example, are now popular sports on our wild rivers. The Grand Canyon of the Colorado River, probably more than any single product of fluvial processes, demonstrates the scenic and recreational value of a great river gorge.

Dam construction also destroys ecosystems adapted to the river environment. In addition, deposition of sediment behind the dam rapidly reduces the holding capacity of the lake and within a century or so may fill the lake basin. This reduces the ability of the lake to provide consistent supplies of water and hydroelectric power. It is small wonder, then, that new dam projects can meet with stiff opposition from concerned local citizens' groups and national environmental organizations.

Figure 15.15 Niagara Falls is formed where the river passes over the eroded edge of a massive limestone layer. Continual undermining of weak shales at the base keeps the fall steep. (Drawn by E. Raisz. Copyright © A. N. Strahler.)

Evolution of a Fluvial Landscape

With the concept of a graded stream in mind, this is a good place to examine some of the broader aspects of fluvial denudation. Just as a stream undergoes a life cycle of change, an entire land mass has its stages of evolution.

a.

b.

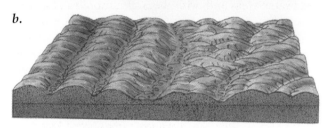

c.

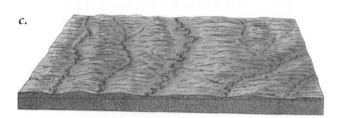

d.

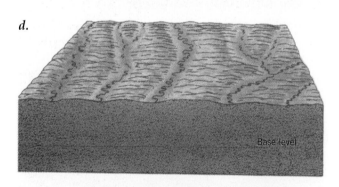

e.

Consider a landscape made up of many drainage basins and their branching stream networks. The region is rugged, with steep mountainsides and high, narrow crests (Figure 15.16*a*). The total volume of rock lying above sea level and available for removal by fluvial action is called the *landmass.* Rock debris is transported out of each drainage basin by the main graded streams at the same average rate as the debris is being contributed from the land surfaces within the basin. Eventually, the export of debris lowers the land surface generally, and the average altitude of the land surface steadily declines. This decline must be accompanied by a reduction in the average gradients of all streams (*b*).

As time passes, the streams and valley-side slopes of the drainage basins undergo gradual change to lower gradients. In theory, the ultimate goal of the denudation process is to reduce the landmass to a featureless plain at sea level. In this process, a sea-level surface imagined to lie beneath the entire landmass represents the lower limiting level, or *base level,* of the fluvial denudation (labeled in Figures 15.8 and 15.16). But because the rate of denudation becomes progressively slower, the land surface approaches the base-level surface of zero elevation at a slower and slower pace. Under this scenario, the ultimate goal can never be reached. Instead, after the passage of some millions of years, the land surface is reduced to a gently rolling surface of low elevation, called a **peneplain** (*c*). You can think of this strange term as meaning an "almost-plain."

If crustal uplift does not occur, land surfaces will eventually be worn down to a low, gently rolling surface, called a peneplain.

Production of a peneplain requires a high degree of crustal and sea-level stability for a period of many millions of years. One region that has been cited as a possible example of a contemporary peneplain is the Amazon-Orinoco basin of South America. This vast region is a stable continental shield of ancient rock.

Figure 15.16 Reduction of a landmass by fluvial erosion. (*a*) In early stages, relief is great, slopes are steep, and the rate of erosion is rapid. (*b*) In an advanced stage, relief is greatly reduced, slopes are gentle, and the rate of erosion is slow. Soils are thick over the broadly rounded hill summits. (*c*) After many millions of years of fluvial denudation, a peneplain is formed. Slopes are very gentle, and the landscape is an undulating plain. Floodplains are broad, and the stream gradients are extremely low. All of the land surface lies close to base level. (*d*) The peneplain is uplifted. (*e*) Streams trench a new system of deep valleys during landmass rejuvenation. (Drawn by A. N. Strahler.)

Uplifted peneplains that are now high-standing land surfaces are numerous within the continental shields. These regions are characterized by an upland surface of nearly uniform elevation. The upland is trenched by stream valleys graded with respect to a lower base level.

Figure 15.16*d*, shows the peneplain of (*c*) uplifted to an elevation of several hundred meters. The base level is now far below the land surface, and a large thickness of new landmass has been created. Soon, however, streams begin to trench the landmass and to carve deep, steep-walled valleys, shown in (*e*). This process is called *landmass rejuvenation*. Landscapes in various stages of rejuvenation are recognized throughout the continental shield area of North America. With the passage of many millions of years, the landscape will be carved into the rugged stage shown in (*a*), and the later stages of (*b*) and (*c*) will follow.

Aggradation and Alluvial Terraces

A graded stream, delicately adjusted to its supply of water and rock waste from upstream sources, is highly sensitive to changes in those inputs. Changes in climate or vegetation cover bring changes in discharge and load at downstream points, and these changes in turn require channel readjustments. One kind of change is the buildup of alluvium in the valley floors.

Consider first what happens when bed load increases, exceeding the transporting capacity of the stream. Along a section of channel where the excess load is introduced, the coarse sediment accumulates on the stream bed in the form of bars of sand, gravel, and pebbles. These deposits raise the elevation of the stream bed, a process called *aggradation*. As more bed materials accumulate, the stream channel gradient is steepened and flow velocity increases. This increase enables bed materials to be dragged downstream and spread over the channel floor at more and more distant downstream sections. In this way, sediment introduced at the head of a stream will be gradually spread along all the length of the stream.

Aggradation typically changes the channel cross section from a narrow and deep form to a wide and shallow one. Because bars are continually being formed, the flow is divided into multiple threads. These rejoin and subdivide repeatedly to give a typical *braided stream* (Figure 15.17). The coarse channel deposits spread across the former floodplain, burying fine-textured alluvium under the coarse material.

What processes cause alluvium to build up on valley floors and induce stream aggradation? One way is for alluvium to accumulate as a result of accelerated soil erosion, which we have already discussed. Other causes for aggradation are related to major changes in global climate, such as the onset of an ice age. Figure 15.17 illustrates one natural cause of aggradation—glacial activity—that has been of major importance in stream systems of North America and Eurasia during the recent Ice Age. In our photo example, a modern valley glacier has provided a large quantity of coarse rock debris at the head of the valley. Valley aggradation of a similar kind was widespread in a broad zone near the edges of the great ice sheets of the Pleistocene Epoch. The accumulated alluvium filled most valleys to depths of several tens of meters. Figure 15.18*a* shows a valley filled in this manner by an aggrading stream. The case could represent any one of a large number of valleys in New England or the Middle West.

Suppose, next, that the source of bed load is cut off or greatly diminished. In the case illustrated in Figure 15.18, the ice sheets have disappeared from the distant headwater areas and, with them, the supplies of coarse rock debris. Reforestation of the landscape restores a protective cover to valley-side and hill slopes of the region, keeping coarse mineral particles from entering the stream through overland flow. Now the streams have abundant water discharges but little bed load. In

Figure 15.17 The braided channel of the Chitina River, Wrangell Mountains, Alaska, shows many separate channels separating and converging on a floodplain filled with glacial debris.

other words, they are operating below their transporting capacity. The result is channel scour. The channel form becomes both deeper and narrower, and it also begins to develop meanders.

Gradually, the stream profile level is lowered, in a process called *degradation*. Because the stream is very close to being in the graded condition at all times, its dominant activity is lateral (sidewise) cutting by growth of meander bends, as shown in Figure 15.18*b*. By this process the valley alluvium is gradually excavated and carried downstream, but it cannot all be removed because the channel encounters hard bedrock masses in many places. These obstructions prevent cutting away of more alluvium. Consequently, as shown in (*c*), steplike alluvial surfaces remain on both sides of the valley. The treads of these steps are called *alluvial terraces*.

Alluvial terraces form when aggrading rivers lose their sediment input and begin degrading their beds, leaving terraces behind as they cut deeper into their sediment-filled valleys.

Alluvial terraces have always attracted human settlement because of the advantages over both the valley-bottom floodplain—which is subject to annual flooding—and the hill slopes beyond—which may be too steep and rocky to cultivate. Besides, terraces are easily tilled and make prime agricultural land (Figure 15.19). Towns are also easily laid out on the flat ground of a terrace. Roads and railroads can be easily run along the terrace surfaces.

Alluvial Rivers and Their Floodplains

We turn now to the graded river with its floodplain. As time passes, the floodplain is widened, so that broad areas of floodplain lie on both sides of the river channel. Civil engineers have given the name **alluvial river** to a large river of very low channel gradient. It flows on a thick floodplain accumulation of alluvium constructed by the river itself in earlier stages of its activity. Characteristically, an alluvial river experiences overbank floods each year or two. These occur during the season of large water surplus over the watershed. Overbank flooding of an alluvial river normally inundates part or all of a floodplain that is bounded on either side by rising steep slopes, called *bluffs*.

Typical landforms of an alluvial river and its floodplain are illustrated in Figure 15.20. Dominating the floodplain is the meandering river channel itself and abandoned stretches of former channels. Meanders develop narrow necks, which are cut through, thus

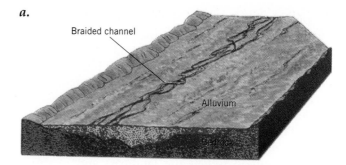

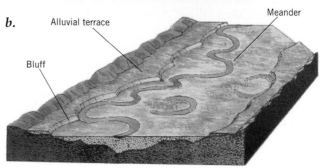

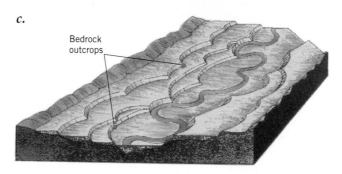

Figure 15.18 Alluvial terraces form when a graded stream slowly cuts away the alluvial fill in its valley. (Drawn by A. N. Strahler.)

shortening the river course and leaving a meander loop abandoned. This event is called a *cutoff*. It is quickly followed by deposition of silt and sand across the ends of the abandoned channel, producing an *oxbow lake*. The oxbow lake is gradually filled in with fine sediment brought in during high floods and with organic matter produced by aquatic plants. Eventually, the oxbows are converted into swamps, but their identity is retained indefinitely (Figure 15.21).

During periods of overbank flooding, when the entire floodplain is inundated, water spreads from the main channel over adjacent floodplain deposits (Figure 15.22). As the current slackens, sand and silt are deposited in a zone adjacent to the channel. The result is an accumulation of higher land on either side of the channel known as a **natural levee**. Because deposition is heavier closest to the channel and decreases away from the channel, the levee surface slopes away from the channel (Figure 15.20). Between the levees and the bluffs is lower ground, called the *backswamp.* In Figure 15.22, the small settlements are located on the higher ground of the levee, next to the river, while the agricultural fields occupy the backswamp area.

Overbank flooding not only results in the deposition of a thin layer of silt on the floodplain, but also brings an infusion of dissolved mineral substances that enter the soil. As a result of the resupply of nutrients, floodplain soils retain their remarkable fertility, even though they are located in regions of rainfall surplus from which these nutrients are normally leached away.

Entrenched Meanders

What happens when a broadly meandering river is uplifted in the process of stream rejuvenation? The uplift increases the river's gradient, so that it cuts downward into the bedrock below. This forms a steep-walled inner gorge. On either side lies the former floodplain, now a flat terrace high above river level. Any river

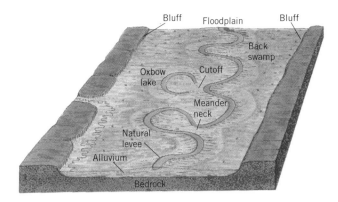

Figure 15.20 Floodplain landforms of an alluvial river. (Drawn by A. N. Strahler.)

deposits left on the terrace are rapidly stripped off by runoff, since no new sediment is accumulating.

Rejuvenation may cause the meanders to become impressed into the bedrock and give the inner gorge a meandering pattern (see chapter opening photo).

Figure 15.19 Alluvial terraces of the Rakaia River gorge on the South Island of New Zealand. The flat terrace surface in the foreground is used as pasture for sheep. Two higher terrace levels can be seen at the left.

These sinuous bends are termed *entrenched meanders* to distinguish them from the floodplain meanders of an alluvial river (Figure 15.23). Although entrenched meanders are not free to shift about as floodplain meanders do, they can enlarge slowly so as to produce cutoffs. Cutoff of an entrenched meander leaves a high, round hill separated from the valley wall by the deep abandoned river channel and the shortened river course (Figure 15.23). As you might guess, such hills formed ideal natural fortifications. Many European fortresses of the Middle Ages were built on such cutoff meander spurs. Under unusual circumstances, where the bedrock includes a strong, massive sandstone formation, meander cutoff leaves a *natural bridge* formed by the narrow meander neck.

Figure 15.21   This aerial photo of the Mississippi River floodplain shows two ox-bow lakes—former river bends that were cut off as the river shifted its course. Vegetation appears bright red in this color-infrared photo.

Figure 15.22   This river floodplain in Bangladesh is largely under water during a 1973 flood. Villages occupy the higher ground of the natural levees bordering the river channel.

Figure 15.23 Rejuvenation of a meandering stream has produced entrenched meanders. One meander neck has been cut through, forming a natural bridge. (Drawn by E. Raisz. Copyright © A. N. Strahler.)

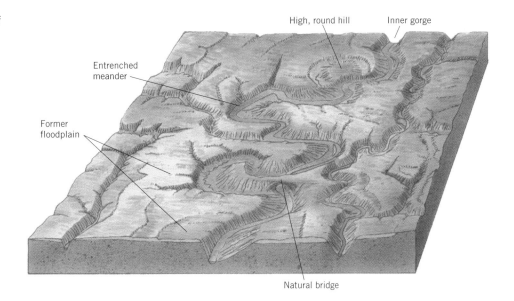

FLUVIAL PROCESSES IN AN ARID CLIMATE

Desert regions look strikingly different from humid regions, in both vegetation and landforms. Obviously, the lower precipitation makes the difference. Vegetation is sparse or absent, and land surfaces are mantled with mineral material—sand, gravel, rock fragments, or bedrock itself.

Although deserts have low precipitation, rain falls in dry climates as well as in moist, and most landforms of desert regions are formed by running water (Figure 15.24). A particular locality in a dry desert may experience heavy rain only once in several years. But when rain does fall, stream channels carry water and perform important work as agents of erosion, transportation, and deposition. Fluvial processes are especially

Figure 15.24 A flash flood has filled this desert channel in the Tucson Mountains of Arizona with raging, turbid waters. A distant thunderstorm produced the runoff.

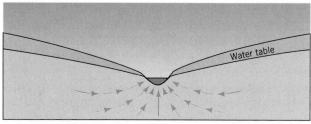

a.

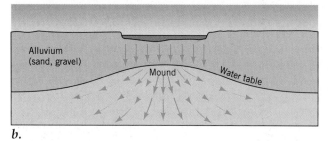

b.

Figure 15.25 In humid regions (*a*), a stream channel receives ground water through seepage. In arid regions (*b*), stream water seeps out of the channel and into the water table below. (Copyright © A. N. Strahler.)

effective in shaping desert landforms because of the sparse vegetation cover. The few small plants that survive offer little or no protection to soil or bedrock. Without a thick vegetative cover to protect the ground and hold back the swift downslope flow of water, large quantities of coarse rock debris are swept into the streams. A dry channel is transformed in a few minutes into a raging flood of muddy water heavily charged with rock fragments.

Although rain is infrequent in desert environments, running water shapes landforms with great effectiveness because of the lack of vegetation cover.

An important contrast between regions of arid and humid climates lies in the way in which the water enters and leaves a stream channel (Figure 15.25). In a humid region (*a*) with a high water table sloping toward a stream channel, ground water moves steadily toward the channel and seeps into the stream bed, producing permanent (perennial) streams. In arid regions (*b*), the water table normally lies far below the channel floor. Where a stream flows across a plain of gravel and sand, water is lost from the channel by seepage. Loss of discharge by seepage and evaporation strongly depletes the flow of streams in alluvium-filled valleys of arid regions. As a result, aggradation occurs and braided channels are common. Streams of desert regions are often short and end in alluvial deposits or on the floors of shallow, dry lakes.

Alluvial Fans

One very common landform built by braided, aggrading streams is the **alluvial fan,** a low cone of alluvial sands and gravels resembling in outline an open Japanese fan (Figure 15.26). The apex, or central point of the fan, lies at the mouth of a canyon or ravine. The fan is built out on an adjacent plain. Alluvial fans are of many sizes. In fact, some desert fans are many kilometers across (Figure 15.27).

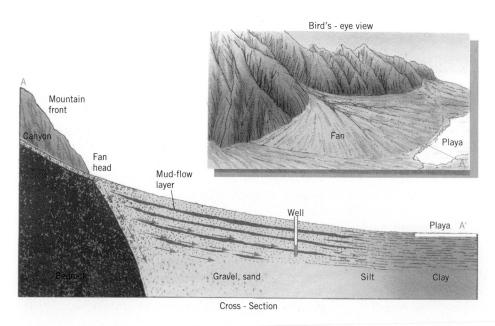

Figure 15.26 Features of an alluvial fan. A cross section shows mudflow layers interbedded with sand layers, providing water (arrows) for a well in the fan. (Copyright © A. N. Strahler.)

Figure 15.27 Great alluvial fans extend out upon the floor of Death Valley. The canyons from which they originate have carved deeply into a great uplifted fault block.

Fans are built by streams carrying heavy loads of coarse rock waste from a mountain or an upland region. The braided channel shifts constantly, but its position is firmly fixed at the canyon mouth. The lower part of the channel, below the apex, sweeps back and forth. This activity accounts for the semicircular fan form and the downward slope in all radial directions away from the apex.

Large, complex alluvial fans also include mudflows (Figure 15.26). Mud layers are interbedded with sand and gravel layers. Water infiltrates the fan at its head, making its way to lower levels along sand layers. The mudflow layers serve as barriers to ground water movement. The trapped ground water is under pressure from water higher in the fan apex. When a well is drilled into the lower slopes of the fan, water rises spontaneously as artesian flow. (See Chapter 16 and Figure 16.8.)

Alluvial fans are the primary sites of ground water reservoirs in the southwestern United States. In many fan areas, sustained heavy pumping of these reserves for irrigation has lowered the water table severely. The rate of recharge is extremely slow in comparison. At some locations, recharge is increased by building water-spreading structures and infiltrating basins on the fan surfaces. A serious side effect of excessive ground water withdrawal is that of subsidence (sinking) of the land surface.

The Landscape of Mountainous Deserts

Where tectonic activity has recently produced block faulting in an area of continental desert, the assemblage of fluvial landforms is particularly diverse. The basin-and-range region of the western United States is such an area. It includes large parts of Nevada and Utah, southeastern California, southern Arizona and New Mexico, and adjacent parts of Mexico. The uplifted and tilted blocks, separated by down-dropped tectonic basins, provide an environment for the development of spectacular erosional and depositional fluvial landforms.

Figure 15.28 demonstrates some landscape features of mountainous deserts. Figure 15.28a shows two uplifted fault blocks with a down-dropped block between them. Although denudation acts on the uplifted blocks as they are being raised, we have shown them as very little modified at the time tectonic activity

Eye on the Environment • Battling the Mighty Mississippi

The violent blast cut the still night air with a force that broke windows and set off car alarms on the far side of the mile-wide river. The muddy waters of the mighty Mississippi, thrown back at first by the force of the dynamite, surged forward through a new gap in the levee just upstream from Prairie du Rocher, Illinois. Fifteen minutes later, a second thundering explosion resounded across the width of the seething river, and floodwaters spilled through another new gash.

At 3:30 that morning, August 4, 1993, the commissioners of the Fort Chartres–Ivy Landing Drainage and Levee District had approved a risky plan to dynamite two new crevasses in the levee upstream from the historic village of Prairie du Rocher. They hoped that the new breaches would skim the top off the crest of high waters approaching the main levee that had so far held the town secure. The waters escaping through the breached levee would take hostage the fields and farms of a large area of upriver floodplain in return for sparing the village. It wasn't long before their vote was affirmed by the two blasts that echoed for miles up and down the river.

In the eyes of the commissioners, Prairie du Rocher was a prize worth saving. The town traces its roots back to its founding as a French trading post in 1722 and considers itself to be the birthplace of Illinois. Its quaint, turn-of-the-century architecture and its reputation for bed-and-breakfast inns and fine French food made the small town famous in the region as an upscale tourist mecca.

The vote of the commissioners to dynamite the levee was against the advice of the U.S. Army Corps of Engineers, the agency charged with building and maintaining the levees, dams, reservoirs, and other flood control structures that normally keep the powerful Mississippi River shackled and controlled. The Corps' engineers were afraid that the shock wave from the blast, transmitted through the river, would damage the levee near the town, flooding the village instead of protecting it. The pressure wave might also cause the levee at the site to lose its strength, turning rapidly into a mixture of sediment and water in a process called liquefaction. The resulting break would be vast enough to send a huge volume of water through the levee that could swamp other levees and dikes and eventually attack the town from the rear. Even without liquefaction, the flooding could be severe enough to engulf areas that would have otherwise remained dry.

On August 3, one day earlier, the Corps had created its own breach in the upstream levee, dispatching a crane shovel on a barge to remove the top 1.2 m (4 ft) of a 120-m (400 ft) section of the levee. According to plan, an area of about 50 sq km (20 sq mi) was flooded sooner rather than later, thus helping to blunt the rapidly approaching flood crest. But the water continued to rise, relentlessly threatening the historic village and leading to the late-night meeting and vote by the district commissioners.

Fortunately, none of the catastrophes foretold by the Corps' engineers came about. Prairie du Rocher was saved from flooding, but whether the extreme measures adopted by the desperate commissioners made the difference in keeping the floodwaters from overtopping the levees is not really known.

The summer of 1993 was a time of extreme measures all along the upper reaches of the Mississippi. In June, unprecedented rainfall in the upper Mississippi basin totaled over 30 cm (12 in.) for large areas of the region. Southern Minnesota and western Wisconsin received even larger amounts. Monthly rainfall totals were the highest on many records dating back more than 100 years. The heavy rainfall continued into July, concentrated in Iowa, Illinois and Missouri. Inundated by huge volumes of rain, the vast wet landscape came to resemble a sixth Great Lake when viewed from the air.

As the water drained from the landscape, so the Mississippi and its tributaries rose, creating a flood crest of swiftly moving turbid water that worked its way slowly downstream through July and early August. At nearly every community along the length of the Mississippi and Missouri rivers, citizens fought to raise and reinforce the levees that protected their lands, homes, and businesses from the ravages of the flood (see Figure 14.12). Where levees failed to keep them in check, the rivers filled their broad floodplains from bluff to bluff. By the time it was over, the waters had crested at 14.8 m (49.4 ft) at St. Louis, nearly 6 m (20 ft) above flood stage.

The Mississippi River Flood of 1993 was an environmental disaster of the first magnitude. An area twice the size of New Jersey was flooded, taking 50 lives and driving nearly 70,000 people from their homes. Property damage, including loss of agricultural crops, was estimated at $12 billion. Of the region's 1400 levees, at least 800 were breached or overtopped. Some flood engineers classed the event as a 500-year flood, meaning that only once in a 500-year interval is a flood of this magnitude likely to occur. However, this does not mean that a similar flood, or an even larger one, could not occur in the near future. For the people living on its banks and bottomlands, the mighty Mississippi is a sleeping giant that awakens with devastating consequences. Only time will tell when the giant is to awaken again.

Questions

1. Why were levees dynamited near Prairie du Rocher? Was the action justified?
2. What caused the Mississippi River Flood of 1993? What were some effects of the flood?

These two Landsat satellite images show St. Louis and vicinity during a normal year and during the flood of 1993. In these images, vegetation appears green, and urban areas appear in pink and purple tones. Some clouds appear in the 1993 image. The Mississippi River (topmost) joins the Missouri River to the north of the city. In 1993, the two rivers left their banks to spread into the bottomlands of their floodplains. The floodwaters crested at an even higher level at St. Louis on August 1, 1993.

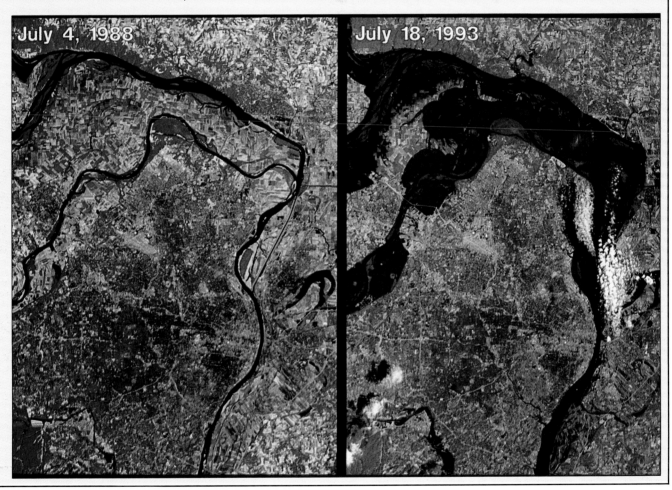

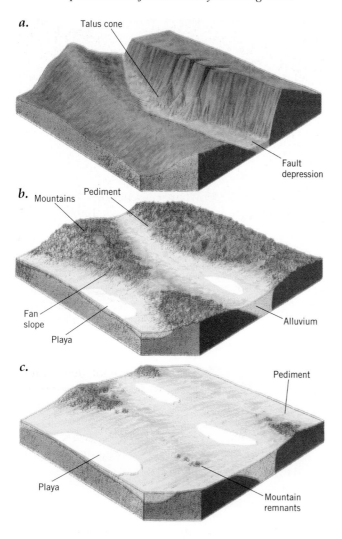

a.
Talus cone

Fault
depression

b.
Mountains Pediment

Fan
slope

Playa

Alluvium

c.
Pediment

Playa

Mountain
remnants

has ceased. At first, the faces of the fault block are extremely steep. They are scored with deep ravines, and talus blocks form cones at the bases of the blocks.

Figure 15.28*b* shows a later stage of denudation in which streams have carved up the mountain blocks into a rugged landscape of deep canyons and high divides. Rock waste furnished by these steep mountain slopes is carried from the mouths of canyons to form numerous large alluvial fans. The fan deposits form a continuous apron extending far out into the basins.

Alluvial fans and playas are features of valleys in mountainous deserts.

In the centers of desert basins lie the saline lakes and dry lake basins mentioned in Chapter 14. Accumulation of fine sediment and precipitated salts produces an extremely flat basin floor, called a **playa** in the southwestern United States and in Mexico. Salt flats are found where an evaporite layer forms the surface. In some playas, shallow water forms a salt lake.

Figure 15.29 is an air photograph of a mountainous desert landscape in the Death Valley region of eastern

Figure 15.28 Idealized diagrams of landforms of the mountainous deserts of the southwestern United States. (*a*) Initial uplift of two blocks, with a downdropped valley between them. (*b*) Stage of rapid filling of tectonic basins with debris from high, rugged mountain blocks. (*c*) Advanced stage with small mountain remnants and broad playa and fan slopes. (Drawn by A. N. Strahler.)

Figure 15.29 Racetrack Playa, a flat, white plain, is surrounded by alluvial fans and rugged mountains. This desert valley lies in the northern part of the Panamint Range, not far west of Death Valley, California. In the distance rises the steep eastern face of the Inyo Mountains, a great fault block.

California. Three environmental zones can be seen: the rugged mountain masses dissected into canyons with steep rocky walls; a zone of alluvial fans along the mountain front; and the white playa occupying the central part of the basin.

In this type of mountainous desert, fluvial processes are limited to local transport of rock particles from a mountain range to the nearest adjacent basin, which receives all the sediment. The basin gradually fills as the mountains diminish in elevation. Because there is no outflow to the sea, the concept of a base level of denudation has no meaning. For that same reason, a peneplain does not represent the next-to-the-last stage of denudation, as it does in the humid environment. Each arid basin becomes a closed system as far as transport of sediment is involved. Only the hydrologic system is open, with water entering as precipitation and leaving as evaporation.

As the desert mountain masses are lowered in height and reduced in extent, a gently sloping rock floor, called a *pediment*, develops close to the receding mountain front (Figure 15.30). As the remaining mountains shrink further in size, the pediment surfaces expand to form wide rock platforms thinly veneered with alluvium. This advanced stage is shown in Figure 15.28*c*. The desert land surface produced in an advanced stage of fluvial activity is an undulating plain consisting largely of areas of pediment surrounded by areas of alluvial fan and playa surfaces.

Figure 15.30   Bedrock is exposed in places on this pediment, carved upon ancient schist. Fragments of schist form a desert pavement atop the rock surface. Cabeza Prieta Range, Arizona.

Running water, the primary agent of denudation, does not act equally on all types of rocks, however. Some rocks are more resistant to erosion, while others are less so. As a result, fluvial erosion can create unique and interesting landscapes that reveal rock structures of various kinds. This will be the subject of our next chapter.

LANDFORMS MADE BY RUNNING WATER IN REVIEW

This chapter has covered the landforms and land-forming processes of running water, one of the four active agents of denudation. Like the other agents, running water erodes, transports, and deposits rock material, forming both erosional and depositional landforms.

The work of running water begins on slopes, producing colluvium where overland flow moves soil particles downslope, and producing alluvium when the particles enter stream channels and are later deposited. In most natural landscapes, soil erosion and soil formation rates are more or less equal, a condition known as the geologic norm. Badlands are an exception in which natural erosion rates are very high.

The work of streams includes stream erosion and stream transportation. Where stream channels are carved into soft materials, large amounts of sediment can be obtained by hydraulic action. Where stream channels flow on bedrock, channels are deepened only by the abrasion of bed and banks by mineral particles, large and small. Both the suspended load and bed load of rivers increase greatly as velocity increases. Velocity, in turn, depends on gradient.

Over time, streams tend to a graded condition, in which their gradients are adjusted to move the average amount of water and sediment supplied to them by slopes. Lakes and waterfalls, created by tectonic, volcanic, or glacial activity, are short-lived events, geologically speaking, that give way to a smooth, graded stream profile. Grade is maintained as landscapes are eroded toward base level. If the land surface is tectonically stable for a very long period, a peneplain can form. When provided with a sudden inflow of rock material, as, for example, by glacial action, streams build up their beds by aggradation. When that inflow ceases, streams resume downcutting, leaving behind alluvial terraces.

Large rivers with low gradients that move large quantities of sediment are termed alluvial rivers. The meandering of these rivers forms cutoff meanders, oxbow lakes, and other typical landforms. Alluvial rivers are sites of intense human activity. Their fertile floodplains yield agricultural crops and provide easy transportation paths. When a region containing a meandering alluvial river is uplifted, the rejuvenation can produce entrenched meanders.

Although rainfall is scarce in deserts, running water is very effective there in producing fluvial landforms. Desert streams, subject to flash flooding, build alluvial fans at the mouths of canyons. Water sinks into the fan deposits, creating local ground water reservoirs. Eventually, desert mountains are worn down into gently sloping pediments. Fine sediments and salts, carried by streams, accumulate in playas, from which water evaporates, leaving sediment and salt behind.

KEY TERMS

fluvial landforms	stream deposition	alluvial river
fluvial processes	stream load	natural levee
alluvium	graded stream	alluvial fan
stream erosion	alluvial meander	playa
stream transportation	peneplain	

REVIEW QUESTIONS

1. List and briefly identify the four flowing substances that serve as agents of denudation.

2. Compare erosional and depositional landforms. Sketch an example of each type.

3. Describe the process of slope erosion. What is meant by the geologic norm?

4. Contrast the two terms colluvium and alluvium. Where on a landscape would you look to find each one?

5. What special conditions are required for badlands to form?

6. When and how does sheet erosion occur? How does it lead to rill erosion and gullying?

7. In what ways do streams erode their bed and banks?

8. What is stream load? Identify its three components. In what form do large rivers carry most of their load?

9. How is velocity related to the ability of a stream to move sediment downstream?

10. What is a graded stream? Sketch the profile of a graded stream.

11. Describe the evolution of a fluvial landscape. What is meant by rejuvenation?

12. How does stream degradation produce alluvial terraces?

13. Define the term alluvial river. Identify some characteristic landforms of alluvial rivers. Why are alluvial rivers important to human civilization?

14. Sketch the floodplain of a graded, meandering river. Identify key landforms on the sketch. How do they form?

15. Why is fluvial action so effective in arid climates, considering that rainfall is scarce? How do streams in arid climates differ from streams of moist climates?

16. Describe the evolution of the landscape in a mountainous desert. Use the terms alluvial fan, playa, and pediment in your answer.

IN-DEPTH QUESTIONS

1. A river originates high in the Rocky Mountains, crosses the high plains, flows through the agricultural regions of the Midwest, and finally reaches the sea. Describe the fluvial processes and landforms you might expect to find on a journey along the river from its headwaters to the ocean.

2. What would be the effects of climate change on a fluvial system? Choose either the effects of cooler temperatures and higher precipitation in a mountainous desert, or warmer temperatures and lower precipitation in a humid agricultural region.

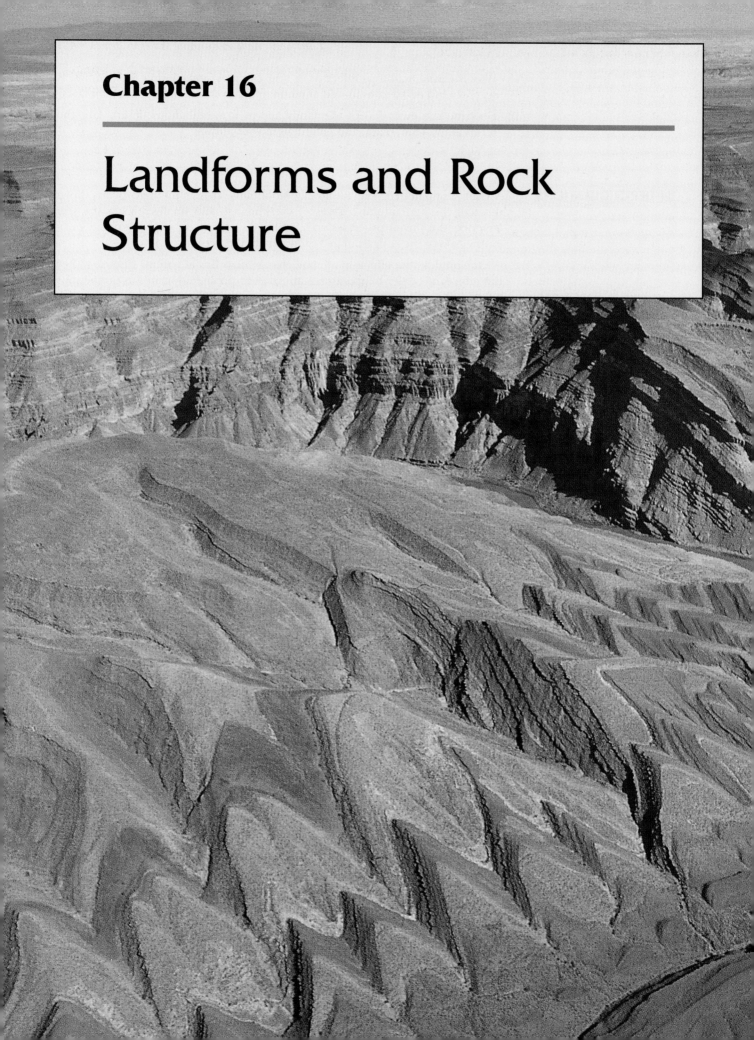

Chapter 16

Landforms and Rock Structure

You are delighted with your window seat, located a few rows in front of the wing. The flight from Denver to Las Vegas will take you across the Rocky Mountain chain and the Colorado Plateau—two of the most spectacular regions of the American landscape. On this clear day in June, the views from the aircraft should be awesome.

You've taken this flight several times before and have always been impressed by the landforms and landscapes below. This time, however, you're prepared to identify the landforms that you will see. While browsing through a pile of dusty maps and charts in a used book store one day, you discovered a marvelous old map of the landforms of the United States by the geographer Erwin Raisz. As you examined its fine black lines in detail, you realized that it showed many of the landscape features that you had seen on this flight. You carefully remove the map from your carry-on bag, folding it so that it shows the first part of your route.

Soon you are rolling down the runway, feeling the force of the powerful jet engines pushing you deeply into your seat. Quickly, the plane is airborne. It banks smoothly to the left, and you head westward toward the magnificent Rockies. The slope of the Front Range looms large, and as you head across the mountain barrier, the ground seems to rise up to meet you. Below is I-70, twisting and turning its way up the narrow valleys toward the summit. There's the entrance to the Eisenhower Tunnel! The tiny cars and trucks below disappear into the mountain slope for the traverse of nearly 3 km (2 mi) beneath the mountain crest.

Looking south, you see the peaks of the Front Range spread out in a long line. Consulting your map, you easily identify Pikes Peak in the far distance, its 4233 m (14,110 ft)

summit silhouetted against the blue. Looking straight down, you see the gray of the ancient granite rock, mantling the slopes between the patches of snow still lying in the low places. The overlying sedimentary rocks have long been eroded away here in the center of this vast elongated dome, exposing the hard granite at the core. Snowpacks in high valley heads below you hint at how the mountains must have looked when their glaciers slowly and majestically marched valleyward, carving broad, U-shaped valleys and leaving the ridges and piles of sediment now resting on the valley floors.

The first part of the flight has been memorable indeed. The Rockies were dramatic, their stark and unyielding summits cut by moving ice into sharp peaks and ridges. As they fell away, the landscape became drier, and the vast high desert of the Colorado Plateau replaced the forested lower slopes of the ranges. Below you now stretches a landscape of nearly flat-lying rocks. Long lines of cliffs, marking the edges of resistant rock beds, advance and recede as the aircraft moves steadily southwestward.

"Well, folks, you're in luck this afternoon." Suddenly, the voice of the pilot cuts the low rushing sound that fills the airplane's cabin. "Air traffic control has vectored us to the south, so we'll be approaching Las Vegas from just over the North Rim of the Grand Canyon. The views should be spectacular!"

And spectacular they are. The gaping chasm is more than 16 km

(10 mi) from rim to rim. Within the Canyon, the flat-lying rock formations form a succession of giant steps leading down to a narrow inner gorge that contains a blue-gray ribbon of water. The resistant cliffs of limestone and sandstone form the vertical risers that offset the sloping surfaces developed on weaker siltstones and shales. The canyons of tributary creeks cut deeply into the rock sequence, recurving to create wedding cakes and turrets of multileveled rock.

But it is the coloring of the rocks that is the most breathtaking. Buff and gray layers of limestone alternate with reds and purples of siltstones and shales. The green tones of vegetation overshadow the earth colors on the wider benches and platforms. The play of light and shade on the land-scape etches the shapes in relief, enhancing and deepening their coloration, making the canyon's landforms look somehow even more grand and monumental.

Descending now, the airplane approaches to Las Vegas. This flight will remain etched in your memory for years to come, not only because of its sheer scenic beauty, but also for the way in which landscape revealed its inner structure to your educated eye.

The adjustment of landforms to the nature of the rock beneath them is a very important factor in shaping regional landscapes. As we will see, weak rocks tend to form valleys or lowlands, while strong rocks tend to stand out as highlands, ridges, or cliffs. By taking a bird's-eye view of the landscape, the pattern of these landforms can often reveal the underlying rock structure—a dome perhaps, or a region of folded rock layers, or even a series of faults in flat-lying strata. The way in which landforms reflect rock type and rock structures is the subject of this chapter.

Erosion of the Rapley Monocline, on the Colorado Plateau. Near Mexican Hat, Utah.

Denudation acts to wear down all rock masses exposed at the land surface. However, different types of rocks are worn down at different rates. Some are easily eroded, while others are highly resistant to erosion. Generally, weak rocks will underlie valleys, while strong rocks will underlie hills, ridges, and uplands. This means that the pattern of landforms on a landscape can reveal the sequence of rock layers underneath it. The sequence, in turn, is determined by the type of rock structure present—for example, a series of folds, or perhaps a dome, in a set of rock layers. Thus, there is often a direct relationship between landforms and rock structure.

Over the world's vast land area, many types of rock and rock structures appear. Repeated episodes of uplift, followed by periods of denudation, can bring to the surface rocks and rock structures formed deep within the crust. In this way, even ancient mountain roots, formed during continental collisions, may eventually appear at the surface and be subjected to denudation. Batholiths and other plutons—igneous rock structures produced by magmas that cool at depth—can similarly be exposed.

As we will see in this chapter, rock structure controls not only the locations of uplands and lowlands, but also the placement of streams and the shapes and heights of the intervening divides. A distinctive assemblage of landforms and stream patterns develops on each of the major types of crustal structures. Recall that in Chapter 12 we classified all landforms as either initial or sequential in origin. Initial landforms are produced directly by volcanic activity, folding, and faulting. Denudation soon converts these initial landforms into sequential landforms, which are controlled in shape, size, and arrangement by the underlying rocks and their structure. The subject of this chapter is therefore the sequential landforms that arise through erosion of rock structures.

ROCK STRUCTURE AS A LANDFORM CONTROL

As denudation takes place, landscape features develop according to patterns of bedrock composition and structure. Figure 16.1 shows an arrangement of five layers of sedimentary rock, together with a mass of much older igneous rock. The sediments were originally deposited in horizontal layers atop the igneous rock, but the block of igneous and sedimentary rock has been bent and tilted in an ancient orogeny. The diagram shows the usual landform shape of each rock type and whether it forms valleys or mountains.

The cross section on the front of the block diagram shows conventional symbols used by geologists for each rock type—for example, fine lines for shale or "bricks" for limestone.

Shale is a weak rock that forms valleys, while sandstone and conglomerate are strong rocks that form ridges.

As indicated by the diagram, shale is a weak rock that is easily eroded and forms the low valley floors of the region. Limestone, subjected to solution by carbonic acid in rainwater and surface water, also forms valleys in humid climates. In arid climates, on the other hand, limestone is a resistant rock and usually stands high to form ridges and cliffs. Sandstone and conglomerate are typically resistant to denudation, and they form ridges or uplands. As a group, the igneous rocks are also resistant—they typically form uplands or mountains rising above adjacent areas of shale and limestone. Although metamorphic rocks in general are more resistant to denudation, individual metamorphic rock types—marble or schist, for example—vary in their resistance.

Strike and Dip

The surfaces of most rock layers are not flat but are tipped away from the horizontal. In addition, internal planes—such as joints—occur within rock layers and are likely to be tilted. We need a system of geometry to enable us to measure and describe the position in which these natural planes are held and to indicate them on maps. Examples of such planes include the bedding layers of sedimentary strata, the sides of a dike, and the joints in granite.

Figure 16.1 Landforms evolve through the slow erosional removal of weaker rock, leaving the more resistant rock standing as ridges or mountains. (Drawn by A. N. Strahler.)

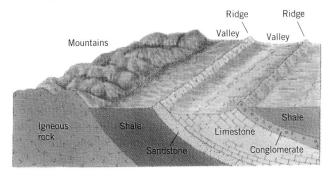

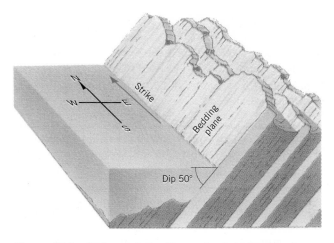

Figure 16.2   Strike and dip. (Drawn by A. N. Strahler.)

The tilt and orientation of a natural rock plane are measured with reference to a horizontal plane. Figure 16.2 shows a water surface resting against tilted sedimentary strata. Since it is horizontal, we can use the water surface as the horizontal plane. The acute angle formed between the rock plane and the horizontal plane is termed the *dip*. The amount of dip is stated in degrees, ranging from 0 for a horizontal rock plane to 90 for a vertical rock plane. Next, we examine the line of intersection between the inclined rock plane and the horizontal water plane. This horizontal line will have an orientation with respect to the compass. This orientation is termed the *strike*. In Figure 16.2, the strike is along a north-south line.

LANDFORMS OF HORIZONTAL STRATA AND COASTAL PLAINS

Extensive areas of the ancient continental shields are covered by thick sequences of horizontal sedimentary rock layers. At various times in the 600 million years following the end of Precambrian time, these strata were deposited in shallow inland seas. Following crustal uplift, with little disturbance other than minor crustal warping or faulting, these areas became continental surfaces undergoing denudation.

Arid Regions

In arid climates, where vegetation is sparse and the action of overland flow especially effective, sharply defined landforms develop on horizontal sedimentary strata (Figure 16.3). The normal sequence of landforms is a sheer rock wall, called a *cliff*, which develops at the edge of a resistant rock layer. At the base of the cliff is an inclined slope, which flattens out into a plain beyond.

In these arid regions, erosion strips away successive rock layers, leaving behind a broad platform capped by hard rock layers. This platform is usually called a **plateau.** Cliffs retreat as near-vertical surfaces because the weak clay or shale formations exposed at the cliff base are rapidly washed away by storm runoff and channel erosion. When undermined, the rock in the upper cliff face repeatedly breaks away along vertical fractures.

Plateaus, mesas, and buttes are landforms of flat-lying strata in arid regions.

Cliff retreat produces a **mesa,** which is a table-topped plateau bordered on all sides by cliffs (Figure 16.3). Mesas represent the remnants of a formerly extensive layer of resistant rock. As a mesa is reduced in area by retreat of the rimming cliffs, it maintains its flat top. Eventually, it becomes a small steep-sided hill known as a **butte.** Further erosion may produce a single, tall column before the landform is totally consumed.

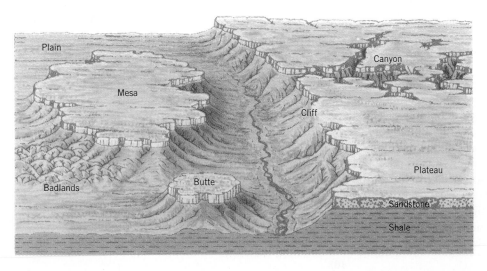

Figure 16.3   In arid climates, distinctive erosional landforms develop in horizontal strata. (Drawn by A. N. Strahler.)

Figure 16.4 Bright Angel Canyon, Grand Canyon National Park, Arizona. Note the "stairstep" landscape, in which resistant sandstones and limestones form flat-topped benches and mesas, while weaker shales form slopes connecting them.

In the walls of the great canyons of the Colorado Plateau region, these landforms are wonderfully displayed. Figure 16.4 shows a view of the Grand Canyon. The flat surface of the canyon rim marks the edge of the vast plateau surrounding the canyon. Within the

canyon, resistant rock layers crop out at different levels, producing a series of steps. Cliffs of resistant rock form the risers, while weaker rocks form the sloping treads below. Benches or platforms develop atop resistant strata, extending the treads and producing a series of levels that break the descent into the canyon. Small mesas and buttes remain within the canyon, isolated by erosion of surrounding rock strata.

Imagine making a map of a region that showed only its stream channels. This map would show the stream pattern, or **drainage pattern,** of the region. The drainage pattern that develops on each land mass type is often related to the underlying rock type and structure.

Regions of horizontal strata show broadly branching stream networks formed into a *dendritic drainage pattern.* Figure 16.5 is a map of the stream channels in a small area within the Appalachian Plateau region of Kentucky. The smaller streams in this pattern take a great variety of compass directions. Because the rock layers lie flat, no one direction is favored.

Figure 16.5 This dendritic drainage pattern is from an area of horizontal strata in the Appalachian Plateau of northern Kentucky. (Derived from data of the U.S. Geological Survey.)

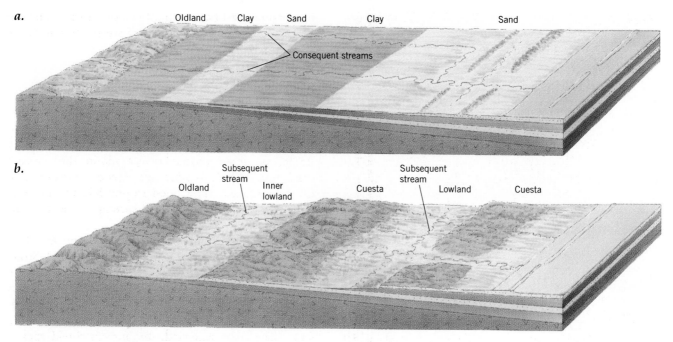

Figure 16.6 Development of a broad coastal plain. (*a*) Early stage—plain recently emerged. (*b*) Advanced stage—cuestas and lowlands developed. (Drawn by A. N. Strahler.)

Coastal Plains

Coastal plains are found along passive continental margins that are largely free of tectonic activity. They are underlain by nearly horizontal strata that slope gently toward the ocean. Figure 16.6*a* shows a coastal zone that has recently emerged from beneath the sea. Formerly, this zone was a shallow continental shelf that accumulated successive layers of sediment brought from the land and distributed by currents. On the newly formed land surface, streams flow directly seaward, down the gentle slope. A stream of this origin is a *consequent stream.* It is a stream whose course is controlled by the initial slope of a new land surface. Consequent streams form on many kinds of initial landforms, such as volcanoes, fault blocks, or beds of drained lakes.

Coastal plains often exhibit a trellis drainage pattern.

In an advanced stage of coastal-plain development, a new series of streams and topographic features has developed (Figures 16.6*b*). Where more easily eroded strata (usually clay or shale) are exposed, denudation is rapid, making *lowlands.* Between them rise broad belts of hills called **cuestas.** Cuestas are commonly underlain by sand, sandstone, limestone, or chalk. The lowland lying between the area of older rock—the oldland—and the first cuesta is called the *inner lowland.*

Streams that develop along the trend of the lowlands, parallel with the shoreline, are of a class known as *subsequent streams.* They take their position along any belt or zone of weak rock, and therefore follow closely the pattern of rock exposure. Subsequent streams occur in many regions, and we will mention them again in the discussion of folds, domes, and faults.

The drainage lines on a fully dissected coastal plain combine to form a *trellis drainage pattern.* In this type of pattern, a main stream has tributaries that are arranged at right angles. You can see this pattern in the streams of Figure 16.6*b*. The subsequent streams trend at about right angles to the consequent streams.

The coastal plain of the United States is a major geographical region, ranging in width from 160 to 500 km (100 to 300 mi) and extending for 3000 km (2000 mi) along the Atlantic and Gulf coasts. In Alabama and Mississippi (Figure 16.7), the coastal plain shows many of the features of Figure 16.6. In this region, the coastal plain sweeps far inland, as the oldland boundary shows. Cuestas and lowlands run in belts curving to follow the ancient coast. The cuestas, named "hills" or "ridges," are underlain by sandy formations and support pine forests. Limestone forms lowlands, such as the Black Belt in Alabama. This belt is named for its dark, fertile soils.

An interesting feature of coastal plain strata is the occurrence of **artesian wells.** As Figure 16.8 shows, cuestas of porous sandstone absorb precipitation, and the water moves down the dip of the sandstone layer. This water-bearing formation is called an *aquifer.* Beds

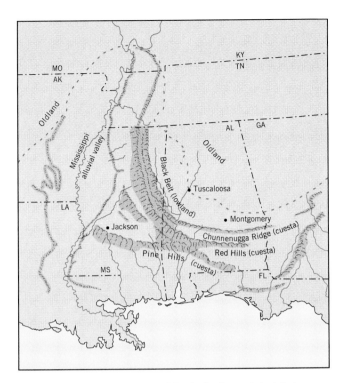

Figure 16.7 The Alabama-Mississippi coastal plain is belted by a series of sandy cuestas and shale lowlands. (After A. K. Lobeck.)

of impervious rock, called *aquicludes,* lie above and below the aquifer. Far out on the coastal plain, at depth, the water is under strong confining pressure. When wells are drilled there, water rises spontaneously to the surface.

LANDFORMS OF WARPED ROCK LAYERS

Although sedimentary rocks are horizontal or gently sloping in some regions, in other areas rock layers may be upwarped or downwarped. In some cases, these

form domes and basins. In others, they form anticlines and synclines—wavelike forms of crests and troughs—as we noted in Chapter 12.

Sedimentary Domes

A distinctive land mass type is the **sedimentary dome,** a circular or oval structure in which strata have been forced upward into a domed shape. Sedimentary domes occur in various places within the covered shield areas of the continents. Igneous intrusions at great depth may have been responsible for some of these uplifts. For others, upthrusting on deep faults may have been the cause.

Erosion features of a sedimentary dome are illustrated in the two block diagrams of Figure 16.9. Strata are first removed from the summit region of the dome, exposing older strata beneath. Eroded edges of steeply dipping strata form sharp-crested sawtooth ridges called *hogbacks* (Figure 16.10). When the last of the strata have been removed, the ancient shield rock is exposed in the central core of the dome, which then develops a mountainous terrain.

The stream network on a deeply eroded dome shows dominant subsequent streams forming a circular system, called an annular drainage pattern (Figure 16.11). ("Annular" means "ring-shaped.") The shorter tributaries make a radial arrangement. The total pattern resembles a trellis pattern bent into a circular form.

The Black Hills Dome

A classic example of a large and rather complex sedimentary dome is the Black Hills dome of western South Dakota and eastern Wyoming (Figure 16.12). The Red Valley is continuously developed around the entire dome and is underlain by a weak shale, which is easily washed away. On the outer side of the Red Valley is a high, sharp hogback of Dakota sandstone, known

Figure 16.8 An artesian well drilled into a dipping sandstone layer overlain by an impervious shale layer. (Drawn by Erwin Raisz. Copyright © by A. N. Strahler.)

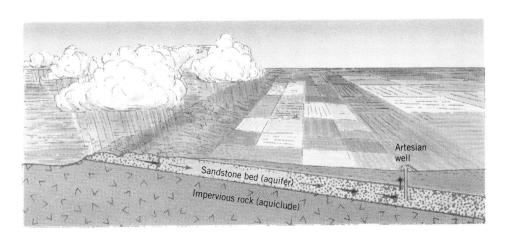

a.

b.

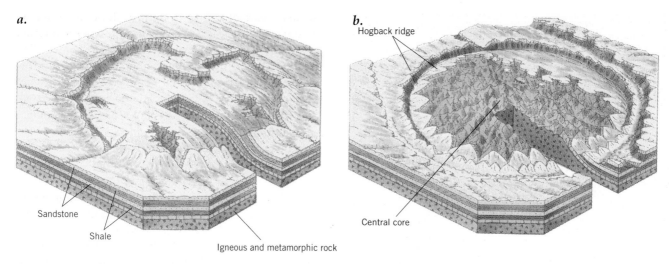

Hogback ridge

Central core

Sandstone

Shale

Igneous and metamorphic rock

Figure 16.9 Erosion of sedimentary strata from the summit of a dome structure. (*a*) The strata are partially eroded, forming an encircling hogback ridge. (*b*) The strata are eroded from the center of dome, revealing a core of older igneous or metamorphic rock. (Drawn by A. N. Strahler.)

Figure 16.10 Hogbacks of sandstone lie along the eastern base of the Colorado Front Range, which is a large domed structure. The view is southward toward the city of Boulder. The rugged forested terrain in the distance is developed on igneous and metamorphic rock exposed in the core of the uplift.

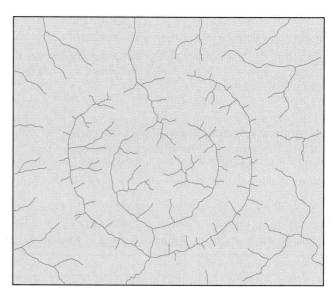

Figure 16.11 The drainage pattern on an eroded dome combines annular and radial elements.

The Black Hills dome of South Dakota is a sedimentary dome that is eroded to produce hogback ridges, a circular valley, and a mountainous central core.

as Hogback Ridge, rising some 150 m (500 ft) above the level of the Red Valley. Farther out toward the margins of the dome, the strata are less steeply inclined and form a series of cuestas.

The eastern central part of the Black Hills consists of a mountainous core of intrusive and metamorphic

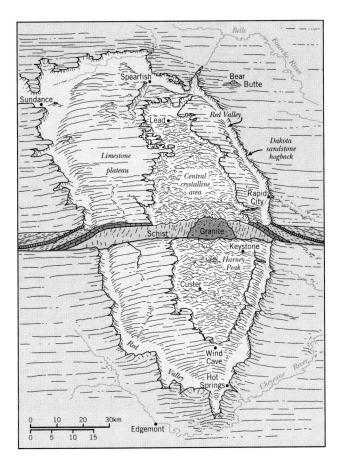

Figure 16.12 The Black Hills consist of a broad, flat-topped dome deeply eroded to expose a core of igneous and metamorphic rocks. (Drawn by A. N. Strahler.)

rocks. These mountains are richly forested, while the surrounding valleys are beautiful open parks. Thus, the region is very attractive as a summer resort area. In the northern part of the central core, there are valuable ore deposits. At Lead is the fabulous Homestake Mine, one of the world's richest gold-producing mines. The west-central part of the Black Hills consists of a limestone plateau deeply carved by streams. The plateau represents one of the last remaining sedimentary rock layers to be stripped from the core of the dome

Fold Belts

According to the model of mountain making developed in Chapter 11, strata of the continental margins are deformed into folds along narrow belts during continental collision. As we explained in Chapter 12, a troughlike downbend of strata is a *syncline,* and the archlike upbend next to it is an *anticline.* In a belt of folded strata, synclines alternate with anticlines, like a succession of wave troughs and wave crests on the ocean surface.

The two block diagrams of Figure 16.13 show some of the distinctive landforms resulting from fluvial denudation of a belt of folded strata. Deep erosion of these simple, open folds in this land mass type produces a **ridge-and-valley landscape.** Weaker formations such as shale and limestone are eroded away, leaving hard strata, such as sandstone, to stand in bold relief as long, narrow ridges (Figure 16.14). Note that anticlines, or upwarps, are not always ridges. If a resistant rock type at the center of the anticline is eroded through to reveal softer rocks underneath, an *anticlinal valley* may form. An anticlinal valley is shown in Figure 16.13*a.* A *synclinal mountain* is also possible. It occurs when a resistant rock type is exposed at the center of a syncline, and the rock stands up as a ridge (Figure 16.13*b*).

Figure 16.13 Stages in the erosional development of folded strata. (*a*) Erosion exposes a highly resistant layer of sandstone or quartzite, which controls much of the ridge-and-valley landscape. (*b*) Continued erosion partly removes the resistant formation but reveals another below it. (Drawn by A. N. Strahler.)

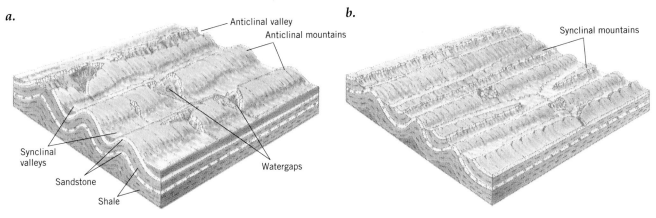

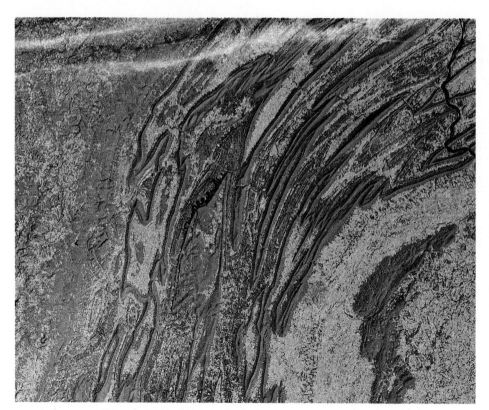

Figure 16.14 Like an old wooden plank deeply etched by drifting sand to reveal the grain, the surface of south-central Pennsylvania shows zigzag ridges formed by bands of hard quartzite. Strata were crumpled into folds during a collision of continents that took place over 200 million years ago to produce the Appalachians. In this September image, the red color depicts vegetation cover, while the blue colors indicate mainly agricultural fields. The cloud bands are formed by condensation from passing aircraft. Note the shadows that the cloud bands cast on the ground.

On eroded folds, the stream network is distorted into a trellis drainage pattern (Figure 16.15). The principal elements of this pattern are long, parallel subsequent streams occupying the narrow valleys of weak rock. In a few places, a larger stream cuts across a ridge in a deep, steep-walled *water gap*.

The folds illustrated in Figure 16.13 are continuous and even-crested. They produce ridges that are approximately parallel in trend and continue for great distances. In some fold regions, however, the folds are not continuous and level-crested. Instead, the fold crests rise or descend from place to place. A descending fold crest is said to "plunge." Plunging folds give rise to a zigzag line of ridges (Figure 16.16). Excellent examples of plunging folds can be seen in Figure 16.14.

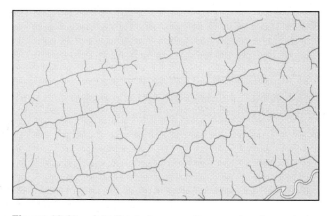

Figure 16.15 A trellis drainage pattern on deeply eroded folds in the Central Appalachians.

LANDFORMS OF HUMID LIMESTONE REGIONS

Recall that limestone is dissolved by carbonic acid action in moist climates, producing lowland areas. Dissolution goes on below the surface as well, producing caves and underground caverns. These features can then collapse, producing a sinking of the ground above and the development of a unique type of landscape called karst. Caverns and karst are the subjects we treat next.

Limestone Caverns

You are probably familiar with the names of caverns, such as Mammoth Cave or Carlsbad Caverns. Millions of Americans have visited these natural wonders. **Limestone caverns** are interconnected subterranean cavities in bedrock formed by the corrosive action of circulating ground water on limestone. The work of carbonic acid is particularly concentrated in the saturated zone just below the water table. Products of solution are carried along in the ground water flow paths

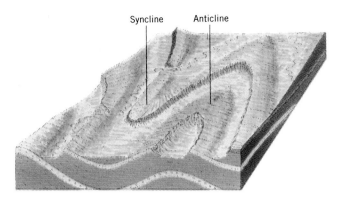

Syncline   Anticline

Figure 16.16 Folds with crests that plunge downward give rise to zigzag ridges following erosion. (Drawn by Erwin Raisz. Copyright © by A. N. Strahler.)

to emerge in streams. This removal process forms many kinds of underground "landforms," such as tortuous tubes and tunnels, great open chambers, and tall chimneys. Subterranean streams can be found flowing in the lowermost tunnels, and these emerge along the banks of surface streams and rivers.

Because the surface streams of the limestone region continually deepen their valleys, the surrounding water table is lowered to new positions. The cavern system previously excavated is now in the unsaturated zone. Deposition of carbonate matter, known as *travertine*, begins to take place on exposed rock surfaces in the caverns. Encrustations of travertine take many beautiful forms—stalactites, stalagmites, columns, drip curtains, and terraces (Figure 16.17).

In limestone caverns, travertine deposits of carbonate matter create beautiful forms.

Karst Landscapes

Where limestone solution is very active, a special landscape with many unique landforms occurs. Such a landscape can be found in regions of horizontal, folded, or domed strata. A classic example is in the Dalmatian coastal area of former Yugoslavia, where the landscape is called **karst.** Geographers apply that term

Figure 16.17 Travertine deposits in the Papoose Room of Carlsbad Caverns include stalactites (slender rods hanging from the ceiling) and sturdy columns.

Figure 16.18   Sinkholes in limestone, near Roswell, New Mexico.

to the topography of any limestone area where sinkholes are numerous and small surface streams are nonexistent. A *sinkhole* is a surface depression in a region of cavernous limestone (Figure 16.18). Some sinkholes are filled with soil washed from nearby hillsides, while others are steep-sided, deep holes.

Development of a karst landscape is shown in Figure 16.19. In an early stage, funnel-like sinkholes are

numerous. Later, the caverns collapse, leaving open, flat-floored valleys. Examples of some important regions of karst or karstlike topography are the Mammoth Cave region of Kentucky, the Yucatan Peninsula, and parts of Cuba and Puerto Rico. In regions such as southern China and west Malaysia, the karst landscape is dominated by steep-sided, conical limestone hills or towers, 100 to 500 m (300 to 1500 ft) high (Figure 16.20). These hills are often riddled with caverns and passageways.

LANDFORMS DEVELOPED ON OTHER LANDMASS TYPES

We now turn to the landforms developed on a few remaining landmass types. First are erosional forms that develop on a landmass exposed to faulting. Then we discuss the landforms that develop on metamorphic rocks and on exposed igneous batholiths. Last are the landforms of eroded volcanoes.

a.

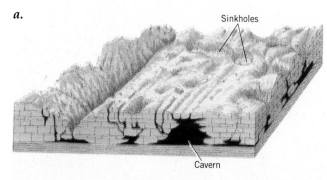

b.

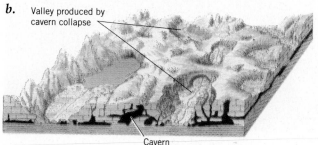

Figure 16.19 (*left*) Features of a karst landscape. (*a*) Rainfall enters the cavern system through sinkholes in the limestone. (*b*) Extensive collapse of caverns reveals surface streams flowing on shale beds beneath the limestone. Some parts of the flat-floored valleys can be cultivated. (Drawn by Erwin Raisz. Copyright © A. N. Strahler.)

Erosion Forms on Fault Structures

In Chapter 12, we found that active normal faulting produces a sharp surface break called a *fault scarp*. Repeated faulting may produce a great rock cliff hundreds of meters high (see Figure 12.14). Erosion quickly modifies a fault scarp, but, because the fault plane extends hundreds of meters down into the bedrock, its effects on erosional landforms persist for long spans of geologic time. Figure 16.21 shows both the original fault scarp (above) and its later landform expression (below), known as a *fault-line scarp*. Even though the cover of sedimentary strata has been completely removed, exposing the ancient shield rock, the fault continues to produce a landform. Because the fault plane is a zone of weak rock that has been crushed during faulting, it is occupied by a subsequent stream. A scarp persists along the upthrown side. Such scarps on ancient fault lines are numerous in the exposed and covered continental shields.

Figure 16.22 shows some erosional features of a large, tilted fault block. The freshly uplifted block has a steep mountain face, but it is rapidly dissected into deep canyons. The upper mountain face becomes less steep as it is removed, and rock debris accumulates in the form of alluvial fans adjacent to the fault block. Vestiges of the fault plane are preserved in a line of triangular facets. Each facet represents the snubbed end of a ridge between two canyon mouths.

Metamorphic Belts

Where strata have been tightly folded and altered into metamorphic rocks during continental collision, denudation eventually develops a landscape with a strong grain of ridges and valleys. These features lack the sharpness and straightness of ridges and valleys in belts of open folds. Even so, the principle that governs the development of ridges and valleys in open folds applies to metamorphic landscapes as well—namely, that rocks resistant to denudation form highlands and ridges, while rocks that are more susceptible to denudation form lowlands and valleys.

Slate and marble are weak metamorphic rocks that underlie valleys. Schist, gneiss, and quartzite are more resistant and underlie uplands and ridges.

Figure 16.20 Tower karst near Guilin (Kweilin), Guanxi Province, in southern China at lat. 25°S. White limestone can be seen exposed in the nearly vertical sides of the towers.

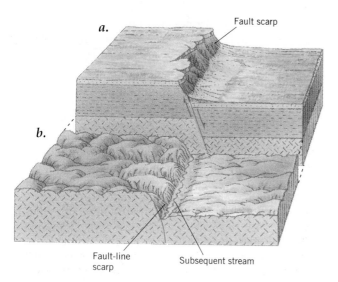

Figure 16.21 (a) A recently formed fault scarp. (b) Despite continental denudation over several million years, the fault plane causes a long narrow valley bounded by a scarp. (Drawn by A. N. Strahler. Copyright © A. N. Strahler.)

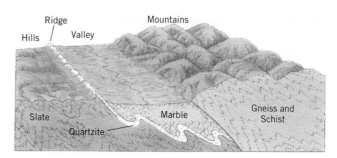

Figure 16.23 Metamorphic rocks form long, narrow parallel belts of valleys and mountains. (Drawn by A. N. Strahler.)

Figure 16.23 shows typical erosional forms associated with parallel belts of metamorphic rocks, such as schist, slate, quartzite, and marble. Marble forms valleys, while slate and schist make hill belts. Quartzite stands out boldly and may produce conspicuous narrow hogbacks. Areas of gneiss form highlands.

Parts of New England, particularly the Taconic and Green mountains of New Hampshire and Vermont,

Figure 16.22 The near-vertical faces of the foreground peaks in this photo mark a fault scarp along which the rocks have been uplifted by many successive motions. Talus slopes of rock debris line the base of the scarp. Eastern Sierra Nevada, California.

illustrate the landforms eroded on an ancient meta-morphic belt. The larger valleys trend north and south and are underlain by marble. These are flanked by ridges of gneiss, schist, slate, or quartzite.

Exposed Batholiths and Monadnocks

Recall from Chapter 12 that batholiths—huge plutons of intrusive igneous rock—are formed deep below the earth's surface. Some are eventually uncovered by erosion and appear at the surface. Because batholiths are typically composed of resistant rock, they are eroded into hilly or mountainous uplands.

Batholiths of granitic composition are a major ingredient in the mosaic of ancient rocks comprising the continental shields. A good example is the Idaho batholith, a granite mass exposed over an area of

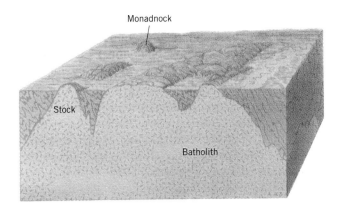

Figure 16.25   Batholiths appear at the land surface only after long-continued erosion has removed thousands of meters of overlying rocks. Small projections of the granite intrusion appear first and are surrounded by older rock. (Drawn by A. N. Strahler.)

Figure 16.24   This dendritic drainage pattern is developed on the dissected Idaho batholith.

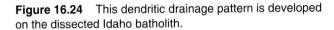

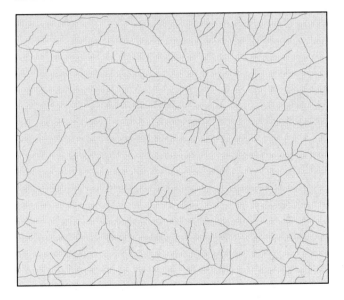

about 40,000 sq km (16,000 sq mi)—a region almost as large as New Hampshire and Vermont combined. Another example is the Sierra Nevada batholith of California, which makes up most of the high central part of that great mountain block.

A dendritic drainage pattern is often well developed on an eroded batholith (Figure 16.24). Note how similar this pattern is to that on horizontal strata (Figure 16.5).

Small bodies of granite, representing domelike projections of batholiths that lie below, are often found surrounded by ancient metamorphic rocks into which the granite was intruded (Figure 16.25). A good example is Stone Mountain, which rises above the Piedmont Upland of Georgia (Figure 16.26). The name **monadnock** has been given to an isolated mountain or hill (such as Stone Mountain) that rises conspicuously above a peneplain. A monadnock develops because the rock within the monadnock is much more resistant to denudation processes than the bedrock of the

Figure 16.26   Stone Mountain, Georgia, is a striking erosion remnant, or monadnock, about 2.4 km (1.5 mi) long and rising 193 m (650 ft) above the surrounding Piedmont peneplain surface. The rock is granite, almost entirely free of joints, and has been rounded into a smooth dome by weathering processes.

Eye on the Environment • Marvelous, Majestic, Monolithic Domes

There is something awesome about a granite dome. Perhaps it's the way it projects above its surroundings, emerging abruptly from the ground and rising so steeply to its rounded summit. Perhaps it's the smooth, barren surface, hardly marred by a crack or crevice. Or maybe it's just the size—a huge, single, uniform object that dwarfs an observer like no other. In any event, these geological curiosities have aroused the interest of observers for centuries—perhaps even millennia.

Probably the most famous of all granite domes is Rio de Janeiro's Sugar Loaf, a monument that serves as a symbol for that cosmopolitan Brazilian city. Of course, Australia has its Ayres Rock, a gigantic rock mass in the desert of

Northern Territory. But, alas, Ayres Rock is mere sandstone, not solid granite. Australia actually has some fine granite domes, located on the Eyre Peninsula of South Australia, but they have not achieved the international fame of Ayres Rock. Splendid granite domes also occur in the Nubian Desert of North Africa. Even Pasadena, California, has its own Little League contender in Eagle Rock, a small but prominent dome of massive conglomerate. These monolithic domes of uniform rock owe their rounded shapes to grain-by-grain disintegration of their rock surfaces. Because the domes lack the joints found in most other rocks, oxidation and hydrolysis act uniformly to weather the rock surface, producing a smooth, rounded form.

And what about the spectacular domes of Yosemite National Park in California's Sierra Nevada Range? Yosemite Valley has several splendid domes high up on either side of the deep canyon. (See photo below.) As we mentioned in Chapter 13, these are exfoliation domes, bearing thick rock shells. As the dome weathers, the outermost shells break apart and slide off as new ones are generated within. Beneath their shells, however, the Yosemite domes seem to be genuine monoliths.

Another fine granite dome is Stone Mountain, near Atlanta (Figure 16.26). Like Sugar Loaf, it is a uniform mass of granite seemingly without joints or fissures. It is the surface remains of an ancient, knoblike granite pluton intruded

These fine examples of monolithic granite domes are found in the Upper Yosemite Valley, Yosemite National Park, California.

Monumental carvings honoring Lee, Jackson, and Davis. Stone Mountain, Georgia.

into older metamorphic rocks of the region. For millions of years, this region has been undergoing denudation, and most of it has been reduced to a rolling upland. Numerous monadnocks of quartzite with elongate, ridgelike forms rise above the upland, but Stone Mountain is unique in terms of its compact outline and smoothly rounded form.

Stone Mountain is unique in yet another way. Carved into its massive granite are huge equestrian sculptures of Robert E. Lee, Stonewall Jackson, and Jefferson Davis (see photo above). These were the artistic vision of Gutzon Borglum, the sculptor who carved the great busts of Washington, Jefferson, Lincoln, and Theodore Roosevelt in the granite of Mount Rushmore. Commissioned in the early 1920s to carve the sculptures by the Daughters of the Confederacy, he began the work with jack hammers and explosives, but in 1924 he resigned in a tiff with his sponsors and destroyed his plans. Eventually, two other sculptors completed his work, and despite long delays, it was dedicated in 1970.

Questions

1. What are the characteristics of domes? What process gives the dome its unique shape?
2. Briefly describe and locate two examples of granite domes.

surrounding region. The name is taken from Mount Monadnock in southern New Hampshire.

Deeply Eroded Volcanoes

Figure 16.27 shows successive stages in the erosion of stratovolcanoes, lava flows, and a caldera. Shown in block (*a*) are active volcanoes in the process of building. These are initial landforms. Lava flows issuing from the volcanoes have spread down into a stream valley, following the downward grade of the valley and forming a lake behind the lava dam.

Figure 16.27 Stages in the erosional development of stratovolcanoes and lava flows. (Drawn by Erwin Raisz. Copyright © by A. N. Strahler.)

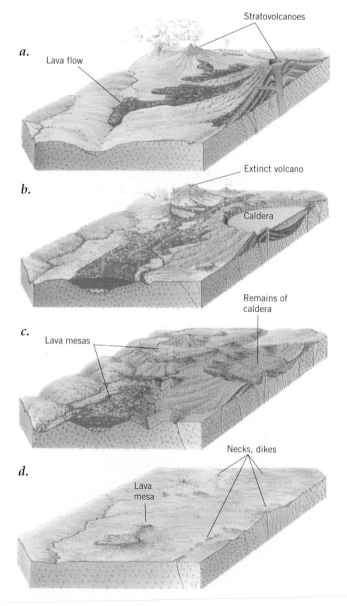

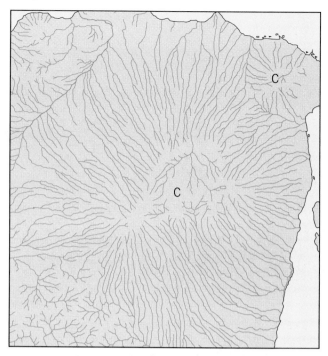

Figure 16.28 Radial drainage patterns of volcanoes in the East Indies. The letter "C" shows the location of a crater.

In block (*b*) some changes have taken place. The most conspicuous change is the destruction of the largest volcano to produce a caldera. A lake occupies the caldera, and a small cone has been built inside. One of the other volcanoes, formed earlier, has become extinct. It has been dissected by streams and has lost its smooth conical form. Smaller, neighboring volcanoes are still active, and the contrast in form is marked.

The system of streams on a dissected volcano cone is a *radial drainage pattern*. Because these streams take their position on a slope of an initial land surface, they are of the consequent variety. It is often possible to recognize volcanoes from a drainage map alone because of the perfection of the radial pattern (Figure 16.28). A good example of a partly dissected volcano is Mount Shasta in California (Figure 16.29).

In Figure 16.27*c* all volcanoes are extinct and have been deeply eroded. The caldera lake has been drained, and the rim has been worn to a low, circular ridge. The lava flows have been able to resist erosion far better than the rock of the surrounding area and have come to stand as mesas high above the general level of the region.

Block (*d*) of Figure 16.27 shows an advanced stage of erosion of stratovolcanoes. All that remains now is a small, sharp peak, called a *volcanic neck*—which is the remains of lava that solidified in the pipe of the volcano. Radiating from it are wall-like dikes, formed of magma, which filled radial fractures around the base

Figure 16.29 Mount Shasta in the Cascade Range, northern California, is a partly dissected composite volcano. Part way up the right-hand side is a more recent subsidiary volcanic cone, Shastina, with a sharp crater rim.

Figure 16.31 Waimaea Canyon, over 760 m (2500 ft) deep, has been eroded into the flank of an extinct shield volcano on Kauai, Hawaii. Gently-sloping layers of basaltic lava flows are exposed in the canyon walls.

of the volcano. Perhaps the finest illustration of a volcanic neck with radial dikes is Ship Rock, New Mexico (Figure 16.30).

Shield volcanoes show erosion features that are quite different from those of stratovolcanoes. Figure 16.31 shows the first stage of erosion of Hawaiian shield volcanoes, with the Mauna Loa volcano and its central crater. The rounded form is an initial land-

form. Eventually, radial consequent streams cut deep canyons into the flanks of the extinct shield volcano, and these canyons are opened out into deep, steep-walled amphitheaters such as Waimea Canyon, on the island of Kauai (Figure 16.30). In the last stages, the original surface of the shield volcano is entirely obliterated, leaving a rugged mountain mass made up of sharp-crested divides and deep canyons.

Figure 16.30 Ship Rock, New Mexico, is a volcanic neck enclosed by a weak shale formation. The peak rises 520 m (1700 ft) above the surrounding plain. In the foreground and to the left, wall-like dikes extend far out from the central peak.

With this chapter, we conclude our series of three chapters on running water as a landforming agent. Because the landscapes of most regions on the earth's land surface are produced by fluvial processes acting on differing rock types, running water is by far the most important agent in creating the variety of landforms that we see around us. The three remaining agents of denudation are waves, wind, and glacial ice. These are the subjects of our final two chapters. In the first, we describe how waves shape the landforms of beaches and coasts, and how wind shapes the dunes of coasts and deserts. In the second, we relate how glacial ice scrapes, pushes, and deposits rock materials to form the distinctive landforms of glaciated regions.

LANDFORMS AND ROCK STRUCTURE IN REVIEW

This chapter has added a new dimension to the realm of landforms produced by fluvial denudation—the variety and complexity introduced by differences in rock composition and crustal structure. Because the various kinds of rocks offer different degrees of resistance to the forces of denudation, they exert a controlling influence on the shapes of landforms. We have seen, too, that streams carving up a landmass are controlled to a high degree by the structure of the rock on which they act. In this way, distinctive drainage patterns evolve.

In arid regions of horizontal strata, resistant rock layers produce vertical cliffs separated by gentler slopes on less resistant rocks. The drainage pattern is dendritic. Where strata are gently dipping, as on a coastal plain, the more resistant rock layers stand out as cuestas, interspersed with lowland valleys on weaker rocks. Consequent streams cut across the cuestas toward the ocean, while subsequent streams follow the valleys. This forms a trellis drainage pattern.

Where rock layers are uparched into a dome, erosion produces a circular arrangement of rock layers outward from the center of the dome. Resistant strata form hogbacks, while weaker rocks form lowlands. Igneous rocks are often revealed in the center. The drainage pattern is annular. In fold belts, the sequence of synclines and anticlines brings a linear pattern of rock layers to the surface. Resistant strata form ridges, while weaker strata form valleys. Like the coastal plain, the drainage pattern will be trellised.

In moist regions, carbonic acid action produces caves and caverns in limestone strata. As the streams draining the region are lowered through time, the water table falls and the caverns are left empty. Eventually, many collapse, providing a surface of numerous sinkholes and streamless valleys that is known as a karst landscape.

Faulting provides an initial surface along which rock layers are moved—the fault scarp. This feature can persist as a fault-line scarp long after the initial scarp is gone. In metamorphic belts, weak layers of shale, slate, and marble underlie valleys, while gneiss and schist form uplands. Quartzite stands out as a ridge-former. Exposed batholiths are often composed of uniform, resistant igneous rock. They erode to form a dendritic drainage pattern. Monadnocks of intrusive igneous rock stand up above a plain of weaker rocks.

Stratovolcanoes produce lava flows that initially follow valleys, but are highly resistant to erosion. After erosion of the surrounding area, they can remain as highlands, lava ridges, or mesas. At the last stages of erosion, all that remains of stratovolcanoes are necks and dikes. Hawaiian shield volcanoes are eroded by streams that form deeply carved valleys with steeply sloping heads. Eventually, these merge to produce a landscape of steep slopes and knifelike ridges.

KEY TERMS

plateau	coastal plain	ridge-and-valley landscape
mesa	cuesta	limestone caverns
butte	artesian well	karst
drainage pattern	sedimentary dome	monadnock

REVIEW QUESTIONS

1. Why is there often a direct relationship between landforms and rock structure? How are geologic structures formed deep within the earth exposed at the surface?

2. Which of the following types of rocks—shale, limestone, sandstone, conglomerate, igneous rocks—tends to form lowlands? Uplands?

3. How are the tilt and orientation of a natural rock plane measured?

4. Identify and sketch a typical landform of flat-lying rock layers found in an arid region.

5. How are coastal plains formed? Identify and describe two landforms found on coastal plains. What drainage pattern is typical of coastal plains?

6. What types of landforms are associated with sedimentary domes? How are they formed? What type of drainage pattern would you expect for a dome and why? Provide an example of an eroded sedimentary dome and describe it briefly.

7. How does a ridge-and-valley landscape arise? Explain the formation of the ridges and valleys. What type of drainage pattern is found on this landscape? Where in the United States can you find a ridge-and-valley landscape?

8. What landforms are associated with limestone rocks in humid regions? Describe what is meant by the term karst.

9. What type of landform(s) might you expect to find in the presence of a fault? Why?

10. Which of the following types of rocks—schist, slate, quartzite, marble, gneiss—tends to form lowlands? Uplands?

11. What types of landforms and drainage pattern develop on batholiths? Provide a sketch of a batholith. What is the difference between a batholith and a monadnock?

12. Describe and sketch the stages through which a landscape of stratovolcanoes is eroded. What distinctive landforms remain as the last remnants?

13. Do shield volcanoes erode differently from stratovolcanoes? How so?

IN-DEPTH QUESTIONS

1. Imagine the following sequence of sedimentary strata—sandstone, shale, limestone, and shale. What landforms would you expect to develop in this structure if the sequence of beds is (a) flat-lying in an arid landscape; (b) slightly tilted as in a coastal plain; (c) folded into a syncline and an anticline in a fold belt; (d) fractured and displaced by a normal fault? Use sketches in your answer.

2. A region of ancient mountain roots, now exposed at the surface, includes a central core of plutonic rocks surrounded on either side by belts of metamorphic rocks. What landforms would you expect for this landscape and why?

Chapter 17

Landforms Made By Waves and Wind

You slow the car and carefully turn right, heading down the narrow, blacktopped road that leads to the lighthouse. The trip wasn't easy, even though the roads were nearly deserted. You saw a number of downed tree limbs along the way and many flooded spots.

What a way to end your Thanksgiving vacation! When your friends offered the use of their cottage on Cape Cod for the long weekend, you had visions of treks along the beach, inhaling the salt air, and watching the sea birds rush to and fro with the swash of the breaking waves. Soon after you arrived, however, the skies turned gray, and the wind and rain began. Instead of a glorious fall weekend, a northeaster was in store. Last night the storm reached its peak. Rain fell in sheets as the wind howled without letup. Although the cottage was snug and the glowing fire was comforting, you were uneasy and slept fitfully. By morning, however, the wind had slackened considerably. Only a fine drizzle emerged from the leaden sky. After breakfast, you decided to venture out and have a look at the beach, and so you headed for the lighthouse.

As you make the turn into the lighthouse road, the car feels the still-strong winds. Correcting the steering, you cautiously follow the narrow black ribbon as it heads toward the beach. Abruptly, the lighthouse comes into view. Although the huge light at the top of the tall, white column is unlit and motionless, the foghorn sounds its low, loud note in measured pulses. The road leads straight into the parking lot, which is perched on top of the coastal bluff overlooking the beach. The far portion of the lot is blocked off with red and white-striped sawhorses anchored by sandbags.

Parking the car, you kill the motor and instantly become aware

of the roaring sound of the surf. Stepping out of the car, you find that the force of the wind is strong, though not overwhelming. Intrigued by the sawhorses, you walk over to see what has happened. A large section of pavement has broken away and slid down the bluff to the beach below. The storm waves, striking the bluff at high tide, must have undermined the pavement, allowing it to fall into the crashing breakers beneath. The wooden stairs down the bluff have fared somewhat better. The sand underneath the bottom part of the lowest flight has been undercut, but access to the beach is not difficult, and you descend. Fortunately, the tide is out.

The beach looks totally different from the way it looked when you visited last summer. Then, the beach had a wide, flat upper portion, ideal for sunbathing and volleyball, and it sloped steeply in the breaker zone. Now the beach is all one long, gentle slope, stretching from the bluffs to the breakers. And the waves are totally different. Instead of approaching the beach, steepening, and then breaking, they are breaking much farther out, where the gray and white of the surf and foam converges with the gray and white of the sky. Disorganized rolls of frothing water move landward, eventually collapsing and running far up the low slope. Clearly, the waves are working the sand from the outer bar all the way to the beach. The roar is intense. A group of gulls flies over the surf zone, swooping low and maneuvering with strong wing beats in the turbulent wind.

Under your feet, the beach feels dense and hard. A few inches above the wet, smooth surface, sheets of wind-blown sand move downwind, braiding and separating into flowing, streamlike currents. The sand strikes your legs and feet, and you bend down to feel the force of the tiny impacts on your hands. Flying grains sting your face.

You turn and look back at the bluff. The layers of sand and silt are clear and distinct. The face of the bluff is almost vertical. Clearly, the wave action has eroded the marine cliff back by several feet. Large, wave-washed lumps of sediment lie at the base of the bluff. Near the lighthouse, the green turf of the lawn with its black topsoil is cleanly cut at the top of the bluff. Suddenly, it dawns on you that it won't be long before the ocean claims the lighthouse itself as an offering to its awesome power.

You linger for a time, enjoying the dark beauty of the scene. Finally, you make your way back to the car and head for home. Although this trip to the beach wasn't exactly what you had in mind when you drove down for the holiday, you are glad to have experienced first-hand the power of the wind and waves.

Wind and waves, the subjects of this chapter, are the primary agents that shape the shoreline. Although storm winds and waves can move a great deal of material and rapidly alter coastal features, gentler waves and winds prevail most of the time. Their work is slower but more consistent, often rebuilding the beach and repairing coastal dunes after severe storms. The problem is that the attractions of the coastline lead to human development, which requires a stable and unchanging coastal configuration. Thus, human and natural forces often conflict in the coastal zone, as we will see in this chapter.

Saharan sand dunes, Algeria.

In Chapter 5, we saw how the rotation of the earth on its axis, combined with unequal solar heating of the earth's surface, produces a system of global winds in which air moves as a fluid across the earth's surface. This motion of air produces a frictional drag that can move surface materials. When winds blow over broad expanses of water, waves are generated. These waves then expend their energy at coastlines, eroding rock, moving sediment, and creating landforms. When winds blow over expanses of soil or alluvium that are not protected by vegetation, fine particles can be picked up and carried long distances before coming to rest. Coarser particles, such as sand, can also be moved to build landforms, such as sand dunes. Because wave action on beaches provides abundant sand, we often find landforms of both wind and waves near coastlines.

This chapter describes the processes of erosion and deposition, as well as the landforms, that result when wind power moves surface materials, either directly as wind or indirectly as waves. Unlike fluvial action or mass wasting, wind and breaking waves can move materials uphill, against the force of gravity. For this reason, wind power is a unique land-forming agent. Another agent that acts in coastal regions is the tide—the rhythmic rise and fall in sea level that is generated by the gravitational attraction of the sun and moon. This subject is also covered in our chapter.

Throughout this chapter we will use the term **shoreline** to mean the shifting line of contact between water and land. The broader term **coastline,** or simply **coast,** refers to a zone in which coastal processes operate or have a strong influence. The coastline includes the shallow water zone in which waves perform their work, as well as beaches and cliffs shaped by waves, and coastal dunes. Also often present are **bays**—bodies of water that are sheltered by the configuration of the coast from strong wave action. Where a river empties into an ocean bay, the bay is termed an **estuary.** In an estuary, fresh and ocean water mix, creating a unique habitat for many plants and animals which is neither fresh water nor ocean.

The continental shoreline is a unique, complex environmental zone of great importance. Here, the saltwater of the oceans receives fresh water from rivers and streams, and contacts the solid mineral base of the continents. Shallow waters and estuaries provide food resources of shellfish, finfish, and waterfowl. The shoreline also provides a base for ocean-going vessels that harvest marine food resources and transport goods between continents. In addition, the coastal zone is a recreational region, with sea breezes, bathing beaches, and opportunities for surfing, skin diving, sport fishing, and boating.

Although the presence of the ocean in the coastal zone provides unique opportunities, it also presents a unique set of restraints and hazards for humans and human development. Some coasts are more suitable for development than others. Where rocky and cliffed, the coast may not provide the sheltered harbors essential for marine commerce. Along other coasts, the enormous energy of storm waves can cut back the shore. Roads and buildings are undermined and eventually claimed by the sea. Storms can raise sea levels, flood low-lying areas, and elevate breaking waves to a level well above normal. Storm surges and seismic sea waves, which cause such inundations, have already been described in earlier chapters. For centuries, humanity has been at war with the sea, building fortresslike walls to keep out ocean waters. Sometimes the battle is won, as land under tidal waters is reclaimed to produce more crops for food and forage. Sometimes it is lost, as storm waves devastate a coastal community.

THE WORK OF WAVES

The most important agent shaping coastal landforms is wave action. The energy of waves is expended primarily in the constant churning of mineral particles and water as waves break at the shore. This churning erodes shoreline materials, moving the shoreline landward. But, as we will see shortly, the action of waves and currents can also move sediment along the shoreline for long distances. This activity can build beaches outward and form barrier islands that shift the land—water contact oceanward.

Wave action is the most important agent in shaping coastal landforms.

Waves travel across the deep ocean with little loss of energy. When waves reach shallow water, the drag of the bottom slows and steepens the wave until it falls and collapses as a *breaker* (Figure 17.1). Many tons of water surge forward, riding up the beach slope.

Where weak or soft materials—various kinds of regolith, such as alluvium—make up the coastline, the force of the forward-moving water alone easily cuts into the coastline. Here, erosion is rapid, and the shoreline may recede rapidly. Under these conditions, a steep bank—a *marine scarp*—is the typical coastal landform (Figure 17.2). It retreats steadily under attack of storm waves.

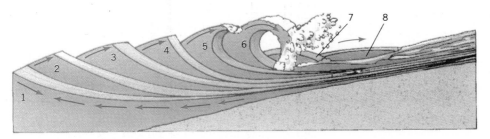

Figure 17.1   A breaking wave. As the wave approaches the beach (1–3), it steepens (4–5) and finally falls forward (6–7), rushing up the beach slope (8). (After W. M. Davis.)

Figure 17.2   Mohegan Bluffs, Block Island, Rhode Island. This marine scarp of Ice Age sediments is being rapidly eroded, threatening historic Southeast Lighthouse. In 1993 the lighthouse was moved to a safer spot 73.5 m (245 ft) away.

Marine Cliffs

Where a **marine cliff** lies within reach of the moving water, it is impacted with enormous force. Rock fragments of all sizes, from sand to cobbles, are carried by the surging water and thrust against bedrock of the cliff. The impact breaks away new rock fragments, and the cliff is undercut at the base. In this way, the cliff erodes shoreward, maintaining its form as it retreats. The retreat of a marine cliff formed of hard bedrock is exceedingly slow, when judged in terms of a human life span.

Figure 17.3 illustrates some details of a typical marine cliff. A deep basal indentation, the *wave-cut notch,* marks the line of most intense wave erosion. The waves find points of weakness in the bedrock and penetrate deeply to form crevices and *sea caves.* Where a more resistant rock mass projects seaward, it may be cut through to form a picturesque *sea arch.* After an arch collapses, a rock column, known as a *stack,* remains. Eventually, the stack is toppled by wave action and is leveled. As a sea cliff retreats landward, continued wave abrasion can form an *abrasion platform* (Figure 17.4). This sloping rock floor is eroded and widened by abrasion beneath the breakers. If a beach is present, it is little more than a thin layer of gravel and cobblestones atop the abrasion platform.

Beaches

Where sand is in abundant supply, it accumulates as a thick, wedge-shaped deposit, or **beach.** Beaches absorb the energy of breaking waves. During short periods of storm activity, the beach is cut back, and sand is carried offshore a short distance by the heavy wave action. However, the sand is slowly returned to the beach dur-

Figure 17.3 Landforms of sea cliffs.

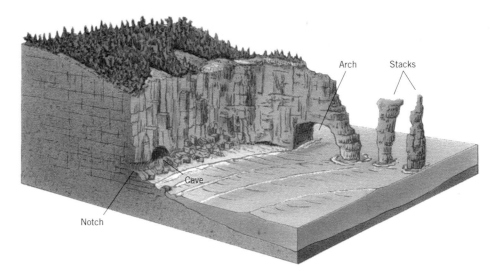

ing long periods when waves are weak. In this way, a beach may retain a fairly stable but alternating configuration over many years' time.

Beaches are shaped by alternate landward and seaward currents of water generated by breaking waves (Figure 17.5). After a breaker has collapsed, a foamy, turbulent sheet of water rides up the beach slope. This *swash* is a powerful surge that causes a landward movement of sand and gravel on the beach. When the force of the swash has been spent against the slope of the beach, a return flow, or *backwash*, pours down the beach (Figure 17.6*a*). Sand and gravel are swept sea-

ward by the backwash. Surf bathers are familiar with the backwash as a strong seaward current felt in the breaker zone. This "undercurrent" or "undertow," as it is popularly called, can be strong enough to sweep unwary bathers off their feet and carry them seaward beneath the next oncoming breaker.

Littoral Drift

The unceasing shifting of beach materials with swash and backwash of breaking waves also results in a side-

Figure 17.4 A wave-cut abrasion platform on the central coast of California, Montagne d'Oro State Park.

wise movement known as *beach drift* (Figure 17.6*a*). Wave fronts usually approach the shore at less than a right angle, so that the swash and its burden of sand ride obliquely up the beach. After the wave has spent its energy, the backwash flows down the slope of the beach in the most direct downhill direction. The particles are dragged directly seaward and come to rest at a position to one side of the starting place. On a particular day, wave fronts usually approach consistently from the same direction, so that this movement is repeated many times. Individual rock particles travel long distances along the shore. Multiplied many thousands of times to include the numberless particles of the beach, beach drift becomes a very significant form of sediment transport.

The swash and backwash of breakers attacking the shore at an angle creates beach drift.

Sediment in the shore zone is also moved along the beach in a related, but different, process. When waves approach a shoreline at an angle to the beach, a current is set up parallel to the shore in a direction away from the wind. This is known as a *longshore current* (Figure 17.6*b*). When wave and wind conditions are favorable, this current is capable of carrying sand along the sea bottom. The process is called *longshore drift*. Beach drift and longshore drift, acting together, move particles in the same direction for a given set of onshore winds. The total process is called **littoral drift**

(Figure 17.6*c*). ("Littoral" means "pertaining to a coast or shore.")

Littoral drift operates to shape shorelines in two quite different situations. Where the shoreline is straight or broadly curved for many kilometers at a stretch, littoral drift moves the sand along the beach in one direction for a given set of prevailing winds. This situation is shown in Figure 17.6*c*. Where a bay exists, the sand is carried out into open water as a long finger, or *sandspit* (Figure 17.7). As the sandspit grows, it forms a barrier, called a *bar*, across the mouth of the bay.

A second situation is shown in Figure 17.8. Here the coastline consists of prominent headlands, projecting seaward, and deep bays. Approaching wave fronts slow when the water becomes shallow, and this slowing effect causes the wave front to wrap around the headland. High, wave-cut cliffs develop shoreward of an abrasion platform. Sediment from the eroding cliffs is carried by littoral drift along the sides of the bay, converging on the head of the bay. The result is a crescent-shaped beach, often called a *pocket beach*.

Littoral Drift and Shore Protection

When sand arrives at a particular section of the beach more rapidly than it is carried away, the beach is widened and built oceanward. This change is called **progradation** (building out). When sand leaves a section of beach more rapidly than it is brought in, the beach is narrowed and the shoreline moves landward. This change is called **retrogradation** (cutting back).

Figure 17.5 Waves breaking on a broad beach on the island of Molokai, Hawaii.

Figure 17.6 How waves move sediment by littoral drift. (*a*) Swash and backwash move particles along the beach in beach drift. (*b*) Waves set up a longshore current that moves particles by longshore drift. (*c*) Littoral drift, produced by these two processes, creates a sandspit.

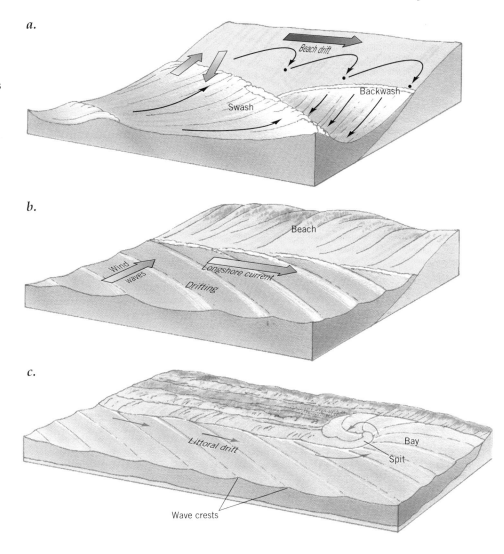

a.

Beach drift

Swash

Backwash

b.

Beach

Wind

waves

Longshore current

Drifting

c.

Littoral drift

Bay

Spit

Wave crests

Figure 17.7 This white sandspit is growing in a direction toward the observer. Monomoy Point, Cape Cod, Massachusetts.

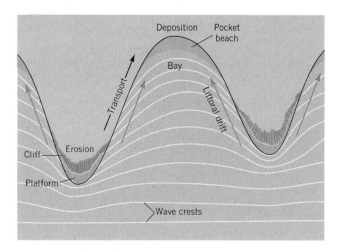

Figure 17.8   On an embayed coast, sediment is carried from eroding headlands to the bayheads, where pocket beaches are formed. (Copyright © A. N. Strahler.)

Along stretches of shoreline affected by retrogradation, the beach may be seriously depleted or even entirely disappear, destroying valuable shore property. In some circumstances, structures can be installed that will cause progradation, and so build a broad, protective beach. This is done by installing groins at close intervals along the beach. A *groin* is simply a wall or embankment built at right angles to the shoreline. It may be constructed of large rock masses, of concrete, or of wooden pilings. The groins act to trap sediment moving along the shore as littoral drift.

The effect of groins is shown in Figure 17.9. The beach toward the back of the scene is protected by groins, while the beach in the foreground is not. The trapping action of the groins has starved the foreground beach of sand, narrowing the beach and leaving the houses unprotected. Storm waves have severely eroded the unprotected beach and cut an inlet across the beach.

Groins trap longshore drift and help prevent beach retrogradation.

In some cases, the source of beach sand is sediment delivered to the coast by a river. Construction of dams far upstream on the river may drastically reduce the sediment load of the river, cutting off the source of sand for littoral drift. Retrogradation can then occur on a long stretch of shoreline.

TIDAL CURRENTS

Most marine coastlines are influenced by the *ocean tide*, a rhythmic rise and fall of sea level under the influ-

Figure 17.9   Erosion by severe storms during the winter of 1993 has carved out an inlet in this barrier beach on the south shore of Long Island, New York. A system of groins has trapped sand, protecting the far stretch of beach. In the foreground, the beach has receded well inland of the houses that were once located on its edge.

ence of changing attractive forces of moon and sun on the rotating earth. Where tides are great, the effects of changing water level and the tidal currents thus set in motion are of major importance in shaping coastal landforms.

The tidal rise and fall of water level is graphically represented by the *tide curve*. We prepare this graph by making observations of water level in a location sheltered from wave action. A measuring stick or scale, known as a tide staff, is attached to a pier or seawall, and the height of the water surface is noted at regular intervals, say, each half-hour. We then plot the changes of water level and draw the tide curve.

Figure 17.10 is a tide curve for Boston Harbor covering a day's time. The water reached its maximum height, or high water, at 3.7 m (12 ft) on the tide staff, and then fell to its minimum height, or low water, at 0.8 m (2.6 ft), occurring about $6\frac{1}{4}$ hours later. A second high water occurred about $12\frac{1}{2}$ hours after the previous high water, completing a single tidal cycle. In this example, the tidal range, or difference between heights of successive high and low waters, is 2.9 m (9.3 ft).

In bays and estuaries, the changing tide sets in motion currents of water known as *tidal currents*. The relationships between tidal currents and the tide curve are shown in Figure 17.11. When the tide begins to fall, an *ebb current* sets in. This flow ceases about the time when the tide is at its lowest point. As the tide begins to rise, a landward current, the *flood current*, begins to flow.

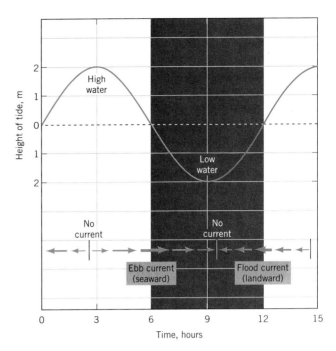

Figure 17.11 The ebb current flows seaward as the tide level falls. The flood current flows landward as the tide level rises. Tidal currents are strongest when water level is changing most rapidly, in the middle of the cycle.

Tidal Current Deposits

Ebb and flood currents generated by tides perform several important functions along a shoreline. First, the currents that flow in and out of bays through narrow inlets are very swift and can scour the inlet strongly. This keeps the inlet open, despite the tendency of shore-drifting processes to close the inlet with sand.

Second, tidal currents carry large amounts of fine silt and clay in suspension. This fine sediment is derived from streams that enter the bays or from bottom muds agitated by storm wave action. It settles to the floors of the bays and estuaries, where it accumulates in layers and gradually fills the bays. Much organic matter is present in this sediment.

In time, tidal sediments fill the bays and produce mud flats, which are barren expanses of silt and clay. They are exposed at low tide but covered at high tide. Next, a growth of salt-tolerant plants takes hold on the mud flat. The plant stems trap more sediment, and the flat is built up to approximately the level of high tide, becoming a *salt marsh* (Figure 17.12). A thick layer of peat is eventually formed at the surface. Tidal currents maintain their flow through the salt marsh by means of a highly complex network of winding tidal streams.

Salt marshes can be drained and used for agriculture. The salt marsh is first cut off from the sea by

Figure 17.10 Height of water at Boston Harbor measured every half hour.

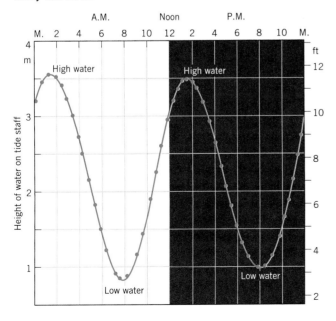

Figure 17.12   A tideland salt marsh. Chincoteague National Wildlife Refuge, Virginia.

construction of an embankment of earth (a dike). Gates are installed in the dike to allow the freshwater drainage of the land to exit when the ocean's tide level is low. Gradually, the saltwater is excluded, and the soil water of the diked land becomes fresh. Such diked lands are intensively developed in Holland (polders) and southeast England (fenlands).

The surface of the reclaimed salt marsh subsides over many decades because of compaction of the underlying peat layers. In fact, the land surface can lie well below mean sea level. The threat of flooding by saltwater, when storm waves break the dikes, constantly hangs over the inhabitants of such low areas

TYPES OF COASTLINES

There are a number of different types of coastlines. Each type is unique because of the distinctive landmass against which the ocean water has come to rest. One group of coastline types derives its qualities from **submergence,** the partial drowning of a coast by a rise of sea level or a sinking of the crust. Another group derives its qualities from **emergence,** the exposure of submarine landforms by a falling of sea level or a rising of the crust. Another group of coastline types results when new land is built out into the ocean by volcanoes and lava flows, by the growth of river deltas, or by the growth of coral reefs.

A few important types of coastlines are illustrated in Figure 17.13. The *ria coast* (*a*) is a deeply embayed coast resulting from submergence of a landmass dissected by streams. The *fiord coast* (*b*) is deeply indented by steep-walled fiords, which are submerged glacial troughs (discussed further in Chapter 18). The *barrier-island coast* (*c*) is associated with a recently emerged coastal plain. Here, the offshore slope is very gentle, and a barrier island of sand, lying a short distance from the coast, is created by wave action. Large rivers build elaborate deltas, producing *delta coasts* (*d*). The *volcano coast* (*e*) is formed by the eruption of volcanoes and lava flows, partly constructed below water level. Reef-building corals create new land and make a *coral-reef coast* (*f*). Down-faulting of the coastal margin of a continent can allow the shoreline to come to rest against a fault scarp, producing a *fault coast* (*g*).

Ria coasts and fiord coasts result from submergence of a landmass, while a barrier-island coast results from emergence.

Ria Coasts and Fiord Coasts

The ria coast has many offshore islands. It takes its name from the Spanish word for estuary, *ria*. A ria coast is formed when a rise of sea level or a crustal sinking (or both) brings the shoreline to rest against the sides of valleys previously carved by streams (Figure 17.13*a*). Wave attack forms cliffs on the exposed seaward sides of islands and headlands. Sediment produced by wave action accumulates in the form of beaches along the cliffed headlands and at the heads of bays. This sediment is carried by littoral drift and is often built into sandspits across bay mouths and as connecting links between islands and mainland.

The fiord coast is similar to the ria coast in that it is a shoreline produced by submergence. However, the submerged valleys were carved by flowing glaciers instead of streams. As a result, the valleys are deep, with straight, steep sides. Because sediment rapidly sinks into the deep water, beaches are rare. Fiords are described further in Chapter 18 (see Figure 18.8).

Barrier-Island Coasts

In contrast to ria and fiord coasts, with their bold relief and deeply embayed outlines produced by submergence, are low-lying coasts from which the land slopes gently beneath the sea. The coastal plain of the Atlantic and Gulf coasts of the United States presents a particularly fine example of such a gently sloping surface. As we explained in Chapter 16, this coastal plain

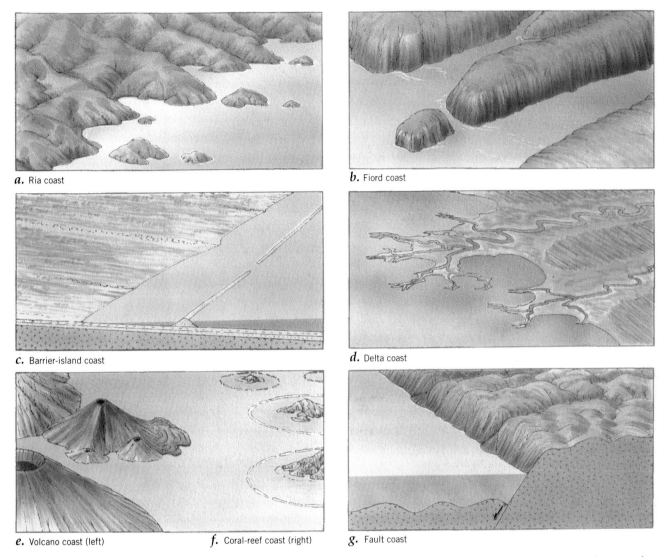

a. Ria coast

b. Fiord coast

c. Barrier-island coast

d. Delta coast

e. Volcano coast (left)   *f.* Coral-reef coast (right)

g. Fault coast

Figure 17.13 Seven common kinds of coastlines. These examples have been selected to illustrate a wide range in coastal features. (Drawn by A. N. Strahler.)

is a belt of relatively young sedimentary strata, formerly accumulated beneath the sea as deposits on the continental shelf. During the latter part of the Cenozoic Era and into recent time, the coastal plain emerged from the ocean as a result of repeated crustal uplifts.

Along much of the Atlantic Gulf coast there exist *barrier islands,* low ridges of sand built by waves and further increased in height by the growth of sand dunes (Figure 17.14). Behind the barrier island lies a *lagoon.* It is a broad expanse of shallow water in places largely filled with tidal deposits.

A characteristic feature of barrier islands is the presence of gaps, known as *tidal inlets.* Strong currents flow alternately landward and seaward through these gaps as the tide rises and falls. In heavy storms, the barrier

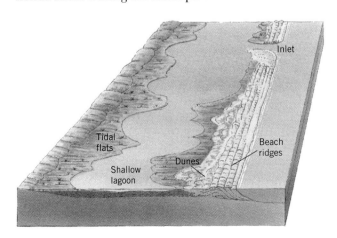

Inlet

Tidal flats

Shallow lagoon

Dunes

Beach ridges

Figure 17.14 A barrier island is separated from the mainland by a wide lagoon. Sediments fill the lagoon, while dune ridges advance over the tidal flats. An inlet allows tidal flows to pass in and out of the lagoon. (Drawn by A. N. Strahler.)

a.

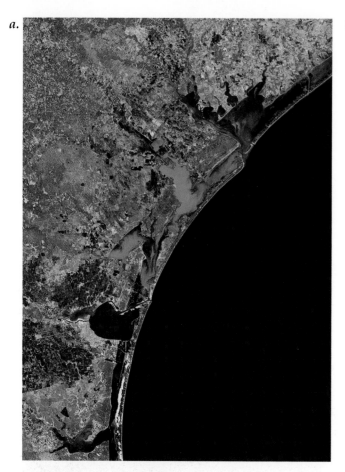

b.

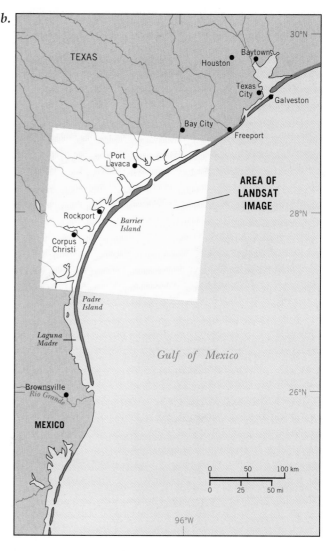

Figure 17.15 (*a*) A Landsat image of the Texas barrier-island coast. (*b*) Index map of the location. Red colors on the Landsat image indicate vegetation. Bright white areas are dunes or beach sand. Lighter blue tones mark the presence of sediment in water inside of the barrier beach.

island may be breached by new inlets (Figure 17.9). After that occurs, tidal current scour will tend to keep a new inlet open. In some cases, the inlet is closed later by shore drifting of beach sand.

Perhaps the finest example of a barrier-island coast is the Gulf coast of Texas (Figure 17.15). Here, the barrier island is unbroken for more than 160 km (100 mi) at a single stretch, and natural passes are few. The lagoon is 8 to 16 km (5 to 10 mi) wide. As is true of most barrier-island coasts, deep natural harbors are lacking. Extensive channel dredging is required. Ocean-going ships must enter and leave the lagoon through passes in the barrier island.

Delta and Volcano Coasts

The deposit of clay, silt, and sand made by a stream or river where it flows into a body of standing water is known as a **delta.** Deposition is caused by rapid reduction in velocity of the current as it pushes out into the

standing water. Typically, the river channel divides and subdivides into lesser channels called *distributaries.* The coarser sand and silt particles settle out first, while the fine clays continue out farthest and eventually come to rest in fairly deep water. Contact of fresh with saltwater causes the finest clay particles to clot together to form larger particles that settle to the sea floor.

A delta results when a river empties into the ocean, depositing sediment that is worked by waves and currents.

Deltas show a wide variety of outlines. The Nile delta has the basic triangular shape of the Greek letter delta. From a spacecraft, the delta appears as a fertile plain bounded by desert land (Figure 17.16). In outline, this delta resembles an alluvial fan. The delta has two major distributaries, one of which terminates in the Damietta mouth and the other in the Rosetta mouth. The mouths project as cusps (tooth-forms) into the

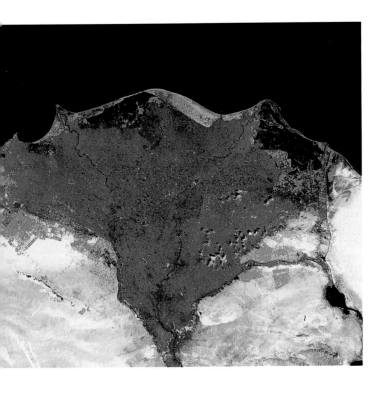

Mediterranean Sea. Sediment from the mouths has been carried along the shore by littoral drift to form curved barrier beaches.

The Mississippi delta has a different shape. Long, branching fingers grow far out into the Gulf of Mexico at the ends of the distributaries. A satellite image of the delta, Figure 17.17, shows the great quantity of suspended sediment—clay and fine silt—being discharged by the river into the Gulf. It amounts to about 1 million metric tons per day.

Delta growth is often rapid, ranging from 3 m (10 ft) per year for the Nile to 60 m (200 ft) per year for the Mississippi delta. Some cities and towns that were at river mouths several hundred years ago are today several kilometers inland. An important engineering problem is to keep an open channel for ocean-going vessels that must enter the delta distributaries to reach port. The mouths of the Mississippi River delta distributaries, known locally as "passes," have been extended by the construction of jetties. The narrowed stream is forced to move faster and thus scours a deep channel suitable for the passage of large vessels.

Figure 17.16   A Landsat image of the Nile delta. The lush vegetation cover of the delta lands appears in deep red tones. Blue colors indicate fallow land. A few puffy clouds are present near the southeast edge of the delta.

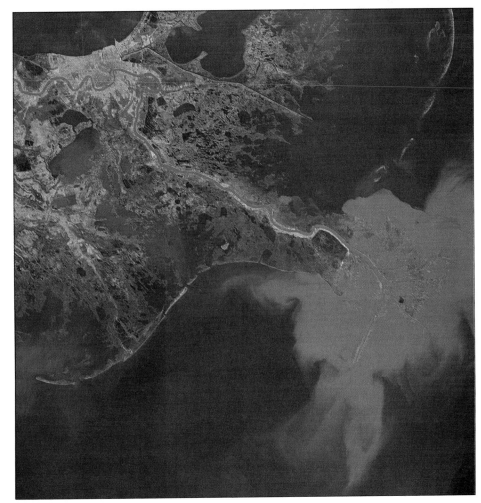

Figure 17.17   The Mississippi delta recorded by Landsat. The natural levees of the bird-foot delta appear as lacelike filaments in a great pool of turbid river water. New Orleans can be seen at the upper left, occupying the region between the natural levees of the Mississippi River and the southern shore of Lake Pontchartrain.

Figure 17.18 The Island of Moorea and its fringing coral reef, Society Islands, South Pacific Ocean. The island is a deeply dissected volcano with a history of submergence. Tahiti lies in the background.

A volcano coast is similar to a delta coast in that it marks the boundary between the ocean and new-created land. However, the new land consists of volcanic deposits—lava and ash—that flow from active volcanoes into the ocean. Low cliffs occur when wave action erodes the fresh deposits. Beaches are typically narrow, steep, and composed of fine particles of the extrusive rock.

Coral-Reef Coasts

Coral-reef coasts are unique in that the addition of new land is made by organisms—corals and algae. Growing together, these organisms secrete rocklike deposits of mineral carbonate, called **coral reefs.** As coral colonies die, new ones are built on them, accumulating as limestone. Coral fragments are torn free by wave attack, and the pulverized fragments accumulate as sand beaches.

Coral-reef coasts occur in warm, tropical and equatorial waters between the limits of lat. 30° N and 25° S. Water temperatures above 20°C (68°F) are necessary for dense growth of coral reefs. Reef corals live near the water surface. The sea water must be free of suspended sediment and well aerated for vigorous coral growth. For this reason, corals thrive in positions exposed to wave attack from the open sea. Because muddy water prevents coral growth, reefs are missing opposite the mouths of muddy streams. Coral reefs are remarkably flat on top. They are exposed at low tide and covered at high tide.

There are three distinctive types of coral reefs— fringing reefs, barrier reefs, and atolls. *Fringing reefs* are built as platforms attached to shore (Figure 17.18). They are widest in front of headlands where wave attack is strongest. At that location, the corals grow best because they are immersed in clear water containing an abundance of food. *Barrier reefs* lie out from shore and are separated from the mainland by a lagoon (Figure 17.18). Narrow gaps occur at intervals in barrier reefs. Through these openings, excess water from breaking waves is returned from the lagoon to the open sea.

Atolls are more or less circular coral reefs enclosing a lagoon but have no land inside (Figure 17.19). On large atolls, parts of the reef have been built up by wave action and wind to form low island chains connected by the reef. Most atolls are built on a foundation of volcanic rock. This foundation is thought to have been a basaltic volcano, originally built on the deep ocean floor and rising above the ocean surface. After being eroded by streams and planed off by wave erosion, the extinct volcano slowly subsided, while the fringing coral reef continued to build upward.

Raised Shorelines and Marine Terraces

The active life of a shoreline is sometimes cut short by a sudden rise of the coast. When this tectonic event occurs, a *raised shoreline* is formed. If present, the marine cliff and abrasion platform are abruptly raised above the level of wave action. The former abrasion

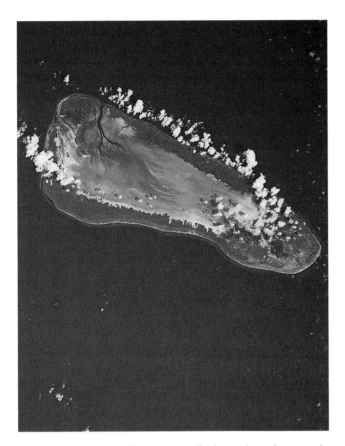

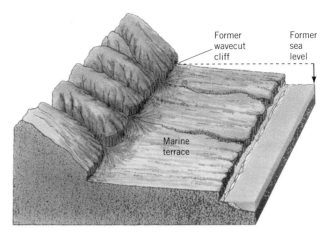

Figure 17.20   A raised shoreline becomes a cliff parallel with the newer, lower shoreline. The former abrasion platform is now a marine terrace. (Drawn by A. N. Strahler.)

platform has now become a *marine terrace* (Figure 17.20). Of course, fluvial denudation acts to erode the terrace as soon as it is formed. The terrace may also undergo partial burial under alluvial fan deposits.

Raised shorelines are common along the continental and island coasts of the Pacific Ocean, because here tectonic processes are active along the mountain and island arcs. Repeated uplifts result in a series of raised shorelines in a steplike arrangement. Fine examples of these multiple marine terraces are seen on the western slope of San Clemente Island, off the California coast (Figure 17.21).

Figure 17.19   A Pacific Island atoll, shown in a photograph made by an orbiting astronaut. The thin white line of beach is the only land above sea level. Outside of the beach, a reef surrounds the island, and an inner lagoon fills its center. A ring of puffy clouds encircles the atoll.

Figure 17.21   Marine terraces on the western slope of San Clemente Island, off the southern California coast. More than twenty terraces have been identified in this series. The highest has an elevation of about 400 m (1300 ft).

WIND ACTION

Transportation and deposition of sand by wind is an important process in shaping certain coastal landforms. We have already mentioned coastal sand dunes, which are derived from beach sand. In the remainder of this chapter, we investigate the transport of mineral particles by wind and the shaping of dune forms. Our discussion also provides information about dune forms far from the coast—in desert environments, where the lack of vegetation cover allows dunes to develop if a source of abundant sand particles is present.

Wind blowing over the land surface is one of the active agents of landform development. Ordinarily, wind is not strong enough to dislodge mineral matter from the surfaces of unweathered rock, or from moist, clay-rich soils, or from soils bound by a dense plant cover. Instead, the action of wind in eroding and transporting sediment is limited to land surfaces where small mineral and organic particles are in a loose, dry state. These areas are typically deserts and semiarid lands (steppes). An exception is the coastal environment, where beaches provide abundant supplies of loose sand. In this environment, wind action shapes coastal dunes, even where the climate is humid and the land surface inland from the coast is well protected by a plant cover.

> **Wind action normally moves mineral particles only when they are in dry state and unprotected by a vegetation cover.**

Landforms shaped and sustained by wind erosion and deposition provide distinctive life environments. The ecosystems that occupy these habitats are often highly specialized. In semidesert climates with little soil water, the growth of plants can be limited due to fierce wind action. Wind action tends to rework the landforms, while plants tend to stabilize the landforms and protect them from wind action. The balance between these two forces can easily be upset by human activities, often with serious consequences of land degradation. Changes in climate, both natural and those induced by human activity, can also upset the balance. To understand these environmental changes, we need to acquire a working knowledge of the physical processes of wind action on the land surfaces.

Erosion by Wind

Wind performs two kinds of erosional work: deflation and abrasion. The removal of loose particles from the ground is termed **deflation.** Loose particles lying on the ground surface may be lifted into the air or rolled along the ground. In the process of *wind abrasion,* wind drives sand and dust particles against an exposed rock or soil surface. This causes the surface to be worn away by the impact of the particles. Abrasion requires cutting tools—mineral particles—carried by the wind, whereas deflation is accomplished by air currents alone.

The sandblasting action of wind abrasion against exposed rock surfaces is limited to the basal meter or two of a rock mass that rises above a flat plain. This height is the limit to which sand grains can rise high into the air. Wind abrasion produces pits, grooves, and hollows in the rock. Wooden utility poles on windswept sandy plains are quickly cut through at the base unless a protective metal sheathing or heap of large stones is placed around the base.

Deflation acts on dry soil or sediment. Dry river courses, beaches, and areas of recently formed glacial deposits are susceptible to deflation. In dry climates, much of the ground surface is subject to deflation because the soil or rock is largely bare of vegetation.

Wind is selective in its deflational action. The finest particles, those of clay and silt sizes, are lifted and raised into the air—sometimes to a height of a thousand meters (3300 ft) or more. Sand grains are moved only when winds are at least moderately strong and usually travel within a meter or two (3 to 6 ft) of the ground. Gravel fragments and rounded pebbles can be rolled or pushed over flat ground by strong winds, but they do not travel far. They become easily lodged in hollows or between other large grains. Consequently, where a mixture of size of particles is present on the ground, the finer sized particles are removed and the coarser particles remain behind.

A landform produced by deflation is a shallow depression called a **blowout.** The size of the depression may range from a few meters (10 to 20 ft) to a kilometer (0.6 mi) or more in diameter, although it is usually only a few meters deep. Blowouts form in plains regions of dry climate. Any small depression in the surface of the plain, especially where the grass cover has been broken or disturbed, can form a blowout. Rains fill the depression and create a shallow pond or lake. As the water evaporates, the mud bottom dries out and cracks, leaving small scales or pellets of dried mud. These particles are lifted out by the wind. In grazing lands, animals trample the margins of the depression into a mass of mud, breaking down the protective grass-root structure and facilitating removal of dried particles. In this way, the depression is enlarged. Blowouts are also found on rock surfaces where the rock is being disintegrated by weathering.

Figure 17.22 This desert pavement is formed of closely-fitted rock fragments. Lying on the surface are fine examples of wind-faceted rocks, which attain their unusual shapes by long-continued sandblasting.

Deflation is also active in semidesert and desert regions. In the southwestern United States, playas often occupy large areas on the flat floors of tectonic basins. Deflation has reduced many playas several meters in elevation.

Rainbeat, overland flow, and deflation may be active for long periods on the gently sloping surface of a desert alluvial fan or alluvial terrace. On these surfaces, rock fragments ranging in size from pebbles to small boulders become concentrated into a surface layer known as a **desert pavement** (Figure 17.22). The large fragments become closely fitted together, concealing the smaller particles—grains of sand, silt, and clay—that lie beneath. The pavement acts as an armor that effectively protects the finer particles from rapid removal by deflation. However, the pavement is easily disturbed by the wheels of trucks and motorcycles, exposing the finer particles and allowing severe deflation and water erosion to follow.

Dust Storms

Strong, turbulent winds blowing over barren surfaces lift great quantities of fine dust into the air, forming a dense, high cloud called a **dust storm.** In semiarid grasslands, a dust storm is generated where ground surfaces have been stripped of protective vegetation cover by cultivation or grazing. Strong winds cause soil particles and coarse sand grains to hop along the ground. This motion breaks down the soil particles and disturbs more soil. With each impact, fine dust is released that can be carried upward by turbulent winds.

A dust storm approaches as a dark cloud extending from the ground surface to heights of several thousand meters (Figure 17.23). Typically, the advancing cloud wall represents a rapidly moving cold front. Within the dust cloud there is deep gloom or even total darkness. Visibility is cut to a few meters, and a fine choking dust penetrates everywhere.

Estimates reveal that as much as 1000 metric tons of dust may be suspended in a cubic kilometer of air (4000 tons per cu mi). On this basis, a large dust storm can carry more than 100 million tons of dust—enough to make a hill 30 m (100 ft) high and 3 km (2 mi) across the base. Dust travels long distances in the air. Dust from a single desert storm can be traced as far as 4000 km (2500 mi).

Figure 17.23 An approaching dust storm in eastern Kenya.

Eye on the Environment • The Threat of Rising Sea Level from Global Warming

As you know, the earth's climate seems to be warming, perhaps in response to human activity. But have you thought about some of the possible effects of climatic warming? One of these effects is the rise of sea level. Imagine, for a moment, the effect of an increase of several meters. Low-lying areas—such as tidal estuaries and deltas—would be flooded. Coastal cities would have to be defended from wave attack. Barrier islands would be swept landward or overtopped. Atolls and other low islands would disappear. The threat of a rise in sea level is real, but like other possible impacts of global warming, it is far from certain how long it will take sea level to rise and how great the effect will be.

A review of the scientific principles involved here may be helpful. First, these changes in sea level are *eustatic*—that is, they refer to worldwide sea-level change. Local relative changes in sea level resulting from the uplift or downsinking of the earth's crust are tectonic effects and not included in eustatic change. Eustatic change can result from a change either in water volume of the world ocean or in the water-holding capacity of the ocean basins. The second possibility relates to plate tectonics and is on a longer time scale than that involved in global warming.

Climate affects the water volume of the oceans in two ways. First, a volume change results from water temperature increase. Global warming will raise the temperature of the uppermost layer of the oceans—at most a few hundred meters thick—and the resulting thermal expansion of that water will cause a small but significant eustatic rise.

Second, global warming may increase the rate of glacial melting, causing mountain glaciers and ice sheets to shrink in volume. Meltwater would then increase the volume of the world ocean. Keep in mind, however, that, by increasing precipitation, global warming will also promote the growth of glaciers. Thus, we must consider the relative strength of the opposite processes of melting and growth. Large swings in world sea level have taken place throughout the past 3 million years as ice sheets have cyclically expanded and contracted. (In Chapter 18 we will return to a discussion of the mechanism of possible sea-level rise by ice melting.)

Has a slow eustatic rise of sea level accompanied the overall global temperature rise that started in the mid-1880s? Apparently so. Scientists examining tide gauge records have concluded that in the 80-year period 1900–1980, the total eustatic rise has been about 80 mm (3 in.). About half of this increase is attributed to thermal expansion. The average rate of rise was about 1.2 mm/yr (0.05 in./yr) through the early and middle 1900s. Recently, the rate has doubled to about 2.4 mm/yr (0.1 in./yr). Although this accelerating rate of rise seems to fit the effects of observed global warming, keep in mind that the linkage between sea level rise and global warming is still uncertain.

Projections of the rate of sea-level rise into the next century vary widely. For example, computer modeling evaluated by the National Academy of Sciences in 1989 showed estimates of sea-level increase by the year 2100 from as small as 0.3 m (1 ft) to as large as 2 m (7 ft). Using a rise of 1 m (3 ft) by the middle or late 2000s, various national and international agencies have estimated the impacts, both physical and financial, on major coastal cities of the world. Large river deltas, along with reclaimed tidal estuaries and marshes and the

SAND DUNES

A **sand dune** is any hill of loose sand shaped by the wind. Active dunes constantly change form under wind currents. Dunes form where there is a source of sand—for example, a sandstone formation that weathers easily to release individual grains, or perhaps a beach supplied with abundant sand from a nearby river mouth. Dunes must be free of a vegetation cover in order to form and move. They become inactive when stabilized by a vegetation cover, or when patterns of wind or sand sources change. Table 17.1 summarizes six important types of sand dunes.

Dune sand is most commonly composed of the mineral quartz, which is extremely hard and largely immune to chemical decay. The grains are beautifully rounded by abrasion (see Figure 10.11). Under the force of strong winds, the grains move in long, low leaps, bouncing after impact with other grains. Rebounding grains rarely rise more than half a

cities built on them, would face catastrophic damage through storm surf and storm surges in concert with high tides. The costs of building protective barriers to meet this threat are enormously high, and the barriers may not be effective in all situations. Along cliffed coasts under direct attack by waves, retrogradation will be intensified, resulting in massive losses of shorefront properties (see photo). Some island nations would lose most or all of their land. The Maldives, a chain of atolls in the Indian Ocean with a population of about a quarter of a million people, is an example.

One note on the positive side comes from biologists specializing in the ecology of coral reefs. They point out that reef-building corals will intensify their rates of growth to maintain the correct level in relation to the sea surface. Their increased production of calcium carbonate, withdrawing carbon dioxide from the atmosphere at a rate as high as 4 to 9 percent of the rate of emission, could substantially reduce the rate of global warming. If, however, reef growth could not keep pace with the rate of rise of sea level, the wholesale extermination of living reefs could result.

Coastal erosion has undermined the bluff beneath these two buildings, depositing them on the beach below.

Questions

1. How might global climate change bring about a rise in sea level? What are the possible effects of such a rise?
2. Is a sea-level rise occurring at present? What are the projections for the future, and what are the implications of the projections?

centimeter above the dune surface. Grains struck by bouncing grains are pushed forward, and, in this way, the surface sand layer creeps downwind. This type of hopping, bouncing movement is termed *saltation*.

Types of Sand Dunes

One common type of sand dune is an isolated heap of free sand called a *barchan*, or *crescentic dune*. This type of dune has the outline of a crescent, and the points of the crescent are directed downwind (Figure 17.24). On the upwind side of the crest, the sand slope is gentle and smoothly rounded. On the downwind side of the dune, within the crescent, is a steep dune slope, the *slip face*. This face maintains an angle of about 35° from the horizontal (Figure 17.25). Sand grains slide down the steep face after being blown over the sharp crest. When a strong wind is blowing, the flying sand makes a visible cloud at the crest.

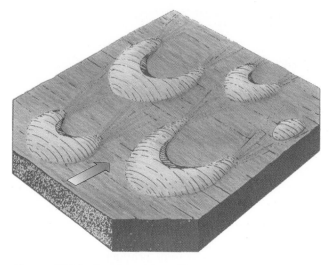

Figure 17.24 Barchan dunes. The arrow indicates wind direction. (Drawn by A. N. Strahler.)

Barchan dunes usually rest on a flat, pebble-covered ground surface. The life of a barchan dune may begin as a sand drift in the lee of some obstacle, such as a small hill, rock, or clump of brush. Once a sufficient mass of sand has formed, it begins to move downwind, taking on the crescent form. For this reason, the dunes are usually arranged in chains extending downwind from the sand source.

Where sand is so abundant that it completely covers the solid ground, dunes take the form of wavelike ridges separated by troughlike furrows. These dunes are called *transverse dunes* because, like ocean waves, their crests trend at right angles to the direction of the dominant wind (Figure 17.26). The entire area may be called a *sand sea*, because it resembles a storm-tossed sea suddenly frozen to immobility. The sand ridges have sharp crests and are asymmetrical, the gentle slope being on the windward and the steep slip face on the lee side. Deep depressions lie between the dune ridges. Sand seas require enormous quantities of sand, supplied by material weathered from sandstone formations or from sands in nearby alluvial plains. Transverse dune belts also form adjacent to beaches that supply abundant sand and have strong onshore winds.

In a barchan dune, the points of the crescent are directed downwind.

Wind is a major agent of landscape development in the Sahara Desert. Enormous quantities of reddish dune sand have been derived from weathering of sandstone formations. The sand is formed into a great expanse of free sand dunes, called an *erg*. Elsewhere, there are vast flat-surfaced sheets of sand that are

Figure 17.25 This aerial view shows a large barchan dune moving from right to left. At its apex is a smaller barchan dune that is overtaking it. Note the network of dry stream channels. Obviously, these dunes have migrated across this channel network since the last water flood occurred. Salton Sea region, California.

Table 17.1   Types of sand dunes.

Type	Shape	Description
Barchan dunes	Crescentic	Crescent-shaped, with points directed downwind. May occur as isolated dunes or as strings or swarms.
Transverse dunes	Wavelike ridges	Resembling storm waves on an ocean. Requires very abundant sand supply.
Star dune	Star with many points	A large hill of sand with radiating arms. Largely fixed in position.
Coastal blowout dunes	Horseshoe	Dunes moving away from a beach. May engulf and smother inland vegetation or development.
Parabolic dunes	Curving arc	Broad, low dune ridges without steep slip faces. May be elongated into hairpin shapes, with points directed upwind.
Longitudinal dunes	Parallel ridges	Long, narrow ridges parallel with wind direction. May cover vast areas of deserts.

armored by a layer of pebbles that forms a desert pavement. A surface of this kind in the Sahara is called a *reg*.

Some of the Saharan dunes are elaborate in shape. For example, the *star dune* (heaped dune), is a large hill of sand whose base resembles a many-pointed star in plan (Figure 17.27). The star dune also occurs in the deserts of the border region between the United States and Mexico. Radial ridges of sand rise toward the dune center and culminate in a sharp peak as high as 100 m (300 ft) or more above the base. The Arabian star dunes remain fixed in position and have served for centuries as reliable landmarks for desert travelers.

Another group of dunes belongs to a family in which the curve of the dune crest is bowed outward in the

Figure 17.26   Transverse dunes of a sand sea, near Yuma, Arizona. Isolated crescent dunes can be seen at the right. The view is eastward; prevailing winds are northerly.

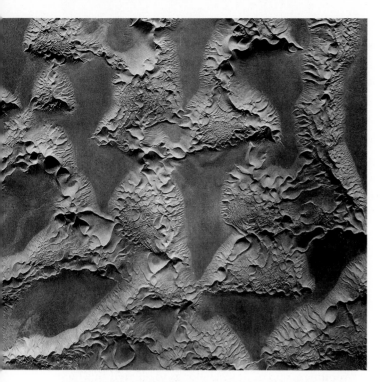

Figure 17.27 Star dunes of the Libyan Desert, seen from an altitude of 10 km (6 mi). Dune peaks rise to heights of 100 to 200 m (about 300 to 600 ft) or more above the intervening level ground.

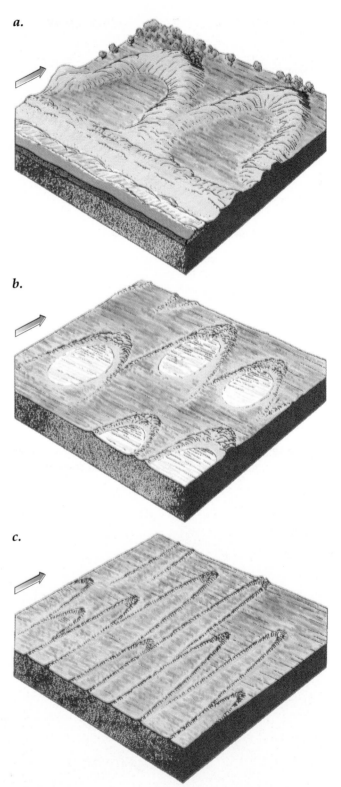

a.

b.

c.

Figure 17.28 Three types of dunes. The prevailing wind direction (arrow) is the same for all three types. (*a*) Coastal blowout dunes. (*b*) Parabolic dunes on a semiarid plain. (*c*) Parabolic dunes drawn out into hairpin forms. (Drawn by A. N. Strahler.)

downwind direction. (This curvature is the opposite of the barchan dune.) These are termed *parabolic dunes.* A common type of parabolic dune is the *coastal blowout dune,* formed adjacent to beaches. Here, large supplies of sand are available, and the sand is blown landward by prevailing winds (Figure 17.28*a*). A saucer-shaped depression is formed by deflation, and the sand is heaped in a curving ridge resembling a horseshoe in plan. On the landward side is a steep slip face that advances over the lower ground and buries forests, killing the trees (Figure 17.29). Coastal blowout dunes are well displayed along the southern and eastern shore of Lake Michigan. Dunes of the southern shore have been protected for public use as the Indiana Dunes State Park.

On semiarid plains, where vegetation is sparse and winds are strong, groups of parabolic blowout dunes develop to the lee of shallow deflation hollows (Figure 17.28*b*). Sand is caught by low bushes and accumulates on a broad, low ridge. These dunes have no steep slip faces and may remain relatively immobile. In some cases, the dune ridge migrates downwind, drawing the dune into a long, narrow form with parallel sides resembling a hairpin in outline (Figure 17.28*c*).

Another class of dunes, described as *longitudinal dunes,* consists of long, narrow ridges oriented parallel

Figure 17.29 This coastal blowout dune is advancing over a coniferous forest, with the slip face gradually burying the tree trunks. Pacific coast, near Florence, Oregon.

Figure 17.30 Longitudinal dunes run parallel to the direction of the wind. (Drawn by A. N. Strahler.)

with the direction of the prevailing wind (Figure 17.30). These dune ridges may be many kilometers long and cover vast areas of tropical and subtropical deserts in Africa and Australia (Figure 17.31).

Figure 17.31 Longitudinal sand dunes in the southern Arabian Peninsula trend northeast-southwest, parallel with the prevailing northeast winds. In the lower right-hand part of the frame is a dissected plateau of the flat-lying sedimentary strata.

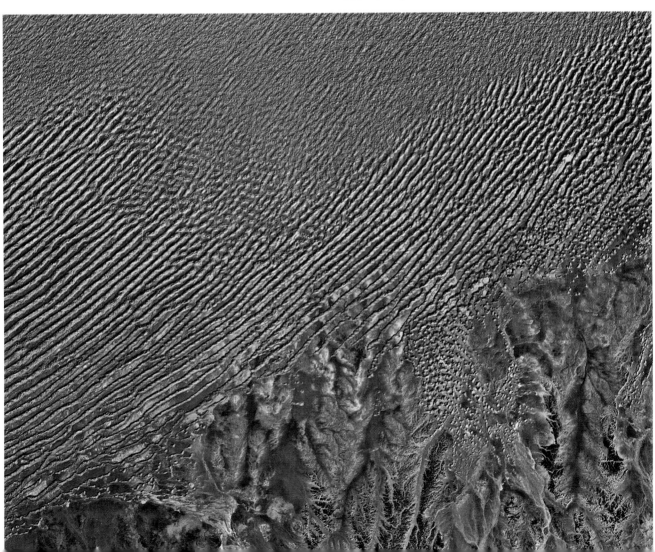

Coastal Foredunes

Landward of sand beaches, we usually find a narrow belt of dunes in the form of irregularly shaped hills and depressions. These are the *foredunes*. They normally bear a cover of beachgrass and a few other species of plants capable of survival in the severe environment (Figure 17.32). On coastal foredunes, the sparse cover of beachgrass and other small plants acts as a baffle to trap sand moving landward from the adjacent beach. As a result, the foredune ridge is built upward to become a barrier standing several meters above high-tide level.

Foredunes form a protective barrier against storm wave action that keeps waves from overwashing a beach ridge or barrier island.

Foredunes form a protective barrier for tidal lands that often lie on the landward side of a beach ridge or barrier island. In a severe storm, the swash of storm waves cuts away the upper part of the beach. Although the foredune barrier may then be attacked by wave action and partly cut away, it will not usually yield. Between storms, the beach is rebuilt, and, in due time, wind action restores the dune ridge, if plants are maintained.

If the plant cover of the dune ridge is reduced by vehicular and foot traffic, a blowout will rapidly develop. The new cavity can extend as a trench across the dune ridge. With the onset of a storm that brings high water levels and intense wave action, swash is funneled through the gap and spreads out on the tidal marsh or tidal lagoon behind the ridge. Sand swept through the gap is spread over the tidal deposits. If eroded, the gap can become a new tidal inlet for ocean water to reach the bay beyond the beach. For many coastal communities of the eastern United States seaboard, the breaching of a dune ridge with its accompanying overwash may bring unwanted change to the tidal marsh or estuary.

Where land, rather than tidal bays, lies behind the beach, blowout dunes can advance on forests, roads,

Figure 17.32 Beachgrass thriving on coastal foredunes has trapped drifting sand to produce a dune ridge. Queen's County, Prince Edward Island, Canada.

Figure 17.33 This thick layer of loess in New Zealand was deposited during the Ice Age. Loess has excellent cohesion and often forms vertical faces as it wastes away.

buildings, and agricultural lands. In the Landes region of the southwestern coast of France, advancing blowout dunes have overwhelmed houses and churches and have even caused entire towns to be abandoned.

LOESS

In several large midlatitude areas of the world, the surface is covered by deposits of wind-transported silt, which has settled out from dust storms over many thousands of years. This material is known as **loess.** (This German word may be pronounced as "lerse" or "luss.") It generally has a uniform yellowish to buff color and lacks any visible layering (Figure 17.33). Loess tends to break away along vertical cliffs wherever it is exposed by the cutting of a stream or grading of a roadway. It is also very easily eroded by running water and subject to rapid gullying when the vegetation cover that protects it is broken.

The thickest deposits of loess are in northern China, where a layer over 30 m (100 ft) thick is common and a maximum thickness of 100 m (300 ft) has been measured. It covers many hundreds of square kilometers and appears to have been brought as dust from the interior of Asia. Loess deposits are also of major importance in the United States, Central Europe, Central Asia, and Argentina.

In the United States, thick loess deposits lie in the Missouri-Mississippi Valley (Figure 17.34). Large areas of the prairie plains region of Indiana, Illinois, Iowa, Missouri, Nebraska, and Kansas are underlain by loess ranging in thickness from 1 to 30 m (3 to 100 ft). Extensive deposits also occur in Tennessee and Mississippi in areas bordering the lower Mississippi River floodplain. Still other loess deposits are in the Palouse region of northeast Washington and western Idaho.

> **Loess is a deposit of wind-blown silt that may be as thick as 30 m (100 ft) in some regions.**

The American and European loess deposits are directly related to the continental glaciers of the Pleistocene Epoch. At the time when the ice covered much of North America and Europe, a generally dry winter climate prevailed in the land bordering the ice sheets. Strong winds blew southward and eastward over the bare ground, picking up silt from the floodplains of braided streams that discharged the meltwater from the ice. This dust settled on the ground between streams, gradually building up a smooth, level ground surface. The loess is particularly thick along the eastern sides of the valleys because of prevailing westerly winds. It is well exposed along the bluffs of most streams flowing through these regions today.

The thick loess deposit covering a large area of north-central China in the province of Shanxi and adjacent provinces poses a difficult problem of severe soil erosion. Although the loess is capable of standing in vertical walls, it also succumbs to deep gullying during the period of torrential summer rains. From the steep walls of these great scars, fine sediment is swept into streams and carried into tributaries of the Huang He (Yellow River). The Chinese government has implemented an intensive program of slope stabilization by using artificial contour terraces (seen in Figure

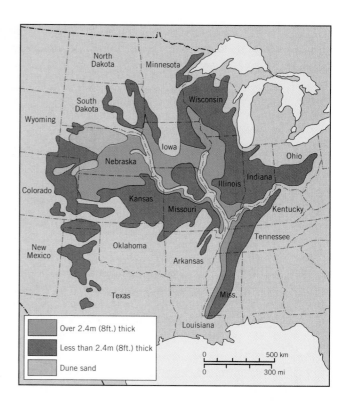

Figure 17.34 Map of loess distribution in the central United States. (Data from *Map of Pleistocene Eolian Deposits of the United States,* Geological Society of America.)

Figure 17.35 A succession of contour terraces, cut into thick loess and covered by tree plantings, is designed to prevent deep gullying and loss of soil. Arched entrances to cave dwellings can be seen at lower left and lower right. Grain fields on the flat valley floor occupy surfaces of thick sediment trapped behind dams. Shanxi Province, near Xian (Sian), People's Republic of China.

17.35) in combination with tree planting. Valley bottoms have been dammed so as to trap the silt to form flat patches of land suitable for cultivation.

Induced Deflation

Induced deflation is a frequent occurrence when short-grass prairie in a semiarid region is cultivated without irrigation. Plowing disturbs the natural soil surface and grass cover, and in drought years, when vegetation dies out, the unprotected soil is easily eroded by wind action. Much of the Great Plains region of the United States has suffered such deflation, experiencing dust storms generated by turbulent winds. Strong cold fronts frequently sweep over this area and lift dust high into the troposphere at times when soil moisture is low.

Human activities in very dry, hot deserts contribute measurably to the raising of high dust clouds. In the desert of northwest India and Pakistan (the Thar Desert bordering the Indus River), the continued trampling of fine-textured soils by hooves of grazing animals and by human feet produces a blanket of dusty, hot air that hangs over the region for long periods. It extends to a height of 9 km (30,000 ft).

One final agent remains to be examined on the list of active agents of erosion, transportation, and deposition—glacial ice. Compared to wind and water, glacial ice moves much more slowly but is far steadier in its motion. Like a vast conveyor belt, glacial ice moves sediment forward relentlessly, depositing the sediment at the ice margin, where the ice melts. By plowing its way over the landscape, glacial ice also shapes the local terrain—bulldozing loose rock from hillsides and plastering sediments underneath its vast bulk. This slow but steady action is very different from that of water, wind, and waves, and produces a set of landforms that is the subject of our last text chapter.

LANDFORMS MADE BY WAVES AND WIND IN REVIEW

This chapter has described the landforms of waves and wind, both of which are indirectly powered by the earth's rotation and the unequal heating of its surface by the sun. Waves act at the shoreline—the boundary between water and land. Waves expend their energy as breakers, which erode hard rock into marine cliffs and create marine scarps in softer materials.

Beaches, usually formed of sand, are shaped by the swash and backwash of waves, which continually work and rework beach sediment. Wave action produces littoral drift, which moves sediment parallel to the beach. This sediment accumulates in bars and sandspits, which further extend the beach. Depending on the nature of longshore currents and the availability of sediment, shorelines can experience progradation or retrogradation.

Tidal forces cause sea level to rise and fall rhythmically, and this change of level produces tidal currents in bays and estuaries. Tidal flows redistribute fine sediments within bays and estuaries, which can accumulate with the help of vegetation to form salt marshes. These are sometimes reclaimed to form new agricultural land.

Coastlines of submergence result when coastal lands sink below sea level or sea level rises rapidly. Scenic ria and fiord coasts are examples. Coastlines of emergence include barrier-island coasts and delta coasts. Coral-reef coasts occur in regions of warm tropical and equatorial waters. Along some coasts, rapid uplift has occurred, creating raised shorelines and marine terraces.

Wind is a landform-creating agent that acts by moving sediment. Deflation occurs when wind removes mineral particles—especially clay and silt, which can be carried long distances. Deflation creates blowouts in semidesert regions and lowers playa surfaces in deserts. In arid regions, deflation produces dust storms.

Sand dunes form when a source, such as a sandstone outcrop or a beach, provides abundant sand that can be moved by wind action. Barchan dunes are arranged individually or in chains leading away from the sand source. Transverse dunes form a sand sea of frozen "wave" forms arranged perpendicular to the wind direction. Parabolic dunes are arc-shaped—coastal blowout dunes are an example. Longitudinal dunes parallel the wind direction and cover vast desert areas. Coastal foredunes are stabilized by dune grass and help protect the coast against storm wave action.

Loess is a surface deposit of fine, wind-transported silt. It can be quite thick, and it typically forms vertical banks. Loess is very easily eroded by water and wind. In eastern Asia, the silt forming the loess was transported by winds from extensive interior deserts located to the north and west. In Europe and North America, the silt was derived from fresh glacial deposits during the Pleistocene epoch. Human activities can hasten the action of deflation by breaking protective surface covers of vegetation and desert pavement.

KEY TERMS

shoreline	littoral drift	atoll
coastline	progradation	deflation
coast	retrogradation	blowout
bay	submergence	desert pavement
estuary	emergence	dust storm
marine cliff	delta	sand dune
beach	coral reef	loess

REVIEW QUESTIONS

1. What is the energy source for wind and wave action?

2. What landforms can be found in areas where bedrock meets the sea?

3. What is littoral drift, and how is it produced by wave action?

4. Identify progradation and retrogradation. How can human activity influence retrogradation?

5. How are salt marshes formed? How can they be reclaimed for agricultural use?

6. What is a ria coast? Under what conditions does it form?

7. Under what conditions do barrier-island coasts form? What are the typical features of this type of coastline? Provide and sketch an example of a barrier-island coast.

8. Describe the features of delta coasts and their formation. Sketch and compare the Mississippi and Nile deltas.

9. What conditions are necessary for the development of coral reefs? Identify three types of coral-reef coastlines.

10. How are marine terraces formed?

11. What is deflation, and what landforms does it produce? What role does the dust storm play in deflation?

12. How do sand dunes form? Describe and compare barchan dunes, transverse dunes, star dunes, coastal blowout dunes, parabolic dunes, and longitudinal dunes.

13. What is the role of coastal dunes in beach preservation? How are coastal dunes influenced by human activity? What problems can result?

14. Define the term loess. What is the source of loess and how are loess deposits formed?

IN-DEPTH QUESTIONS

1. Consult an atlas to identify a good example of each of the following types of coastlines: ria coast, fiord coast, barrier-island coast, delta coast, coral-reef coast, and fault coast. For each example, provide a brief description of the key features you used to identify the coastline type.

2. Wind action moves sand close to the ground in a bouncing motion, whereas silt and clay are lifted and carried longer distances. Compare landforms and deposits that result from wind transportation of sand with those that result from wind transportation of silt and finer particles.

Glacial Landforms
And the Ice Age

The huge cruise ship moves silently and effortlessly between the two steep, forest-covered slopes. The green walls of spruce and hemlock glide by slowly, the spires of the forest giants fading into the low ceiling of gray clouds overhead. Tiny droplets of drizzle cool your face as you lean against the rail, observing the somber scene. The trees seem almost close enough to reach out and touch. Below you, the ship's bow parts the gray-green water with little resistance, providing only a low rushing noise that barely breaks the stillness. Ahead, the sea and forest merge, fade, and then blend into the clouds and fog.

So this is Alaska! The travel brochure showed your ship cruising in a crystal-blue ocean past magnificent vistas of snow-capped peaks, but so far the trip along the inland passage from Vancouver to Juneau has provided only low clouds, fog, and spruce-clad slopes. Not that it has been all that unpleasant. The world of gray and green has been strangely cozy and comforting. Even though there were plenty of other things to do in this huge floating hotel, you returned time after time to your place on the rail, communing with the quiet majesty of this unspoiled landscape.

The weather may soon change, however. The forecast calls for slow clearing later in the day, about the time that your ship will reach your ultimate destination, Glacier Bay. This steep-walled fiord is a drowned valley that was once filled with a huge tongue of glacial ice. During the Ice Age, the moving ice broadened and deepened the valley to a depth below present sea level. As late as 200 years ago, the valley was still filled with ice. But then the forward motion of the glacier slowed considerably, and the ice front retreated. As ocean water entered the valley, the retreat uncovered a fiord over 100 km (60 mi) in length receiving 16 streams of glacial ice.

It is not long before you notice a change in the ship's rhythm. A slow curving turn, and you leave Icy Strait to enter the mouth of Glacier Bay. It will take several hours to make the journey to the head of the Bay, where the spectacular Grand Pacific Glacier enters.

Miraculously, the overcast skies have yielded to a blue and white patchwork of sky and cloud, and the sun has appeared. The timing is perfect. Your ship is moving slowly down Tarr Inlet, the last narrow passage leading to the Grand Pacific Glacier. To either side are low, rocky peaks, sculpted by the ice into stark angular shapes composed of points and edges. Streaked with snow, the peaks alternately glisten and glow as cloud shadows drift across their flanks. Ahead are the high peaks of the St. Elias Mountains, entirely capped with white.

As the glacier comes into view, the ship slows further and approaches the steep ice face carefully. It is fractured into tall vertical columns with steplike horizontal facets at the top. The water below is littered with white chunks of floating ice of all sizes and shapes. Soon the ship lies motionless alongside the huge glistening wall, closer than you thought possible. Your view from the ship's rail is incredibly beautiful.

The color of the glacial ice is what strikes you most. Beneath a top layer of white decaying ice and snow, the ice is a bright greenish-blue. The color is so intense that the ice seems to glow from within. Above the clear aqua color of the ice are the gray and brown colors of the rocks, the white of the snow-capped peaks, and the blue and white patchwork of sky. Below is the gray-green of the water, dotted with white ice fragments.

The ice is also noisy. It groans, screeches, clicks, and pops. A loud crack startles you, the sound echoing off the walls of the fiord. A monumental column of ice breaks free and falls, as if in slow motion. A great splash wells upward as it breaks the surface of the water. Next to the huge ice front, the bulk of the cruise ship seems insignificant indeed. Suddenly, you connect the force of the moving ice with the shaping of the vast fiord that has contained your ship for the past hours. The power that carved this huge, water-filled canyon is awesome indeed.

Perhaps the best way to appreciate the work of moving ice in shaping the landscape is at first hand, by a visit to a glacier. An Alaska cruise to the front of a tidewater glacier is certainly an easy way to do it. However, for occupants of regions that were covered by ice during the last Ice Age, the evidence of glaciation is all around. From ponds formed by the melting of buried ice blocks to the great hills of sediments dumped by glacial ice sheets at their margins, the action of glaciers has drastically modified much of the landscape of northern regions. The landform-making activities of glaciers are the subject of this chapter.

Gorner Glacier, Swiss Alps. Near Zermatt, Switzerland.

In this chapter, we turn to the last of the active agents that create landforms—glacial ice. Not long ago, during the Ice Age, much of northern North America and Eurasia was covered by massive sheets of glacial ice. As a result, glacial ice has played a dominant role in shaping landforms of large areas in midlatitude and subarctic zones. Yet, glacial ice still exists today in two great accumulations of continental dimensions—the Greenland and Antarctic Ice Sheets—and in many smaller masses in high mountains.

The glacial ice sheets of Greenland and Antarctica strongly influence the radiation and heat balance of the globe. Because of their intense whiteness, they reflect much of the solar radiation they receive. Their intensely cold surface air temperatures contrast with temperatures at more equatorial latitudes. This temperature difference helps drive the system of meridional heat transport that we described in Chapters 2 and 4. In addition, these enormous ice accumulations represent water in storage in the solid state. They figure as a major component of the global water balance. When the volume of glacial ice increases, as during an ice age, sea levels must fall. When ice sheets melt away, sea level rises. Today's coastal environments evolved during the rising sea level that followed the melting of the last ice sheets of the Ice Age.

GLACIERS

Most of us know ice only as a brittle, crystalline solid because we are accustomed to seeing it in small quantities. Where a great thickness of ice exists, the pressure on the ice at the bottom makes the ice lose its rigidity. That is, the ice becomes plastic. This allows the ice mass to flow in response to gravity, slowly spreading out over a larger area or moving downhill. On steep mountain slopes, the ice can also move by sliding. Movement is the key characteristic of a *glacier*, defined as any large natural accumulation of land ice affected by present or past motion.

Glacial ice accumulates when the average snowfall of the winter exceeds the amount of snow that is lost in summer by ablation. The term **ablation** means the loss of snow and ice by evaporation and melting. When winter snowfall exceeds summer ablation, a layer of snow is added each year to what has already accumulated. As the snow compacts by surface melting and refreezing, it turns into a granular ice and is then compressed by overlying layers into hard crystalline ice.

When the ice mass is so thick that the lower layers become plastic, outward or downhill flow starts, and the ice mass is now an active glacier.

Glacial ice forms where temperatures are low and snowfall is high. These conditions can occur both at high elevations and at high latitudes. In mountains, glacial ice can form even in tropical and equatorial zones if the elevation is high enough to keep average annual temperatures below freezing. Orographic precipitation encourages the growth of glacial ice. In high mountains, glaciers flow from small high-elevation collecting grounds down to lower elevations, where temperatures are warmer. Here the ice disappears by ablation. Typically, mountain glaciers are long and narrow because they occupy former stream valleys. These **alpine glaciers** are a distinctive type (see chapter opening photo).

In arctic and polar regions, prevailing temperatures are low enough that snow can accumulate over broad areas, eventually forming a vast layer of glacial ice. Accumulation starts on uplands that intercept heavy snowfall. The uplands become buried under enormous volumes of ice, which can reach a thickness of several thousand meters. The ice then spreads outward, over surrounding lowlands, and covers all landforms it encounters. This extensive type of ice mass is called an **ice sheet.** As already noted, ice sheets exist today in Greenland and Antarctica.

Glacial ice normally contains abundant rock fragments ranging from huge angular boulders to pulverized rock flour. Some of this material is eroded from the rock floor on which the ice moves. In alpine glaciers, rock debris is also derived from material that slides or falls from valley walls onto the ice.

Alpine glaciers and ice sheets are two forms of moving glacial ice.

Glaciers are capable of eroding and depositing great quantities of sediment. *Glacial abrasion* is a glacial erosion process caused by rock fragments that are held within the ice and scrape and grind against bedrock (Figure 18.1). Erosion also occurs by *plucking*, as moving ice lifts out blocks of bedrock that have been loosened by the freezing and expansion of water in joint fractures. Rock debris brought into a glacier is eventually deposited at the lower end of a glacier, where the ice melts. Both erosion and deposition result in distinctive glacial landforms.

Figure 18.1   This grooved and polished surface, now partly eroded, marks the former path of glacial ice. Cathedral Lakes, Yosemite National Park, California.

ALPINE GLACIERS

Figure 18.2 illustrates a number of features of alpine glaciers. The illustration shows a simple glacier occupying a sloping valley between steep rock walls. Snow collects at the upper end in a bowl-shaped depression, the **cirque.** The upper end lies in a zone of accumulation. Layers of snow in the process of compaction and recrystallization are called *firn.*

The smooth firn field is slightly bowl-shaped in profile. Flowage in the glacial ice beneath the firn carries the ice downvalley out of the cirque. The rate of ice flow is accelerated at a steep rock step, where deep *crevasses* (gaping fractures), mark an ice fall. The lower part of the glacier lies in the zone of ablation. In this area, the rate of ice wastage is rapid, and old ice is exposed at the glacier surface. This surface may be quite rough, with deep crevasses. At its lower end, or terminus, the glacier carries abundant rock debris. As the downward-flowing ice melts, the debris accumulates.

Although the uppermost layer of a glacier is brittle and fractures readily into crevasses, the ice beneath behaves as a plastic substance and moves by slow flowage (Figure 18.3). Like stream flow, glacier flow is most rapid far from the glacier's bed—near the midline and toward the top of the glacier's surface. Alpine glaciers also move by basal sliding. In this process, the ice slides downhill, lubricated by meltwater and mud at its base.

A glacier establishes a dynamic balance in which the rate of accumulation at the upper end balances the rate of ablation at the lower end. This balance is easily upset by changes in the average annual rates of accumulation or ablation, causing the glacier's terminus to move forward or melt.

Figure 18.2 Cross section of an alpine glacier. Ice accumulates in the glacial cirque, then flows downhill, abrading and plucking the bedrock. Glacial debris accumulates at the glacier terminus. (After A. N. Strahler.)

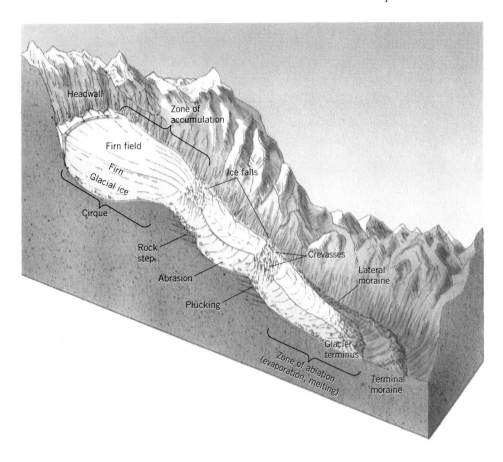

Glacial flow is usually very slow. It amounts to a few centimeters per day for large ice sheets and the more sluggish alpine glaciers, but as fast as several meters per day for an active alpine glacier. However, some alpine glaciers experience episodes of very rapid move-

ment, termed *surges*. A surging glacier may travel downvalley at speeds of more than 60 m (200 ft) per day for several months. The reasons for surging are not well understood, but probably involve mechanisms that increase the amount of meltwater beneath the ice, enhancing basal sliding. Most glaciers do not experience surging.

Landforms Made by Alpine Glaciers

Landforms made by alpine glaciers are shown in a series of diagrams in Figure 18.4. Mountains are eroded and shaped by glaciers, and after the glaciers melt, the remaining landforms are exposed to view. Diagram *a* shows a region sculptured entirely by weathering, mass wasting, and streams. The mountains have a smooth, rounded appearance. Soil and regolith are thick.

Imagine now that a climatic change results in the accumulation of snow in the heads of the higher valleys. An early stage of glaciation is shown at the right side of diagram *b*, where snow is collecting and cirques are being carved by the grinding motion of the ice. Deepening of the cirques is aided by intensive frost shattering of the bedrock near the masses of compacted snow. On the left side of the diagram, the landscape remains unglaciated.

Figure 18.3 Motion of glacial ice. Ice moves most rapidly on the glacier's surface at its midline. Movement is slowest near the bed, where the ice contacts bedrock or sediment.

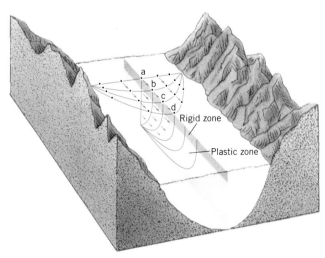

a.

Figure 18.4 Landforms produced by alpine glaciers. (*a*) Before glaciation sets in, the region has smoothly rounded divides and narrow, V-shaped stream valleys. (*b*) After glaciation has been in progress for thousands of years, new erosional forms are developed. (*c*) With the disappearance of the ice, a system of glacial troughs is exposed. (Drawn by A. N. Strahler.)

b.

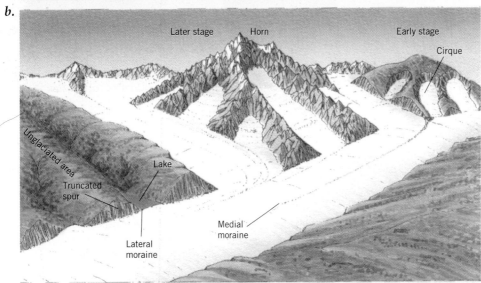

c.

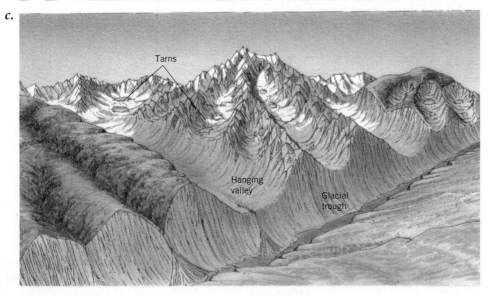

Figure 18.5 Aerial view of the Grand Plateau Glacier, St. Elias Mountains, Glacier Bay National Park. The dark stripes are medial moraines.

In Figure 18.4*b*, glaciers have filled the valleys and are integrated into a system of tributaries that feed a trunk glacier. Tributary glaciers join the main glacier smoothly. The cirques grow steadily larger. Their rough, steep walls soon replace the smooth, rounded slopes of the original mountain mass. Where two cirque walls intersect from opposite sides, a jagged, knifelike ridge, called an *arête*, is formed. Where three or more cirques grow together, a sharp-pointed peak is formed. Such peaks in the Swiss Alps are called "horns."

A ridge or pile of rock debris left by glacial action that marks the edge of a glacier is termed a **moraine.** A *lateral moraine* is a debris ridge formed along the edge of the ice adjacent to the trough wall (Figure 18.2). Where two ice streams join, this marginal debris is dragged along to form a narrow band riding on the ice in midstream (Figures 18.4*b*, 18.5), called a *medial moraine.* At the terminus of a glacier, rock debris accumulates in a *terminal moraine,* an embankment curving across the valley floor and bending upvalley along each wall of the trough (Figure 18.6).

Glacial Troughs and Fiords

Glacier flow constantly deepens and widens its rock channel, so that after the ice has finally melted, a deep, steep-walled **glacial trough** remains (Figure 18.4*c*). The trough typically has a U-shape in cross-profile (Figure

Figure 18.6 A terminal moraine, shaped like the bow of a great canoe, lies at the mouth of a deep glacial trough on the east face of the Sierra Nevada. Cirques can be seen in the distance. Near Lee Vining, California.

18.7). Tributary glaciers also carve U-shaped troughs, but they are smaller in cross section. Because their floors lie high above the level of the main trough, they are called *hanging valleys*. Streams later occupy the abandoned valleys, providing a scenic waterfall that cascades over the lip of the hanging valley to the main trough below. High up in the smaller troughs, the bedrock is unevenly excavated, so that the floors of troughs and cirques contain rock basins and rock steps. The rock basins are occupied by small lakes, called *tarns* (Figure 18.4*c*). Major troughs sometimes hold large, elongated trough lakes.

When the floor of a trough open to the sea lies below sea level, the sea water enters as the ice front recedes. The result is a deep, narrow estuary known as a **fiord** (Figure 18.8). Fiords are opening up today along the Alaskan coast, where some glaciers are melting back rapidly and ocean waters are filling their troughs. Fiords are found largely along mountainous coasts between lat. 50° and 70° N and S. On these coasts, glaciers were nourished by heavy orographic snowfall, associated with the marine west-coast climate ⑧.

Cirques are rounded rock basins that contain the heads of alpine glaciers. They form high on glaciated peaks.

Figure 18.8 Geirangerfjord, Norway, is a deeply carved glacial trough occupied by an arm of the sea.

Figure 18.7 This U-shaped glacial trough in the Beartooth Plateau, near Red Lodge, Montana, was carved by an alpine glacier during the Ice Age. Talus cones have been built out from the steep trough walls.

ICE SHEETS OF THE PRESENT

In contrast to alpine glaciers are the enormous ice sheets of Antarctica and Greenland. These are huge plates of ice, thousands of meters thick in the central areas, resting on land masses of subcontinental size. The Greenland Ice Sheet has an area of 1.7 million sq km (670,000 sq mi) and occupies about seven-eighths of the entire island of Greenland (Figure 18.9). Only a narrow, mountainous coastal strip of land is exposed. The Antarctic Ice Sheet covers 13 million sq km (5 million sq mi) (Figure 18.10). Both ice sheets are devel-

Figure 18.9 The Greenland Ice Sheet. (Based on data of R. F. Flint, *Glacial and Pleistocene Geology,* John Wiley & Sons, New York.)

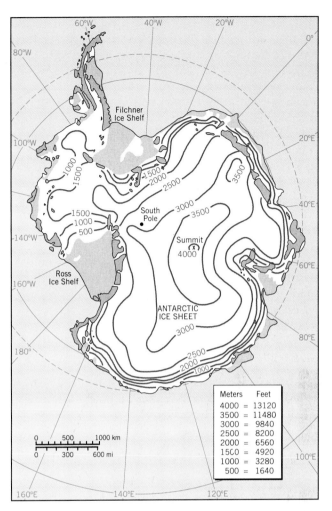

Figure 18.10 The Antarctic Ice Sheet and its ice shelves. (Based on data of American Geophysical Union.)

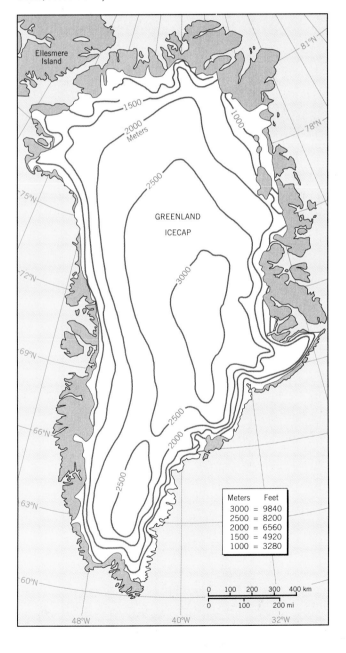

oped on large, elevated land masses in high latitudes. No ice sheet exists near the north pole, which is positioned in the vast Arctic Ocean. Ice there occurs only as floating sea ice.

The surface of the Greenland Ice Sheet has the form of a very broad, smooth dome. Underneath the ice sheet, the rock floor lies near or slightly below sea level under the central region but is higher near the edges. The Antarctic Ice Sheet is thicker than the Greenland Ice Sheet—as much as 4000 m (13,000 ft) at maximum. At some locations, ice sheets extend long tongues, called outlet glaciers, to reach the sea at the heads of fiords (Figure 18.11). From the floating edge of the glacier, huge masses of ice break off and drift out to open sea with tidal currents to become icebergs. An important glacial feature of Antarctica is the presence of great plates of floating glacial ice, called *ice shelves* (Figure 18.10). Ice shelves are fed by the ice sheet, but they also accumulate new ice through the compaction of snow.

Figure 18.11   This outlet glacier on the coast of Antarctica moves ice rapidly away from the interior to the ocean. Ambers Island, Antarctic Peninsula.

SEA ICE AND ICEBERGS

Free-floating ice on the sea surface is of two types—sea ice and icebergs. *Sea ice* is formed by direct freezing of ocean water. In contrast, *icebergs* are bodies of land ice that have broken free from glaciers that terminate in the ocean. Aside from differences in origin, a major difference between sea ice and icebergs is thickness. Sea ice does not exceed 5 m (15 ft) in thickness, while icebergs may be hundreds of meters thick.

Pack ice is sea ice that completely covers the sea surface (Figure 18.12). Under the forces of wind and currents, pack ice breaks up into individual patches called ice floes. The narrow strips of open water between such floes are known as *leads*. Where ice floes are forcibly brought together by winds, the ice margins buckle and turn upward into pressure ridges that often resemble walls of ice. Travel on foot across the polar sea ice is extremely difficult because of such obstacles. The surface zone of sea ice is composed of fresh water, while the deeper ice is salty.

When a valley glacier or tongue of an ice sheet terminates in sea water, blocks of ice break off to form icebergs (Figure 18.13). An iceberg may be as thick as several hundred meters. Because it is only slightly less dense than sea water, the iceberg floats very low in the

Figure 18.12   This Landsat image shows sea ice in the Canadian arctic region.

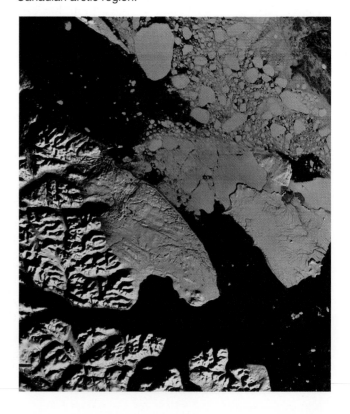

water, with about five-sixths of its bulk submerged. The ice is composed of fresh water since it is formed from compacted and recrystallized snow.

In the northern hemisphere, icebergs are derived largely from glacier tongues of the Greenland Ice Sheet. They drift slowly south with the Labrador and Greenland currents and may find their way into the North Atlantic in the vicinity of the Grand Banks of Newfoundland. They are irregular in shape and present rather peaked outlines above water. In contrast, antarctic icebergs are commonly tabular in form, with flat tops and steep clifflike sides (Figure 18.13). This shape is not surprising, since these tabular bergs are parts of ice shelves. A large tabular berg of the Antarctic may be tens of kilometers broad and over 600 m (2000 ft) thick, with an ice wall rising 100 m (300 ft) above sea level.

THE ICE AGE

The period during which continental ice sheets grow and spread outward over vast areas is known as a **glaciation.** Glaciation is associated with a general cooling of average air temperatures over the regions where the ice sheets originate. At the same time, ample snowfall must persist over the growth areas to allow the ice masses to build in volume.

When the climate warms or snowfall decreases, ice sheets become thinner and cover less area. Eventually, the ice sheets may melt completely. This period is called a *deglaciation.* Following a deglaciation, but preceding the next glaciation, is a period in which a mild climate prevails—an **interglaciation.** The last interglaciation began about 140,000 years ago, and ended between 120,000 and 110,000 year ago. A succession of alternating glaciations and interglaciations, spanning a period of 1 to 10 million years or more, constitutes an *ice age.*

An ice age includes cycles of glaciation, deglaciation, and interglaciation.

Throughout the past 3 million years or so, the earth has been experiencing the **Late-Cenozoic Ice Age** (or, simply, the **Ice Age**). As you may recall from Chapter 11, the Cenozoic Era has seven epochs (see Table 11.1). The Ice Age falls within the last three epochs: Pliocene, Pleistocene, and Holocene. These three epochs comprise only a small fraction—about one-twelfth—of the total duration of the Cenozoic Era.

During the first half of this century, most geologists associated the Ice Age with the Pleistocene Epoch, which began about 1.6 million years ago. However, new evidence obtained from deep-sea sediments shows that the glaciations of the Ice Age began in late Pliocene time, perhaps 2.5 to 3.0 million years ago.

At present, we are within an interglaciation of the Late-Cenozoic Ice Age, following a deglaciation that set in quite rapidly about 15,000 years ago. In the preceding glaciation, called the *Wisconsinan Glaciation,* ice sheets covered much of North America and Europe, as well as parts of northern Asia and southern South America. The maximum ice advance of the Wisconsinan Glaciation was reached about 18,000 years ago.

Figure 18.13 An iceberg off the coast of Antarctica near Elephant Island.

Glaciation During the Ice Age

Figures 18.14 and 18.15 show the maximum extent to which North America and Europe were covered during the last advance of the ice. Most of Canada was engulfed by the vast Laurentide Ice Sheet. It spread south into the United States, covering most of the land lying north of the Missouri and Ohio rivers, as well as northern Pennsylvania and all of New York and New England. Alpine glaciers of the western ranges coalesced into a single ice sheet that spread to the Pacific shores and met the Laurentide sheet on the east. Notice that an area in southwestern Wisconsin escaped inundation. Known as the Driftless Area, it was apparently bypassed by glacial lobes moving on either side.

In Europe, the Scandinavian Ice Sheet centered on the Baltic Sea, covering the Scandinavian countries. It spread south into central Germany and far eastward to cover much of Russia. In north-central Siberia, large ice caps formed over the northern Ural Mountains and highland areas farther east. Ice from these centers grew into a large sheet covering much of central Siberia. The European Alps were capped by enlarged

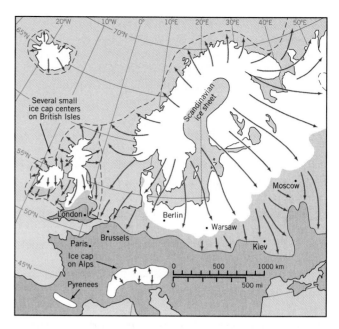

Figure 18.15 The Scandinavian ice sheet dominated northern Europe during the Ice Age glaciations. As noted in Figure 18.14, the present coastline is far inland from the coastline that prevailed during glaciations. (Based on data of R. F. Flint, *Glacial and Pleistocene Geology,* John Wiley & Sons, New York.)

Figure 18.14 Continental glaciers of the Ice Age in North America at their maximum extent reached as far south as the present Ohio and Missouri rivers. Note that during glaciations sea level was much lower. The present coastline is shown for reference only. (Based on data of R. F. Flint, *Glacial and Pleistocene Geology,* John Wiley & Sons, New York.)

alpine glaciers. The British Isles were mostly covered by a small ice sheet that had several centers on highland areas and spread outward to coalesce with the Scandinavian Ice Sheet.

At the maximum spread of these ice sheets, sea level was as much as 125 m (410 ft) lower than today, exposing large areas of the continental shelf on both sides of the Atlantic basin. The shelf supported a vegetated landscape populated with animal life, including Pleistocene elephants (mastodons and mammoths). The drawdown of sea level explains why the ice sheets shown on our maps extend far out into what is now the open ocean.

South America, too, had an ice sheet. It grew from ice caps on the southern Andes Range south of about latitude 40° S and spread westward to the Pacific shore, as well as eastward, to cover a broad belt of Patagonia. It covered all of Tierra del Fuego, the southern tip of the continent. The South Island of New Zealand, which today has a high spine of alpine mountains with small relict glaciers, developed a massive ice cap in late Pleistocene time. All high mountain areas of the world underwent greatly intensified alpine glaciation at the time of maximum ice sheet advance. Today, most remaining alpine glaciers are small ones. In less favorable locations, the Ice Age alpine glaciers are entirely gone.

LANDFORMS MADE BY ICE SHEETS

Landforms made by the last ice advance and recession are very fresh in appearance and show little modification by erosion processes. It is to these landforms that we now turn our attention.

Erosion by Ice Sheets

Like alpine glaciers, ice sheets are highly effective eroding agents. The slowly moving ice scraped and ground away much solid bedrock, leaving behind smoothly rounded rock masses. These bear countless grooves

---

Although the largest ice sheets covered North America and Eurasia, South America also developed an ice sheet.

---

and scratches trending in the general direction of ice movement (see Figure 18.1). Sometimes the ice polishes the rock to a smooth, shining surface. The evidence of ice abrasion is common throughout glaciated regions of North America and may be seen on almost any hard rock surface that is freshly exposed. Conspicuous knobs of solid bedrock shaped by the moving ice are also common features (Figure 18.16). The side from which the ice approached is usually smoothly rounded. The lee side, where the ice plucked out angular joint blocks, is irregular and blocky.

The ice sheets also excavated enormous amounts of rock at locations where the bedrock was weak and the flow of ice was channeled by the presence of a valley trending in the direction of ice flow. Under these conditions, the ice sheet behaved like a valley glacier, scooping out a deep, U-shaped trough. The Finger Lakes of western New York State are fine examples (Figure 18.17). Here, a set of former stream valleys lay largely parallel to the southward spread of the ice, and a set of long, deep basins was eroded. Blocked at their north ends by glacial debris, the basins now hold lakes. Many hundreds of lake basins were created by glacial erosion and deposition over the glaciated portion of North America.

Figure 18.17 The Finger Lakes region of New York is shown in this photo taken by astronauts aboard the Space Shuttle. The lakes occupy stream valleys that were eroded and deepened by glacial ice.

Deposits Left by Ice Sheets

The term **glacial drift** includes all varieties of rock debris deposited in close association with glaciers. Drift is of two major types. *Stratified drift* consists of layers of sorted and stratified clays, silts, sands, or gravels. These materials were deposited by meltwater streams or in bodies of water adjacent to the ice. **Till** is an unstratified mixture of rock fragments, ranging in size from clay to boulders, that is deposited directly from the ice without water transport.

Figure 18.16 A glacially abraded rock knob. Glacial action abrades the rock into a smooth form as it rides over the rock summit, then plucks bedrock blocks from the lee side, producing a steep, rocky slope. (Copyright © A. N. Strahler.)

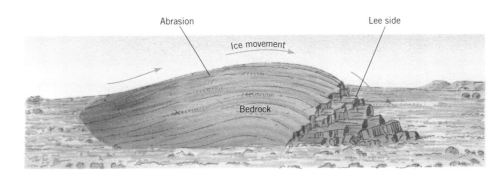

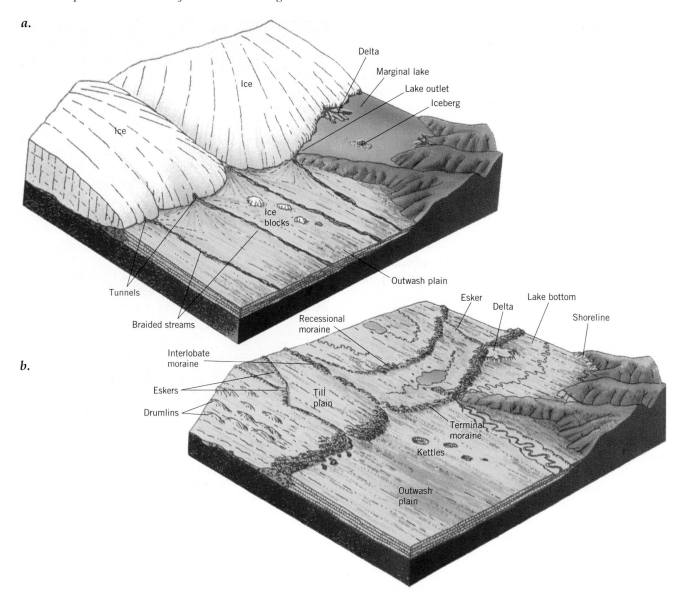

Figure 18.18 Marginal landforms of continental glaciers. (*a*) With the ice front stabilized and the ice in a wasting, stagnant condition, various depositional features are built by meltwater. (*b*) The ice has wasted completely away, exposing a variety of new landforms made under the ice. (Drawn by A. N. Strahler.)

Over those parts of North America formerly covered by late-Cenozoic ice sheets, glacial drift thickness averages from 6 m (20 ft) over mountainous terrain, such as New England, to 15 m (50 ft) and more over the lowlands of the north-central United States. Over Iowa, drift thickness is from 40 to 60 m (150 to 200 ft), and over Illinois it averages more than 30 m (100 ft). In some places where deep stream valleys existed prior to glacial advance, as in parts of Ohio, drift is much thicker.

To understand the form and composition of deposits left by ice sheets, it will help to examine the conditions prevailing at the time of the ice sheet's existence. Figure 18.18*a* shows a region partly covered by an ice sheet with a stationary front edge. This condition occurs when the rate of ice ablation balances the amount of ice brought forward by spreading of the ice sheet. Although the ice fronts of the Ice Age advanced and receded in many minor and major fluctuations, there were long periods when the front was essentially stable and thick deposits of drift accumulated.

Moraines

The transportational work of an ice sheet resembles that of a huge conveyor belt. Anything carried on the belt is dumped off at the end and, if not constantly removed, will pile up in increasing quantity. Rock fragments brought within the ice are deposited at its outer edge as the ice evaporates or melts. There is no possibility of return transportation.

Glacial till that accumulates at the immediate ice edge forms an irregular, rubbly heap—the terminal moraine. After the ice has disappeared (Figure 18.18*b*), the moraine appears as a belt of knobby hills interspersed with basinlike hollows, or kettles, some of which hold small lakes. The name *knob-and-kettle* is often applied to morainal belts.

Moraines are built of rock debris that is deposited at the edge of a melting glacier.

Terminal moraines form great curving patterns. The outward curvature is southward and indicates that the ice advanced as a series of great *ice lobes,* each with a curved front (Figure 18.19). Where two lobes come together, the moraines curve back and fuse together into a single *interlobate moraine* pointed northward. In its general recession accompanying disappearance, the ice front paused for some time along a number of positions, causing morainal belts similar to the terminal moraine belt to be formed. These belts are known as recessional moraines (Figure 18.18*b*.). They run roughly parallel with the terminal moraine but are often thin and discontinuous.

Outwash and Eskers

Figure 18.18 shows a smooth, sloping plain lying in front of the ice margin. This is the *outwash plain,* formed of *outwash*—stratified drift left by braided streams issuing from the ice. The plain is built of layer upon layer of sands and gravels.

Large streams carrying meltwater issue from tunnels in the ice. These form when the ice front stops moving for many kilometers back from the front. After the ice has gone, the position of a former ice tunnel is marked by a long, sinuous ridge of sediment known as an *esker* (Figure 18.20). The esker is the deposit of sand and gravel laid on the floor of the former ice tunnel. After the ice has melted away, only the streambed deposit remains, forming a ridge. Many eskers are several kilometers long.

Drumlins and Till Plains

Another common glacial form is the *drumlin,* a smoothly rounded, oval hill resembling the bowl of an inverted teaspoon. It consists in most cases of glacial till (Figure 18.21). Drumlins invariably lie in a zone behind the terminal moraine. They commonly occur in groups or swarms and may number in the hundreds. The long axis of each drumlin parallels the direction

Figure 18.19 Moraine belts of the northcentral United States have a curving pattern left by ice lobes. Some regions of interlobate moraines are shown by the color overlay. (Based on data of R. F. Flint and others, *Glacial Map of North America,* Geological Society of America.)

of ice movement. The origin of drumlins is not well understood. They seem to have been formed under moving ice by a plastering action in which layer upon layer of bouldery clay was spread on the drumlin.

Between moraines, the surface overridden by the ice is covered by glacial till. This cover is often inconspicuous since it forms no prominent landscape feature. The till layer may be thick and obscure, or it may entirely bury the hills and valleys that existed before glaciation. Where thick and smoothly spread, the layer forms a level *till plain.* Plains of this origin are widespread throughout the central lowlands of the United States and southern Canada.

Marginal Lakes and Their Deposits

When the ice advanced toward higher ground, valleys that may have opened out northward were blocked by ice. Under such conditions, marginal glacial lakes formed along the ice front (see Figure 18.18*a*). These lakes overflowed along the lowest available channel

Figure 18.20 The curving ridge of sand and gravel in this photo is an esker, marking the bed of a river of meltwater flowing underneath a continental ice sheet near its margin. Kettle-Moraine State Park, Wisconsin.

between the ice and the rising ground slope, or over some low pass along a divide. Streams of meltwater from the ice built *glacial deltas* into these marginal lakes.

When the ice withered away, the lakes drained, leaving a flat floor exposed. Here, layers of fine clay and silt had accumulated. Glacial lake plains often contain extensive areas of marshland. The deltas are now curiously isolated, flat-topped landforms composed of well-washed and well-sorted sands and gravels (Figure 18.8*b*).

Environmental Aspects of Glacial Deposits

Because much of Europe and North America was glaciated by the Pleistocene ice sheets, landforms asso-

ciated with the ice are of major environmental importance. Agricultural influences of glaciation are both favorable and unfavorable, depending on preglacial topography and the degree and nature of ice erosion and deposition.

In hilly or mountainous regions, such as New England, the glacial till is thinly distributed and extremely stony. Till cultivation is difficult because of countless boulders and cobbles in the clay soil. Till accumulations on steep mountain or roadside slopes are subject to mass movement as earthflows when clay in the till becomes weakened after absorbing water from melting snows and spring rains. Along moraine belts, the steep slopes, the irregularity of knob-and-kettle topography, and the abundance of boulders conspired to prevent crop cultivation but invited use as pasture.

Flat till plains, outwash plains, and lake plains, on the other hand, comprise some of the most productive agricultural land in the world. Fertile soils have formed on these till plains and on exposed lakebeds bordering the Great Lakes. This fertility is enhanced by a blanket of wind-deposited silt (loess) that covers these plains (Chapter 17).

Stratified drift deposits are of great commercial value. The sands and gravels of outwash plains, deltas, and eskers provide necessary materials for both concrete manufacture and highway construction. Where it is thick, stratified drift forms an excellent aquifer and is a major source of groundwater supplies.

INVESTIGATING THE ICE AGE

A great scientific breakthrough in the study of Ice Age glacial history came in the 1960s. First, scientists learned how to measure the absolute age of certain types of water-laid sediments by means of ancient magnetism. The earth's magnetic field experienced many sudden reversals of polarity in Cenozoic time, and the absolute ages of these reversals have been firmly established. Second, techniques were developed to take long sample cores of undisturbed fine-textured sediments of the deep ocean floor. Within each core, scientists could determine the age of sediment layers at various control points by identifying magnetic polarity reversals. By further studying the composition and chemistry of the layers within the core, a record of ancient temperature cycles in the air and ocean could be established.

Figure 18.21 This small drumlin, located south of Sodus, New York, shows a tapered form from upper right to lower left, indicating that the ice moved in that direction (north to south).

Deep-sea cores reveal a long history of alternating glaciations and interglaciations going back at least as far as 2 million years and possibly 3 million years before present. The cores show that in late-Cenozoic time more than 30 glaciations occurred, spaced at time intervals of about 90,000 years. How much longer this sequence will continue into the future is not known, but perhaps for 1 or 2 million years, or even longer.

Possible Causes of the Late-Cenozoic Ice Age

What caused the earth to enter into an Ice Age with its numerous cycles of glaciation and interglaciation? Three causes seem possible: (1) a change in the placement of continents on the earth's surface through plate tectonic activity; (2) an increase in the number and severity of volcanic eruptions; and (3) a reduction in the sun's energy output.

Perhaps the answer lies in plate tectonics, through the motions of lithospheric plates following the breakup of Pangaea. Recall from Chapter 11 that in Permian time, only the northern tip of the Eurasian continent projected into the polar zone. As the Atlantic basin opened up, however, North America moved westward and poleward to a position opposite Eurasia, while Greenland took up a position between North America and Europe (see Figure 11.19).

The effect of these plate motions was to bring an enormous landmass area to a high latitude and to surround a polar ocean with land. Because the flow of warm ocean currents into the polar ocean was greatly reduced, or at times totally cut off, this arrangement was favorable to growth of ice sheets. The polar ocean was ice-covered much of the time, and average air temperatures in high latitudes were at times lowered enough to allow ice sheets to grow on the encircling continents. Also, Antarctica moved southward during the breakup of Pangaea and took up a position over the south pole. In that location, it was ideally situated to develop a large ice sheet. Some scientists have also proposed that the uplift of the Himalayan Plateau, a result of the collision of the Austral-Indian and Eurasian plates, could have modified weather patterns sufficiently to trigger the ice age.

During the Cenozoic era, increased volcanic activity may have helped to trigger the Ice Age.

The second geological mechanism suggested as a basic cause of the Ice Age is increased volcanic activity on a global scope in late-Cenozoic time. Volcanic eruptions produce dust veils that linger in the stratosphere and reduce the intensity of solar radiation reaching the ground (Chapter 2). Temporary cooling of near-surface air temperatures follows such eruptions. Although the geologic record shows periods of high levels of volcanic activity in the Miocene and Pliocene epochs, their role in initiating the Ice Age has not been convincingly demonstrated on the basis of evidence now available.

Another possible cause of the Ice Age is a slow decrease in the sun's energy output over the last several million years, perhaps as part of a cycle of slow increase and decrease that lasts many million years. As yet, data are insufficient to identify this mechanism as a possible basic cause. However, research on this topic is being stepped up as new knowledge of the sun is acquired from satellites that probe the sun's atmosphere and map its surface features.

Possible Causes of Glaciation Cycles

What timing and triggering mechanisms are responsible for the many cycles of glaciation and interglaciation that the earth is experiencing during the present Ice Age? Although many causes for glacial cycles have been proposed, we will limit our discussion here to one major contender called the **astronomical hypothesis**. It has been under consideration for about 40 years and is based on well-established motions of the earth in its orbit around the sun.

The mechanism that links ice sheet growth and decline with cycles of insolation is complex and not well understood.

Two factors are involved—the changing distance between earth and sun, and the changing angle of tilt of the earth's axis of rotation. As to the first factor, the distance that separates the earth and sun at summer solstice, June 21, undergoes a cyclic variation. A single cycle lasts 21,000 years. Next, we take into account that the earth's axial tilt (now $23\frac{1}{2}°$) undergoes a 40,000-year cycle of change in which the tilt angle may increase to as much as about 24° and decrease to as little as 22°. If we pick a point on the earth at, say, lat. 65° N, we can calculate the cycle of total change in the intensity of incoming solar radiation, or insolation, on a day in summer (June 21), resulting from changing earth–sun distance and the cycle of axial tilt.

The changing insolation, plotted against time on a graph, yields a series of peaks and valleys (see graph in *Eye on the Environment • Earth–Sun Cycles and Ice Ages*, Chapter 1). The major peaks are repeated at intervals of about 80,000 to 90,000 years. Each peak seems to have triggered a deglaciation, ending a glaciation and bringing on an interglaciation. During the longer periods of reduced insolation, ice sheets formed and grew in volume. However, the glacial events may have

Eye on the Environment • Ice Sheets and Global Warming

What effects will global warming have on the ice sheets of Greenland and Antarctica? The Antarctic Ice Sheet (see photo) holds 91 percent of the earth's ice. If it were to melt entirely, the earth's mean sea level would rise by about 40 m (200 ft). Additional water could be released by the melting of the Greenland Ice Sheet, which holds most of the remaining volume of land ice. It seems unlikely that global warming will melt these ice sheets completely, unless the warming is much greater than anticipated. Still, the ice sheets might be greatly affected by only a small warming.

In the 1970s climatologist John Mercer pointed out that a portion of the Antarctic Ice Sheet lying west

The Antarctic Ice Sheet, here flowing into the ocean, is shedding a huge iceberg. In the distance, rock masses project through the ice sheet.

lagged behind the insolation peaks and valleys by some thousands of years.

The actual mechanisms by which insolation changes cause ice sheets to grow or to disappear are unclear. The entire subject is so complex that it is difficult for even a research scientist to grasp fully. Interactions between the atmosphere, the oceans, and the continental surfaces (including the ice sheets) are numerous and closely interrelated with many threads of cause and effect. Changes that occur in one earth realm are fed back to the other realms in a most complex manner.

HOLOCENE ENVIRONMENTS

The elapsed time span of about 10,000 years since the Wisconsinan Glaciation ended is called the *Holocene Epoch*. It began with rapid warming of ocean sur-

face temperatures. Continental climate zones shifted rapidly poleward, and soil-forming processes began to act on new parent matter of glacial deposits in midlatitudes. Plants became reestablished in glaciated areas in a succession of climate stages.

One method of reconstructing the history of climate and vegetation throughout Holocene time is to study spores and pollen grains found in layered order from bottom to top in postglacial bogs. Because many types of plants can be identified from these tiny remains, the general nature of the vegetation in the vicinity of the bog can be determined. Samples can also be dated, so that we can establish the time of appearance of a vegetation cover type.

Studies of this nature have identified three major climatic periods leading up to the last 2000 years. The first of these is known as the Boreal stage. ("Boreal" refers to the present subarctic region where needleleaf

of the Transantarctic Mountains was in a potentially unstable condition and that it might surge rapidly into the ocean, causing a sea-level rise of about 5 m (16 ft). Mercer's theory depends heavily on the role of the Ross ice shelf of western Antarctica, which merges on its landward sides with the ice sheet (see Figure 18.11). Because the ice shelf is already afloat, its detachment and/or melting would not change sea level. It does, however, hold back the downslope flow of ice from the western Antarctic Ice Sheet.

The shelf ice is in contact with the bottom (grounded) at a number of points. If sea level were to rise slightly, owing to thermal expansion, it could lift the ice shelf and allow sea water to circulate more freely underneath it, melting the underside of the floating ice shelf. The ice shelf would then move out to sea, allowing the unsupported land ice to quickly surge forward and enter the ocean.

Thus far, there are no indications that this activity has begun. On the other hand, if Mercer's prediction should materialize, it could result in a sea-level rise as much as 5 m (16 ft)—a disaster of monumental proportions.

In 1989 other scientists active in glaciological research—among them geophysicist Charles R. Bentley—presented a different view of the effects of global warming on ice sheets. They reasoned that, although global warming will accelerate glacier melting, increased precipitation over the ice sheets will produce a net growth in their thickness. Measurements made by radar on satellites have shown that the Greenland Ice Sheet has increased in thickness by 23 cm (9 in.) in the past decade. Growth in bulk of the Antarctic Ice Sheet, by itself, could cause a lowering of global sea level. In balance with total melting, ice sheet growth would result in only a modest rate of rise of sea level—perhaps no

faster than the rate that has been documented over the past decade.

The possible effects of climate change on global ice sheets are thus far from certain. We can only hope that human-induced global warming will result in slow, progressive changes to which human civilization can readily adapt.

Questions

1. Explain Mercer's theory that global warming could create a sudden rise in sea level of as much as 5 m (16 ft).
2. Explain Bentley's theory that global warming would only increase sea level slightly, if at all.
3. What new evidence has complicated the understanding of the role of ice shelves in retaining unstable ice on west Antarctica?

---

forests dominate the vegetation.) A dominant tree of the Boreal stage was spruce. Interpretation of pollen indicates that the Boreal stage in midlatitudes had a vegetation similar to that now found in the region of Boreal forest climate.

A general warming of climate followed the Boreal stage until the Atlantic climatic stage was reached about 8000 years ago (–8000 years). The Atlantic stage lasted about 3000 years and had average air temperatures somewhat higher than those of today—perhaps on the order of 2.5° C (4.5° F) higher. We call such a period a climatic optimum, and it applies mostly to the midlatitude zone of North America and Europe.

Next came a period of temperatures that were below average, the Subboreal climatic stage, in which alpine glaciers showed a period of readvance. This stage spanned the age range –5000 to –2000 years. At that time sea level, drawn far down from present levels dur-

ing glaciation, had returned to a position close to that of the present, and the stable continent margins were inundated and submerged.

Through the availability of historical records and of more detailed evidence, we have been able to describe the climate of the past 2000 years on a finer scale than other Holocene stages. A secondary climatic optimum occurred in the period A.D. 1000 to 1200 (–1000 to –800 years). This warm episode was followed by the Little Ice Age, A.D. 1450–1850 (–550 to –50 years). During the Little Ice Age, valley glaciers made new advances and extended to lower elevations. In the process, the ice overrode nearby forests and so left a mark of its maximum extent.

In Chapter 3, we examined conflicting views of the role of human activity in causing global climate change. Cycles of glaciation and interglaciation that have proceeded without human influence for millions of years demonstrate the power of natural forces to make drastic swings from cold to warm climates. The lesser climatic cycles of most of the Holocene epoch were also produced by natural causes. Only following the Industrial Revolution have we recognized possible linkages between the burning of hydrocarbon fuels on a massive scale and global air temperature change. There is general agreement that increased carbon dioxide causes a rise in average temperatures. But we do not yet know to what extent recently observed changes in global temperatures are part of a natural cycle and to what extent these changes are influenced by the impacts of an industrial society.

The group of chapters we have now completed has reviewed landform-making processes that operate on the surface of the continents. These have ranged from mass wasting to the erosion and deposition brought about by glacial ice. The great variety and complexity of landforms we have described are not difficult to understand when each agent of denudation is examined in turn.

Human influence on landforms is felt most strongly on surfaces of fluvial denudation because of the severity of surface changes caused by agriculture and urbanization. Landforms shaped by wind and by waves and currents are also highly sensitive to changes induced by human activity. Only glaciers maintain their integrity and are thus far largely undisturbed by human activity. Perhaps even this last realm of nature's superiority will eventually fall prey to human interference through climate changes induced by industrial activity.

GLACIAL LANDFORMS AND THE ICE AGE IN REVIEW

Glaciers form when snow accumulates to a great depth, creating a mass of ice that is plastic in lower layers and flows outward or downhill from a center in response to gravity. Glaciers require cold temperatures and abundant snow to form and grow. These conditions create both alpine glaciers and continental ice sheets. Glaciers can deeply erode bedrock as they move, by abrasion and plucking. The eroded fragments, incorporated into the flowing ice, leave depositional landforms when the ice melts.

Alpine glaciers develop in cirques in high mountain locations. Alpine glaciers flow downvalley on steep slopes, picking up rock debris and depositing it in lateral and terminal moraines. Through erosion, glaciers carve U-shaped glacial troughs that are distinctive features of glaciated mountain regions. They become fiords if later submerged by rising sea level.

Ice sheets are huge plates of ice that cover vast areas. They are present today in Greenland and Antarctica. The Antarctic Ice Sheet includes ice shelves—great plates of floating glacial ice. Icebergs form when glacial ice flowing into an ocean breaks into great chunks and floats free. Sea ice, which is much thinner and more continuous, is formed by direct freezing of ocean water and accumulation of snow on the sea-ice surface.

An ice age includes alternating periods of glaciation, deglaciation, and interglaciation. During the past 2 to 3 million years, the earth has experienced the Late-Cenozoic Ice Age. During this ice age, continental ice sheets have grown and melted as many as 30 times. The most recent glaciation is the Wisconsinan Glaciation, in which ice sheets covered much of North America and Europe, as well as parts of northern Asia and southern South America.

Moving ice sheets create many types of landforms. Bedrock is grooved and scratched. Where rocks are weak, long valleys can be excavated to depths of hundreds of meters. The melting of glacial ice deposits glacial drift, which may be stratified by water flow or deposited directly as till. Moraines accumulate at ice edges. Outwash plains are built up by meltwater streams. Tunnels within the ice leave streambed deposits as eskers. Till may be spread smooth and thick under an ice sheet, leaving a till plain. This may be studded with elongated till mounds, termed drumlins. Meltwater streams build deltas into lakes formed at the ice margin and line lake bottoms with clay and silt. When the lakes drain, these features remain.

Several factors have been proposed to explain the cause of present ice age glaciations and interglaciations. These include ongoing change in the global position of continents, an increase in volcanism, and a reduction in the sun's energy output. Individual cycles of glaciation seem strongly related to small, cyclic changes in earth–sun distance and axial tilt.

KEY TERMS

ablation
alpine glacier
ice sheet
cirque
moraine

glacial trough
fiord
glaciation
interglaciation

Late-Cenozoic Ice Age
glacial drift
till
astronomical hypothesis

REVIEW QUESTIONS

1. How does a glacier form? What factors are important? Why does a glacier move?

2. Distinguish between alpine glaciers and ice sheets.

3. What are some typical features of an alpine glacier? Sketch a cross section along the length of an alpine glacier and label it.

4. What is a glacial trough and how is it formed? What is its basic shape? In what ways can a glacial trough appear after glaciation is over?

5. Where are ice sheets present today? How thick are they?

6. Contrast sea ice and icebergs, including the processes by which they form.

7. Identify the Late-Cenozoic Ice Age. When did it begin? What was the last glaciation in this cycle? When did it end?

8. What areas were covered with ice sheets by the last glaciation? How was sea level affected?

9. What are moraines? How are they formed? What types of moraines are there?

10. Identify the landforms and deposits associated with stream action at or near the front of an ice sheet.

11. Identify the landforms and deposits associated with deposition underneath a moving ice sheet.

12. Identify the landforms and deposits associated with lakes that form at ice sheet margins.

13. How have environments changed during the Holocene Epoch? What periods are recognized, and what are their characteristics?

IN-DEPTH QUESTIONS

1. Imagine that you are planning a car trip to the Canadian Rockies. What glacial landforms would you expect to find there? Where would you look for them?

2. At some time during the latter part of the Pliocene Epoch, the earth entered an ice age. Describe the nature of this ice age and the cycles that occur within it. What explanations are proposed for causing an ice age and its cycles? What cycles have been observed since the last ice sheets retreated?

EPILOGUE

News from the Future

Our text began with a Prologue describing a trip you might take as a lunar astronaut, traveling home from the moon to earth in 2050. What will life be like in the middle of the twenty-first century? Imagine turning on a futuristic television set and watching the news for June 26, 2050. It might go something like this.

"Good evening, America, and welcome to the Nightly News. This is Dan Rogers in Washington. Tonight's big stories are from the environment. First, we'll go to Galveston, where the story is the flooding from the first storm of the season, Hurricane Alma. Here's our correspondent, Lorraine Jackson, on the scene. Lorraine, what's it like there now?"

"Thanks, Dan. The storm has slackened off in the last few hours, but it seems that the system of dams and pumps they have here has worked, at least for now, to keep the floodwaters out of the center of the city. As you can see, the rain is still falling and the winds are still strong. But the real problem with this storm has not been the wind and rain. Alma is one of the weaker storms to reach the Texas coast in the past several hurricane seasons. The problem is with the ocean flooding that this coastal outpost endures when a storm like Alma passes close by.

"The ocean dikes have been taking a terrible beating from the wave action. Here's some footage of the waves breaking on the dikes that we shot earlier. All that wave water spilling over the dikes has to be pumped out, and fast. We understand from the local authorities that several smaller dikes west of the city were breached this morning, adding to the floodwaters. Apparently, several subdivisions out there are well under water now.

Hurricane Alicia lashes the Texas coast at Galveston.

476

Floating market, Bangkok, Thailand.

There was a scare when the power went out earlier in the day, and one of the huge backup generators supplying power to the pumps refused to start. But the other generators seem to be keeping up with the load of the pumps.

"When the citizens of this Texas city decided 30 years ago to fight against the rise in sea level rather than move inland like most other beach communities, they knew that the battle against the ocean would be costly and could eventually be a losing one. But for now, the valiant city of Galveston seems to be holding its own, Dan."

"Lorraine, what about the death toll?"

"So far, Dan, the authorities have not reported any deaths from the storm, although there were some injuries in the evacuation. We had a report that one of the huge buses they use for the evacuation hit the rear end of another bus, but only the driver and two passengers were injured. Apparently, the accident didn't slow down the river of huge vehicles streaming across the viaduct to the mainland. As I reported this morning, last night's evacuation went pretty smoothly. The residents here have been through so many evacuations in the last few years that they are all familiar with the emergency routine."

"Thanks, Lorraine. That seems to be relatively good news."

"Our next story marks a momentous event for the human race—the birth of the 7 billionth citizen of earth. For the story, we go to Beijing, to our correspondent Li Feng Sun. Where are you, Feng Sun?"

"Dan, I'm here in the corridor outside the maternity ward in this small birthing center in a suburb west of Beijing, waiting word about the birth of Zhu Peng, destined to be the world's 7 billionth citizen. Of course, we're not precisely certain that young Zhu Peng will be the world's 7 billionth citizen, but scientists have predicted that in precisely 4 minutes and 14 seconds, the earth's population will reach 7 billion. They also tell us that the 7 billionth citizen is most likely to be Chinese.

"Economists and population planners have been marking the event as a real milestone. Before the turn of the century, they projected that the world's population would be over 9 billion by 2050. You may remember some years ago the many predictions that the plans for sustainable world development would crash along with populations of Africa and Southeast Asia, which were decimated by the AIDS epidemic. China was only little impacted because AIDS spread quite slowly here, and because of China's crash program to inoculate its entire population with the AIDS vaccine shortly after it was developed. That left this huge country to continue its slow population growth coupled with rapid economic growth fueled by this country's vast natural and human resources. The result is the success story we're all familiar with—how China's economy surpassed the economies of Europe, the Americas, and finally the mighty East Asian trading bloc of Japan, Taiwan, Korea, and Singapore."

"Feng Sun, I seem to hear some chanting in the background. Are those demonstrators I hear?"

Earthquake damage, Golden State Freeway, Los Angeles.

"That's right, Dan. The militant arm of the Chinese Green party has mounted a demonstration here. In fact, we have one of the leaders here, Huang Chung, ready to talk with us. Huang Chung, what are the demonstrators here for?"

"Well, Feng Sun, we're protesting this event because we don't think the world needs any more people. Every consumer requires more food and uses more resources. We've already cut down almost half of the world's forests. Global warming has made the deserts spread and our cropland become less productive. We all know that global warming was generated by consumers demanding products and lifestyles that depend on fossil fuel. . . ."

"Feng Sun, I hate to break in, but we've got an interrupt here. Apparently there's been a breakthrough on the biodiversity talks at the U.N. For the story, let's go to Maria Remarquez in New York. Maria?"

"Hi, Dan. We've just gotten word here that the rainforest nations will back down from their threat to withdraw from the biodiversity treaty. Apparently, the U.S., the European Union, and Chinese-East Asian bloc have agreed to make some of the payments demanded for the use of the rainforest genetic materials that the rainforest nations contend were stolen from them by biotechnology companies.

"Here to tell us about it is Professor Kate Carver of New York's City University. Professor Carver, why is this agreement important?"

"Well, Maria, because with this agreement in place, we can go ahead with the plan to set up rainforest preservation zones, perhaps as early as next year. As you know, the rainforest ecosystems are our most complex ecological communities, and they hold the greatest diversity of life of all habitats on earth. They are a reservoir of genetic material that is irreplaceable, and they need to be protected

as vast areas forever. Without these genetic reserves, we'd be missing many of our cures for diseases ranging from cancer to AIDS.

"You know, Maria, I might also point out that this agreement is good for the environment in another way. The money will be used by the rainforest nations to stimulate the use of nonpolluting technologies for sustainable development. That is, the funds will be helping these nations promote more use of solar power, wind power, and methane gas production. In this way, they will reduce their reliance on old-fashioned fossil fuels and with it the release of carbon dioxide that produces climate change."

"Thanks for that perspective, Professor. It looks like a win for the environment here at the United Nations this evening, Dan."

"And thanks for that fine story, Maria. For our next story, remember that it was exactly one month ago when the disastrous El Cajon earthquake struck Los Angeles. We go now to Pasadena for a report on the cleanup and rebuilding. Our reporter with the story is Keith Mancini. Keith?"

"It was early on a Sunday morning just one month ago when Angelinos were thrown from their beds by the violent shaking that devastated so much of Los Angeles. It was the largest earthquake to hit state of South California in modern times, measuring magnitude 8.1 on the Richter Scale, flattening thousands of buildings and rendering thousands more unusable. The death toll, estimated at over 16,000 with more than 100,000 injured, was also by far the largest in this century for an American natural disaster.

"Today Los Angeles is still digging out. The magnitude of the task ahead is mind-boggling, Dan. The electric freeway system is virtually closed for all but a few segments. Nearly every viaduct and

bridge experienced some damage, and engineers are struggling to inspect them all for safety and set the priorities for repair. Most of the city's skyscrapers are still unoccupied, since structural repairs are expected to take many more months to complete. In the hardest-hit areas containing old masonry buildings, almost nothing was left standing. Now that the searches for casualties are over, bulldozers and dump trucks are removing the debris, collecting it in huge piles for recycling.

"In the Long Beach area, the cleanup is beginning from the fires that started in the oil processing plants and spread to the chemical factories. The toxic compounds carried by the smoke and ash contaminated a huge area. It has taken several weeks for the National Guard to replace their cordon of troops and vehicles with the fencing and barbed wire that will keep looters out while the area is being detoxified, which is expected to take six to nine months.

"Meanwhile, homeless residents are flocking to the hundreds of cities constructed of portable geodomes now being set up by the National Guard. The demand increases daily as troops and local authorities evict inhabitants still trying to live in the dangerous, crumbling buildings. The geodome cities at least provide shelter, hot food and sanitation for the thousands of displaced residents, luxuries that are not easy to come by here.

"Communications systems are still in disarray here, Dan. Signal channels are scarce, and it can take 5 to 10 minutes to use a payment card for food at the grocery store or for a recharge at the filling station. Cash is at a premium, particularly for businesses still without power. The banks that are open are struggling to meet the demand for currency. Most merchants are offering discounts for cash, and using that money to pay their employees at the same discounts.

"But the amazing thing, Dan, is the dedication of these residents to rebuilding their city. Everywhere you go in this vast metropolis, there is activity. People are pulling together to rebuild their homes, their businesses, and their lives. There's a real 'can-do' spirit here, and L.A. should emerge from this tragedy all the stronger for it."

"Thanks, Keith, for that fine report. It won't be easy for the folks in Los Angeles in the weeks and months ahead, but at least they know that the rest of the nation is pulling for them.

"And, that's it for the top news stories today. Stay with us for the local report, coming up after these messages."

Danum River gorge, Sabah, Borneo.

PERSPECTIVE ON THE NEWS

This news report dramatizes some of the concerns about the environment that have arisen in our survey of physical geography. The first segment calls attention to the rise in sea level that might be created by global warming. Conservative projections are that, by 2050, sea level will rise by about one-third of a meter (1 foot), partly depending on how much warming occurs. The problem with rising sea level is not so much with day-to-day levels, but rather the levels that occur when tides combine with storms to produce intense wave action on low-lying coasts. An even greater rise in sea level could occur if the trend continues for a century or more. The Republic of Maldives, which is a nation of 1190 small islands, would be entirely submerged by a rise of 2 meters (6.6 ft). Even with a rise of only 1 meter (3.3 ft), the country could be wiped out by a severe storm surge.

A rise in sea level is just one possible effect of climatic warming. Even though the extent of warming is uncertain, scientists have concluded that warming would bring not only warmer temperatures, but also more global precipitation. However, this precipitation increase will not be evenly distributed. Computer models predict that high latitudes will experience more precipitation, while midlatitude continental interiors will experience drier summers. These types of changes would have important effects on crop growth, shifting the varieties and types of crops grown within large regions. Increased irrigation could also be needed.

Because the effects of climatic warming could be so far-reaching, scientists have managed to alert the public about the danger of continued increases of CO_2 and other gases that contribute to global warming. At the 1992 Earth Summit in Rio de Janiero, most of the world's nations signed a treaty to reduce human impact on climate. The industrialized countries agreed to reduce the levels of greenhouse gases they release, thus slowing the rapid increases in atmospheric CO_2, methane, chlorofluorocarbons, and nitrous oxide (N_2O) that have been measured. Although the actual targets for these reductions are still under negotiation, the world is nonetheless united in reducing the buildup in greenhouse gases.

The Rio Earth Summit also adopted a treaty on biodiversity, which is the subject of the third news story. In the treaty, nations pledged to protect the genetic resources in their plant and animal populations by shielding them from extinction. Forest preservation was addressed as a need as well. The low-latitude rainforests cover only about 7 percent of the earth's surface, but they contain over half of the world's species. They are especially rich in insect species and flowering plants. Plants of these regions are a particularly valuable resource. They have provided most of the species on which humans rely for food. Many highly effective medical drugs are produced from these plants.

Not only are low-latitude rainforests valuable resources for human exploitation, they are also very complex ecosystems that have evolved over millions of years. Many environmentalists believe that humans have a responsibility to preserve our planet's natural systems and to maintain their diversity. It is encouraging that the world's nations have recognized the importance of ecological resources such as the low-latitude rainforests and are working toward their preservation.

The last future news story, about an earthquake in Los Angeles, demonstrates the impact of natural disasters on human activities. The story is only slightly exaggerated—an earthquake of magnitude 8.1 on the San Andreas fault would most certainly have devastating effects on a nearby city the size of Los Angeles.

By nature, the human species has a short-term outlook, preferring to prepare for dangers that are likely to occur within a time span of a few years to a decade or so. Many natural processes work on far longer time scales, however. While a major earthquake on the San Andreas fault may not happen more than once every few generations, tens of thousands of severe earthquakes probably occurred on that fault in the Los Angeles region during the Ice Age, with hundreds of these taking place since the last advance of glacial ice. A major flood on the Mississippi River, like that of 1993, is a still more likely event. Although infrequent in human terms, these rare events have shaped much of the earth's landscape. For the geologist or physical geographer trained in earth processes, the evidence of these past events is everywhere.

The second event of the future news broadcast, which dramatizes the birth of earth's 7 billionth citizen, is probably the one with the most important implications.

It is actually quite an optimistic scenario, for present estimates of the world population in 2050 are closer to 9.5 billion. The story refers to future population crashes in Africa and Southeast Asia produced by the AIDS epidemic that now rages unchecked throughout the world, and it advances the idea that China will be impacted less severely than other developing nations. Whether this future will come to pass is uncertain. In any event, the human population will continue to increase until it either regulates itself or destroys itself.

Why is population growth so important? Humans are the most powerful force transforming the surface of the earth today. About 15 to 20 percent of the world's forest area has been cleared since human civilization began to exploit forest resources. In the last three hundred years, humans have increased the area of cropland by 450 percent. Populations of the largest urban areas have increased 25 times in that same time span. These changes have directly influenced the global energy balance, the hydrologic cycle, and the dynamics of the atmosphere, with effects that are unknown in the long term. Furthermore, as economic development proceeds, the technological demands of each person increase. This translates to an increasing consumption of natural resources and energy that, in turn, places increasing demands on the environment. Thus, as population expands, the impact on natural environments increases.

Recognizing these trends, the nations of the world have recently come together in an effort to promote sustainable development. Under their newly formulated policy, the needs and aspirations of present and future generations will be satisfied without impacting the ability of future generations to meet their needs. The idea is to focus on social and economic progress that is compatible with the environment. The concept draws on the premise that human knowledge, sophisticated technology, and access to resources are now so well developed that they can be used to allow developing countries to increase the well-being of their populations in a way that does not damage the environment. In this effort, developed nations need to provide not only the funds to stimulate sustainable development, but also the necessary technological base.

The social and economic aspects of continued population growth and policies such as sustainable development are far beyond the scope of physical geography, but they are concerns of which every responsible citizen on our planet must be aware. This book has provided a background and context on the workings of the earth and the environment that we, as authors, hope will be valuable to you, its readers, as you face the challenges of world citizenship in the twenty-first century.

More About Maps

Maps play an essential role in the study of physical geography, because much of the information content of geography is stored and displayed on maps. Map literacy—the ability to read and understand what a map shows—is a basic requirement for day-to-day functioning in our society. Maps appear in almost every issue of a newspaper and in nearly every television newscast. Most people routinely use highway maps and street maps. The purpose of this appendix is to supplement the map-reading skills you have already acquired with some additional information on the subject of *cartography*, the science of maps and their construction. In understanding the following material, it will be helpful to review the section "Map Projections" in Chapter 1 of the text.

MAP PROJECTIONS

Recall from Chapter 1 that a *map projection* is an orderly system of parallels and meridians used as a base on which to draw a map on a flat surface. A projection is needed because the earth's surface is not flat but rather is curved in a shape that is very close to the surface of a sphere. All map projections misstate the shape of the earth in some way. It is simply impossible to transform a spherical surface to a flat (planar) surface without some violation of the true surface as a result of cutting, stretching, or otherwise distorting the information that lies on the sphere.

Perhaps the simplest of all map projections is a grid of perfect squares. In this simple map, horizontal lines are parallels and vertical lines are meridians. This grid is often used in modern computer-generated world maps displaying data that consist of a single number

for each square, representing, for example, 1, 5, or 10 degrees of latitude and longitude. A grid of this kind can show the true spacing (approximately) of the parallels, but it fails to show how the meridians converge toward the two poles. This grid fails dismally in high latitudes, and the map usually has to be terminated at about 70° to 80° north and south.

Early attempts to find satisfactory map projections made use of a simple concept. Imagine the spherical earth grid as a cage of wires (a kind of bird cage). A tiny light source is placed at the center of the sphere, and the image of the wire grid is cast on a surface outside the sphere. This situation is like a reading lamp with a lampshade. Basically, three kinds of "lampshades" can be used, as shown in Figure A1.1.

First is a flat paper disk balanced on the north pole. The shadow of the wire grid on this plane surface will appear as a combination of concentric circles (parallels) and radial straight lines (meridians). Here we have a polar-centered projection. Second is a cone of paper resting point-up on the wire grid. The cone can be slit down the side, unrolled, and laid flat to produce a map that is some part of a full circle. This is called a conic projection. Parallels are arcs of circles, and meridians are radiating straight lines. Third, a cylinder of paper can be wrapped around the wire sphere so as to be touching all around the equator. When slit down the side along a meridian, the cylinder can be unrolled to produce a cylindrical projection, which is a true rectangular grid.

Take note that none of these three projection methods can show the entire earth grid, no matter how large a sheet of paper is used to receive the image. Obviously, if the entire earth grid is to be shown, some quite different system must be devised. Many

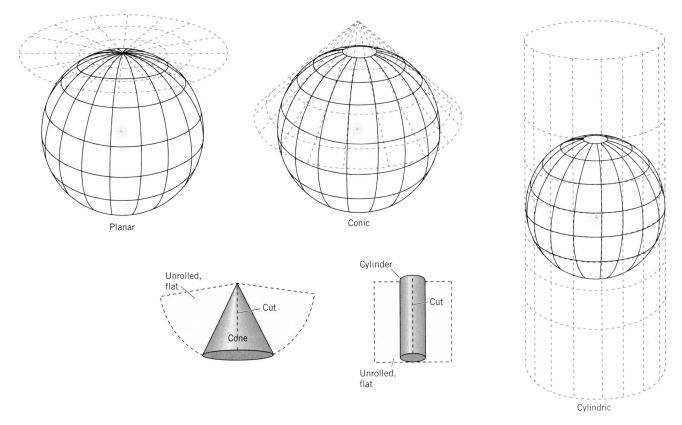

Figure A1.1 Simple ways to generate map projections. Rays from a central light source cast shadows of the spherical geographic grid on target screens. The conical and cylindrical screens can be unrolled to become flat maps. (A. N. Strahler.)

such alternative solutions have been proposed. In Chapter 1, we described the three types of projections used throughout the book—the *polar projection;* the *Mercator projection,* which is a cylindrical projection; and the *Goode projection,* which uses a special mathematical principle.

SCALES OF GLOBES AND MAPS

All globes and maps depict the earth's features in much smaller size than the true features they represent. In principle, globes are intended to be perfect scale models of the earth itself, differing from the earth only in size. The *scale* of a globe is therefore the ratio between the size of the globe and the size of the earth, where "size" is some measure of length or distance (but not of area or volume).

Take, for example, a globe 20 cm (about 8 in.) in diameter, representing the earth, with a diameter of about 13,000 km. The scale of the globe is the ratio between 20 cm and 13,000 km. This ratio reduces to a scale stated as follows: 1 centimeter on the globe

represents 650 kilometers on the earth. The relationship holds true for distances between any two points on a globe.

Scale is more usefully stated as a simple fraction, termed the *fractional scale,* or *representative fraction.* It can be obtained by reducing both earth and globe distances to the same unit of measure, which in this case is centimeters. (There are 100,000 centimeters in one kilometer.) The advantage of the representative fraction is that it is entirely free of any specified units of measure, such as the foot, mile, meter, or kilometer. Persons of any nationality can understand the fraction, regardless of their language or units of measure.

Being a true-scale model of the earth, a globe has a constant scale everywhere on its surface, but this is not true of a map projection drawn on a flat surface. In flattening the curved surface of the sphere to conform to a plane surface, all map projections stretch the earth's surface in a nonuniform manner, so that the map scale changes from place to place. Thus, we can't say about any world map: "Everywhere on this map the scale is 1:65,000,000." It is, however, possible to select a meridian or parallel—the equator, for example—for which a fractional scale can be given, relating the map to the globe it represents. For example, in Figure 1.8, the scale of the Mercator projection along its equator is about 1:325,000,000.

SMALL-SCALE AND LARGE-SCALE MAPS

From maps on the global scale, showing the geographic grid of the entire earth or a full hemisphere, we turn to maps that show only small sections of the earth's surface. These large-scale maps are capable of carrying the enormous amount of geographic information that is available and must be shown in a convenient and effective manner. For practical reasons, maps are printed on sheets of paper that are usually less than 1 meter (3 ft) wide, as in the case of the ordinary highway map or navigation chart. Bound books of maps—atlases, that is—consist of pages that are usually no larger than 30 by 40 cm (12 by 16 in.), whereas maps found in textbooks and scientific journals are even smaller.

Actually, it is not simply the size of map sheet or page that determines how much information a map can carry. Instead, it is the scale on which the surface is depicted. Keep in mind that the relative magnitude of

two different map scales is determined according to which representative fraction is the larger and which the smaller. For example, a scale of 1:10,000 is twice as large as a scale of 1:20,000. Don't make the mistake of supposing that the fraction with the larger denominator represents the larger scale. If in doubt, ask yourself "Which fraction is the larger: $\frac{1}{4}$ or $\frac{1}{2}$?"

In the case of globes and maps of the entire earth, the fractional scale may range from 1:100,000,000 to 1:10,000,000; these are *small-scale maps*. *Large-scale maps* have fractional scales that are generally greater than 1:100,000. Stated in words, this fraction means: "One centimeter on the map represents one kilometer on the ground." Most large-scale maps carry a *graphic scale*, which is a line marked off into units representing kilometers or miles. Figure A1.2 shows a portion of a large-scale map on which sample graphic scales in miles, feet, and kilometers are superimposed. Graphic scales make it easy to measure ground distances.

INFORMATIONAL CONTENT OF MAPS

Up to this point, our investigation of cartography has yielded a geographic grid with only the barest outlines of the continents. Information conveyed by a map

Figure A1.2 A portion of a modern, large-scale map for which three graphic scales have been provided. (U.S. Geological Survey.)

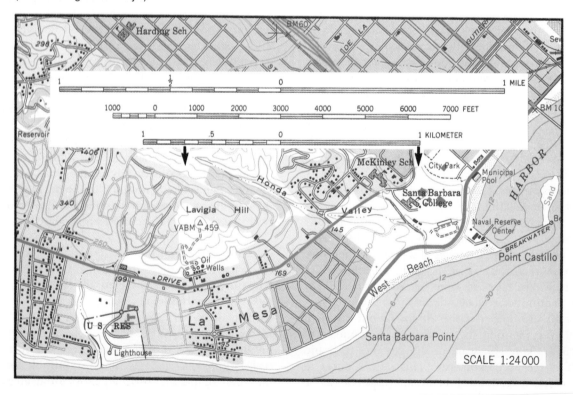

projection grid system is limited to one category only: absolute location of points on the earth's surface. What other categories of information can be shown on the flat map sheet? The number of categories is almost infinite, since the map is capable of carrying information about anything that occupies a particular location at a specified time. For the needs of physical geography, which is focused primarily on the physical environment of life in the surface layer, the list is greatly reduced but still enormously long.

Multipurpose maps show a number of different types of information. A simple example is shown in Figure A1.3. Map sheets published by national governments are usually multipurpose maps. Using a great variety of symbols, patterns, and colors, these maps carry a high information content available on demand to the general user, who may select and use only one information category, or theme, at any one time. A real example of a multipurpose map is shown on the endpapers of this book, located on the inside of the back cover. It is a portion of a U.S. Geological Survey topographic quadrangle map for San Rafael, California.

In contrast to the multipurpose map is the *thematic map*, which shows only one type of information. Some examples are distribution of the human population, boundaries of election districts, geologic age of exposed rock, elevation of the land surface above sea level, number of days of frost in the year, and speed of winds in the upper atmosphere.

MAP SYMBOLS

Symbols on maps associate information with points, lines, and areas. To express points as symbols, they can be renamed "dots." Broadly defined, a dot can be any small device to show point location. It might be a closed circle, an open circle, a letter, a numeral, or a little picture of the object it represents (see "church with tower" in Figure A1.3). A line can vary in width and can be single or double. The line can also consist of a string of dots or dashes. A specific area of surface can be referred to as a "patch." The patch can be shown simply by a line marking its edge, or it can be depicted by a distinctive pattern or a solid color. Patterns are highly varied. Some consist of tiny dots and others of parallel, intersecting, or wavy lines.

A map consisting of dots, lines, and patches can carry a great deal of information, as Figure A1.3 shows. In this case, the map uses two kinds of dot symbols (both symbolic of churches), three kinds of line symbols, and three kinds of patch symbols. Altogether, eight types of information are offered. Line symbols freely cross patches, and dots can appear within

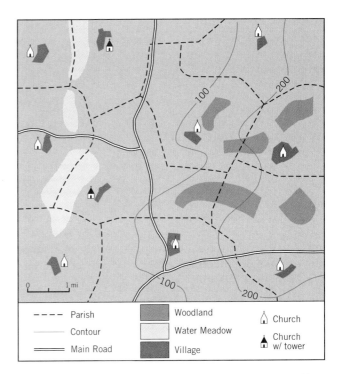

Figure A1.3 Multipurpose map of an imaginary area with 10 villages. (After J.P. Cole and C.A.M. King, *Quantitative Geography,* copyright © John Wiley & Sons, London. Used by permission.)

patches. Two different kinds of patches can overlap.

The relationship of map symbols to map scale is of prime importance in cartography. Maps of very large scale, along with architectural and engineering plans, can show objects to their true outline form. As map scale is decreased, representation becomes more and more generalized. In physical geography an excellent example is the depiction of a river, such as the lower Mississippi. Figure A1.4 shows the river channel at three scales, starting with a detailed plan, progressing to a double-line analog that generalizes the channel form, and ending with a single-line symbol. As generalization develops, the details of the river banks and channel bends are simplified as well. The level of depiction of fine details is described by the term *resolution.* Maps of large scale have much greater resolving power than maps of small scale.

PRESENTING NUMERICAL DATA ON THEMATIC MAPS

How are numerical data shown on maps? In physical geography, much of the information collected about particular areas is in the form of numbers. The numbers might represent readings taken from a scientific

a.

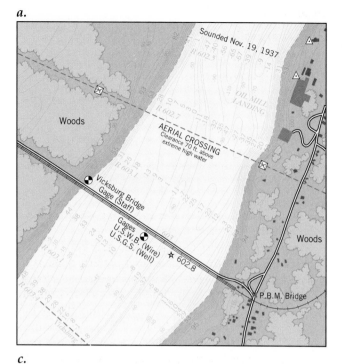

b.

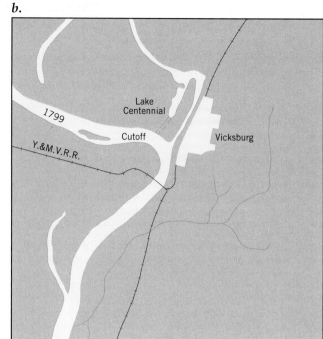

c.

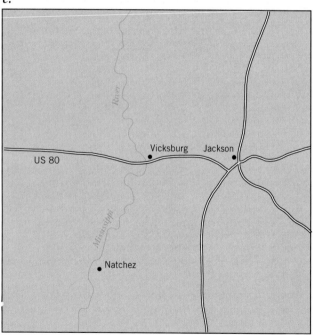

Figure A1.4 Maps of the Mississippi River on three scales. (*a*) 1:20,000. Channel contours give depth below mean water level. (*b*) 1:250,000. Waterline only shown to depict channel. (*c*) 1:3,000,000. Channel shown as solid line symbol. (Maps slightly enlarged for reproduction.) (Modified from U.S. Army Corps of Engineers.)

instrument at various places throughout the study area. A simple example is the collection of weather data, such as air temperature, air pressure, wind speed, and amount of rainfall. Another example is the set of measurements of the elevation of a land surface, given in meters above sea level. Another category of information consists merely of the presence or absence of a quantity or an attribute. In such cases, we can simply place a dot to mean "present," so that when entries are completed, we have before us a field of scattered dots (Figure A1.5).

In some scientific programs, measurements are taken uniformly, for example, at the centers of grid squares laid over a map. For many classes of data, how-

Table A1.1   Examples of Isopleths

Name of Isopleth	Greek Root	Property Described	Examples in Figures
Isobar	*baros,* weight	Barometric pressure	5.17
Isotherm	*therme,* heat	Temperature of air, water, or soil	3.18
Isotach	*tachos,* swift	Fluid velocity	5.22
Isohyet	*hyetos,* rain	Precipitation	4.18
Isohypse	*hypso,* height	Elevation	18.9, 18.10

ever, the locations of the observation points are predetermined by a fixed and nonuniform set of observing stations. For example, data of weather and climate are collected at stations typically located at airports. Whatever the sampling method used, we end up with an array of numbers and dots indicating their location on the field of the base map.

While the numbers and locations may be accurate, it may be difficult to see the spatial pattern present in the data being displayed. For this reason, cartographers often simplify arrays of point values into isopleth maps. An *isopleth* is a line of equal value (from the Greek *isos,* "equal," and *plethos,* "fullness" or "quantity"). Figure 3.17 shows how an isopleth map is constructed for temperature data. In this case, the isopleth is an *isotherm,* or a line of constant temperature. In drawing an isopleth, the line is routed among the points in a way that best indicates a uniform value, given the observations at hand.

Isopleth maps are important in various branches of physical geography. Table A1.1 gives a partial list of isopleths of various kinds used in the earth sciences, together with their special names and the kinds of information they display. Examples are cited from our text.

A special kind of isopleth, the *topographic contour* (or isohypse), is shown on the maps in Figures A1.2, A1.4*a* and in the back endpapers. Topographic contours show the configuration of land surface features, such as hills, valleys, and basins.

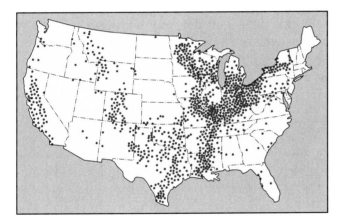

Figure A1.5   A dot map showing the distribution of soils of the soil order Alfisols in the United States. (From P. Gersmehl, *Annals of the Assoc. of Amer. Geographers,* vol. 67. Copyright © Association of American Geographers. Used by permission.)

APPENDIX II

Remote Sensing for Physical Geography

In various branches of physical geography, as in other fields of the earth sciences, a new technical discipline called *remote sensing* has expanded rapidly within the past decade and is adding greatly to our ability to perceive and analyze the physical, chemical, biological, and cultural character of the earth's surface. Remote sensing refers to gathering information from great distances and over broad areas, usually through instruments mounted on aircraft or orbiting space vehicles. All substances, whether naturally occurring or synthetic, are capable of reflecting, absorbing, and emitting energy in forms that can be detected by instruments known collectively as *remote sensors*.

RADAR—AN ACTIVE REMOTE SENSING SYSTEM

There are two classes of remote sensor systems: active systems and passive systems. *Active systems* use a beam of wave energy as a source, sending the beam toward an object or surface. Part of the energy is reflected back to the source, where it is recorded by a detector. A simple analogy would be the use of a spotlight on a dark night to illuminate a target, which reflects light back to the eye.

Radar is an example of an active sensing system. Radar uses radiation from the *microwave* portion of the electromagnetic spectrum, so named because the waves have a short wavelength. Radar systems emit short

pulses of microwave radiation and then "listen" for a returning microwave echo. Many radar systems can penetrate clouds to provide images of the earth's surface in any weather. At short wavelengths, however, microwaves

Figure A2.1 Radar image of a portion of the folded Appalachians in south-central Pennsylvania. The area shown is about 40 km (25 mi) wide. Compare with Figure 16.15, which is a Landsat image of this region.

488

Figure A2.2 High-altitude infrared photograph of an area near Bakersfield, California, in the southern San Joaquin Valley, taken by a NASA U-2 aircraft flying at approximately 18 km (60,000 ft). Photos such as this can be used to study problems associated with agriculture. Problems affecting crop yields arise in (A) perched ground water areas, which appear dark, and (B) areas of high soil salinity, which appear light. The various red tones are associated with different types of crops. (Photo by NASA, compiled and annotated by John E. Estes and Leslie W. Senger.)

can be scattered by water droplets and produce a return signal sensed by the radar apparatus. This effect is the basis for weather radars, which can detect rain and hail and are used in local weather forecasting.

Figure A2.1 shows a radar image of the folded Appalachian Mountains in south central Pennsylvania. It is produced by an airborne radar instrument that sends pulses of radio waves downward and sideward as the airplane flies forward. Surfaces oriented most

Figure A2.3 This color infrared photograph was taken with a large-format camera carried on a flight of the NASA Space Shuttle. It gives an extremely detailed picture of the terrain and is suitable for precision mapping. The area shown, about 125 km in width, includes the Sulaiman Range in West Pakistan (upper left). The Indus and Sutlej rivers, fed by snowmelt from the distant Hindu Kush and Himalaya ranges, cross the scene flowing from northeast to southwest. A mottled pattern of green fields (red) interspersed with barren patches of saline soil (white) covers much of the lower fourth of the area.

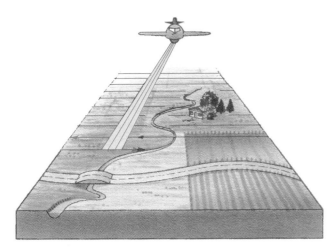

Figure A2.4 Scanning from aircraft. As the aircraft flies forward, the scanner's field of view sweeps from side to side. The result is a digital image covering the overflight area. (A. H. Strahler.)

nearly at right angles to the slanting radar beam will return the strongest echo and therefore appear lightest in tone. In contrast, those surfaces facing away from the beam will appear darkest. The effect is to produce an image resembling a shaded relief map. Various types of surfaces, such as forest, rangeland, and agricultural fields, can also be identified by variations in image tone and pattern.

PASSIVE REMOTE SENSING SYSTEMS

Passive systems measure radiant energy that is reflected or emitted by an object. Reflected energy falls mostly in the visible light and near-infrared regions, whereas emitted energy lies in the longer thermal infrared region. The most familiar instrument of the passive type is the camera, which uses film that is sensitive to the energy reflected from the scene. Aerial photographs, taken by downward-pointing cameras from aircraft, have been in wide use since before World War II. Commonly, the field of one photograph overlaps the next along the plane's flight path, so that the photographs can be viewed stereoscopically for a three-dimensional effect. Because of its high resolution (degree of sharpness), aerial photography remains one of the most valuable of the older remote-sensing techniques.

Photography can also extend into infrared wavelengths. Within the infrared spectrum, there is a region in the near-infrared, immediately adjacent to the visible red region, in which reflected rays can be recorded by cameras with suitable film and filter combinations. One example is color infrared film (Figure A2.2). In this type of film, the red color is produced as a response to infrared light, the green color is produced by red light, and the blue color by green light. Because healthy, growing vegetation reflects much more strongly in the infrared than in the red or green regions of the spectrum, vegetation has a characteristic red appearance. Figure 15.21, which shows a portion of the Mississippi River and its floodplain, is an example of an air photo taken with color-infrared film.

Satellite images are often processed to give the appearance of color-infrared film. Examples are Figures 16.14 and 17.15. On these images, agricultural crops appear as color shades ranging from pink to orange-red to deep red. Mature crops and dried vegetation (dormant grasses) appear yellow or brown. Urbanized areas typically appear in tones of blue and gray. Shallow water areas appear blue; deep water appears dark blue to blue-black.

Photography has been extended to greater distances through the use of cameras on orbiting space vehicles. A recent Space Shuttle flight included a specially constructed large-format camera, designed to produce very large, very detailed transparencies of the earth's surface suitable for precise topographic mapping. An excellent example is shown in Figure A2.3.

SCANNING SYSTEMS

Passive remote sensing also includes images acquired by scanning systems, which may be mounted in aircraft or orbiting space vehicles. *Scanning* is the process of receiving information instantaneously from only a very small portion of the area being imaged (Figure A2.4). The scanning instrument senses a very small field of view that runs rapidly across the ground scene. Light from the field of view is focused on a detector that measures its intensity at very short time intervals and records the intensities as a series of numbers that can be processed by a computer. Later, the computer reconstructs an image of the ground scene from the measurements acquired by the scanning system.

Most scanning systems in common use are *multispectral scanners*. These devices measure brightness in several wavelength regions simultaneously. An example is the Multispectral Scanning System (MSS) used aboard the Landsat series of earth-observing satellites. This instrument simultaneously collects reflectance data in four spectral bands. Two are bands in the visible light spectrum (one green, one red), and two in the infrared region. A successor to this system, the Landsat Thematic Mapper (TM), collects data in seven spectral bands. The text contains a number of fine examples of Landsat images. (See Figures 16.14, 17.15, 17.16, 17.31, and 18.12.)

Another type of scanner provides images from the *thermal infrared* portion of the spectrum, which is sensitive to the temperature of the objects in the scene. Since warmer objects emit more infrared radiation than do cooler ones, the warmer objects will appear lighter on thermal infrared imagery. An example is shown in Figure 2.3. Here, a color spectrum has been used so that the warmest surfaces are shown in red, while cooler surfaces appear in blue, violet, or black.

ORBITING EARTH SATELLITES

With the development of orbiting earth satellites carrying remote-sensing systems, remote sensing has expanded into a major branch of geographic research, going far beyond the limitations of conventional aerial photography. One reason for this advance is that most satellite remote sensors are scanners that provide data instead of photographs. These data can be processed

Figure A2.5 Digital image of New York Harbor, acquired by the SPOT satellite in June, 1986. The data are presented as a color infrared picture. The tip of Manhattan is at the top center; the blocky, dark structures at the tip are Wall Street skyscrapers and their shadows. Just below the tip is Governor's Island, a Coast Guard military reservation, and immediately to the west (left) is small Liberty Island, home of the Statue of Liberty. The long container-loading piers of Port Newark are visible on the left, as are piers on the Brooklyn shorelines to the right. At the center bottom, the Verrazano Narrows bridge connects Brooklyn with Staten Island. The fine detail of the street grid pattern, and even of the wakes of ships in the harbor, show the fine resolution of the SPOT's imaging instrument.

and enhanced by computers much more effectively than photographic images. Another reason is the ability of orbiting satellites to monitor nearly all the earth's surface. As a result, global and regional studies that could not have been carried out with air photographs have become possible.

How does a satellite sensor cover the entire earth? Nearly all satellites used for remote sensing are placed in a *sun-synchronous orbit.* In this orbit, all the satellite images of a location are acquired at about the same time of day. For the Landsat series, this is normally between 9:30 and 10:30 A.M. local time. This means that images collected on different dates can easily be compared, since they are all illuminated in a similar way by the sun. The orbit is also designed so that images are acquired from all locations on the globe before the orbit repeats. This can take from two to twenty-seven days, depending on how large an area is covered by each image.

Although in the past NASA has dominated the field of earth-observing satellites with its Landsat series, new satellites have been developed and launched by other nations. An example is the French satellite system, *SPOT* (an acronym for the French title Système Probatoire d'Observation de la Terre). The first SPOT satellite was launched in 1986, and the series continues to the present. Figure A2.5 shows a SPOT color image of the New York harbor area. The resolution, which is finer than that of the Landsat series, shows how much small detail can be seen with new-generation sensing systems.

Many technical advances will continue to be made in remote sensing. An international effort, led by NASA, is developing the *Earth Observing System,* which includes many satellites and sensing instruments to be used to study global change during the next few decades. Using these instruments, research in physical geography will be significantly enhanced. Remote sensing will be extended not only into remote areas of the earth that have never been mapped in detail, but also to global studies of the earth as a whole, utilizing the vantage point of earth orbit as never before in the history of science.

APPENDIX III

The Canadian System of Soil Classification*

The formation in 1940 of the National Soil Survey Committee of Canada was a milestone in the development of soil classification and of pedology generally in Canada. Prior to that time, the 1938 USDA system was used in Canada, but although the Canadian experience showed that the concept of zonal soils was useful in the western plains, it proved less applicable in eastern Canada where parent materials and relief factors had a dominant influence on soil properties and development in many areas.

Canadian pedologists observed closely the evolution of the U.S. Comprehensive Soil Classification System (CSCS) during the 1950s and 1960s and ultimately adopted several important features of that system. Nevertheless, the special needs of a workable Canadian national system required that a completely independent classification system be established. Because Canada lies entirely in a latitude zone poleward of the 40th parallel, there was no need to incorporate those soil orders found only in lower latitudes. Furthermore, the vast expanse of Canadian territory lying within the boreal forest and tundra climates necessitated the recognition at the highest taxonomic level (the order) of soils of cold regions that appear only as suborders and even as great groups in the CSCS—for example, the Cryaquents and Cryorthents that occupy much of the soils map of northern Canada above the 50th parallel (Figure 9.10).

The overall philosophy of the Canadian system is pragmatic: the aim is to organize the knowledge of soils in a reasonable and usable way. The system is a natural or taxonomic one in which the classes (taxa) are based on properties of the soils themselves and not on interpretations of the soils for various uses. Thus, the taxa are concepts based on generalization of properties of real bodies of soils rather than idealized concepts of the kinds of soils that would result from the action of presumed genetic processes. In this respect, the philosophy agrees with that used in the CSCS. Although taxa in the Canadian system are defined on the basis of actual soil properties that can be observed and measured, the system has a genetic bias in that properties or combinations of properties that reflect genesis are favored in distinguishing among the higher taxa. Thus, the soils brought together under a single soil order are seen as the product of a similar set of dominant soil-forming processes resulting from broadly similar climatic conditions.

The Canadian system recognizes the pedon as the basic unit of soils; it is defined as in the CSCS. Major mineral horizons of the soil (A, B, C) are defined in much the same way as in the U.S. system. Thus, the Canadian system of soil taxonomy is more closely related to the U.S. system than to any other. Both are hierarchical, and the taxa are defined on the basis of measurable soil properties. However, they differ in several respects. The Canadian system is designed to classify only the soils that occur in Canada and is not a comprehensive system. The U.S. system includes the suborder, a taxon not recognized in the Canadian system. Because 90 percent of the area of Canada is not likely to be cultivated, the Canadian system does not recognize as diagnostic those horizons strongly affected by plowing and application of soil conditioners and fertilizers.

---

*Throughout this section numerous sentences and phrases are taken verbatim or paraphrased from the following work: Canada Soil Survey Committee, *The Canadian System of Soil Classification*, Research Branch, Canada Department of Agriculture, Publication 1646, 1978. Table A3.1 and Figure A3.2 are also compiled from this source.

SOIL HORIZONS AND OTHER LAYERS

The definitions of classes in the Canadian system are based mainly on kinds, degrees of development, and the sequence of soil horizons and other layers in pedons. The major mineral horizons are A, B, and C. The major organic horizons are L, F, and H, which are mainly forest litter at various stages of decomposition, and O, which is derived mainly from bog, marsh, or swamp vegetation. Subdivisions of horizons are labeled by adding lowercase suffixes to the major horizon symbols; for example, Ah or Ae.

Besides the horizons, nonsoil layers are recognized. Two such layers are R, rock, and W, water. Lower mineral layers not affected by pedogenic processes are also identified. In organic soils, layers are described as *tiers*.

The principal mineral horizons, A, B, and C, are defined as follows:

A   Mineral horizon found at or near the surface in the zone of leaching or eluviation of materials in solution or suspension, or of maximum *in situ* accumulation of organic matter or both.

B   Mineral horizon characterized by enrichment in organic matter, sesquioxides, or clay; or by the development of soil structure; or by change of color denoting hydrolysis, reduction, or oxidation.

C   Mineral horizon comparatively unaffected by the pedogenic processes operative in A and B horizons. The processes of gleying and the accumulation of calcium and magnesium and more soluble salts can occur in this horizon.

Lowercase suffixes, used to designate subdivisions of horizons, are shown in Table A3.1.

Table A3.1   Subhorizons and Organic Horizons of the Canadian System of Soil Classification

Subhorizons; Lowercase Suffixes

b   Buried soil horizon.

c   Cemented (irreversible) pedogenic horizon.

ca   Horizon of secondary carbonate enrichment in which the concentration of lime exceeds that in the unenriched parent material.

e   Horizon characterized by the eluviation of clay, Fe, Al, or organic matter alone or in combination.

f   Horizon enriched with amorphous material, principally Al and Fe combined with organic matter; reddish near upper boundary, becoming yellower at depth.

g   Horizon characterized by gray colors, or prominent mottling, or both, indicative of permanent or intense reduction.

h   Horizon enriched with organic matter.

j   Used as a modifier of suffixes e, f, g, n, and t to denote an expression of, but failure to meet, the specified limits of the suffix it modifies.

k   Denotes presence of carbonate as indicated by visible effervescence when dilute HCl is added.

m   Horizon slightly altered by hydrolysis, oxidation, or solution, or all three to give a change in color or structure, or both.

n   Horizon in which the ratio of exchangeable Ca to exchangeable Na is 10 or less. It must also have the following distinctive morphological characteristics: prismatic or columnar structure, dark coatings on ped surfaces, and hard to very hard consistence when dry.

p   Horizon disturbed by human activities such as cultivation, logging, and habitation.

s   Horizon of salts, including gypsum, which may be detected as crystals or veins or as surface crusts of salt crystals.

sa   Horizon with secondary enrichment of salts more soluble than Ca and Mg carbonates; the concentration of salts exceeds that in the unenriched parent material.

t   Illuvial horizon enriched with silicate clay.

u   Horizon that is markedly disrupted by physical or faunal processes other than cryoturbation.

x   Horizon of fragipan character. (A fragipan is a loamy subsurface horizon of high bulk density and very low organic matter content. When dry it is hard and seems to be cemented.)

y   Horizon affected by cryoturbation as manifested by disrupted or broken horizons, incorporation of materials from other horizons, and mechanical sorting.

z   A frozen layer.

Organic Horizons

O   Organic horizon developed mainly from mosses, rushes, and woody materials.

L   Organic horizon characterized by an accumulation of organic matter derived mainly from leaves, twigs and woody materials in which the organic structures are easily discernible.

F   Same as L, above, except that original structures are difficult to recognize.

H   Organic horizon characterized by decomposed organic matter in which the original structures are indiscernible.

SOIL ORDERS OF THE CANADIAN SYSTEM

Nine soil orders make up the highest taxon of the Canadian System of Soil Classification. Listed in alphabetical order, they are as follows:

Brunisolic   Gleysolic   Podzolic
Chernozemic   Luvisolic   Regosolic
Cryosolic   Organic   Solonetzic

Table A3.2 lists the great groups within each order.

Brunisolic Order

The central concept of the *Brunisolic order* is that of soils under forest having brownish-colored Bm horizons. Most Brunisolic soils are well to imperfectly drained. They occur in a wide range of climatic and vegetative environments, including boreal forest; mixed forest, shrubs, and grass; and heath and tundra.

Table A3.2   Great Groups of the Canadian Soil Classification System

Order	Great Group
Brunisolic	Melanic Brunisol
	Eutric Brunisol
	Sombric Brunisol
	Dystric Brunisol
Chernozemic	Brown
	Dark Brown
	Black
	Dark Gray
Cryosolic	Turbic Cryosol
	Static Cryosol
	Organic Cryosol
Gleysolic	Humic Gleysol
	Gleysol
	Luvic Gleysol
Luvisolic	Gray Brown Luvisol
	Gray Luvisol
Organic	Fibrisol
	Mesisol
	Humisol
	Folisol
Podzolic	Humic Podzol
	Ferro-Humic Podzol
	Humo-Ferric Podzol
Regosolic	Regosol
	Humic Regosol
Solonetzic	Solentz
	Solodized Solonetz
	Solod

As compared with the Chernozemic soils, the Brunisolic soils show a weak B horizon of accumulation attributable to their moister environment. Brunisolic soils lack the diagnostic podzolic B horizon of the Podzolic soils, in which accumulation in the B horizon is strongly developed. The Melanic Brunisol shown in profile Figure A3.1 can be found in the St. Lawrence Lowlands, surrounded by Podzolic soils (Figure A3.2).

Chernozemic Order

The general concept of the *Chernozemic order* is that of well to imperfectly drained soils having surface horizons darkened by the accumulation of organic matter from the decomposition of xerophytic or mesophytic grasses and forms representative of grassland communities or of grassland-forest communities with associated shrubs and forbs. The major area of Chernozemic soils is the cool, subarid to subhumid Interior Plains of western Canada. Most Chernozemic soils are frozen during some period each winter, and the soil is dry at some period each summer. The mean annual temperature is higher than 0°C and usually less than 5.5°C. The associated climate is typically the semiarid (steppe) variety of the dry midlatitude climate (Figure A3.2).

Essential to the definition of soils of the Chernozemic order is that they must have an A horizon (typically, Ah) in which organic matter has accumulated and they must meet several other requirements. The A horizon is at least 10 cm thick; its color is dark brown to black. It usually has sufficiently good structure that it is neither massive and hard nor single-grained when dry. The profile shown in Figure A3.1 is that of the Orthic subgroup of the Brown great group; it shows a Bm horizon that is typically of prismatic structure. The C horizon is one of lime accumulation (Cca). Clearly, the Chernozemic soils can be closely correlated with the Mollisols of the CSCS.

Cryosolic Order

Soils of the *Cryosolic order* occupy much of the northern third of Canada where permafrost remains close to the surface of both mineral and organic deposits. Cryosolic soils predominate north of the tree line, are common in the subarctic forest area in fine-textured soils, and extend into the boreal forest in some organic materials and into some alpine areas of mountainous regions. Cryoturbation (intense disturbance by freeze–thaw activity) of these soils is common, and it may be indicated by patterned ground features such as sorted

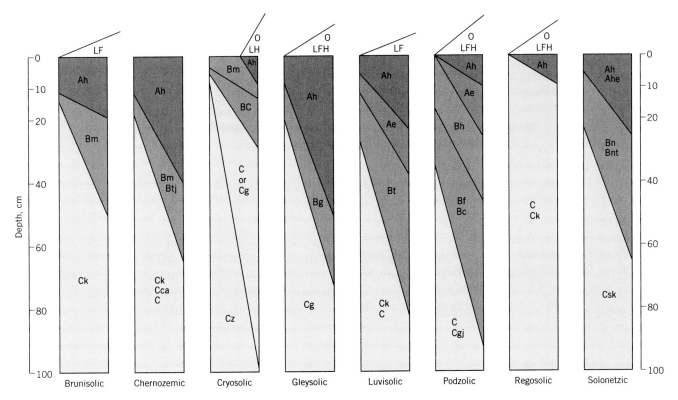

Figure A3.1 Representative schematic profiles of eight of the nine orders of the Canadian system of soil classification. Slanting lines show the range in depth and thickness of each horizon. (The horizon planes are actually approximately horizontal within the pedon.) See Table A3.1 for explanation of symbols. (From Canada Soil Survey Committee, Research Branch, Canada Department of Agriculture, 1978.)

and nonsorted nets, circles, polygons, stripes, and earth hummocks.

Cryosolic soils are found in either mineral or organic materials that have permafrost either within 1 m of the surface or within 2 m if more than one-third of the pedon has been strongly cryoturbated, as indicated by disrupted, mixed, or broken horizons. The profile shown in Figure A3.1 is that of the Orthic subgroup of the Static Cyrosol great group. Note the presence of organic L, H, and O surface horizons and the thin Ah horizon. The Cryosolic soils are closely correlated with the Cryaquepts of the CSCS.

Gleysolic Order

Soils of the *Gleysolic order* have features indicative of periodic or prolonged saturation with water and reducing conditions. They commonly occur in patchy association with other soils in the landscape. Gleysolic soils are usually associated with either a high ground water table at some period of the year or temporary saturation above a relatively impermeable layer. Some Gleysolic soils may be submerged under shallow water throughout the year. The profile shown in Figure A3.1

is that of the Gleysol great group. It has a thick Ah horizon. The underlying Bg horizon is grayish and shows mottling typical of reducing conditions.

Luvisolic Order

Soils of the *Luvisolic order* generally have light-colored, eluvial horizons (Ae), and they have illuvial B horizons (Bt) in which silicate clay has accumulated. These soils develop characteristically in well to imperfectly drained sites, in sandy loam to clay base-saturated parent materials under frost vegetation in subhumid to humid, mild to very cold climates. The genesis of Luvisolic soils is thought to involve the suspension of clay in the soil solution near the soil surface, downward movement of the suspended clay with the soil solution, and deposition of the translocated clay at a depth where downward motion of the soil solution ceases or becomes very slow. The representative profile shown in Figure A3.1 is that of the Orthic subgroup of the Gray Brown Luvisol great group.

Luvisolic soils occur from the southern extremity of Ontario to the zone of permafrost and from Newfoundland to British Columbia. The largest area of

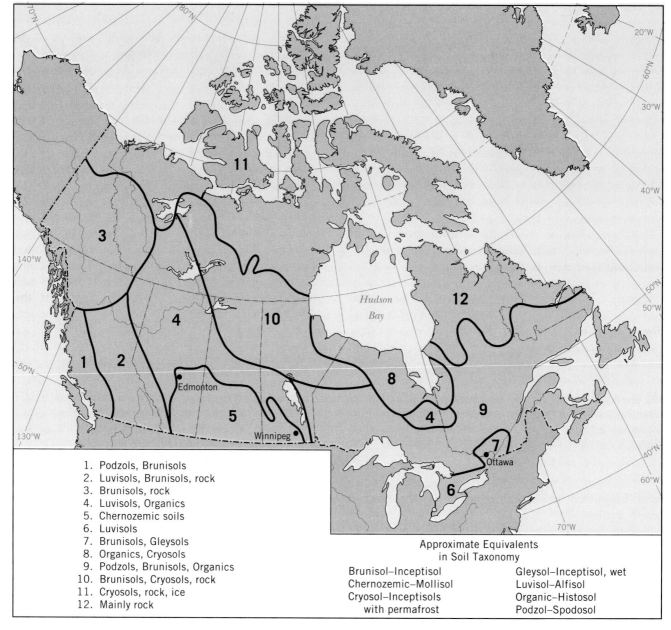

Figure A 3.2 Generalized map of soil regions of Canada. (Courtesy of Land Resource Research Institute, Agriculture Canada.) (Illustration is taken from *Fundamentals of Soil Science,* 7th ed., by Henry D. Foth, John Wiley & Sons.)

these soils are Gray Luvisols occurring in the central to northern Interior Plains under deciduous, mixed, and coniferous forest. In this location they appear to correlate with the Boralfs of the Alfisol order in the CSCS. Gray-Brown Luvisolic soils of southern Ontario would correlate with the suborder of Udalfs.

Organic Order

Soils of the *Organic order* are composed largely of organic materials. They include most of the soils commonly known as peat, muck, or bog soils. Organic soils contain 17 percent or more organic carbon (30 percent organic matter) by weight. Most Organic soils are saturated with water for prolonged periods. They occur widely in poorly and very poorly drained depressions and level areas in regions of subhumid to humid climate and are derived from vegetation that grows in such sites. However, one group of Organic soils consists of leaf litter overlying rock or fragmental material; soils of this group may occur on steep slopes and rarely be saturated with water. (No profile of the Organic soils is shown in Figure A3.1.) Organic soils can be correlated with the Histosols of the CSCS.

Podzolic Order

Soils of the *Podzolic order* have B horizons in which the dominant accumulation product is amorphous material composed mainly of humified organic matter in varying degrees with Al and Fe. Typically, Podzolic soils occur in coarse- to medium-textured, acid parent materials, under forest and heath vegetation in cool to very cold humid to very humid climates. Podzolic soils can usually be readily recognized in the field. Generally, they have organic surface horizons that are commonly L, F, and H. Most Podzolic soils have a reddish brown to black B horizon (Bh) with an abrupt upper boundary. The profile shown in Figure A3.1 is that of the Orthic subgroup of the Humic Podzol great group.

The Podzolic soils correspond closely with the Spodosols (Orthods) of the CSCS.

Regosolic Order

Regosolic soils have weakly developed horizons. The lack of development of genetic horizons may be due to any number of factors: youthfulness of the parent material, recent alluvium; instability of the material, colluvium on slopes subject to mass wasting; nature of the material, nearly pure quartz sand; climate, dry cold conditions. Regosolic soils are generally rapidly to imperfectly drained. They occur in a wide range of vegetation and climates. The profile shown in Figure A3.1 is that of the Orthic subgroup of the Regosol great group. It has only a thin humic A horizon (Ah) and a surface horizon of organic materials.

Regosolic soils correspond with the Entisols of the CSCS.

Solonetzic Order

Soils of the *Solonetzic order* have B horizons that are very hard when dry and swell to a sticky mass of very low permeability when wet. Typically, the Solonetzic B horizon has prismatic or columnar macrostructure that breaks into hard to extremely hard, blocky peds with dark coatings. Solonetzic soils occur on saline parent materials in some areas of the semiarid to subhumid Interior Plains in association with Chernozemic soils and to a lesser extent with Luvisolic and Gleysolic soils. Most Solonetzic soils are associated with a vegetative cover of grasses and forbs. The profile shown in Figure A3.1 is that of the Brown subgroup of the Solonetz great group.

Solonetzic soils are thought to be have developed from parent materials that were more or less uniformly salinized with salts high in sodium. Leaching of salts by descending rainwater presumably mobilizes the sodium-saturated colloids. The colloids are apparently carried downward and deposited in the B horizon. Further leaching results in depletion of alkali cations in the A horizon, which becomes acidic, and a platy Ahe horizon usually develops. The underlying Solonetzic B horizon (Bn, Bnt) usually consists of darkly stained, fused, intact columnar peds. This stage is followed by the structural breakdown of the upper part of the B horizon and eventually its complete destruction in the most advanced stage, known as *solodization*. Solonetzic soils are correlated with the suborder of Argids in the order of Aridisols under the CSCS.

GLOSSARY

This glossary contains definitions of terms shown in the text in italics or boldface. Terms that are italicized within the definitions will be found as individual entries elsewhere in the glossary.

A horizon   *mineral* horizon of the *soil*, overlying the *E* and *B* horizons.

ablation   a wastage of glacial ice by both *melting* and *evaporation*.

abrasion   erosion of *bedrock* of a *stream channel* by impact of particles carried in a *stream* and by rolling of larger rock fragments over the stream bed; abrasion is also an activity of glacial ice, waves, and *wind*.

abrasion platform   sloping, nearly flat *bedrock* surface extending out from the foot of a *marine cliff* under the shallow water of the breaker zone.

absorption of radiation   transfer of *electromagnetic energy* into heat energy within a *gas* or *liquid* through which the radiation is passing.

abyssal plain   large expanse of very smooth, flat ocean floor found at depths of 4600 to 5500 m (15,000 to 18,000 ft).

accelerated erosion   *soil erosion* occurring at a rate much faster than *soil horizons* can be formed from the parent *regolith*.

accretion of lithosphere   production of new *oceanic lithosphere* at an active *spreading plate boundary* by the rise and solidification of *magma* of basaltic composition.

acid deposition   the *deposition* of acid raindrops and/or dry acidic dust particles on vegetation and ground surfaces.

acid mine drainage   sulfuric acid effluent from *coal* mines, mine tailings, or spoil ridges made by *strip mining*.

acid rain   rainwater having an abnormally low *pH*, between 2 and 5, as a result of air pollution by sulfur oxides and nitrogen oxides.

active continental margins   continental margins that coincide with tectonically active plate boundaries. (See also *continental margins; passive continental margins*.)

active systems   *remote sensing* systems that emit a beam of wave energy at a source and measure the intensity of that energy reflected back to the source.

adiabatic lapse rate   (See *dry adiabatic lapse rate; wet adiabatic lapse rate*.)

adiabatic process   change of temperature within a *gas* because of compression or expansion, without gain or loss of heat from the outside.

advection fog   *fog* produced by *condensation* within a moist basal air layer moving over a cold land or water surface.

aerosols   tiny dust particles present in the *atmosphere*, so small and light that the slightest movements of air keep them aloft.

aggradation   raising of *stream channel* altitude by continued *deposition* of bed load.

air mass   extensive body of air within which upward gradients of temperature and moisture are fairly uniform over a large area.

air pollutant   an unwanted substance injected into the *atmosphere* from the earth's surface by either natural or human activities; includes *aerosols, gases,* and *particulates*.

albedo   percentage of *electromagnetic energy* reflected from a surface.

albic horizon   pale, often sandy *soil horizon* from which *clay* and free iron oxides have been removed; found in the profile of the *Spodosols*.

Alfisols   *soil order* consisting of *soils* of humid and subhumid climates, with high *base status* and an *argillic horizon*.

alluvial fan   gently sloping, conical accumulation of coarse *alluvium* deposited by a *braided stream* undergoing

500

aggradation below the point of emergence of the channel from a narrow *gorge* or *canyon.*

alluvial meanders   sinuous bends of a *graded stream* flowing in the alluvial deposit of a *floodplain.*

alluvial river   *stream* of low *gradient* flowing upon thick deposits of *alluvium* and experiencing approximately annual overbank flooding of the adjacent *floodplain.*

alluvial terrace   benchlike *landform* carved in *alluvium* by a *stream* during *degradation.*

alluvium   any stream-laid *sediment* deposit found in a *stream channel* and in low parts of a stream valley subject to flooding.

alpine chains   high mountain ranges that are narrow belts of *tectonic activity* severely deformed by *folding* and thrusting in comparatively recent geologic time.

alpine glacier   long, narrow, mountain *glacier* on a steep downgrade, occupying the floor of a troughlike valley.

alpine tundra   a plant *formation class* within the *tundra biome*, found at high altitudes above the limit of *tree* growth.

amphibole group   *silicate minerals* rich in calcium, magnesium, and iron, dark in color, high in *density*, and classed as *mafic minerals.*

andesite   *extrusive igneous rock* of diorite composition, dominated by *plagioclase feldspar*; the extrusive equivalent of *diorite.*

Andisols   a *soil order* that includes *soils* formed on volcanic ash; often enriched by organic matter, yielding a dark soil color.

anemometer   weather instrument used to indicate *wind* speed.

aneroid barometer   *barometer* using a mechanism consisting of a partially evacuated air chamber and a flexible diaphragm.

annuals   plants that live only a single growing season, passing the unfavorable season as a seed or spore.

annular drainage pattern   a stream network dominated by concentric (ringlike) major *subsequent streams.*

antarctic circle   *parallel of latitude* at $66\frac{1}{2}°$ S.

antarctic front zone   frontal zone of interaction between antarctic *air masses* and polar air masses.

antarctic zone   *latitude* zone in the latitude range 60° to 75° S (more or less), centered on the *antarctic circle* and lying between the *subantarctic zone* and the *polar zone.*

anticlinal valley   valley eroded in weak *strata* along the central line or axis of an eroded *anticline.*

anticline   upfold of *strata* or other layered rock in an arch-like structure; a class of *folds.* (See also *syncline.*)

anticyclone   center of high *atmospheric pressure.*

aphelion   point on the earth's elliptical orbit at which the earth is farthest from the sun.

aquatic ecosystem   *ecosystem* of a *lake*, bog, pond, river, *estuary*, or other body of water.

Aquepts   *suborder* of the *soil order Inceptisols*; includes Inceptisols of wet places, seasonally saturated with water.

aquiclude   rock mass or layer that impedes or prevents the movement of *ground water.*

aquifer   rock mass or layer that readily transmits and holds *ground water.*

arc   curved line that forms a portion of a circle.

arctic circle   *parallel of latitude* at $66\frac{1}{2}°$ N.

arctic front zone   frontal zone of interaction between arctic *air masses* and polar air masses.

arctic tundra   a plant *formation class* within the *tundra biome*, consisting of low, mostly herbaceous plants, but with some very small stunted *trees*, associated with the *tundra climate* [12].

arctic zone   *latitude* zone in the latitude range 60° to 75° N (more or less), centered about on the *arctic circle*, and lying between the *subarctic zone* and the *polar zone.*

arête   sharp, knifelike divide or crest formed between two *cirques* by alpine glaciation.

argillic horizon   *soil horizon*, usually the *B horizon*, in which *clay minerals* have accumulated by *illuviation.*

arid (dry climate subtype)   subtype of the dry climates that is extremely dry and supports little or no vegetation cover.

Aridisols   *soil order* consisting of soils of dry climates, with or without *argillic horizons*, and with accumulations of *carbonates* or soluble salts.

artesian well   drilled well in which water rises under hydraulic pressure above the level of the surrounding *water table* and may reach the surface.

aspect   compass orientation of a *slope* as an inclined element of the ground surface.

asthenosphere   soft layer of the upper *mantle*, beneath the rigid *lithosphere.*

astronomical hypothesis   explanation for *glaciations* and *interglaciations* making use of cyclic variations in the form of solar energy received at the earth's surface.

atmosphere   envelope of gases surrounding the earth, held by *gravity.*

atmospheric pressure   pressure exerted by the *atmosphere* because of the force of *gravity* acting on the overlying column of air.

atoll   circular or closed-loop *coral reef* enclosing an open *lagoon* with no island inside.

autumnal equinox   *equinox* occurring on September 22 or 23.

axial rift   narrow, trenchlike depression situated along the center line of the *midoceanic ridge* and identified with active seafloor spreading.

axis of rotation   center line around which a body revolves, as the earth's axis of rotation.

B horizon   mineral *soil horizon* located beneath the *A horizon* and usually characterized by a gain of *mineral matter* (such as *clay minerals* and oxides of aluminum and iron) and organic matter (*humus*).

backswamp   area of low, swampy ground on the *floodplain* of an *alluvial river* between the *natural levee* and the *bluffs.*

backwash   return flow of *swash* water under influence of *gravity.*

badlands   rugged land surface of steep *slopes*, resembling miniature mountains, developed on weak *clay* formations or clay-rich *regolith* by fluvial erosion too rapid to permit plant growth and *soil* formation.

bar   low ridge of *sand* built above water level across the mouth of a *bay* or in shallow water paralleling the *shoreline.* May also refer to embankment of *sand* or gravel on the floor of a *stream channel.*

bar (pressure)   unit of pressure or *capillary tension* equal to 1 million dynes per sq cm; approximately equal to the pressure of the earth's *atmosphere* at sea level.

barchan dunes   *sand dune* of crescentic base outline with a sharp crest and a steep lee *slip face*, with crescent points (horns) pointing downwind.

barometer   instrument for measuring *atmospheric pressure*.

barrier island   long, narrow island, built largely of beach *sand* and dune sand, parallel with the mainland and separated from it by a *lagoon*.

barrier-island coast   *coastline* with broad zone of shallow water offshore (a *lagoon*) shut off from the ocean by a *barrier island*.

barrier reef   *coral reef* separated from mainland *shoreline* by a *lagoon*.

basalt   *extrusive igneous rock* of *gabbro* composition; occurs as *lava*.

base level   lower limiting surface or level that can ultimately be attained by a *stream* under conditions of stability of the *earth's crust* and sea level; an imaginary surface equivalent to sea level projected inland.

bases   certain positively charged *ions* in the *soil* that are also plant nutrients; the most important are calcium, magnesium, potassium, and sodium.

base status of soils   quality of a *soil* as measured by the presence or absence of *clay minerals* capable of holding large numbers of *bases*. Soils of high *base status* are rich in base-holding clay minerals; soils of low base status are deficient in such minerals.

batholith   large, deep-seated body of *intrusive igneous rock*, usually with an area of surface exposure greater than 100 km² (40 mi²).

bauxite   mixture of several *clay minerals*, consisting largely of aluminum oxide and water with impurities; a principal ore of aluminum.

bay   a body of water sheltered from strong wave action by the configuration of the *coast*.

beach   thick, wedge-shaped accumulation of *sand*, gravel, or cobbles in the zone of breaking waves.

beach drift   transport of *sand* on a beach parallel with a *shoreline* by a succession of landward and seaward water movements at times when *swash* approaches obliquely.

bed load   that portion of the *stream load* moving close to the stream bed by rolling and sliding.

bedrock   solid *rock* in place with respect to the surrounding and underlying rock and relatively unchanged by *weathering processes*.

biome   largest recognizable subdivision of *terrestrial ecosystems*, including the total assemblage of plant and animal life interacting within the *life layer*.

biosphere   all living organisms of the earth and the environments with which they interact.

block mountains   class of mountains produced by block faulting and usually bounded by *normal faults*.

blowout   shallow depression produced by continued *deflation*.

bluffs   steeply rising ground slopes marking the outer limits of a *floodplain*.

Boralfs   *suborder* of the *soil order Alfisols*; includes Alfisols of *boreal forests* or high mountains.

boreal forest   variety of *needleleaf forest* found in the *boreal forest climate* ⑪ regions of North America and Eurasia.

boreal forest climate ⑪   cold climate of the *subarctic zone* in the northern *hemisphere* with long, extremely severe winters and several consecutive months of zero *potential evapotranspiration (water need)*.

Borolls   *suborder* of the *soil order Mollisols*; includes Mollisols of cold-winter semiarid plants (*steppes*) or high mountains.

braided stream   *stream* with shallow channel in coarse *alluvium* carrying multiple threads of fast flow that subdivide and rejoin repeatedly and continually shift in position.

breaker   sudden collapse of a steepened water wave as it approaches the *shoreline*.

broadleaf evergreen forest   *formation class* in the *forest biome* consisting of broadleaf evergreen *trees* and found in the *moist subtropical climate* ⑥ and in parts of the *marine west-coast climate* ⑧.

butte   prominent, steep-sided hill or peak, often representing the final remnant of a resistant layer in a region of flat-lying *strata*.

C horizon   *soil horizon* lying beneath the *soil solum* (*A*, *E*, and *B horizons*); a layer of *sediment* or *regolith* that is the *parent material* of the solum.

calcification   accumulation of *calcium carbonate* in a *soil*, usually occurring in the *B horizon* or in the *C horizon* below the *soil solum*.

calcite   *mineral* having the composition *calcium carbonate*.

calcium carbonate   compound consisting of calcium (Ca) and carbonate (CO_3) *ions*, formula $CaCO_3$, occurring naturally as the mineral *calcite*.

caldera   large, steep-sided circular depression resulting from the explosion and subsidence of a *stratovolcano*.

canyon   (See *gorge*.)

capillary tension   a cohesive force among surface molecules of a *liquid* that gives a droplet its rounded shape.

carbon dioxide   the chemical compound CO_2, formed by the union of two atoms of oxygen and one atom of carbon; normally, a *gas* present in low concentration in the *atmosphere*.

carbonates (carbonate minerals, carbonate rocks)   *minerals* that are carbonate compounds of calcium or magnesium or both, that is, *calcium carbonate* or magnesium carbonate. (See also *calcite*.)

carbonic acid   a weak acid created when CO_2 *gas* dissolves in water.

carbonic acid action   chemical reaction of *carbonic acid* in rainwater, *soil water*, and *ground water* with *minerals*; most strongly affects carbonate minerals and *rocks*, such as *limestone* and marble; an activity of *chemical weathering*.

cartography   the science and art of making maps.

Celsius scale   temperature scale in which the *freezing* point of water is 0° and the boiling point 100°.

Cenozoic Era   last (youngest) of the *eras* of geologic time.

channel   (See *stream channel*.)

chaparral   sclerophyll *scrub* and dwarf *forest* plant *formation class* found throughout the coastal mountain ranges and hills of central and southern California.

chemical weathering   chemical change in rock-forming *minerals* through exposure to atmospheric conditions in the presence of water; mainly involving *oxidation*, *hydrolysis*, *carbonic acid action*, or direct solution.

chemically precipitated sediment   *sediment* consisting of *mineral matter* precipitated from a water solution in which the matter has been transported in the dissolved state as *ions*.

chernozem type of *soil order* closely equivalent to *Mollisol*; an order of the Canadian Soil Classification System.

chert *sedimentary rock* composed largely of silicon dioxide and various impurities, in the form of nodules and layers, often occurring with *limestone* layers.

chinook wind a *local wind* occurring at certain times to the lee of the Rocky Mountains; a very dry wind with a high capacity to evaporate *snow*.

chlorofluorocarbons (CFCs) synthetic chemical compounds containing chlorine, fluorine, and carbon atoms that are widely used as coolant fluids in refrigeration systems.

cinder cone conical hill built of coarse *tephra* ejected from a narrow volcanic vent; a type of *volcano*.

circle of illumination *great circle* that divides the globe at all times into a sunlit *hemisphere* and a shadowed hemisphere.

circum-Pacific belt chains of andesite *volcanoes* making up mountain belts and *island arcs* surrounding the Pacific Ocean basin.

cirque bowl-shaped depression carved in *rock* by glacial processes and holding the *firn* of the upper end of an *alpine glacier*.

clastic sediment *sediment* consisting of particles broken away physically from a parent *rock* source.

clay *sediment* particles smaller than 0.004 mm in diameter.

clay minerals class of *minerals* produced by alteration of *silicate minerals*, having plastic properties when moist.

claystone *sedimentary rock* formed by lithification of *clay* and lacking *fissile* structure.

cliff sheer, near-vertical rock wall formed from flat-lying resistant layered rocks, usually *sandstone, limestone,* or *lava* flows. Cliff may refer to any near-vertical rock wall. (See also *marine cliff*.)

climate generalized statement of the prevailing *weather* conditions at a given place, based on statistics of a long period of record and including mean values, departures from those means, and the probabilities associated with those departures.

climatic frontier a geographical boundary that marks the limit of survival of a plant species subjected to climatic stress.

climax stable community of plants and animals reached at the endpoint of *ecological succession.*

climograph a graph on which two or more climatic variables, such as monthly mean temperature and monthly mean precipitation, are plotted for each month of the year.

cloud forest a type of low evergreen rainforest that occurs high on mountain slopes, where *clouds* and *fog* are frequent.

clouds dense concentrations of suspended water or ice particles in the diameter range 20 to 50 μm. (See *cumuliform clouds; stratiform clouds.*)

coal *rock* consisting of hydrocarbon compounds, formed of compacted, lithified, and altered accumulations of plant remains (*peat*).

coarse textured (rock) having *mineral* crystals sufficiently large that they are at least visible to the naked eye or with low magnification.

coast (See *coastline.*)

coastal blowout dune high *sand dune* of the *parabolic dunes* class formed adjacent to a *beach*, usually with a deep *deflation* hollow (*blowout*) enclosed within the dune ridge.

coastal forest subtype of *needleleaf evergreen forest* found in the humid coastal zone of the northwestern United States and western Canada.

coastal plain coastal belt, emerged from beneath the sea as a former *continental shelf*, underlain by *strata* with gentle *dip* seaward.

coastline (coast) zone in which coastal processes operate or have a strong influence.

cold front moving weather *front* along which a cold *air mass* moves underneath a warm air mass, causing the warm air mass to be lifted.

colloids particles of extremely small size, capable of remaining indefinitely in suspension in water. May be mineral or organic in nature.

colluvium deposit of *sediment* or *rock* particles accumulating from overland flow at the base of a *slope* and originating from higher slopes where *sheet erosion* is in progress. (See also *alluvium*.)

compression (tectonic) squeezing together, as horizontal compression of crustal layers by *tectonic processes*.

condensation process of change of matter in the gaseous state (*water vapor*) to the liquid state (liquid water) or solid state (ice).

condensation nucleus a tiny bit of solid matter (*aerosol*) in the *atmosphere* on which *water vapor* condenses to form a tiny water droplet.

conduction of heat transmission of *sensible heat* through matter by transfer of energy from one atom or molecule to the next in the direction of decreasing temperature.

cone of depression conical configuration of the lowered *water table* around a well from which water is being rapidly withdrawn.

conformal projection *map projection* that preserves without shearing the true shape or outline of any small surface feature of the earth.

conglomerate a *sedimentary rock* composed of pebbles in a matrix of finer *rock* particles.

conic projections a group of *map projections* in which the *geographic grid* is transformed to lie on the surface of a developed cone.

consequent stream *stream* that takes its course down the slope of an *initial landform*, such as a newly emerged *coastal plain* or a *volcano*.

consumption of plate destruction or disappearance of a subducting *lithospheric plate* in the *asthenosphere*, in part by *melting* of the upper surface, but largely by softening because of heating to the temperature of the surrounding *mantle* rock.

continental collision event in *plate tectonics* in which subduction brings two segments of the *continental lithosphere* into contact, leading to formation of a *continental suture*.

continental crust crust of the continents, of felsic composition in the upper part; thicker and less dense than *oceanic crust*.

continental drift hypothesis, introduced by Alfred Wegener and others early in the 1900s, of the breakup of a parent continent, *Pangaea*, starting near the close of the *Mesozoic*

Era and resulting in the present arrangement of *continental shields* and intervening *ocean-basin floors.*

continental lithosphere   *lithosphere* bearing *continental crust* of *felsic igneous rock.*

continental margins   (1) Topographic: one of three major divisions of the ocean basins, being the zones directly adjacent to the continent and including the *continental shelf, continental slope,* and *continental rise.* (2) Tectonic: marginal belt of continental crust and lithosphere that is in contact with *oceanic crust* and *lithosphere,* with or without an active plate boundary being present at the contact. (See also *active continental margins; passive continental margins.*)

continental rise   gently sloping seafloor lying at the foot of the *continental slope* and leading gradually into the *abyssal plain.*

continental rupture   crustal spreading apart affecting the *continental lithosphere,* so as to cause a *rift valley* to appear and to widen, eventually creating a new belt of *oceanic lithosphere.*

continental shelf   shallow, gently sloping belt of seafloor adjacent to the continental shoreline and terminating at its outer edge in the *continental slope.*

continental shields   ancient crustal rock masses of the continents, largely *igneous rock* and *metamorphic rock,* and mostly of *Precambrian age.*

continental slope   steeply descending belt of seafloor between the *continental shelf* and the *continental rise.*

continental suture   long, narrow zone of crustal deformation, including underthrusting and intense *folding,* produced by a *continental collision.* Examples: Himalayan Range, European Alps.

convection (atmospheric)   air motion consisting of strong updrafts taking place within a *convection cell.*

convection cell   individual column of strong updrafts produced by atmospheric *convection.*

convectional precipitation   a form of *precipitation* induced when warm, moist air is heated at the ground surface, rises, cools, and condenses to form water droplets, raindrops, and, eventually, rainfall.

converging boundary   boundary between two crustal plates along which *subduction* is occurring and *lithosphere* is being consumed.

coral reef   rocklike accumulation of *carbonates* secreted by corals and algae in shallow water along a marine shoreline.

coral-reef coast   *coast* built out by accumulations of *limestone* in *coral reefs.*

core of earth   spherical central mass of the earth composed largely of iron and consisting of an outer liquid zone and an interior solid zone.

Coriolis effect   effect of the earth's *rotation* tending to turn the direction of motion of any object or *fluid* toward the right in the northern *hemisphere* and to the left in the southern hemisphere.

corrosion   erosion of *bedrock* of a *stream channel* (or other *rock* surface) by chemical reactions between solutions in stream water and *mineral* surfaces.

counterradiation   *longwave radiation* of *atmosphere* directed downward to the earth's surface.

covered shields   areas of *continental shields* in which the ancient *rocks* are covered beneath a thin layer of sedimentary *strata.*

crater   central summit depression associated with the principal vent of a *volcano.*

crescentic dune   (See *barchan dunes.*)

crevasse   gaping crack in the brittle surface ice of a *glacier.*

crude oil   liquid fraction of *petroleum.*

crust of earth   outermost solid shell or layer of the earth, composed largely of *silicate minerals.*

Cryaquepts   great group within the soil *suborder* of *Aquepts;* includes Aquepts of cold climate regions and particularly the *tundra climate* ⑫.

cuesta   *erosional landform* developed on resistant *strata* having low to moderate *dip* and taking the form of an asymmetrical low ridge or hill belt with one side a steep *slope* and the other a gentle slope; usually associated with a *coastal plain.*

cumuliform clouds   *clouds* of globular shape, often with extended vertical development.

cumulonimbus cloud   large, dense *cumuliform cloud* yielding *precipitation.*

cumulus   *cloud* type consisting of low-lying, white cloud masses of globular shape well separated from one another.

cutoff   cutting through of a narrow neck of land, so as to bypass the stream flow in an *alluvial meander* and cause it to be abandoned.

cycle of rock change   total cycle of changes in which *rock* of any one of the three major rock classes—*igneous rock, sedimentary rock,* and *metamorphic rock*—is transformed into rock of one of the other classes.

cyclone   center of low *atmospheric pressure.* (See *tropical cyclone; wave cyclone.*)

cyclonic precipitation   a form of *precipitation* that occurs as warm moist air is lifted by air motion occurring in a *cyclone.*

cyclonic storm   intense weather disturbance within a moving *cyclone* generating strong winds, cloudiness, and *precipitation.*

cylindric projections   group of *map projections* in which the *geographic grid* is transformed to lie on the surface of a developed cylinder.

daylight saving time   time system under which time is advanced by one hour with respect to the *standard time* of the prevailing *standard meridian.*

debris flood (debris flow)   streamlike flow of muddy water heavily charged with *sediment* of a wide range of size grades, including boulders, generated by sporadic torrential rains on steep mountain watersheds.

decalcification   removal of *calcium carbonate* from a *soil horizon* or *soil solum* as *carbonic acid* reacts with *carbonate mineral* matter.

December solstice   (See *winter solstice.*)

deciduous plant   *tree* or *shrub* that sheds its leaves seasonally.

deep-sea cone   a fan-shaped accumulation of undersea *sediment* on the *continental rise* produced by sediment-rich currents flowing down the *continental slope.*

deflation   lifting and transport in *turbulent suspension* by *wind* of loose particles of *soil* or *regolith* from dry ground surfaces.

deglaciation   widespread recession of *ice sheets* during a period of warming global climate, leading to an interglaciation. (See also *glaciation; interglaciation.*)

degradation   lowering or downcutting of a *stream channel* by *stream erosion* in *alluvium* or *bedrock.*

degree of arc   measurement of the angle associated with an *arc*, in degrees.

delta   *sediment* deposit built by a *stream* entering a body of standing water and formed of the *stream load.*

delta coast   *coast* bordered by a *delta.*

dendritic drainage pattern   *drainage pattern* of treelike branched form in which the smaller *streams* take a wide variety of directions and show no parallelism or dominant trend.

density of matter   quantity of mass per unit of volume, stated in gm/cc.

denudation   total action of all processes whereby the exposed *rocks* of the continents are worn down and the resulting *sediments* are transported to the sea by the *fluid agents*; also includes *weathering* and *mass wasting.*

deposition   (See *stream deposition.*)

deposition (atmosphere)   the change of state of a substance from a *gas* (*water vapor*) to a *solid* (ice); in the science of *meteorology*, the term sublimation is used to describe both this process and the change of state from solid to vapor. (See *sublimation.*)

depositional landform   *landform* made by *deposition* of *sediment.*

desert biome   *biome* of the dry climates consisting of thinly dispersed plants that may be *shrubs*, grasses, or perennial *herbs*, but lacking in *trees.*

desertification   (See *land degradation.*)

desert pavement   surface layer of closely fitted pebbles or coarse *sand* from which finer particles have been removed.

dew-point temperature   temperature of an *air mass* at which the air holds its full capacity of *water vapor.*

diagnostic horizons   *soil horizons*, rigorously defined, that are used as diagnostic criteria in classifying *soils.*

diffuse radiation   solar radiation that has been *scattered* (deflected or reflected) by minute dust particles or *cloud* particles in the *atmosphere.*

digital image   numeric representation of a picture consisting of a collection of numeric brightness values (pixels) arrayed in a fine grid pattern.

dike   thin layer of *intrusive igneous rock*, often near-vertical or with steep *dip*, occupying a widened fracture in the surrounding *rock* and typically cutting across older rock planes.

diorite   *intrusive igneous rock* consisting dominantly of *plagioclase feldspar* and pyroxene; a *felsic igneous rock.*

dip   acute angle between an inclined natural *rock* plane or surface and an imaginary horizontal plane of reference; always measured perpendicular to the *strike.* Also a verb, meaning to incline toward.

discharge   volume of flow moving through a given cross section of a *stream* in a given unit of time; commonly given in cubic meters (feet) per second.

distributary   branching *stream channel* that crosses a *delta* to discharge into open water.

diurnal   adjective meaning "daily."

doldrums   belt of calms and variable *winds* occurring at times along the *equatorial trough.*

dolomite   carbonate *mineral* or *sedimentary rock* having the composition calcium magnesium carbonate.

dome   (See *sedimentary dome.*)

drainage basin   total land surface occupied by a *drainage system*, bounded by a *drainage divide* or watershed.

drainage divide   imaginary line following a crest of high land such that overland flow on opposite sides of the line enters different *streams.*

drainage pattern   the plan of a network of interconnected *stream channels.*

drainage system   a branched network of *stream channels* and adjacent land *slopes*, bounded by a *drainage divide* and converging to a single channel at the outlet.

drainage winds   *winds*, usually cold, that flow from higher to lower regions under the direct influence of *gravity.*

drawdown (of a well)   difference in height between base of cone of depression and original *water table* surface.

drought   occurrence of substantially lower-than-average *precipitation* in a season that normally has ample precipitation for the support of food-producing plants.

drumlin   hill of glacial *till*, oval or elliptical in basal outline and with smoothly rounded summit, formed by plastering of till beneath moving, debris-laden glacial ice.

dry adiabatic lapse rate   rate at which rising air is cooled by expansion when no *condensation* is occurring; 1.0 C°/100 m (5.5 F°/1000 ft).

dry desert   plant *formation class* in the *desert biome* consisting of widely dispersed xerophytic plants that may be small, hardleaved, or spiny *shrubs*, succulent plants (cacti), or hard grasses.

dry midlatitude climate ⑨   dry climate of the *midlatitude zone* with a strong annual cycle of *potential evapotranspiration (water need)* and cold winters.

dry subtropical climate ⑤   dry climate of the *subtropical zone*, transitional between the *dry tropical climate* ④ and the *dry midlatitude climate* ⑨.

dry tropical climate ④   dry climate of the *tropical zone* with large total annual *potential evapotranspiration (water need).*

dune   (See *sand dune.*)

dust storm   heavy concentration of dust in a turbulent *air mass*, often associated with a *cold front.*

E horizon   mineral horizon of the *soil solum* lying below the *A horizon* and characterized by the loss of *clay minerals* and oxides of iron and aluminum; may show a concentration of *quartz* grains and is often pale in color.

earthflow   moderately rapid downhill flowage of masses of water-saturated *soil*, *regolith*, or weak *shale*, typically forming a steplike terrace at the top and a bulging toe at the base.

earthquake   a trembling or shaking of the ground produced by the passage of *seismic waves.*

earthquake focus   point within the earth at which the energy of an *earthquake* is first released by rupture and from which *seismic waves* emanate.

earth's crust   (See *crust of earth.*)

easterly wave   weak, slowly moving trough of low pressure within the belt of *tropical easterlies*; causes a weather disturbance with rain showers.

ebb current   oceanward flow of *tidal current* in a *bay* or tidal stream.

ecological succession   time-succession (sequence) of distinctive plant and animal communities occurring within a given area of newly formed land or land cleared of plant cover by burning, clear cutting, or other agents.

ecosystem group of organisms and the environment with which the organisms interact.

electromagnetic radiation (electromagnetic energy) wave-like form of *energy* radiated by any substance possessing heat; travels through space at the speed of light.

electromagnetic spectrum the total *wavelength* range of *electromagnetic energy.*

El Niño episodic cessation of the typical *upwelling* of cold, deep water off the coast of Peru; literally, "The Christ Child," for its occurrence in the Christmas season once every few years.

eluviation soil-forming process consisting of the downward transport of fine particles, particularly the *soil colloids* (both mineral and organic), carrying them out of an upper *soil horizon.*

emergence exposure of submarine *landforms* by a lowering of sea level or a rise of the crust, or both.

energy the capacity to do work, that is, to bring about a change in the state or motion of matter.

Entisols *soil order* consisting of mineral soils lacking *soil horizons* that would persist after normal plowing.

entrenched meanders winding, sinuous valley produced by *degradation* of a *stream* with trenching into the *bedrock* by downcutting.

environmental temperature lapse rate rate of temperature decrease upward through the *troposphere*; standard value is 6.4 C°/km (3 F°/1000 ft).

epipedon *soil horizon* that forms at the surface.

epiphytes plants that live above ground level out of contact with the *soil*, usually growing on the limbs of *trees* or *shrubs*; also called "air plants."

epoch a subdivision of geologic time.

equal-area projections class of *map projections* on which any given area of the earth's surface is shown to correct relative areal extent, regardless of position on the globe.

equator *parallel of latitude* occupying a position midway between the earth's poles of *rotation*; the largest of the parallels, designated as *latitude* 0°.

equatorial easterlies upper-level easterly airflow over the *equatorial zone.*

equatorial rainforest plant *formation class* within the *forest biome*, consisting of tall, closely set broadleaf *trees* of evergreen or semideciduous habit.

equatorial trough atmospheric low-pressure trough centered more or less over the *equator* and situated between the two belts of *trade winds.*

equatorial zone *latitude* zone lying between lat. 10° S and 10° N (more or less) and centered on the *equator.*

equinox instant in time when the *subsolar point* falls on the earth's *equator* and the *circle of illumination* passes through both poles. *Vernal equinox* occurs on March 20 or 21; *autumnal equinox* on September 22 or 23.

era major subdivision of geologic time consisting of a number of geologic periods. The three *eras* following *Precambrian time* are *Paleozoic, Mesozoic,* and *Cenozoic.*

erg large expanse of active *sand dunes* in the Sahara Desert of North Africa.

erosional landforms class of the *sequential landforms* shaped by the removal of *regolith* or *bedrock* by agents of erosion. Examples: *gorge*, glacial *cirque*, *marine cliff.*

esker narrow, often sinuous embankment of coarse gravel and boulders deposited in the bed of a meltwater *stream* enclosed in a tunnel within stagnant ice of an *ice sheet.*

estuary *bay* that receives fresh water from a river mouth and saltwater from the ocean.

Eurasian-Indonesian belt mountain arc system extending from southern Europe across southern Asia and Indonesia.

eustatic referring to a true change in sea level, as opposed to a local change created by upward or downward *tectonic* motion of land.

eutrophication excessive growth of algae and other related organisms in a *stream* or *lake* as a result of the input of large amounts of nutrient *ions*, especially phosphate and nitrate.

evaporation process in which water in liquid state or solid state passes into the vapor state.

evaporites class of *chemically precipitated sediment* and *sedimentary rock* composed of soluble salts deposited from saltwater bodies.

evapotranspiration combined water loss to the *atmosphere* by *evaporation* from the *soil* and *transpiration* from plants.

evergreen plant *tree* or *shrub* that holds most of its green leaves throughout the year.

exfoliation dome smoothly rounded rock knob or hilltop bearing rock sheets or shells produced by spontaneous expansion accompanying *unloading.*

exotic river *stream* that flows across a region of dry climate and derives its *discharge* from adjacent uplands where a *water surplus* exists.

exposed shields areas of *continental shields* in which the ancient basement *rock*, usually of *Precambrian* age, is exposed to the surface.

extension (tectonic) drawing apart of crustal layers by *tectonic activity* resulting in faulting.

extrusion release of molten rock *magma* at the surface, as in a flow of *lava* or shower of volcanic ash.

extrusive igneous rock rock produced by the solidification of *lava* or ejected fragments of *igneous rock* (*tephra*).

Fahrenheit scale temperature scale in which the *freezing* point of water is 32° and the boiling point is 212°.

fallout *gravity* fall of atmospheric particles of *particulates* reaching the ground.

fault sharp break in rock with a displacement (slippage) of the block on one side with respect to an adjacent block. (See *normal fault; overthrust fault; strike-slip fault; transform fault.*)

fault coast *coast* formed when a *shoreline* comes to rest against a *fault scarp.*

fault creep more or less continuous slippage on a *fault plane*, relieving some of the accumulated strain.

fault-line scarp erosion scarp developed upon an inactive *fault* line.

fault plane surface of slippage between two earth blocks moving relative to each other during faulting.

fault scarp clifflike surface feature produced by faulting and exposing the *fault plane*, commonly associated with a *normal fault.*

feldspar group of *silicate minerals* consisting of silicate of aluminum and one or more of the metals potassium, sodium, or calcium. (See *plagioclase feldspar; potash feldspar.*)

felsic igneous rock *igneous rock* dominantly composed of *felsic minerals*.

felsic minerals (felsic mineral group) *quartz* and *feldspars* treated as a mineral group of light color and relatively low density. (See also *mafic minerals.*)

fine textured (rock) having *mineral* crystals too small to be seen by eye or with low magnification.

fiord narrow, deep ocean embayment partially filling a *glacial trough.*

fiord coast deeply embayed, rugged coast formed by partial *submergence* of *glacial troughs.*

firn granular old *snow* forming a surface layer in the zone of accumulation of a *glacier.*

fissile adjective describing a *rock*, usually *shale*, that readily splits up into small flakes or scales.

flood stream flow at a stream *stage* so high that it cannot be accommodated within the *stream channel* and must spread over the banks to inundate the adjacent *floodplain.*

flood basalts large-scale outpourings of basalt *lava* to produce thick accumulations of *basalt* over large areas.

flood current landward flow of a *tidal current.*

floodplain belt of low, flat ground, present on one or both sides of a *stream channel*, subject to inundation by a *flood* about once annually and underlain by *alluvium.*

flood stage designated stream-surface level for a particular point on a *stream*, higher than which overbank flooding may be expected.

fluid substance that flows readily when subjected to unbalanced stresses; may exist as a *gas* or a *liquid.*

fluid agents *fluids* that erode, transport, and deposit *mineral matter* and organic matter; they are running water, waves and currents, glacial ice, and *wind.*

fluvial landforms *landforms* shaped by running water.

fluvial processes geomorphic processes in which running water is the dominant *fluid* agent, acting as *overland flow* and *stream flow.*

focus (See *earthquake focus.*)

fog *cloud* layer in contact with land or sea surface, or very close to that surface. (See *advection fog; radiation fog.*)

folding process by which *folds* are produced; a form of *tectonic activity.*

folds wavelike corrugations of *strata* (or other layered rock masses) as a result of crustal *compression.*

forb broad-leaved *herb*, as distinguished from the grasses.

foredunes ridge of irregular *sand dunes* typically found adjacent to *beaches* on low-lying *coasts* and bearing a partial cover of plants.

foreland folds *folds* produced by *continental collision* in *strata* of a *passive continental margin.*

forest assemblage of *trees* growing close together, their crowns forming a layer of foliage that largely shades the ground.

forest biome *biome* that includes all regions of *forest* over the lands of the earth.

formation classes subdivisions within a *biome* based on the size, shape, and structure of the plants that dominate the vegetation.

fossil fuels naturally occurring hydrocarbon compounds that represent the altered remains of organic materials enclosed in *rock*; examples are *coal, petroleum (crude oil)*, and *natural gas.*

fractional scale (See *scale fraction.*)

freezing change from liquid state to solid state accompanied by release of *latent heat*, becoming *sensible heat.*

fringing reef *coral reef* directly attached to land with no intervening *lagoon* of open water.

front surface of contact between two unlike *air masses.* (See *cold front; occluded front; polar front; warm front.*)

frost action *rock* breakup by forces accompanying the *freezing* of water.

gabbro *intrusive igneous rock* consisting largely of pyroxene and *plagioclase feldspar*, with variable amounts of *olivine*; a *mafic igneous rock.*

gas (gaseous state) *fluid* of very low density (as compared with a liquid of the same chemical composition) that expands to fill uniformly any small container and is readily compressed.

geographic grid complete network of parallels and meridians on the surface of the globe, used to fix the locations of surface points.

geologic norm stable natural condition in a moist climate in which slow *soil erosion* is paced by maintenance of *soil horizons* bearing a plant community in an equilibrium state.

geology science of the solid earth, including the earth's origin and history, materials comprising the earth, and the processes acting within the earth and on its surface.

geomorphology science of *landforms*, including their history and processes of origin.

geostrophic wind *wind* at high levels above the earth's surface blowing parallel with a system of straight, parallel *isobars.*

geyser periodic jetlike emission of hot water and steam from a narrow vent at a geothermal locality.

glacial abrasion *abrasion* by a moving *glacier* of the *bedrock* floor beneath it.

glacial delta *delta* built by meltwater streams of a *glacier* into standing water of a marginal glacial lake.

glacial drift general term for all varieties and forms of *rock* debris deposited in close association with *ice sheets* of the *Pleistocene Epoch.*

glacial plucking removal of masses of *bedrock* from beneath an *alpine glacier* or *ice sheets* as ice moves forward suddenly.

glacial trough deep, steep-sided rock trench of U-shaped cross section formed by *alpine glacier* erosion.

glaciation (1) General term for the total process of glacier growth and *landform* modification by *glaciers.* (2) Single episode or time period in which *ice sheets* formed, spread, and disappeared.

glacier large natural accumulation of land ice affected by present or past flowage. (See *alpine glacier.*)

gneiss variety of *metamorphic rock* showing banding and commonly rich in *quartz* and *feldspar.*

Goode projection an equal-area *map projection*, often used to display areal thematic information, such as *climate* or *soil* type.

gorge (canyon)   steep-sided *bedrock* valley with a narrow floor limited to the width of a *stream channel.*

graben   trenchlike depression representing the surface of a crustal block dropped down between two opposed, infacing *normal faults.* (See *rift valley.*)

graded profile   smoothly descending profile displayed by a *graded stream.*

graded stream   *stream* (or *stream channel*) with *stream gradient* so adjusted as to achieve a balanced state in which average *bed load* transport is matched to average bed load input; an average condition over periods of many years' duration.

gradient   degree of *slope,* as the gradient of a river or a flowing *glacier.*

granite   *intrusive igneous rock* consisting largely of *quartz, potash feldspar,* and *plagioclase feldspar,* with minor amounts of biotite and hornblende; a *felsic igneous rock.*

granitic rock   general term for rock of the upper layer of the *continental crust,* composed largely of *felsic igneous* and *metamorphic rock;* rock of composition similar to that of *granite.*

graphic scale   map scale as shown by a line divided into equal parts.

grassland biome   *biome* consisting largely or entirely of *herbs,* which may include grasses, grasslike plants, and *forbs.*

gravitation   mutual attraction between any two masses.

gravity   gravitational attraction of the earth upon any small mass near the earth's surface. (See *gravitation.*)

great circle   circle formed by passing a plane through the exact center of a perfect sphere; the largest circle that can be drawn on the surface of a sphere.

greenhouse effect   accumulation of heat in the lower *atmosphere* through the absorption of *longwave radiation* from the earth's surface.

greenhouse gases   atmospheric gases such as CO_2 and *chlorofluorocarbons (CFCs)* that absorb outgoing *longwave radiation,* contributing to the *greenhouse effect.*

groin   wall or embankment built out into the water at right angles to the *shoreline.*

ground water   *subsurface water* occupying the *saturated zone* and moving under the force of *gravity.*

gullies   deep, V-shaped trenches carved by newly formed *streams* in rapid headward growth during advanced stages of *accelerated soil erosion.*

gyres   large circular *ocean current* systems centered upon the oceanic subtropical *high-pressure cells.*

habitat   subdivision of the plant environment having a certain combination of *slope,* drainage, *soil* type, and other controlling physical factors.

Hadley cell   atmospheric circulation cell in low latitudes involving rising air over the *equatorial trough* and sinking air over the *subtropical high-pressure belts.*

hail   form of *precipitation* consisting of pellets or spheres of ice with a concentric layered structure.

hanging valley   stream valley that has been truncated by marine erosion so as to appear in cross section in a *marine cliff,* or truncated by glacial erosion so as to appear in cross section in the upper wall of a *glacial trough.*

haze   minor concentration of *pollutants* or natural forms of *aerosols* in the *atmosphere* causing a reduction in visibility.

heat   (See *latent heat; sensible heat.*)

heat island   persistent region of higher air temperatures centered over a city.

hemisphere   half of a sphere; that portion of the earth's surface found between the *equator* and a pole.

herbs   tender plants, lacking woody stems, usually small or low; may be annual or *perennial.*

high base status   (See *base status of soils.*)

high-latitude climates   group of climates in the *subarctic zone, arctic zone,* and *polar zone,* dominated by arctic *air masses* and polar air masses.

high-level temperature inversion   condition in which a high-level layer of warm air overlies a layer of cooler air, reversing the normal trend of cooling with altitude.

high-pressure cell   center of high barometric pressure; an *anticyclone.*

Histosols   *soil order* consisting of *soils* with a thick upper layer of organic matter.

hogbacks   sharp-crested, often sawtooth ridges formed of the upturned edge of a resistant *rock* layer of *sandstone, limestone,* or *lava.*

Holocene Epoch   last *epoch* of geologic time, commencing about 10,000 years ago; it followed the *Pleistocene Epoch* and includes the present.

horse latitudes   *subtropical high-pressure belt* of the North Atlantic Ocean, coincident with the central region of the Azores High; a belt of weak, variable *winds* and frequent calms.

horst   crustal block uplifted between two *normal faults.*

hot spot   center of intrusive igneous and volcanic activity thought to be located over a rising *mantle plume.*

hot springs   springs discharging heated *ground water* at a temperature close to the boiling point; found in geothermal areas and thought to be related to a *magma* body at depth.

human habitat   the lands of the earth that support human life.

human-influenced vegetation   vegetation that has been influenced in some way by human activity, for example, through cultivation, grazing, timber cutting, or urbanization.

humidity   general term for the amount of *water vapor* present in the air. (See *relative humidity; specific humidity.*)

humification   *pedogenic process* of transformation of plant tissues into *humus.*

humus   dark brown to black organic matter on or in the *soil,* consisting of fragmented plant tissues partly digested by organisms.

hurricane   *tropical cyclone* of the western North Atlantic and Caribbean Sea.

hydraulic action   *stream erosion* by impact force of the flowing water on the bed and banks of the *stream channel.*

hydrologic cycle   total plan of movement, exchange, and storage of the earth's free water in gaseous state, liquid state, and solid state.

hydrology   science of the earth's water and its motions through the *hydrologic cycle.*

hydrolysis   chemical union of water molecules with *minerals* to form different, more stable mineral compounds.

hydrosphere   total water realm of the earth's surface zone, including the oceans, surface waters of the lands, *ground water,* and water held in the *atmosphere.*

hygrometer instrument that measures the *water vapor* content of the *atmosphere;* some types measure *relative humidity* directly.

ice age span of geologic time, usually on the order of 1 to 3 million years, or longer, in which glaciations alternate with interglaciations repeatedly in rhythm with cyclic global climate changes. (See also *glaciation; interglaciation.*)

iceberg mass of glacial ice floating in the ocean, derived from a *glacier* that extends into tidal water.

ice lobes (glacial lobes) broad tonguelike extensions of an *ice sheet* resulting from more rapid ice motion where terrain was more favorable.

ice sheet large thick plate of glacial ice moving outward in all directions from a central region of accumulation.

ice sheet climate ⑬ severely cold climate, found on the Greenland and Antarctic *ice sheets,* with *potential evapotranspiration (water need)* effectively zero throughout the year.

ice shelf thick plate of floating glacial ice attached to an *ice sheet* and fed by the ice sheet and by *snow* accumulation.

ice storm occurrence of heavy glaze of ice on solid surfaces.

ice wedge vertical, wall-like body of ground ice, often tapering downward, occupying a shrinkage crack in *silt* of *permafrost* areas.

ice-wedge polygons polygonal networks of *ice wedges.*

igneous rock *rock* solidified from a high-temperature molten state; rock formed by cooling of *magma.* (See *extrusive igneous rock; felsic igneous rock; intrusive igneous rock; mafic igneous rock; ultramafic igneous rock.*)

illuviation accumulation in a lower *soil horizon* (typically, the *B horizon*) of materials brought down from a higher horizon; a soil-forming process.

image processing mathematical manipulation of digital images, for example, to enhance contrast or edges.

Inceptisols *soil order* consisting of soils having weakly developed *soil horizons* and containing weatherable *minerals.*

induced deflation loss of *soil* by wind erosion that is triggered by human activity such as cultivation or overgrazing.

induced mass wasting *mass wasting* that is induced by human activity, such as creation of waste *soil* and *rock* piles or undercutting of *slopes* in construction.

infiltration absorption and downward movement of *precipitation* into the *soil* and *regolith.*

infrared imagery images formed by *infrared radiation* emanating from the ground surface as recorded by a remote sensor.

infrared radiation *electromagnetic energy* in the *wavelength* range of 0.7 to about 200 μm.

initial landforms *landforms* produced directly by internal earth processes of *volcanism* and *tectonic activity.* Examples: *volcano, fault scarp.*

inner lowland on a *coastal plain,* a shallow valley lying between the first *cuesta* and the area of older rock (oldland).

insolation interception of solar energy (*shortwave radiation*) by an exposed surface.

inspiral horizontal inward spiral or motion, such as that found in a *cyclone.*

interglaciation within an *ice age,* a time interval of mild global climate in which continental *ice sheets* were largely absent or were limited to the Greenland and Antarctic ice sheets; the interval between two glaciations. (See also *deglaciation; glaciation.*)

interlobate moraine *moraine* formed between two adjacent lobes of an *ice sheet.*

International Date Line the 180° *meridian of longitude,* together with deviations east and west of that meridian, forming the time boundary between adjacent *standard time zones* that are 12 hours fast and 12 hours slow with respect to Greenwich standard time.

interrupted projection projection subdivided into a number of sectors (gores), each of which is centered on a different central meridian.

intertropical convergence zone (ITC) zone of convergence of *air masses* of *tropical easterlies* (*trade winds*) along the axis of the *equatorial trough.*

intrusion body of *igneous rock* injected as *magma* into preexisting crustal rock. Example: *dike* or *sill.*

intrusive igneous rock *igneous rock* body produced by solidification of *magma* beneath the surface, surrounded by preexisting rock.

inversion (See *temperature inversion.*)

ion atom or group of atoms bearing an electrical charge as the result of a gain or loss of one or more electrons.

island arcs curved lines of volcanic islands associated with active *subduction* zones along the boundaries of *lithospheric plates.*

isobars lines on map passing through all points having the same *atmospheric pressure.*

isohyet line on a map drawn through all points having the same numerical value of *precipitation.*

isopleth line on a map or globe drawn through all points having the same value of a selected property or entity.

isotherm line on a map drawn through all points having the same air temperature.

jet stream high-speed airflow in narrow bands within the *upper-air westerlies* and along certain other global *latitude* zones at high levels.

joints fractures within *bedrock,* usually occurring in parallel and intersecting sets of planes.

June solstice (See *summer solstice.*)

karst landscape or topography dominated by surface features of *limestone* solution and underlain by a *limestone cavern* system.

Kelvin scale (K) temperature scale on which the starting point is absolute zero, equivalent to —273°C.

kinetic energy form of energy represented by matter (mass) in motion.

knob and kettle terrain of numerous small knobs of *glacial drift* and deep depressions usually situated along the *moraine* belt of a former *ice sheet.*

lagoon shallow body of open water lying between a *barrier island* or a *barrier reef* and the mainland.

lahar rapid downslope or downvalley movement of a tonguelike mass of water-saturated *tephra* (volcanic ash) originating high up on a steep-sided volcanic cone; a variety of *mudflow.*

lake terrestrial body of standing water surrounded by land or glacial ice.

land breeze   local *wind* blowing from land to water during the night.

land degradation   *degradation* of the quality of plant cover and *soil* as a result of overuse by humans and their domesticated animals, especially during periods of *drought*.

landforms   configurations of the land surface taking distinctive forms and produced by natural processes. Examples: hill, valley, plateau. (See *depositional landforms; erosional landforms; initial landforms; sequential landforms*.)

landmass   large area of *continental crust* lying above sea level (base level) and thus available for removal by *denudation*.

landmass rejuvenation   episode of rapid fluvial *denudation* set off by a rapid crustal rise, increasing the available *landmass*.

landslide   rapid sliding of large masses of *bedrock* on steep mountain slopes or from high *cliffs*.

lapse rate   rate at which temperature decreases with increasing altitude. (See *dry adiabatic lapse rate; environmental temperature lapse rate; wet adiabatic lapse rate*.)

large-scale map   map with *fractional scale* greater than 1:100,000; usually shows a small area.

Late-Cenozoic Ice Age   the series of *glaciations, deglaciations,* and *interglaciations* experienced during the late *Cenozoic Era*.

latent heat   heat absorbed and held in storage in a *gas* or *liquid* during the processes of *evaporation*, or *melting*, or *sublimation;* distinguished from *sensible heat*.

latent heat transfer   flow of *latent heat* that results when water absorbs heat to change from a *liquid* or *solid* to a *gas* and then later releases that heat to new surroundings by *condensation* or *deposition*.

lateral moraine   *moraine* forming an embankment between the ice of an *alpine glacier* and the adjacent valley wall.

laterite   rocklike layer rich in *sequioxides* and iron, including the minerals *bauxite* and *limonite*, found in low latitudes in association with *Ultisols* and *Oxisols*.

latitude   *arc* of a meridian between the *equator* and a given point on the globe.

lava   *magma* emerging on the earth's solid surface, exposed to air or water.

leaching   *pedogenic process* in which material is lost from the *soil* by downward washing out and removal by percolating surplus *soil water*.

leads   narrow strips of open ocean water between ice floes.

level of condensation   elevation at which an upward-moving parcel of moist air cools to the *dewpoint* and *condensation* begins to occur.

liana   woody vine supported on the trunk or branches of a *tree*.

lichens   plant forms in which algae and fungi live together (in a symbiotic relationship) to create a single structure; typically form tough, leathery coatings or crusts attached to *rocks* and *tree* trunks.

life-form   characteristic physical structure, size, and shape of a plant or of an assemblage of plants.

life layer   shallow surface zone containing the *biosphere;* a zone of interaction between *atmosphere* and land surface, and between atmosphere and ocean surface.

life zones   series of vegetation zones describing vegetation types that are encountered with increasing elevation, especially in the southwestern United States.

limestone   nonelastic *sedimentary rock* in which *calcite* is the predominant *mineral*, and with varying minor amounts of other minerals and *clay*.

limestone caverns   interconnected subterranean cavities formed in *limestone* by *carbonic acid action* occurring in slowly moving *ground water*.

limonite   mineral or group of *minerals* consisting largely of iron oxide and water, produced by *chemical weathering* of other iron-bearing minerals.

liquid   *fluid* that maintains a free upper surface and is only very slightly compressible, as compared with a *gas*.

lithosphere   (1) General term for the entire solid earth realm. (2) In *plate tectonics*, it is the strong, brittle outermost rock layer lying above the *asthenosphere*.

lithospheric plate   segment of *lithosphere* moving as a unit, in contact with adjacent lithospheric plates along plate boundaries.

littoral drift   transport of *sediment* parallel with the *shoreline* by the combined action of *beach drift* and *longshore current* transport.

loam   soil-texture class in which no one of the three size grades (*sand, silt, clay*) dominates over the other two.

local winds   general term for *winds* generated as direct or immediate effects of the local terrain.

loess   accumulation of yellowish to buff-colored, fine-grained *sediment*, largely of *silt* grade, upon upland surfaces after transport in the air in *turbulent suspension* (i.e., carried in a *dust storm*).

longitude   *arc* of a *parallel* between the *prime meridian* and a given point on the globe.

longitudinal dunes   class of *sand dunes* in which the dune ridges are oriented parallel with the prevailing *wind*.

longshore current   current in the breaker zone, running parallel with the *shoreline* and set up by the oblique approach of waves.

longshore drift   *littoral drift* caused by action of a *longshore current*.

longwave radiation   *electromagnetic energy* emitted by the earth, largely in the range from 3 to 50 μm.

low-angle overthrust fault   *overthrust fault* in which the *fault plane* or fault surface has a low angle of *dip* or may be horizontal.

low base status   (See *base status of soils*.)

lowlands   broad, open valleys between two *cuestas* of a *coastal plain*. (The term may refer to any low areas of land surface.)

low-latitude climates   group of climates of the *equatorial zone* and *tropical zone* dominated by the *subtropical high-pressure belt* and the *equatorial trough*.

low-latitude rainforest   evergreen broadleaf forest of the wet equatorial and tropical climate zones.

low-latitude rainforest environment   low-latitude environment of warm temperatures and abundant *precipitation* that characterizes rainforest in the *wet equatorial* ① and *monsoon and trade-wind coastal* ② climates.

low-level temperature inversion   atmospheric condition in which temperature near the ground increases, rather than decreases, with elevation.

mafic igneous rock   *igneous rock* dominantly composed of *mafic minerals*.

mafic minerals (mafic mineral group) *minerals*, largely *silicate minerals*, rich in magnesium and iron, dark in color, and of relatively great density.

magma mobile, high-temperature molten state of *rock*, usually of *silicate mineral* composition and with dissolved *gases*.

mantle *rock* layer or shell of the earth beneath the *crust* and surrounding the *core*, composed of *ultramafic igneous rock* of *silicate mineral* composition.

mantle plume a columnlike rising of heated *mantle* rock, thought to be the cause of a *hot spot* in the overlying *lithospheric plate*.

map projection any orderly system of parallels and meridians drawn on a flat surface to represent the earth's curved surface.

marble variety of *metamorphic rock* derived from *limestone* or *dolomite* by recrystallization under pressure.

marine cliff *rock* cliff shaped and maintained by the undermining action of breaking waves.

marine scarp steep seaward *slope* in poorly consolidated *alluvium*, *glacial drift*, or other forms of *regolith*, produced along a *coastline* by the undermining action of waves.

marine terrace former *abrasion platform* elevated to become a steplike coastal *landform*.

marine west-coast climate ⑧ cool moist climate of west coasts in the *midlatitude zone*, usually with a substantial annual *water surplus* and a distinct winter *precipitation* maximum.

mass wasting spontaneous downhill movement of *soil*, *regolith*, and *bedrock* under the influence of *gravity*, rather than by the action of *fluid* agents.

mean annual temperature mean of daily air temperature means for a given year or succession of years.

mean daily temperature sum of daily maximum and minimum air temperature readings divided by two.

mean monthly temperature mean of daily air temperature means for a given calendar month.

mean velocity mean, or average, speed of flow of water through an entire stream cross section.

meanders (See *alluvial meanders*.)

medial moraine long, narrow deposit of fragments on the surface of a *glacier*, created by the merging of *lateral moraines* when two glaciers join into a single stream of ice flow.

Mediterranean climate ⑦ climate type of the *subtropical zone*, characterized by the alternation of a very dry summer and a mild, rainy winter.

melting change from solid state to liquid state, accompanied by absorption of *sensible heat* to become *latent heat*.

Mercator projection conformal *map projection* with horizontal parallels and vertical meridians and with map scale rapidly increasing with increase in *latitude*.

mercury barometer *barometer* using the Torricelli principle, in which *atmospheric pressure* counterbalances a column of mercury in a tube.

meridian of longitude north-south line on the surface of the global *oblate ellipsoid*, connecting the *north pole* and *south pole*.

meridional transport flow of energy (heat) or matter (water) across the *parallels of latitude*, either poleward or equatorward.

mesa table-topped *plateau* of comparatively small extent bounded by *cliffs* and occurring in a region of flat-lying *strata*.

Mesozoic Era second of three geologic *eras* following *Precambrian time*.

metamorphic rock *rock* altered in physical structure and/or chemical (*mineral*) composition by action of heat, pressure, *shearing* stress, or infusion of elements, all taking place at substantial depth beneath the surface.

meteorology science of the *atmosphere*, particularly the physics of the lower or inner atmosphere.

mica group aluminum-silicate *mineral* group of complex chemical formula having perfect cleavage into thin sheets.

micrometer metric unit of length equal to one-millionth of a meter (0.000001 m); abbreviated µm.

microwaves waves of the *electromagnetic radiation* spectrum in the *wavelength* band from about 0.03 cm to about 1 cm.

midlatitude climates group of climates of the *midlatitude zone* and *subtropical zone*, located in the *polar front zone* and dominated by both tropical *air masses* and polar air masses.

midlatitude deciduous forest plant *formation class* within the *forest biome* dominated by tall, broadleaf deciduous *trees*, found mostly in the *moist continental climate* ⑩ and *marine west-coast climate* ⑧.

midlatitude zones latitude zones occupying the *latitude* range 35° to 55° N and S (more or less) and lying between the *subtropical zones* and the *subarctic (subantarctic) zones*.

midoceanic ridge one of three major divisions of the ocean basins, being the central belt of submarine mountain topography with a characteristic *axial rift*.

millibar unit of *atmospheric pressure*, one-thousandth of a bar. *Bar* is a force of 1 million dynes per square centimeter.

mineral naturally occurring inorganic substance, usually having a definite chemical composition and a characteristic atomic structure. (See *felsic minerals*; *mafic minerals*; *silicate minerals*.)

mineral alteration chemical change of *minerals* to more stable compounds upon exposure to atmospheric conditions; same as *chemical weathering*.

mineral matter (soils) component of *soil* consisting of weathered or unweathered mineral grains.

mineral oxides (soils) secondary *minerals* found in *soils* in which original minerals have been altered by chemical combination with oxygen.

minute (of arc) 1/60 of a degree.

mistral local drainage *wind* of cold air affecting the Rhone Valley of southern France.

Moho contact surface between the earth's *crust* and *mantle*; a contraction of Mohorovic, the name of the seismologist who discovered this feature.

moist continental climate ⑩ moist climate of the *midlatitude zone* with strongly defined winter and summer seasons, adequate *precipitation* throughout the year, and a substantial annual *water surplus*.

moist subtropical climate ⑥ moist climate of the *subtropical zone*, characterized by a moderate to large annual *water surplus* and a strongly seasonal cycle of *potential evapotranspiration* (*water need*).

mollic epipedon   relatively thick, dark-colored surface *soil horizon*, containing substantial amounts of organic matter (*humus*) and usually rich in *bases*.

Mollisols   *soil order* consisting of *soils* with a *mollic horizon* and high *base status*.

monadnock   prominent, isolated mountain or large hill rising conspicuously above a surrounding *peneplain* and composed of a *rock* more resistant than that underlying the peneplain; a *landform* of *denudation* in moist climates.

monsoon and trade-wind coastal climate ②   moist climate of low latitudes showing a strong rainfall peak in the season of high sun and a short period of reduced rainfall.

monsoon forest   *formation class* within the *forest biome* consisting in part of deciduous *trees* adapted to a long dry season in the *wet-dry tropical climate* ③.

monsoon system   system of low-level *winds* blowing into a continent in summer and out of it in winter, controlled by *atmospheric pressure* systems developed seasonally over the continent.

montane forest   plant *formation class* of the *forest biome* found in cool upland environments of the *tropical zone* and *equatorial zone*.

moraine   accumulation of *rock* debris carried by an *alpine glacier* or an *ice sheet* and deposited by the ice to become a *depositional landform*. (See *lateral moraine; terminal moraine*.)

mountain arc   curving section of an *alpine chain* occurring on a *converging boundary* between two crustal plates.

mountain roots   erosional remnants of deep portions of ancient *continental sutures* that were once *alpine chains*.

mountain winds   daytime movements of air up the *gradient* of valleys and mountain slopes; alternating with nocturnal *valley winds*.

mucks   organic *soils* largely composed of fine, black, sticky organic matter.

mud   *sediment* consisting of a mixture of *clay* and *silt* with water, often with minor amounts of *sand* and sometimes with organic matter.

mudflow   a form of *mass wasting* consisting of the downslope flowage of a mixture of water and *mineral* fragments (*soil, regolith*, disintegrated *bedrock*), usually following a natural drainage line or *stream channel*.

mudstone   *sedimentary rock* formed by the lithification of *mud*.

multipurpose map   map containing several different types of information.

multispectral image   image consisting of two or more images, each of which is taken from a different portion of the spectrum (e.g., blue, green, red, infrared).

multispectral scanner   *remote sensing* instrument, flown on an aircraft or spacecraft, that simultaneously collects multiple *digital images* (*multispectral images*) of the ground. Typically, images are collected in four or more spectral bands.

nappe   overturned recumbent *fold* of *strata*, usually associated with *thrust sheets* in a collision *orogen*.

natural bridge   natural *rock* arch spanning a *stream channel*, formed by cutoff of an *entrenched meander* bend.

natural gas   naturally occurring mixture of hydrocarbon compounds (principally methane) in the gaseous state held within certain porous *rocks*.

natural levee   belt of higher ground paralleling a meandering *alluvial river* on both sides of the *stream channel* and built up by *deposition* of fine *sediment* during periods of overbank flooding.

natural vegetation   stable, mature plant cover characteristic of a given area of land surface largely free from the influences and impacts of human activities.

needleleaf evergreen forest   needleleaf forest composed of evergreen *tree* species, such as spruce, fir, and pine.

needleleaf forest   plant *formation class* within the *forest biome*, consisting largely of needleleaf evergreen *trees*. (See also *boreal forest*.)

net radiation   difference in intensity between all incoming energy (positive quantity) and all outgoing energy (negative quantity) carried by both *shortwave radiation* and *longwave radiation*.

noon   (See *solar noon*.)

normal fault   variety of *fault* in which the *fault plane* inclines (*dips*) toward the downthrown block and a major component of the motion is vertical.

north pole   point at which the northern end of the earth's *axis of rotation* intersects the earth's surface.

nuclei (atmospheric)   minute particles of solid matter suspended in the *atmosphere* and serving as cores for *condensation* of water or ice.

O_1 horizon   surface *soil horizon* containing decaying organic matter that is recognizable as leaves, twigs, or other organic structures.

O_a horizon   *soil horizon* below the O_1 horizon containing decaying organic matter that is too decomposed to recognize as specific plant parts, such as leaves or twigs.

oblate ellipsoid   geometric solid resembling a flattened sphere, with polar axis shorter than the equatorial diameter.

occluded front   weather *front* along which a moving *cold front* has overtaken a *warm front*, forcing the warm *air mass* aloft.

ocean-basin floors   one of the major divisions of the ocean basins, comprising the deep portions consisting of *abyssal plains* and low hills.

ocean current   persistent, dominantly horizontal flow of ocean water.

ocean tide   periodic rise and fall of the ocean level induced by gravitational attraction between the earth and moon in combination with earth *rotation*.

oceanic crust   crust of basaltic composition beneath the ocean floors, capping *oceanic lithosphere*. (See also *continental crust*.)

oceanic lithosphere   *lithosphere* bearing *oceanic crust*.

oceanic trench   narrow, deep depression in the seafloor representing the line of *subduction* of an oceanic *lithospheric plate* beneath the margin of a continental lithospheric plate; often associated with an *island arc*.

olivine   *silicate mineral* with magnesium and iron but no aluminum, usually olive-green or grayish-green; a *mafic mineral*.

organic matter (soils)   material in *soil* that was originally produced by plants or animals and has been subjected to decay.

organic sediment   *sediment* consisting of the organic remains of plants or animals.

orogen   the mass of tectonically deformed rocks and related *igneous rocks* produced during an *orogeny*.

orogeny   major episode of *tectonic activity* resulting in *strata* being deformed by *folding* and faulting.

orographic precipitation   *precipitation* induced by the forced rise of moist air over a mountain barrier.

outcrop   surface exposure of *bedrock*.

outspiral   horizontal outward spiral or motion, such as that found in an *anticyclone*.

outwash   glacial deposit of stratified drift left by *braided streams* issuing from the front of a *glacier*.

outwash plain   flat, gently sloping plain built up of *sand* and gravel by the *aggradation* of meltwater *streams* in front of the margin of an *ice sheet*.

overland flow   motion of a surface layer of water over a sloping ground surface at times when the *infiltration* rate is exceeded by the *precipitation* rate; a form of *runoff*.

overthrust fault   *fault* characterized by the overriding of one crustal block (or *thrust sheet*) over another along a gently inclined *fault plane*; associated with crustal *compression*.

oxbow lake   crescent-shaped *lake* representing the abandoned channel left by the *cutoff* of an *alluvial meander*.

oxidation   chemical union of free oxygen with metallic elements in *minerals*.

oxide   chemical compound containing oxygen; in *soils*, iron oxides and aluminum oxides are examples.

Oxisols   *soil order* consisting of very old, highly weathered *soils* of low latitudes, with an oxic horizon and low *base status*.

ozone   a form of oxygen with a molecule consisting of three atoms of oxygen, O_3.

ozone layer   layer in the *stratosphere*, mostly in the altitude range 20 to 35 km (12 to 31 mi), in which a concentration of *ozone* is produced by the action of solar *ultraviolet radiation*.

pack ice   floating *sea ice* that completely covers the sea surface.

Paleozoic Era   first of three geologic *eras* comprising all geologic time younger than *Precambrian time*.

Pangaea   hypothetical parent continent, enduring until near the close of the *Mesozoic Era*, consisting of the *continental shields* of Laurasia and Gondwana joined into a single unit. (See also *continental drift*.)

parabolic dunes   isolated low *sand dunes* of parabolic outline, with points directed into the prevailing *wind*.

parallel of latitude   east-west circle on the earth's surface, lying in a plane parallel with the *equator* and at right angles to the *axis of rotation*.

parent material   inorganic, *mineral* base from which the *soil* is formed; usually consists of *regolith*.

particulates   *solid* and *liquid* particles capable of being suspended for long periods in the *atmosphere*.

passive continental margins   continental margins lacking active plate boundaries at the contact of *continental crust* with *oceanic crust*. A passive margin thus lies within a single *lithospheric plate*. Example: Atlantic continental margin of North America. (See also *active continental margins; continental margins*.)

passive systems   electromagnetic *remote sensing* systems that measure radiant energy reflected or emitted by an object or surface.

peat   partially decomposed, compacted accumulation of plant remains occurring in a bog environment.

ped   individual natural *soil* aggregate.

pediment   gently sloping, rock-floored land surface found at the base of a mountain mass or *cliff* in an arid region.

pedogenic processes   group of recognized basic soil-forming processes, mostly involving the gain, loss, *translocation*, or transformation of materials within the *soil* body.

pedology   science of the *soil* as a natural surface layer capable of supporting living plants; synonymous with *soil science*.

peneplain   land surface of low elevation and slight relief produced in the late stages of *denudation* of a *landmass*.

percolation   slow, downward flow of water by *gravity* through *soil* and subsurface layers toward the *water table*.

perennials   plants that live for more than one growing season.

peridotite   *igneous rock* consisting largely of *olivine* and *pyroxene*; an *ultramafic igneous rock* occurring as a *pluton*, also thought to compose much of the upper *mantle*.

perihelion   point on the earth's elliptical orbit at which the earth is nearest to the sun.

period of geologic time   time subdivision of the *era*, each ranging in duration between about 35 and 70 million years.

permafrost   condition of permanently frozen water in the *soil*, *regolith*, and *bedrock* in cold climates of subarctic and arctic regions.

petroleum (crude oil)   natural liquid mixture of many complex hydrocarbon compounds of organic origin, found in accumulations (oil pools) within certain *sedimentary rocks*.

pH   measure of the concentration of hydrogen ions in a solution. (The number represents the logarithm to the base 10 of the reciprocal of the weight in grams of hydrogen ions per liter of water.) Acid solutions have pH values less than 6, and basic solutions have pH values greater than 6.

photosynthesis   production of carbohydrate by the union of water with *carbon dioxide* while absorbing light energy.

physical geography   the study and synthesis of selected subject areas from the natural sciences—especially atmospheric science, *hydrology*, physical oceanography, *geology*, *geomorphology*, *soil science*, and *plant ecology*—in order to gain a complete picture of the physical environment of humans and to examine the interactions of humans with that environment.

physical weathering   breakup of massive *rock* (*bedrock*) into small particles through the action of physical forces acting at or near the earth's surface. (See *weathering*.)

pioneer plants   plants that first invade an environment of new land or a *soil* that has been cleared of vegetation cover; often these are annual *herbs*.

plagioclase feldspar   aluminum-silicate *mineral* with sodium or calcium or both.

plane of the ecliptic   imaginary plane in which the earth's orbit lies.

plant ecology   the study of the relationships between plants and their environment.

plant nutrients   *ions* or chemical compounds that are needed for plant growth.

plate tectonics   theory of *tectonic activity* dealing with *lithospheric plates* and their activity.

plateau   upland surface, more or less flat and horizontal, upheld by resistant beds of *sedimentary rock* or *lava* flows and bounded by a steep *cliff*.

playa flat land surface underlain by fine *sediment* or evap-orite minerals deposited from shallow *lake* waters in a dry climate in the floor of a closed topographic depression.

Pleistocene Epoch *epoch* of the *Cenozoic Era*, often identified as the Ice Age; it preceded the *Holocene Epoch*.

plinthite iron-rich concentrations present in some kinds of *soils* in deeper *soil horizons* and capable of hardening into rocklike material with repeated wetting and drying.

plucking (See *glacial plucking*.)

pluton any body of *intrusive igneous rock* that has solidified below the surface, enclosed in preexisting *rock*.

pocket beach *beach* of crescentic outline located at a *bay* head.

podzol type of *soil order* closely equivalent to *Spodosol*; an order of the Canadian Soil Classification System.

point bar deposit of coarse bed-load *alluvium* accumulated on the inside of a growing *alluvial meander*.

polar easterlies system of easterly surface winds at high latitude, best developed in the southern *hemisphere*, over Antarctica.

polar front *front* lying between cold polar *air masses* and warm tropical air masses, often situated along a *jet stream* within the *upper-air westerlies*.

polar front jet stream *jet stream* found along the *polar front*, where cold polar air and warm tropical air are in contact.

polar front zone broad zone in midlatitudes and higher latitudes, occupied by the shifting *polar front*.

polar high persistent low-level center of high *atmospheric pressure* located over the *polar zone* of Antarctica.

polar outbreak tongue of cold polar air, preceded by a *cold front*, penetrating far into the *tropical zone* and often reaching the *equatorial zone;* it brings rain squalls and unusual cold.

polar projection *map projection* centered on earth's *north pole* or *south pole*.

polar zones *latitude* zones lying between 75° and 90° N and S.

poleward heat transport movement of heat from equatorial and tropical regions toward the poles, occurring as *latent* and *sensible heat transfer*.

pollutants in air pollution studies, foreign matter injected into the lower *atmosphere* as *particulates* or as chemical pollutant *gases*.

pollution dome broad, low dome-shaped layer of polluted air, formed over an urban area at times when *winds* are weak or calm prevails.

pollution plume (1) The trace or path of pollutant substances, moving along the flow paths of *ground water*. (2) Trail of polluted air carried downwind from a pollution source by strong *winds*.

polypedon smallest distinctive geographic unit of the *soil* of a given area.

potash feldspar aluminum-silicate *mineral* with potassium the dominant metal.

potential evapotranspiration (water need) ideal or hypothetical rate of *evapotranspiration* estimated to occur from a complete canopy of green foliage of growing plants continuously supplied with all the *soil water* they can use; a real condition reached in those situations where *precipitation* is sufficiently great or irrigation water is supplied in sufficient amounts.

pothole cylindrical cavity in hard *bedrock* of a *stream channel* produced by *abrasion* of a rounded *rock* fragment rotating within the cavity.

prairie plant *formation class* of the *grassland biome*, consisting of dominant tall grasses and subdominant *forbs*, widespread in subhumid continental climate regions of the *subtropical zone* and *midlatitude zone*. (See *short-grass prairie; tall-grass prairie*.)

Precambrian time all of geologic time older than the beginning of the Cambrian period, that is, older than 600 million years.

precipitation particles of *liquid* water or ice that fall from the atmosphere and may reach the ground. (See *convectional precipitation; cyclonic precipitation; orographic precipitation*.)

pressure gradient change of *atmospheric pressure* measured along a line at right angles to the *isobars*.

pressure gradient force force acting horizontally, tending to move air in the direction of lower *atmospheric pressure*.

prevailing westerly winds (westerlies) surface winds blowing from a generally westerly direction in the *midlatitude zone*, but varying greatly in direction and intensity.

primary minerals in *pedology* (*soil science*), the original, unaltered *silicate minerals* of *igneous rocks* and *metamorphic rocks*.

prime meridian reference meridian of zero *longitude;* universally accepted as the Greenwich meridian.

progradation shoreward building of a *beach, bar,* or *sandspit* by addition of coarse *sediment* carried by *littoral drift* or brought from deeper water offshore.

pyroxene group complex aluminum-silicate *minerals* rich in calcium, magnesium, and iron, dark in color, high in density, classed as *mafic minerals*.

quartz mineral of silicon dioxide composition.

quartzite *metamorphic rock* consisting largely of the mineral *quartz*.

radar an active *remote sensing* system in which a pulse of *radiation* is emitted by an instrument and the strength of the echo of the pulse is recorded.

radial drainage pattern stream pattern consisting of *streams* radiating outward from a central peak or highland, such as a *sedimentary dome* or a *volcano*.

radiation (See *electromagnetic radiation*.)

radiation balance condition of balance between incoming energy of solar *shortwave radiation* and outgoing *longwave radiation* emitted by the earth into space.

radiation fog *fog* produced by radiational cooling of the basal air layer.

rain form of *precipitation* consisting of falling water drops, usually 0.5 mm or larger in diameter.

rain gauge instrument used to measure the amount of *rain* that has fallen.

rain-green vegetation vegetation that puts out green foliage in the wet season but becomes largely dormant in the dry season; found in the *tropical zone*, it includes the *savanna biome* and *monsoon forest*.

rainshadow belt of arid climate to lee of a mountain barrier, produced as a result of adiabatic warming of descending air.

raised shoreline former *shoreline* lifted above the limit of wave action; also called an elevated shoreline.

rapids steep-*gradient* reaches of a *stream channel* in which *stream* velocity is high.

recumbent overturned, as a folded sequence of *rock* layers in which the folds are doubled back on themselves.

reflection outward scattering of *radiation* toward space by the *atmosphere* and/or earth's surface.

reg desert surface armored with a pebble layer, resulting from long-continued *deflation*; found in the Sahara Desert of North Africa.

regolith layer of *mineral* particles overlying the *bedrock*; may be derived by *weathering* of underlying bedrock or be transported from other locations by *fluid* agents. (See *residual regolith; transported regolith*.)

relative humidity ratio of *water vapor* present in the air to the maximum quantity possible for *saturated air* at the same temperature.

remote sensing measurement of some property of an object or surface by means other than direct contact; usually refers to the gathering of scientific information about the earth's surface from great heights and over broad areas, using instruments mounted on aircraft or orbiting space vehicles.

remote sensor instrument or device measuring *electromagnetic radiation* reflected or emitted from a target body.

representative fraction (R.F.) (See *scale fraction*.)

residual regolith *regolith* formed in place by alteration of the *bedrock* directly beneath it.

resolution on a map, power to resolve small objects present on the ground.

retrogradation cutting back (retreat) of a *shoreline, beach, marine cliff,* or *marine scarp* by wave action.

reverse fault type of *fault* in which one fault block rides up over the other on a steep *fault plane*.

revolution motion of a planet in its orbit around the sun, or of a planetary satellite around a planet.

rhyolite *extrusive igneous rock* of *granite* composition; occurs as *lava* or *tephra*.

ria coast deeply embayed *coast* formed by partial *submergence* of a *land mass* previously shaped by fluvial *denudation*.

Richter scale scale of magnitude numbers describing the quantity of energy released by an *earthquake*.

ridge-and-valley landscape assemblage of *landforms* developed by *denudation* of a system of open *folds* of *strata* and consisting of long, narrow ridges and valleys arranged in parallel or zigzag patterns.

rift valley trenchlike valley with steep, parallel sides; essentially a *graben* between two *normal faults*; associated with crustal spreading.

rill erosion form of *accelerated erosion* in which numerous, closely spaced miniature channels (rills) are scored into the surface of exposed *soil* or *regolith*.

rock natural aggregate of *minerals* in the solid state; usually hard and consisting of one, two, or more mineral varieties.

rock terrace terrace carved in *bedrock* during the *degradation* of a *stream channel* induced by the crustal rise or a fall of the sea level. (See also *alluvial terrace; marine terrace*.)

Rossby waves horizontal undulations in the flow path of the *upper-air westerlies;* also known as upper-air waves.

rotation spinning of an object around an axis.

runoff flow of water from continents to oceans by way of *stream flow* and *ground water* flow; a term in the water balance of the *hydrologic cycle*. In a more restricted sense, runoff refers to surface flow by *overland flow* and channel flow.

Sahel (Sahelian zones) belt of *wet-dry tropical* ③ and *semiarid dry tropical* ④ climate in Africa in which *precipitation* is highly variable from year to year.

salic horizon *soil horizon* enriched by soluble salts.

salinization precipitation of soluble salts within the *soil*.

saltation leaping, impacting, and rebounding of sand grains transported over a *sand* or pebble surface by *wind*.

salt-crystal growth a form of *weathering* in which *rock* is disintegrated by the expansive pressure of growing salt crystals during dry weather periods when *evaporation* is rapid.

salt flat shallow basin covered with salt deposits formed when *stream* input to the basin is subjected to severe *evaporation*; may also form by evaporation of a saline *lake* when climate changes.

salt marsh *peat*-covered expanse of *sediment* built up to the level of high tide over a previously formed tidal mud flat.

sand *Sediment* particles between 0.06 and 2 mm in diameter.

sand dune hill or ridge of loose, well-sorted *sand* shaped by *wind* and usually capable of downwind motion.

sand sea field of *transverse dunes*.

sandspit narrow, fingerlike embankment of *sand* constructed by *littoral drift* into the open water of a *bay*.

sandstone variety of *sedimentary rock* consisting largely of *mineral* particles of *sand* grade size.

Santa Ana easterly *wind*, often hot and dry, that blows from the interior desert region of southern California and passes over the coastal mountain ranges to reach the Pacific Ocean.

saturated air air holding the maximum possible quantity of *water vapor at* a given temperature and pressure.

saturated zone zone beneath the land surface in which all pores of the *bedrock* or *regolith* are filled with *ground water*.

savanna a vegetation cover of widely spaced *trees* with a grassland beneath.

savanna biome *biome* that consists of a combination of *trees* and grassland in various proportions.

savanna woodland plant *formation class* of the *savanna biome* consisting of a *woodland* of widely spaced *trees* and a grass layer, found throughout the *wet-dry tropical climate* ③ regions in a belt adjacent to the *monsoon forest* and *low-latitude rainforest*.

scale fraction ratio that relates distance on the earth's surface to distance on a map or surface of a globe.

scale of globe ratio of size of a globe to size of the earth, where size is expressed by a measure of length or distance.

scale of map ratio of distance between two points on a map and the same two points on the ground.

scanning systems *remote sensing* systems that make use of a scanning beam to generate images over the frame of surveillance.

scarification general term for artificial excavations and other land disturbances produced for purposes of extracting or processing *mineral* resources.

scattering turning aside by reflection of solar *shortwave radiation* by *gas* molecules of the *atmosphere*.

schist foliated *metamorphic rock* in which mica flakes are typically found oriented parallel with foliation surfaces.

sclerophyll forest plant *formation class* of the *forest biome,* consisting of low sclerophyll *trees* and often including sclerophyll *woodland* or *scrub,* associated with regions of *Mediterranean climate* ⑦.

sclerophylls hardleaved evergreen *trees* and *shrubs* capable of enduring a long, dry summer.

sclerophyll woodland plant *formation class* of the *forest biome* composed of widely spaced sclerophyll *trees* and *shrubs.*

scoria *lava* or *tephra* containing numerous cavities produced by expanding *gases* during cooling.

scrub plant *formation class* or subclass consisting of *shrubs* and having a canopy coverage of about 50 percent.

sea arch archlike *landform* of a rocky, cliffed *coast* created when waves erode through a narrow headland from both sides.

sea breeze local *wind* blowing from sea to land during the day.

sea cave cave near the base of a *marine cliff,* eroded by breaking waves.

sea fog *fog* layer formed at sea when warm moist air passes over a cool ocean current and is chilled to the *condensation* point.

sea ice floating ice of the oceans formed by direct *freezing* of ocean water.

second of arc 1/60 of a minute, or 1/3600 of a degree.

secondary minerals in *soil science, minerals* that are stable in the surface environment, derived by *mineral alteration* of the *primary minerals.*

sediment finely divided *mineral matter* and organic matter derived directly or indirectly from preexisting *rock* and from life processes. (See *chemically precipitated sediment; organic sediment.*)

sedimentary dome up-arched *strata* forming a circular structure with domed summit and flanks with moderate to steep outward *dip.*

sedimentary rock *rock* formed from accumulation of *sediment.*

seismic sea wave (tsunami) train of sea waves set off by an *earthquake* (or other seafloor disturbance) traveling over the ocean surface.

seismic waves waves sent out during an *earthquake* by faulting or other crustal disturbance from an *earthquake focus* and propagated through the solid earth.

semiarid (dry climate subtype) subtype of the dry climates exhibiting a short wet season supporting the growth of grasses and *annual* plants.

semiarid (steppe) climate subtype subtype of the dry climate in which *soil water* storage equals or exceeds 6 cm in at least two months of the year.

semidesert plant *formation class* of the *desert biome,* consisting of xerophytic *shrub* vegetation with a poorly developed herbaceous lower layer; subtypes are semidesert scrub and *woodland.*

sensible heat heat measurable by a *thermometer;* an indication of the intensity of *kinetic energy* of molecular motion within a substance.

sensible heat transfer flow of heat from one substance to another by direct contact.

sequential landforms *landforms* produced by external earth processes in the total activity of *denudation.* Examples: *gorge, alluvial fan, floodplain.*

sesquioxides oxides of aluminum or iron with a ratio of two atoms of aluminum or iron to three atoms of oxygen.

shale *fissile, sedimentary rock* of *mud* or *clay* composition, showing lamination.

shearing (of rock) slipping motion between very thin *rock* layers, like a deck of cards fanned with the sweep of a palm.

sheet erosion type of *accelerated soil erosion* in which thin layers of *soil* are removed without formation of rills or *gullies.*

sheet flow overland flow taking the form of a continuous thin film of water over a smooth surface of *soil, regolith,* or *rock.*

sheeting structure thick, subparallel layers of massive *bedrock* formed by spontaneous expansion accompanying *unloading.*

shield volcano low, often large, domelike accumulation of basalt *lava* flows emerging from long radial fissures on flanks.

shoreline shifting line of contact between water and land.

short-grass prairie plant *formation class* in the *grassland biome* consisting of short grasses sparsely distributed in clumps and bunches and some *shrubs,* widespread in areas of *semiarid climate* in continental interiors of North America and Eurasia; also called *steppe.*

shortwave infrared *infrared radiation* with *wavelengths* shorter than 3 µm.

shortwave radiation *electromagnetic energy* in the range from 0.2 to 3 µm, including most of the energy spectrum of solar *radiation.*

shrubs woody *perennial* plants, usually small or low, with several low-branching stems and a foliage mass close to the ground.

silica silicon dioxide in any of several *mineral* forms.

silicate minerals (silicates) *minerals* containing silicon and oxygen atoms, linked in the crystal space lattice in units of four oxygen atoms to each silicon atom.

sill *intrusive igneous rock* in the form of a plate where *magma* was forced into a natural parting in the *bedrock,* such as a bedding surface in a sequence of *sedimentary rocks.*

silt *sediment* particles between 0.004 and 0.06 mm in diameter.

sinkhole surface depression in *limestone,* leading down into *limestone caverns.*

slash-and-burn agricultural system, practiced in the *low-latitude rainforest,* in which small areas are cleared and the *trees* burned, forming plots that can be cultivated for brief periods.

slate compact, fine-grained variety of *metamorphic rock,* derived from *shale,* showing well-developed cleavage.

sleet form of *precipitation* consisting of ice pellets, which may be frozen raindrops.

sling psychrometer form of *hygrometer* consisting of a wet-bulb *thermometer* and a dry-bulb thermometer.

slip face steep face of an active *sand dune,* receiving *sand* by *saltation* over the dune crest and repeatedly sliding because of oversteepening.

slope (1) Degree of inclination from the horizontal of an element of ground surface, analogous to *dip* in the geologic sense. (2) Any portion or element of the earth's solid surface. (3) Verb meaning "to incline."

small-scale map map with *fractional scale* of less than 1:100,000; usually shows a large area.

smog mixture of *aerosols* and chemical *pollutants* in the lower *atmosphere*, usually found over urban areas.

snow form of *precipitation* consisting of ice particles.

soil natural terrestrial surface layer containing living matter and supporting or capable of supporting plants.

soil colloids *mineral* particles of extremely small size, capable of remaining suspended indefinitely in water; typically, they have the form of thin plates or scales.

soil creep extremely slow downhill movement of *soil* and *regolith* as a result of continued agitation and disturbance of the particles by such activities as *frost action*, temperature changes, or wetting and drying of the soil.

soil enrichment additions of materials to the *soil* body; one of the *pedogenic processes*.

soil erosion erosional removal of material from the *soil* surface.

soil horizon distinctive layer of the *soil*, more or less horizontal, set apart from other soil zones or layers by differences in physical and chemical composition, organic content, structure, or a combination of those properties, produced by soil-forming processes.

soil orders those eleven *soil* classes forming the highest category in the classification of soils.

soil profile display of *soil horizons* on the face of a freshly cut vertical exposure through the *soil*.

soil science (See *pedology*.)

soil solum that part of the *soil* made up of the *A*, *E*, and *B soil horizons*; the soil zone in which living plant roots can influence the development of soil horizons.

soil structure presence, size, and form of aggregations (lumps or clusters) of *soil* particles.

soil texture descriptive property of the *mineral* portion of the *soil* based on varying proportions of *sand*, *silt*, and *clay*.

soil water water held in the *soil* and available to plants through their root systems; a form of *subsurface water*.

soil water balance balance among the component terms of the *soil water budget*, namely, *precipitation*, *evapotranspiration*, change in soil water storage, and *water surplus*.

soil water belt *soil* layer from which plants draw *soil water*.

soil water budget accounting system evaluating the daily, monthly, or yearly amounts of *precipitation*, *evapotranspiration*, soil water storage, water deficit, and water surplus.

soil water recharge restoring of depleted *soil water* by *infiltration* of *precipitation*.

solar constant intensity of solar radiation falling on a unit area of surface held at right angles to the sun's rays at a point outside the earth's *atmosphere*; equal to an energy flow of about 1400 W/m^2.

solar day average time required for the earth to complete one *rotation* with respect to the sun; time elapsed between one solar noon and the next, averaged over the period of one year.

solar noon instant at which the *subsolar point* crosses the *meridian of longitude* of a given point on the earth; instant at which the sun's shadow points exactly due north or due south at a given location.

solids substances in the solid state; they resist changes in shape and volume, are usually capable of withstanding large unbalanced forces without yielding, but will ultimately yield by sudden breakage.

solifluction tundra (arctic) variety of *earthflow* in which the saturated thawed layer over *permafrost* flows slowly downhill to produce multiple terraces and solifluction lobes.

sorting separation of one grade size of *sediment* particles from another by the action of currents of air or water.

source region extensive land or ocean surface over which an *air mass* derives its temperature and moisture characteristics.

south pole point at which the southern end of the earth's *axis of rotation* intersects the earth's surface.

specific heat property of a substance that governs the temperature change of the substance with a given input of heat energy.

specific humidity mass of *water vapor* contained in a unit mass of air.

spit (See *sandspit*.)

splash erosion *soil erosion* caused by direct impact of falling raindrops on a wet surface of *soil* or *regolith*.

spodic horizon *soil horizon* containing precipitated amorphous materials composed of organic matter and *sesquioxides* of aluminum, with or without iron.

Spodosols *soil order* consisting of *soils* with a *spodic horizon*, an *albic horizon*, with low *base status*, and lacking in *carbonate* materials.

spreading plate boundary *lithospheric plate* boundary along which two plates of *oceanic lithosphere* are undergoing separation, while at the same time new *lithosphere* is being formed by *accretion*. (See also *transform plate boundary*.)

stable air mass *air mass* in which the *environmental temperature lapse rate* is less than the *dry adiabatic lapse rate*, inhibiting *convectional* uplift and mixing.

stack (marine) isolated columnar mass of *bedrock* left standing in front of a retreating *marine cliff*.

stage height of the surface of a river above its bed or a fixed level near the bed.

standard meridians *standard time* meridians separated by 15° of *longitude* and having values that are multiples of 15°. (In some cases the meridians used are multiples of $7\frac{1}{2}°$.)

standard time system time system based on the local time of a *standard meridian* and applied to belts of *longitude* extending $7\frac{1}{2}°$ (more or less) on either side of that meridian.

standard time zone zone of the earth in which all inhabitants keep the same time, which is that of a *standard meridian* within the zone.

star dune large, isolated *sand dune* with radial ridges culminating in a peaked summit; found in the deserts of North Africa and the Arabian Peninsula.

steppe (See *short-grass prairie*.)

steppe climate (See *semiarid (steppe) climate subtype*.)

stone rings linked ringlike ridges of cobbles or boulders lying at the surface of the ground in arctic and alpine tundra regions.

storage capacity maximum capacity of *soil* to hold water against the pull of *gravity*.

storm surge rapid rise of coastal water level accompanying the onshore arrival of a *tropical cyclone*.

strata layers of *sediment* or *sedimentary rock* in which individual beds are separated from one another along bedding planes.

stratified drift   *glacial drift* made up of sorted and layered *clay, silt, sand,* or gravel deposited from meltwater in *stream channels,* or in marginal *lakes* close to the ice front.

stratiform clouds   *clouds* of layered, blanketlike form.

stratosphere   layer of *atmosphere* lying directly above the *troposphere.*

stratovolcano   *volcano* constructed of multiple layers of *lava* and *tephra* (volcanic ash).

stratus   *cloud* type of the low-height family formed into a dense, dark gray layer.

stream   long, narrow body of flowing water occupying a *stream channel* and moving to lower levels under the force of *gravity.* (See *consequent stream; graded stream; subsequent stream.*)

stream capacity   maximum *stream load* of solid matter that can be carried by a *stream* for a given *discharge.*

stream channel   long, narrow, troughlike depression occupied and shaped by a *stream* moving to progressively lower levels.

stream deposition   accumulation of transported particles on a *stream* bed, upon the adjacent *floodplain,* or in a body of standing water.

stream erosion   progressive removal of *mineral* particles from the floor or sides of a *stream channel* by drag force of the moving water, or by *abrasion,* or by *corrosion.*

stream flow   water flow in a *stream channel;* same as channel flow.

stream gradient   rate of descent to lower elevations along the length of a *stream channel,* stated in m/km, ft/mi, degrees, or percent.

stream load   solid matter carried by a *stream* in dissolved form (as *ions*), in *turbulent suspension,* and as *bed load.*

stream profile   a graph of the elevation of a *stream* plotted against its distance downstream.

stream transportation   downvalley movement of eroded particles in a *stream channel* in solution, in *turbulent suspension,* or as *bed load.*

strike   compass direction of the line of intersection of an inclined rock plane and a horizontal plane of reference. (See also *dip.*)

strike-slip fault   variety of *fault* on which the motion is dominantly horizontal along a near-vertical *fault plane.*

strip mining   mining method in which overburden is first removed from a seam of *coal,* or a sedimentary ore, allowing the coal or ore to be extracted.

subantarctic low-pressure belt   persistent belt of low *atmospheric pressure* centered about at lat. 65° S over the Southern Ocean.

subantarctic zone   *latitude* zone lying between lat. 55° and 60° S (more or less) and occupying a region between the *midlatitude zone* and the *antarctic zone.*

subarctic zone   *latitude* zone between lat. 55° and 60° N (more or less), occupying a region between the *midlatitude zone* and the *arctic zone.*

subduction   descent of the downbent edge of a *lithospheric plate* into the *asthenosphere* so as to pass beneath the edge of the adjoining plate.

sublimation   process of change of ice (solid state) to *water vapor* (gaseous state); in *meteorology,* sublimation also refers

to the change of state from water vapor (*liquid*) to ice (*solid*), which is referred to as *deposition* in this text.

submergence   inundation or partial drowning of a former land surface by a rise of sea level or a sinking of the *crust,* or both.

suborder   a unit of *soil* classification representing a subdivision of the *soil order.*

subsequent stream   *stream* that develops its course by *stream erosion* along a band or belt of weaker *rock.*

subsolar point   point on the earth's surface at which solar rays are perpendicular to the surface.

subsurface water   water of the lands held in *soil, regolith,* or *bedrock* below the surface.

subtropical broadleaf evergreen forest   a *formation class* of the *forest biome* composed of broadleaf evergreen *trees;* occurs primarily in the regions of the *moist subtropical climate* ⑥.

subtropical evergreen forest   a subdivision of the *forest biome* composed of both broadleaf and needleleaf evergreen *trees.*

subtropical high-pressure belts   belts of persistent high *atmospheric pressure* trending east-west and centered about on lat. 30° N and S.

subtropical jet stream   *jet stream* of westerly winds forming at the *tropopause,* just above the *Hadley cell.*

subtropical needleleaf evergreen forest   a *formation class* of the *forest biome* composed of needleleaf evergreen *trees* occurring in the *moist subtropical climate* ⑥ of the southeastern United States; also referred to as the southern pine forest.

subtropical zones   *latitude* zones occupying the region of lat. 25° to 35° N and S (more or less) and lying between the *tropical zone* and the *midlatitude zone.*

summer monsoon   inflow of maritime air at low levels from the Indian Ocean toward the Asiatic low-pressure center in the season of high sun; associated with the rainy season of the *wet-dry tropical climate* ③ and the Asiatic monsoon climate.

summer solstice   solstice occurring on June 21 or 22, when the *subsolar point* is located at $23\frac{1}{2}°$ N.

sun-synchronous orbit   satellite orbit in which the orbital plane remains fixed in position with respect to the sun.

supercooled water   water existing in the liquid state at a temperature lower than the normal *freezing* point.

surface water   water of the lands flowing freely (as *streams*) or impounded (as ponds, *lakes,* marshes).

surges   episodes of very rapid downvalley movement within an *alpine glacier.*

suspended load   that part of the *stream load* carried in *turbulent suspension.*

suspension   (See *turbulent suspension.*)

suture   (See *continental suture.*)

swash   surge of water up the *beach* slope (landward) following collapse of a breaker.

synclinal mountain   steep-sided ridge or elongate mountain developed by erosion of a *syncline.*

synclinal valley   valley eroded on weak *strata* along the central trough or axis of a *syncline.*

syncline   downfold of *strata* (or other layered *rock*) in a troughlike structure; a class of *folds.* (See also *anticline.*)

taiga plant *formation class* consisting of *woodland* with low, widely spaced *trees* and a ground cover of *lichens* and mosses, found along the northern fringes of the region of *boreal forest climate* ⑪; also called cold woodland.

tall-grass prairie a *formation class* of the *grassland biome* that consists of tall grasses with broad-leaved *herbs*.

talus accumulation of loose rock fragments derived by fall of *rock* from a *cliff*.

talus slope slope formed of *talus*.

tarn small *lake* occupying a rock basin in a *cirque* of glacial *trough*.

tectonic activity process of bending (*folding*) and breaking (faulting) of crustal mountains, concentrated on or near active *lithospheric plate* boundaries.

tectonics branch of *geology* relating to tectonic activity and the features it produces. (See also *plate tectonics; tectonic activity*.)

temperature gradient rate of temperature change along a selected line or direction.

temperature inversion upward reversal of the normal *environmental temperature lapse rate*, so that the air temperature increases upward. (See *high-level temperature inversion; low-level temperature inversion*.)

tephra collective term for all size grades of solid *igneous rock* particles blown out under *gas* pressure from a volcanic vent.

terminal moraine *moraine* deposited as an embankment at the terminus of an *alpine glacier* or at the leading edge of an *ice sheet*.

terrestrial ecosystems *ecosystems* of land plants and animals found on upland surfaces of the continents.

thematic map map showing a single type of information.

thermal infrared a portion of the *infrared radiation wavelength* band, from approximately from 3 to 20 μm, in which objects at temperatures encountered on the earth's surface (including fires) emit *electromagnetic radiation*.

thermal pollution form of water pollution in which heated water is discharged into a *stream* or *lake* from the cooling system of a power plant or other industrial heat source.

thermister electronic device that measures (air) temperature.

thermometer instrument measuring temperature.

thermometer shelter louvered wooden cabinet of standard construction used to hold *thermometers* and other weather-monitoring equipment.

thorntree semidesert *formation class* within the *desert biome*, transitional from *grassland biome* and *savanna biome* and consisting of xerophytic *trees* and *shrubs*.

thorntree–tall-grass savanna plant *formation class*, transitional between the *savanna biome* and the *grassland biome*, consisting of widely scattered *trees* in an open grassland.

thrust sheet sheetlike mass of *rock* moving forward over a *low-angle overthrust fault*.

thunderstorm intense, local convectional storm associated with a *cumulonimbus cloud* and yielding heavy *precipitation*, also with lightning and thunder, and sometimes the fall of *hail*.

tidal current current set in motion by the *ocean tide*.

tidal inlet narrow opening in a *barrier island* or baymouth *bar* through which *tidal currents* flow.

tide (See *ocean tide*.)

tide curve graphical presentation of the rhythmic rise and fall of ocean water because of *ocean tides*.

till heterogeneous mixture of *rock* fragments ranging in size from *clay* to boulders, deposited beneath moving glacial ice or directly from the *melting* in place of stagnant glacial ice.

till plain undulating, plainlike land surface underlain by glacial *till*.

time zones zones or belts of given east-west (*longitudinal*) extent within which *standard time* is applied according to a uniform system.

topographic contour *isopleth* of uniform elevation appearing on a map.

tornado small, very intense wind vortex with extremely low air pressure in center, formed beneath a dense *cumulonimbus cloud* in proximity to a *cold front*.

trade winds (trades) surface *winds* in low *latitudes*, representing the low-level airflow within the *tropical easterlies*.

transcurrent fault *fault* on which the relative motion is dominantly horizontal, in the direction of the *strike* of the fault; also called a *strike-slip fault*.

transformation (soils) a class of soil-forming processes that transform materials within the *soil* body; examples include *mineral alteration* and *humification*.

transform fault special case of a *strike-slip fault* making up the boundary of two moving *lithospheric plates*; usually found along an offset of the *midoceanic ridge* where seafloor spreading is in progress.

transform plate boundary *lithospheric plate* boundary along which two plates are in contact on a *transform fault*; the relative motion is that of a *strike-slip fault*.

transform scar linear topographic feature of the ocean floor taking the form of an irregular scarp or ridge and originating at the offset *axial rift* of the *midoceanic ridge*; it represents a former *transform fault* but is no longer a plate boundary.

translocation a soil-forming process in which materials are moved within the *soil* body, usually from one horizon to another.

transpiration evaporative loss of water to the *atmosphere* from leaf pores of plants.

transportation (See *stream transportation*.)

transported regolith *regolith* formed of *mineral matter* carried by *fluid* agents from a distant source and deposited on the *bedrock* or on older regolith. Examples: floodplain silt, lake clay, beach sand.

transverse dunes field of wavelike *sand dunes* with crests running at right angles to the direction of the prevailing *wind*.

traveling cyclone center of low pressure and *inspiraling* winds that travels over the earth's surface; includes *wave cyclones, tropical cyclones*, and *tornadoes*.

travertine *carbonate mineral* matter, usually *calcite*, accumulating on *limestone cavern* surfaces situated in the *unsaturated zone*.

tree large, erect, woody *perennial* plant typically having a single main trunk, few branches in the lower part, and a branching crown.

trellis drainage pattern *drainage pattern* characterized by a dominant parallel set of major *subsequent streams*, joined at right angles by numerous short tributaries; typical of *coastal plains* and belts of eroded *folds*.

tropic of cancer *parallel of latitude* at $23\frac{1}{2}°$ N.

tropic of capricorn *parallel of latitude* at $23\frac{1}{2}°$ S.

tropical cyclone intense *traveling cyclone* of tropical and subtropical latitudes, accompanied by high *winds* and heavy rainfall.

tropical easterlies low-latitude *wind* system of persistent airflow from east to west between the two *subtropical high-pressure belts*.

tropical easterly jet stream upper-air *jet stream* of seasonal occurrence, running east to west at very high altitudes over Southeast Asia.

tropical high-pressure belt a high-pressure belt occurring in tropical latitudes at a high level in the *troposphere*; extends downward and poleward to form the *subtropical high-pressure belt*, located at the surface.

tropical-zone rainforest plant *formation class* within the *forest biome* similar to *equatorial rainforest*, but occurring farther poleward in tropical regions.

tropical zones *latitude* zones centered on the *tropic of cancer* and the *tropic of capricorn*, within the latitude ranges 10° to 25° N and 10° to 25° S, respectively.

tropopause boundary between *troposphere* and *stratosphere*.

troposphere lowermost layer of the *atmosphere* in which air temperature falls steadily with increasing altitude.

tsunami (See *seismic sea wave*.)

tundra biome *biome* of the cold regions of *arctic tundra* and *alpine tundra*, consisting of grasses, grasslike plants, flowering *herbs*, dwarf *shrubs*, mosses, and *lichens*.

tundra climate ⑫ cold climate of the *arctic zone* with eight or more consecutive months of zero *potential evapotranspiration (water need)*.

tundra soils soils of the arctic *tundra climate* ⑫ regions.

turbulence in *fluid* flow, the motion of individual water particles in complex eddies, superimposed on the average downstream flow path.

turbulent suspension *stream transportation* in which particles of *sediment* are held in the body of the *stream* by turbulent eddies. (Also applies to *wind* transportation.)

typhoon *tropical cyclone* of the western North Pacific and coastal waters of Southeast Asia.

Udalfs suborder of the *soil order Alfisols*; includes Alfisols of moist regions, usually in the *midlatitude zone*, with deciduous forest as the natural vegetation.

Udolls suborder of the *soil order Mollisols*; includes Mollisols of the moist soil water regime in the *midlatitude zone* and with no horizon of *calcium carbonate* accumulation.

Ultisols *soil order* consisting of *soils* of warm soil temperatures with an *argillic horizon* and low *base status*.

ultramafic igneous rock *igneous rock* composed almost entirely of *mafic minerals*, usually *olivine* or *pyroxene group*.

ultraviolet radiation *electromagnetic energy* in the *wavelength* range of 0.2 to 0.4 µm.

unloading process of removal of overlying *rock* load from *bedrock* by processes of *denudation*, accompanied by expansion and often leading to the development of *sheeting structure*.

unsaturated zone subsurface *water* zone in which pores are not fully saturated, except at times when *infiltration* is very rapid; lies above the *saturated zone*.

unstable air air with substantial content of *water vapor*, capable of breaking into spontaneous convectional activity leading to the development of heavy showers and *thunderstorms*.

upper-air westerlies system of westerly *winds* in the upper *atmosphere* over middle and high latitudes.

upwelling upward motion of cold, nutrient-rich ocean waters, often associated with cool equatorward currents occurring along *continental margins*.

Ustalfs suborder of the *soil order Alfisols*; includes Alfisols of semiarid and seasonally dry climates in which the *soil* is dry for a long period in most years.

Ustolls suborder of the *soil order Mollisols*; includes Mollisols of the semiarid climate in the *midlatitude zone*, with a horizon of *calcium carbonate* accumulation.

valley winds air movement at night down the *gradient* of valleys and the enclosing mountainsides; alternating with daytime *mountain winds*.

veins small, irregular, branching network of *intrusive rock* within a preexisting rock mass.

vernal equinox *equinox* occurring on March 20 or 21, when the *subsolar point* is at the *equator*.

Vertisols *soil order* consisting of *soils* of the *subtropical zone* and the *tropical zone* with high *clay* content, developing deep, wide cracks when dry, and showing evidence of movement between aggregates.

visible light *electromagnetic energy* in the *wavelength* range of 0.4 to 0.7 µm.

volcanic bombs boulder-sized, semisolid masses of *lava* that are ejected from an erupting *volcano*.

volcanic neck isolated, narrow steep-sided peak formed by erosion of *igneous rock* previously solidified in the feeder pipe of an extinct *volcano*.

volcanism general term for *volcano* building and related forms of extrusive igneous activity.

volcano conical, circular structure built by accumulation of *lava* flows and *tephra*. (See *shield volcano*; *stratovolcano*.)

volcano coast *coast* formed by *volcanoes* and *lava* flows built partly below and partly above sea level.

warm front moving weather *front* along which a warm *air mass* is sliding up over a cold air mass, leading to production of *stratiform clouds* and *precipitation*.

washout downsweeping of atmospheric *particulates* by *precipitation*.

water gap narrow transverse *gorge* cut across a narrow ridge by a *stream*, usually in a region of eroded *folds*.

water need (See *potential evapotranspiration*.)

water surplus water disposed of by *runoff* or *percolation* to the *ground water* zone after the *storage capacity* of the *soil* is full.

water table upper boundary surface of the *saturated zone*; the upper limit of the *ground water* body.

water vapor the gaseous state of water.

waterfall abrupt descent of a *stream* over a *bedrock* downstep in the *stream channel*.

waterlogging rise of a *water table* in *alluvium* to bring the zone of saturation into the root zone of plants.

watt unit of power equal to the quantity of work done at the rate of 1 joule per second, or 10^7 ergs per second.

wave-cut notch   *rock* recess at the base of a *marine cliff* where wave impact is concentrated.

wave cyclone   traveling, vortexlike *cyclone* involving the interaction of cold and warm *air masses* along sharply defined *fronts*.

wavelength   distance separating one wave crest from the next in any uniform succession of traveling waves.

weak equatorial low   weak, slowly moving low-pressure center (*cyclone*) accompanied by numerous convectional showers and *thunderstorms;* it forms close to the *intertropical convergence zone* in the rainy season, or *summer monsoon*.

weather   physical state of the *atmosphere* at a given time and place.

weather system   organized state of the *atmosphere* associated with a characteristic weather pattern, such as a *cyclone* or *anticyclone*.

weathering   total of all processes acting at or near the earth's surface to cause physical disruption and chemical decomposition of *rock*. (See *chemical weathering; physical weathering*.)

west-wind drift   ocean drift current moving eastward in zone of *prevailing westerlies*.

westerlies   (See *prevailing westerly winds; upper-air westerlies*.)

wet adiabatic lapse rate   reduced *adiabatic lapse rate* when *condensation* is taking place in rising air; value ranges from 3 to 6C°/1000m (2 to 3F°/1000 ft).

wet-dry tropical climate ③   climate of the *tropical zone* characterized by a very wet season alternating with a very dry season.

wet equatorial climate ①   moist climate of the *equatorial zone* with a large annual *water surplus* and with uniformly warm temperatures throughout the year.

wetlands   land areas of poor surface drainage, such as marshes and swamps.

wilting point   quantity of stored *soil water*, less than which the foliage of plants not adapted to *drought* will wilt.

wind   air motion, dominantly horizontal relative to the earth's surface.

wind abrasion   mechanical wearing action of *wind*-driven *mineral* particles striking exposed *rock* surfaces.

wind vane   weather instrument used to indicate *wind* direction.

winter monsoon   outflow of continental air at low levels from the Siberian High, passing over Southeast Asia as a dry, cool northerly *wind*.

winter solstice   solstice occurring on December 21 or 22, when the *subsolar point* is at $23\frac{1}{2}°$ S.

Wisconsinan glaciation   last glaciation of the *Pleistocene Epoch*.

woodland   plant *formation class*, transitional between *forest biome* and *savanna biome*, consisting of widely spaced *trees* with canopy coverage between 25 and 60 percent.

Xeralfs   suborder of the *soil order Alfisols;* includes Alfisols of the *Mediterranean climate* ⑦.

Xerolls   suborder of the *soil order Mollisols;* includes Mollisols of the *Mediterranean climate* ⑦.

xerophytes   plants adapted to a dry environment.

PHOTO CREDITS

Prologue
Page 3: Courtesy NOAA. **Figure P.1:** Courtesy NASA. **Figure P.2:** Earth Satellite Corporation. **Figure P.5:** Earth Satellite Corporation, Science Photo Library/Photo Researchers. **Figure P.6:** Courtesy NASA.

Chapter 1
Page 10: Ron Watts/Black Star. **Figure 1.1:** Dale Knuepfer/Bruce Coleman, Inc. **Figure 1.6:** ©National Maritime Museum, Greenwich, London.

Chapter 2
Page 28: John Marshall/AllStock, Inc. **Figure 2.3:** Courtesy Daedalus Enterprises, Inc. & National Geographic Magazine. **Figure E2.1:** Courtesy NASA.

Chapter 3
Page 48: Carr Clifton/AllStock, Inc. **Figure 3.2A:** /Steve McCutcheon. **Figure 3.2B:** Arthur N. Strahler. **Figure 3.8:** /George Wuerthner. **Figure 3.11:** David Falconer/STOP Pictures. **Figure 3.22:** Shawn Henry/SABA.

Chapter 4
Page 76: Barbara Filet/Tony Stone Worldwide. **Figure 4.6:** Arthur N. Strahler. **Figure 4.12A:** Van Bucher/Photo Researchers. **Figure 4.12B:** Kees van den Berg/Photo Researchers. **Figure 4.13:** Michael S. Renner/Bruce Coleman, Inc. **Figure 4.14:** Larry West/Bruce Coleman, Inc. **Figure 4.15:** Arthur N. Strahler. **Figure 4.20:** Keith Kent/Peter Arnold, Inc. **Figure 4.22:** Nuridsany et Perennou/Photo Researchers. **Figure 4.24:** John H. Hoffman/Bruce Coleman, Inc. **Figure 4.26:** Radarsat. **Figure 4.27:** Paul Barton/The Stock Market. **Figure 4.30:** John S. Shelton.

Chapter 5
Page 106: Grapes, Michaud/Photo Researchers. **Figure 5.7:** Courtesy Taylor Instrument Company/Wards Natural Science Establishment, Rochester, New York. **Figure 5.18:** P. F. Bentley/Black Star. **Figure 5.24:** Courtesy NASA. **Figure 5.26:** Courtesy NOAA/Satellite Data Services Division.

Chapter 6
Page 132: Courtesy NASA Goddard Space Flight Center/A. F. Hasler, H. Pierce, K. Palaniappan, and M. Manyin. **Figure 6.3:** Steve Pace/Envision. **Figure 6.11:** Keith Brewster/Weatherstock. **Figure 6.13:** Alex S. Maclean/Landslides. **Figure E6.1:** Courtesy NOAA. **Figure 6.18:** John Lopinot/Sygma.

Chapter 7
Page 152: M. Thonig/AllStock, Inc. **Figure 7.10:** Will & Deni McIntyre/Photo Researchers. **Figure E7.2:** Alain Nogues/ Sygma. **Figure 7.13:** ©Boyd Norton. **Figure 7.16:** Frans Lanting/Minden Pictures, Inc. **Figure 7.19:** /Fred Hirschmann. **Figure 7.24:** Tom McHugh/Photo Researchers. **Figure 7.26:** David Muench Photography. **Figure 7.30:** ©Porterfield/Chickering/Photo Researchers. **Figure 7.36:** Warren Garst/Tom Stack & Associates.

Chapter 8
Page 196: John M. Roberts/The Stock Market. **Figure 8.2:** Steve McCutcheon. **Figure 8.4:** Josef Muench. **Figure 8.7:** M. Freeman/Bruce Coleman, Inc. **Figure 8.10:** /Tom Bean. **Figure 8.9:** Ferrero/Labat/Auscape International. **Figure 8.12:** John S. Shelton. **Figure 8.15:** Kenneth Murray/Photo Researchers. **Figure 8.16:** Jake Rajs/The Image Bank. **Figure 8.19:** Michael Townsend/AllStock, Inc. **Figure 8.20A:** Arthur N. Strahler. **Figure 8.20B:** Alan H. Strahler. **Figure 8.22:** Arthur N. Strahler. **Figure 8.23:** David Muench Photography. **Figure 8.25:** Annie Griffiths Belt/DRK Photo. **Figure E8.1:** Jacques Jangoux/Peter Arnold, Inc. **Figure E8.2:** Toby Molenaar/Woodfin Camp & Associates. **Figure 8.26:** Joe Viesti/Viesti Associates, Inc. **Figure 8.27:** Brian A. Vikander. **Figure 8.28:** Arthur N. Strahler. **Figure 8.30:** Arthur N. Strahler.

Chapter 9
Page 232: George Holton/Photo Researchers. **Figure 9.4:** Harry GoldCourtesy R. B. Krone, San Francisco District Corps of Engineers, U.S. Army. **Figure 9.5:** R. Schaetzel. **Figure 9.9:** Arthur N. Strahler. **Figure 9.11a:** Henry D. Foth. **Figure 9.11b:** Henry Foth. **Figure 9.11c:** Henry D. Foth. **Figure 9.11d:** Henry D. Foth. **Figure 9.11e:** Henry D. Foth. **Figure 9.11f:** Henry D. Foth. **Figure 9.11g:** Henry D. Foth. **Figure 9.11h:** Henry D. Foth. **Figure 9.11i:** Henry D. Foth. **Figure 9.11j:** Henry D. Foth. **Figure 9.11k:** Henry D. Foth. **Figure 9.11l:** Soil Conservation Service. **Figure 9.12:** Alan H. Strahler. **Figure 9.13:** Henry D. Foth. **Figure 9.14:** Soil Conservation Service. **Figure 9.15:** ©William E. Ferguson. **Figure 9.17:** Henry D. Foth. **Figure 9.18:** Ric Ergenbright/AllStock, Inc. **Figure 9.19:** Robin White/Fotolex Associates. **Figure 9.21:** R. Schaetzl. **Figure 9.22:** Courtesy Soil Conservation Service. **Figure 9.23:** Ned L. Reglein. **Figure 9.24:** Henry D. Foth.

Chapter 10

Page 262: J. M. La Roque/Auscape International. **Figure 10.2:** Roberto de Gugliemo/Science Photo Library, Photo Researchers. **Figure 10.6:** G. R. Roberts. **Figure 10.7:** /Travelpix, FPG International. **Figure 10.8a:** Ward's Natural Science Establishment. **Figure 10.8b:** Ward's Natural Science Establishment. **Figure 10.10:** Arthur N. Strahler. **Figure E10.1:** Courtesy of Richard S. Williams, Jr., U.S. Geological Survey. **Figure 10.11:** Andrew McIntyre/Columbia University. **Figure 10.12:** Panoramic Images. **Figure 10.13:** Arthur N. Strahler. **Figure 10.14:** Victor Englebert/Photo Researchers. **Figure 10.15:** Charles R. Belinky/Photo Researchers. **Figure 10.16:** Arthur N. Strahler. **Figure 10.17:** Freeman Patterson/Masterfile.

Chapter 11

Page 282: Courtesy NASA. **Figure 11.8:** Hammond Incorporated.

Chapter 12

Page 310: Douglas Peebles Photography. **Figure 12.3:** James Mason/Black Star. **Figure 12.4:** K. Hamdorf/Auscape International. **Figure 12.5:** Greg Vaughn/Tom Stack & Associates. **Figure 12.6:** Werner Stoy/Camera Hawaii. **Figure 12.7:** Arthur N. Strahler. **Figure 12.8:** Arthur N. Strahler. **Figure 12.9:** ©Larry Ulrich. **Figure 12.10:** Steve Vidler/Leo de Wys, Inc. **Figure 12.14:** Arthur N. Strahler. **Figure 12.17:** James Balog/Black Star. **Figure 12.19:** Georg Gerster/COMSTOCK, Inc. **Figure 12.20:** ©Sipa Press. **Figure E12.1:** G. Hall/Woodfin Camp & Associates. **Figure 12.23:** J. Eyerman/ Black Star.

Chapter 13

Page 332: Douglas Peebles Photography. **Figure 13.2:** Susan Rayfield/Photo Researchers. **Figure 13.3:** Arthur N. Strahler. **Figure 13.4:** ©Steve McCutcheon. **Figure 13.5:** Bernard Hallet, Periglacial Laboratory, Quaternary Research Center. **Figure 13.6:** Stephen J. Krasemann/DRK Photo. **Figure 13.8:** Kunio Owaki/The Stock Market. **Figure 13.9:** ©George Wuerthner. **Figure 13.10:** David Muench Photography. **Figure 13.11:** ©Tom Bean. **Figure E13.2:** Courtesy Montana Historical Society. **Figure 13.12A:** Alan H. Strahler. **Figure 13.12B:** Alan H. Strahler. **Figure 13.14:** Mark A. Melton. **Figure 13.16:** ©Steve McCutcheon. **Figure 13.18:** Orlo E. Childs. **Figure 13.19:** Courtesy NOAA. **Figure 13.20:** Billy Davis/Black Star.

Chapter 14

Page 351: Laurence Parent. **Figure 14.7:** Mark A. Melton. **Figure E14.1:** M. Smith/Sipa Press. **Figure 14.12:** Chris Stewart/Black Star. **Figure 14.15:** Tom Bean/DRK Photo. **Figure 14.16:** Tom Till Photography. **Figure 14.17:** National Audubon Society/Photo Researchers.

Chapter 15

Page 374: W. P. Fleming/Viesti Associates, Inc. **Figure 15.2:** Official U. S. Navy Photographs. **Figure 15.3:** Mark A. Melton. **Figure 15.4:** Joe Englander/Viesti Associates, Inc. **Figure 15.5:** Alan H. Strahler. **Figure 15.7:** Frans Lanting/Minden Pictures, Inc. **Figure 15.10:** Dick Dietrich/Dietrich Photography. **Figure 15.13:** Courtesy Harm J. deBlij. **Figure 15.17:** ©Tom Bean. **Figure 15.19:** ©F. Kenneth Hare. **Figure 15.21:** /NASA, Science Source, Photo Researchers. **Figure 15.22:** Albert Moldvay/Eriako Associates. **Figure 15.24:** T.A. Wiewandt/DRK Photo. **Figure 15.27:** Mark A. Melton. **Figure E15.1:** Earth Observation Satellite Company. **Figure 15.29:** John S. Shelton. **Figure 15.30:** Mark A. Melton.

Chapter 16

Page 402: Tom Bean/DRK Photo. **Figure 16.4:** Larry Ulrich/DRK Photo. **Figure 16.10:** John S. Shelton. **Figure 16.14:** Earth Satellite Corporation. **Figure 16.17:** ©Laurence Parent. **Figure 16.18:** John S. Shelton. **Figure 16.20:** Bruno Barbey/Magnum Photos, Inc. **Figure 16.22:** J. A. Kraulis/Masterfile. **Figure 16.26:** Landis Aerial Photo. **Figure E16.1:** ©Carr Clifton. **Figure E16.2:** Nancy S. Dituri/Envision. **Figure 16.29:** John S. Shelton. **Figure 16.30:** Alex S. MacLean/Landslides. **Figure 16.31:** ©Larry Ulrich.

Chapter 17

Page 424: Geoff Renner/Robert Harding Picture Library. **Figure 17.2:** Alex S. Maclean/Landslides. **Figure 17.4:** Arthur N. Strahler. **Figure 17.5:** ©Mike Yamashita. **Figure 17.7:** Steve Dunnell/The Image Bank. **Figure 17.9:** Vic DeLucia/ NYT Pictures. **Figure 17.12:** David Muench Photography. **Figure 17.15A:** Courtesy NASA. **Figure 17.16:** Earth, Satellite Corporation. **Figure 17.17:** Courtesy NASA. **Figure 17.18·** David Hiser/Photographers Aspen. **Figure 17.19:** Courtesy NASA/Lyndon B. Johnson Space. **Figure 17.21:** John S. Shelton. **Figure 17.22:** ©Tom Bean. **Figure 17.23:** M. J. Coe/Earth Scenes. **Figure E17.1:** Stephen Rose/Gamma Liaison. **Figure 17.25:** John S. Shelton. **Figure 17.26:** John S. Shelton. **Figure 17.27:** Aero Service Corporation, Litton Industries. **Figure 17.29:** ©William E. Ferguson. **Figure 17.31:** ©ERIM, Ann Arbor MI. **Figure 17.32:** J. A. Kraulis/Masterfile. **Figure 17.33:** G. R. Roberts. **Figure 17.35:** Alan H. Strahler.

Chapter 18

Page 454: Ric Ergenbright Photography. **Figure 18.1:** Carr Clifton/AllStock, Inc. **Figure 18.5:** Fred Hirschmann Wilderness Photography. **Figure 18.6:** John S. Shelton. **Figure 18.8:** Floyd L. Norgaard/Ric Engenbright Photography. **Figure 18.7:** Alan H. Strahler. **Figure 18.11:** D. Parer & E. Parer-Cook/Auscape International. **Figure 18.12:** Courtesy NASA. **Figure 18.13:** ©Wolfgang Kaehler. **Figure 18.17:** Courtesy NASA. **Figure 18.20:** C. Wolinsky/Stock, Boston. **Figure 18.21:** Arthur N. Strahler. **Figure E18.1:** Jonathan Chester© Natural History Photographic Agency.

Epilogue

Figure EP.1: Alicia Hurr/Weatherstock. **Page 477:** Bob Krist/Black Star. **Figure EP.2:** Tom McHugh/Photo Researchers. **Figure EP.3:** Frans Lanting/Minden Pictures, Inc.

Appendix

Figure A2.1: SAR image courtesy of Intera Technologies Corporation, Calgary, Alberta, Canada. **Figure A2.2:** Courtesy NASA, compiled and annotated by John E. Estes and Leslie W. Senger. **Figure A2.3:** Courtesy NASA. **Figure A2.5:** Copyright ©1993 by CNES; Courtesy SPOT Image Corporation, Reston, Virginia.

INDEX

CONVERSION FACTORS
Metric to English

Metric Measure	*Multiply by**	*English Measure*
LENGTH		
Millimeters (mm)	0.0394	Inches (in)
Centimeters (cm)	0.394	Inches (in)
Meters (m)	3.28	Feet (ft)
Kilometers (km)	0.621	Miles (mi)
AREA		
Square centimeters (cm^2)	0.155	Square inches (in^2)
Square meters (m^2)	10.8	Square feet (ft^2)
Square meters (m^2)	1.12	Square yards (yd^2)
Square kilometers (km^2)	0.386	Square miles (mi^2)
Hectares (ha)	2.47	Acres
VOLUME		
Cubic centimeters (cm^3)	0.0610	Cubic inches (in^3)
Cubic meters (m^3)	35.3	Cubic feet (ft^3)
Cubic meters (m^3)	1.31	Cubic yards (yd^3)
Milliliters (ml)	0.0338	Fluid ounces (fl oz)
Liters (l)	1.06	Quarts (qt)
Liters (l)	0.264	Gallons (gal)
MASS (weight)		
Grams (g)	0.0353	Ounces (oz)
Kilograms (kg)	2.20	Pounds (lb)
Kilograms (kg)	0.00110	Tons (2000 lb)
Tonnes (t)	1.10	Tons (2000 lb)

*Conversion factors shown to 3 decimal-digit precision.

CONVERSION FACTORS
English to Metric

English Measure	Multiply by*	Metric Measure
LENGTH		
Inches (in)	2.54	Centimeters (cm)
Feet (ft)	0.305	Meters (m)
Yards (yd)	0.914	Meters (m)
Miles (mi)	1.61	Kilometers (km)
AREA		
Square inches (in^2)	6.45	Square centimeters (cm^2)
Square feet (ft^2)	0.0929	Square meters (m^2)
Square yards (yd^2)	0.836	Square meters (m^2)
Square miles (mi^2)	2.59	Square kilometers (km^2)
Acres	0.405	Hectares (ha)
VOLUME		
Cubic inches (in^3)	16.4	Cubic centimeters (cm^3)
Cubic feet (ft^3)	0.0283	Cubic meters (m^3)
Cubic yards (yd^3)	0.765	Cubic meters (m^3)
Fluid ounces (fl oz)	29.6	Milliliters (ml)
Pints (pt)	0.473	Liters (l)
Quarts (qt)	0.946	Liters (l)
Gallons (gal)	3.79	Liters (l)
MASS (weight)		
Ounces (oz)	28.4	Grams (g)
Pounds (lb)	0.454	Kilograms (kg)
Tons (2000 lb)	907	Kilograms (kg)
Tons (2000 lb)	0.907	Tonnes (t)

*Conversion factors shown to 3 decimal-digit precision.

TOPOGRAPHIC MAP SYMBOLS

VARIATIONS WILL BE FOUND ON OLDER MAPS

Hard surface, heavy duty road, four or more lanes ......

Hard surface, heavy duty road, two or three lanes ......

Hard surface, medium duty road, four or more lanes ....

Hard surface, medium duty road, two or three lanes .....

Improved light duty road ....

Unimproved dirt road and trail ....

Dual highway, dividing strip 25 feet or less ....

Dual highway, dividing strip exceeding 25 feet ....

Road under construction ....

Railroad, single track and mult ple track ....

Railroads in juxtaposition ....

Narrow gage, single track and multiple track ....

Railroad in street and carline ....

Bridge road and railroad ....

Drawbridge, road and railroad ....

Footbridge ....

Tunnel, road and railroad ....

Overpass and underpass ....

Important small masonry or earth dam ....

Dam with lock ....

Dam with road ....

Canal with lock ....

Buildings (dwelling, place of employment, etc.) ....

School, church, and cemetery ....

Buildings (barn, warehouse, etc.) ....

Power transmission line ....

Telephone line, pipeline, etc. (labeled as to type) ....

Wells other than water (labeled as to type) .... o Oil ..... o Gas

Tanks; oil, water, etc. (labeled as to type) .... ● ● ● ⊘ Water

Located or landmark object; windmill ....

Open pit, mine, or quarry; prospect ....

Shaft and tunnel entrance ....

Horizontal and vertical control station:

  Tablet, spirit level elevation .... BM△ 5653

  Other recoverable mark, spirit level elevation .... △ 5455

Horizontal control station: tablet, vertical angle elevation VABM△ 9519

  Any recoverable mark, vertical angle or checked elevation △3775

Vertical control station: tablet, spirit level elevation ..... BM ✕ 957

  Other recoverable mark, spirit level elevation .... ✕ 954

Checked spot elevation .... ✕4675

Unchecked spot elevation and water elevation .... ✕ 5657 870

Boundary, national ....

  State ....

  County, parish, municipio ....

  Civil township, precinct, town, barrio ....

  Incorporated city, village, town, hamlet ....

  Reservation, national or state ....

  Small park, cemetery, airport, etc. ....

  Land grant ....

Township or range line, United States land survey ......

Township or range line, approximate location ....

Section line, United States land survey ....

Section line, approximate location ....

Township line, not United States land survey ....

Section line, not United States land survey ....

Section corner, found and indicated .... + ... +

Boundary monument: land grant and other .... □ .... □

United States mineral or location monument .... ▲

Index contour	Intermediate contour..
Supplementary contour	Depression contours ..
Fill............	Cut............
Levee............	Levee with road.....
Mine dump........	Wash..........
Tailings........	Tailings pond........
Strip mine........	Distorted surface....
Sand area........	Gravel beach.......

Perennial streams	Intermittent streams..
Elevated aqueduct....	Aqueduct tunnel......
Water well and spring o..... ～	Disappearing stream..
Small rapids.........	Small falls........
Large rapids.......	Large falls.......
Intermittent lake.....	Dry lake........
Foreshore flat......	Rock or coral reef...
Sounding, depth curve. 10	Piling or dolphin.....
Exposed wreck......	Sunken wreck.......

Rock, bare or awash; dangerous to navigation ....

Marsh (swamp).......	Submerged marsh ...
Wooded marsh......	Mangrove........
Woods or brushwood..	Orchard........
Vineyard.........	Scrub.........
Inundation area.......	Urban area........

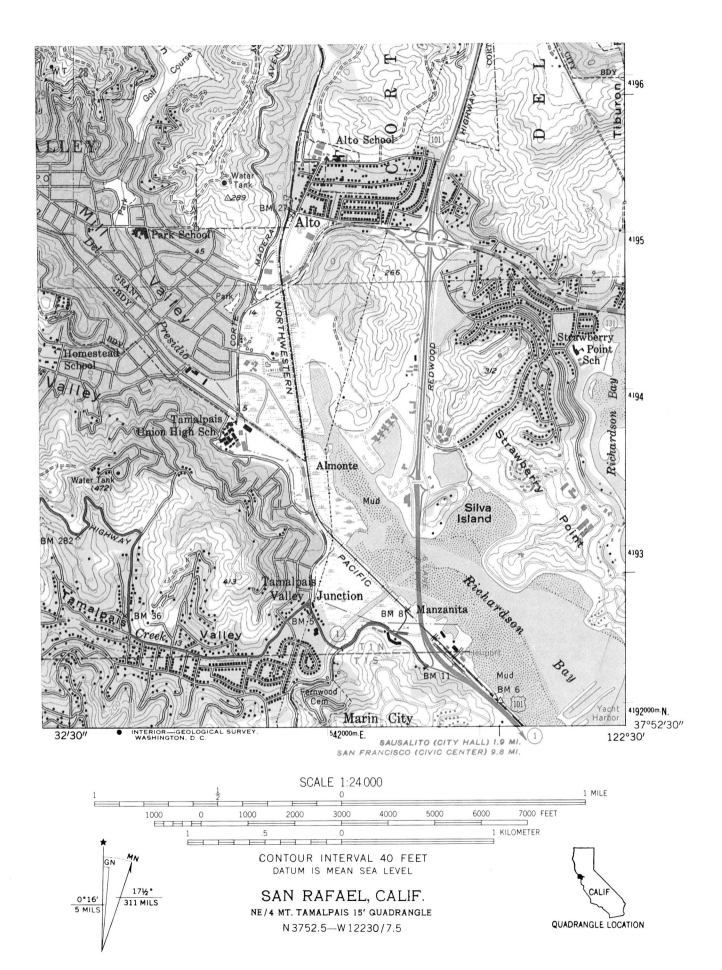

SCALE 1:24 000

CONTOUR INTERVAL 40 FEET
DATUM IS MEAN SEA LEVEL

SAN RAFAEL, CALIF.

NE/4 MT. TAMALPAIS 15' QUADRANGLE

N 3752.5—W 12230/7.5

GN
MN

★

0°16'
5 MILS

17½°
311 MILS

CALIF

QUADRANGLE LOCATION